Smart Innovation, Systems and Technologies

Volume 290

The Smart Innovation, Systems and Technologies book series encompasses the topics of knowledge, intelligence, innovation and sustainability. The aim of the series is to make available a platform for the publication of books on all aspects of single and multi-disciplinary research on these themes in order to make the latest results available in a readily-accessible form. Volumes on interdisciplinary research combining two or more of these areas is particularly sought.

The series covers systems and paradigms that employ knowledge and intelligence in a broad sense. Its scope is systems having embedded knowledge and intelligence, which may be applied to the solution of world problems in industry, the environment and the community. It also focusses on the knowledge-transfer methodologies and innovation strategies employed to make this happen effectively. The combination of intelligent systems tools and a broad range of applications introduces a need for a synergy of disciplines from science, technology, business and the humanities. The series will include conference proceedings, edited collections, monographs, handbooks, reference books, and other relevant types of book in areas of science and technology where smart systems and technologies can offer innovative solutions.

High quality content is an essential feature for all book proposals accepted for the series. It is expected that editors of all accepted volumes will ensure that contributions are subjected to an appropriate level of reviewing process and adhere to KES quality principles.

Indexed by SCOPUS, EI Compendex, INSPEC, WTI Frankfurt eG, zbMATH, Japanese Science and Technology Agency (JST), SCImago, DBLP.

All books published in the series are submitted for consideration in Web of Science.

More information about this series at https://link.springer.com/bookseries/8767

Sanjay Yadav · Abid Haleem · P. K. Arora ·
Harish Kumar
Editors

Proceedings of Second International Conference in Mechanical and Energy Technology

ICMET 2021, India

Springer

Editors
Sanjay Yadav
Physico-Mechanical Metrology
CSIR—National Physical Laboratory
New Delhi, Delhi, India

P. K. Arora
Galgotias College of Engineering
and Technology
Greater Noida, Uttar Pradesh, India

Abid Haleem
Department of Mechanical Engineering
Jamia Millia Islamia
New Delhi, Delhi, India

Harish Kumar
Department of Mechanical Engineering
National Institute of Technology Delhi
New Delhi, India

ISSN 2190-3018 ISSN 2190-3026 (electronic)
Smart Innovation, Systems and Technologies
ISBN 978-981-19-0107-2 ISBN 978-981-19-0108-9 (eBook)
https://doi.org/10.1007/978-981-19-0108-9

This Springer imprint is published by the registered company Springer Nature Singapore Pte Ltd.
The registered company address is: 152 Beach Road, #21-01/04 Gateway East, Singapore 189721, Singapore

Preface

International Conference on Mechanical and Energy Technologies (ICMET-2021) has been the second international conference of its series organized by the Department of Mechanical Engineering of Galgotias College of Engineering and Technology, Greater Noida, Uttar Pradesh, India. It was considered as a forum to bring together scientists, university professors, graduate students, and mechanical engineers, presenting new science, technology, and engineering ideas and achievements.

The conference attracted many participants working in various fields of engineering: design, mechanics, materials, energy, etc. More than 200 manuscripts were submitted to the conference, topics ranging from the latest research in the field of aerodynamics and fluid mechanics, artificial intelligence, rapid manufacturing technologies, remanufacturing, refrigeration and air conditioning, renewable energies technology, I.C. engines, turbo machinery metrology, and computer-aided design and inspection, etc.

Furthermore, we thank the management and director of Galgotias College of Engineering and Technology, Greater Noida, India, for their cooperation and support. We are thankful to all the members of the organizing committee for their contribution in organizing the conference. Last but not least, we thank Springer for its professional assistance and particularly Miss Priya Vyas who supported this publication.

New Delhi, India — Sanjay Yadav
New Delhi, India — Abid Haleem
Greater Noida, India — P. K. Arora
New Delhi, India — Harish Kumar

Contents

About the Editors

Sanjay Yadav is Senior Principal Scientist and Head of the Pressure, Vacuum and Ultrasonic Metrology Section and Prof. Academy of Scientific and Innovative Research (AcSIR), National Physical Laboratory, New Delhi, India. He received his Ph.D. in Physics from Meerut University in 1990. He has published more than 280 papers in international journals and conference proceedings and is Editor-in-Chief of the journal *MAPAN/Journal of Metrology Society of India*, published by Springer.

Dr. Abid Haleem is Professor of Mechanical Engineering, Jamia Millia Islamia, New Delhi, India. Presently, he is Regional Editor, Global Journal of Flexible Systems Management, Springer, and Associate Editor with *JCOT* (Elsevier) and i-Scholar. He has conferred with the fellowship of AGBA and has professional & life membership of IIIE, ISME, ISTE, GIFT, SCL, IICC, and ISC. With h-index of 39 in Google Scholar (26 in SCOPUS) and more than 6100 citations, he has published more than 400 research articles of which more than 185 articles indexed in SCOPUS and three books to his credit. His latest book is *Handbook of Materials Management*, published by Prentice Hall of India, while the latest edited book is *An update on Medical 3D Printing* from Solubris. He has more than 33 years of teaching and research experience at graduate & higher level and has produced more than 28 Ph.D.

P. K. Arora is Professor at the Mechanical Engineering Department of Galgotias College of Engineering and Technology, Greater Noida, India. He received his Ph.D. in Mechanical Engineering from Jamia Millia Islamia, New Delhi, in 2014. With 20 years of academic experience, his main research areas are industrial engineering and optimization. Dr. Arora has published more than 35 papers in international journals and conference proceedings.

Harish Kumar is currently working as Assistant Professor at the National Institute of Technology, Delhi. With more than 17 years of research and academic experience, he is Active Researcher in the area of mechanical measurement and metrology. He served as Scientist at the National Physical Laboratory, India, for more than a decade, and as Guest Researcher at the National Institute of Standards and Technology, USA, in 2016. He has authored more than 90 publications in peer-reviewed journals and conference proceedings. He is Active Reviewer for many prominent journals on measurement, metrology, and related areas and is Associate Editor of the *MAPAN/Journal of Metrology Society of India*, published by Springer.

Chapter 1
Redesign and Analysis of Cargo Containers for Delivery Drone Applications

Shivanshu Mishra, Vishal Kumar, Abdul Gani, and Faisal Shameem

Abstract Drone or octocopter drone is a new generation innovation that can perform operations like surveillance, media, etc. with ease setup and cost-efficient. The delivery drone is the upcoming evolution in the field of engineering. The design of cargo containers attached to drones affects the parameters like the speed, efficiency, dynamics, and controls. The present work proposes a modified elliptical-shaped design instead of a regular rectangular cargo container. For lightweight and durability, carbon fiber as a material is also proposed. The shape was analyzed through computational fluid dynamics technique by using Ansys. The results of the study showed that the new ellipse shape design of cargo container is more practical to use in high-speed operation and results in increased efficiency of the system like drone.

1.1 Introduction

The advancement in today's technology has extended the boundaries of drone technologies. The major application of drones is in military, civil, and agriculture services. Due to its wide applications, drones are now used for delivery purposes to solve efficient and reduced delivery time problems [1]. The major advantage of drone is they can cut short 80% of delivery time compared to ground vehicles based on their flexible routes and parallelized operations [2]. Due to their cheap, flexible, and fast delivery drones can bring revolution to delivery department.

The shape and size of the cargo container in delivery drone play a vital role in overall efficiency of the operation as most of the cargo containers are rectangular in shape which occupies a huge air resistance at its surface making overall delivery drone design less efficient. Some studies did airflow analysis on different structures of freight vehicles and it has been observed that the streamline shape is most efficient as air flows smoothly over the surfaces at high wind speed operation [3]. The main demerit of an ideal rectangular shape cargo container is that at high-speed operations

S. Mishra (✉) · V. Kumar · A. Gani · F. Shameem
School of Mechanical Engineering, Galgotias University, Gr. Noida, India
e-mail: mishrashivanshu.26@gmail.com

S. Yadav et al. (eds.), *Proceedings of Second International Conference in Mechanical and Energy Technology*, Smart Innovation, Systems and Technologies 290,
https://doi.org/10.1007/978-981-19-0108-9_1

the container becomes heavily unstable and limits the operational speed [4]. Therefore, we redesign the structure of cargo container into an ellipse shape, changed the material of construction to make it durable- lightweight, and carried out CFD analysis to know the results with respect to efficiency, control, and dynamics. The new design is separated into three different storing sections to utilize space and maintain equal weight distribution during operation.

1.2 Literature Review

The advancement in today's technology has pushed the boundaries of drone application; drones when attached to a cargo container can deliver various products in less time and more efficiently [5]. If we consider suburban areas and compare deliveries by a truck-only delivery and truck-drone delivery we see that the truck-only delivery is very advantageous economically [6]. Thus, it is important to study the drone characteristics and airflow analysis to be economical. Changing the design of light UAVS has resulted in more stability and performance [7]. These days drones can also perform their operations in automated modes with high velocity where their operation paths are predefined, Path generation is achieved by a unit quaternion curve and an associated parallel transport frame in the interactive process. [8]. But, it is important to select the specific type of drone for the delivery operations with respect to weight of cargo as more the number of motors and blades it has the more it consumes power, therefore, a study was conducted on octocopter drone and dodecacopter drone with, without periodic disturbance and it was observed that octocopter is more stable than dodecacopter without disturbance [9]. Hence, we can choose octocopter application for delivery drones.

The aerodynamic characteristic of the cargo container plays a major role in stability and performance parameters in delivery drones. A study on freight wagon was done where airflow analysis was done on the freight wagon the Reynolds number of the flow came out to be 10^5 [10]. Such shape containers with slung load are subject to massively separated unsteady flow and are limited by stability to operational airspeeds well below the power-limited speed of the configuration [4]. Thus, we believe there needs to be change in design of cargo containers for high-speed delivery applications. The shape can withstand the high-speed drag, giving more stability and dynamic controls. Cargo can be highly unstable while in motions and needed to be tied up for so but, locking and unlocking cargo can consume more time hence, we can use baffle at different intersections of container to reduce the instability of cargo similar to baffle used in liquid containers which reduces the amplitude of fluid slosh in partly filled tanks [11].

1.2.1 Proposed Design

The design is developed on Autodesk software, the conventional design of cargo section is rectangular shaped so we modified it to give the cargo section a shape of an ellipse (Fig. 1.1). This design is aimed to minimize resistance and drag caused by air thus making cargo section more aerodynamic and will eventually enhance the productivity of the drone to carry out the delivery operation in a less time and power.

The exterior of the design is developed like an ellipse shape as it has very less air resistance, the front and back edges have smooth (Fig. 1.2), which helps in creating aerodynamic path and increases controls due to less air resistance.

The interior of the cargo section is divided into three parts namely cargo1, cargo2, cargo3 separated by baffle (Fig. 1.3). We have divided it into three parts because by separating each compartment the weight distribution will be balanced and there will be less chances of instability while in motion.

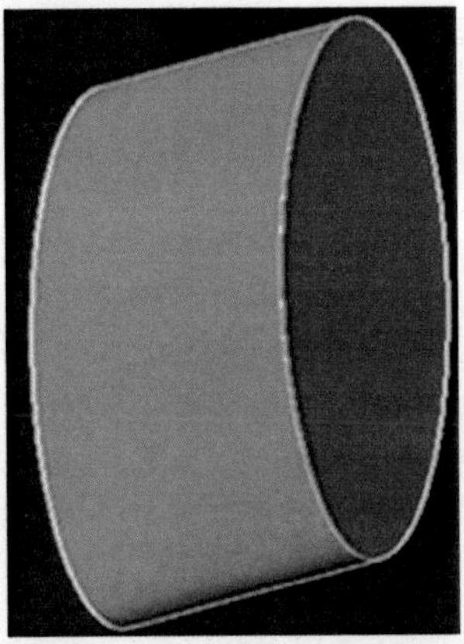

Fig. 1.1 3D view of proposed cargo container

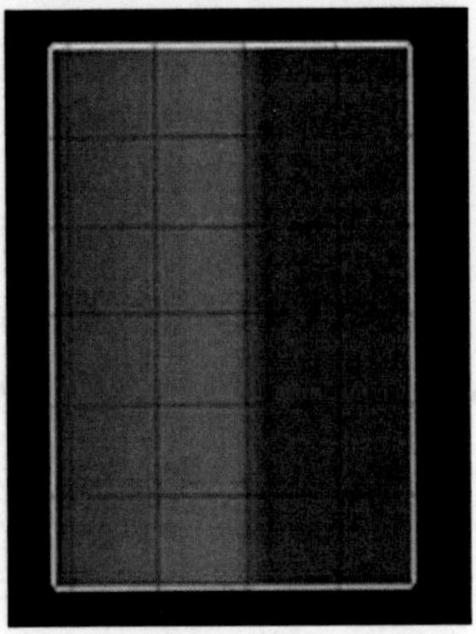

Fig. 1.2 Top view of proposed cargo container for drone application

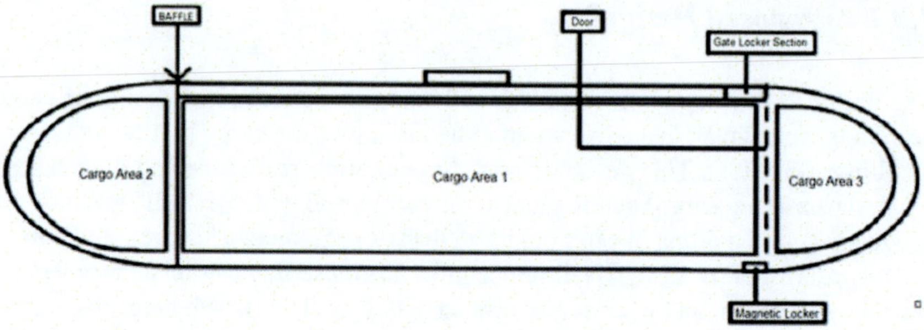

Fig. 1.3 Interior of proposed cargo container for drone application

Fig. 1.4 Bottom view of proposed cargo container for drone application

The cargo areas 2 and 3 are built to store small or lightweight goods whereas cargo area 1 has large storage area for large heavy goods. They both have separate doors for loading unloading cargo situated at top. The interior of the new design consists of single door situated between the cargo area 1 and 3 and also performs as baffle between the surfaces (Fig. 1.3). The door lock mechanism and stopper will be situated at top and bottom of cargo area 1. Additionally, user can construct an area on top of cargo area 1 to mount the container and drone. The top view of the design is replica of bottom view shown in Fig. 1.4 and the back view of the design is replica of front view of design shown in Fig. 1.5.

1.2.2 *CFD Analysis*

We carried out computation fluid dynamics (CFD) analysis on ANSYS 2021 software. Table 1.1 shows the analysis parameters. The design was exported to ANSYS for simulation results. The airflow analysis was performed with Air as inlet material, we first exported the design and various geometry were set which provided the

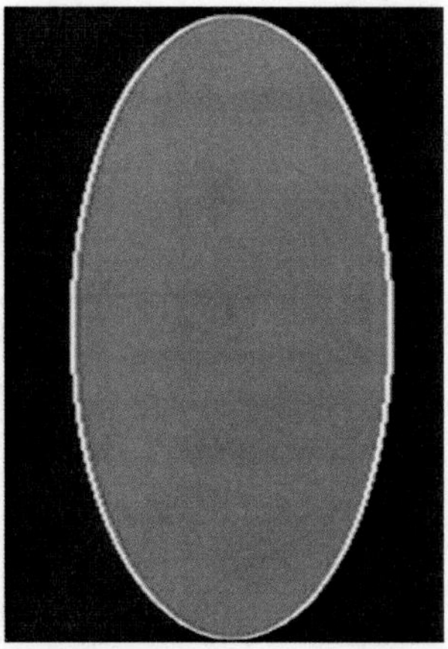

Fig. 1.5 Front view of proposed cargo container for drone application

Table 1.1 Analysis parameters

Velocity formulation	30 m/s
Viscous model	k-omega
Time	Steady
Flow	Fluent
Material	Air
Boundary conditions	Inlet velocity—30 m/s Pressure outlet—1 atm
Number of nodes	44,758
Number of elements	187,788
Enclosure dimensions	Length 12000 mm Breadth 4000 mm Height 2000 mm
Inflation	10 layers
Element size	Minimum 0.2 m Maximum 0.5 m
Converge	140 iterations

enclosure. Meshing was done and various input parameters were given mentioned below for simulation result.

In computational fluid dynamics, the k-omega (k–ω) turbulence model is a common two-equation turbulence model that is used as an approximation for the Reynolds-averaged Navier–Stokes equations (RANS equations). The model attempts to predict turbulence by two partial differential equations for two variables, k, and ω, with the first variable being the turbulence kinetic energy (k) while the second (ω) is the specific rate of dissipation (of the turbulence kinetic energy k into internal thermal energy). This design required SST k-omega turbulence model to calculate the value of lift force and drag force.

The two-equation model (written in conservation form) is given by the following:

$$\partial\left(\frac{\rho k}{\partial t}\right) + \frac{\partial(\rho ujk)}{\partial xj} = P - \beta^{*}\partial\omega k + \frac{\partial}{\partial xj}\left[\left(\mu + \sigma_k \frac{\rho k}{\omega}\right)\frac{\partial k}{\rho xj}\right]$$

where,

$$Sij = \frac{1}{2}\left(\frac{\partial uj}{\partial xi} + \frac{\partial uj}{\partial xi}\right)$$

1.3 Result and Discussion

Figures 1.6, 1.7, 1.8, 1.9, 1.10, 1.11, and 1.12 are derived from airflow simulation and we can see that the graph stables up after peak of air which shows the lower air resistance on surfaces.

Fig. 1.6 Pressure contour line with respect to Pascal

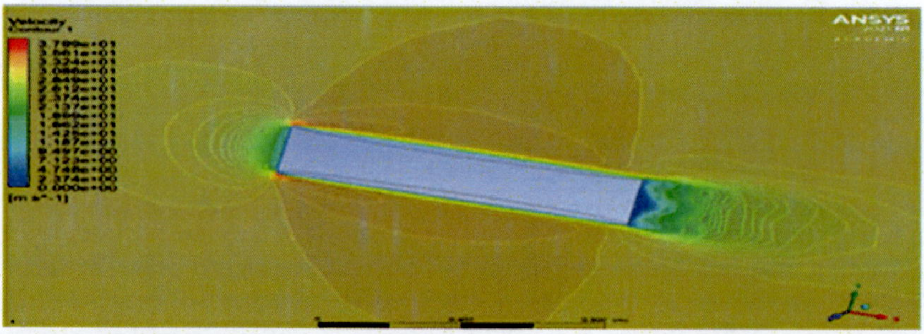

Fig. 1.7 Velocity contour lines with respect to kilohertz

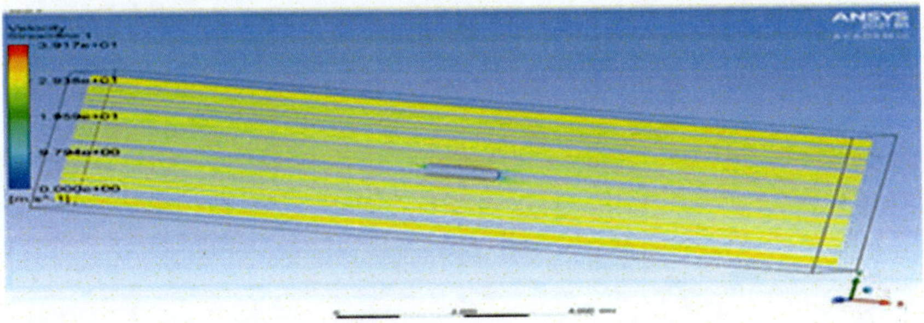

Fig. 1.8 Velocity streamlines with respect to kilohertz

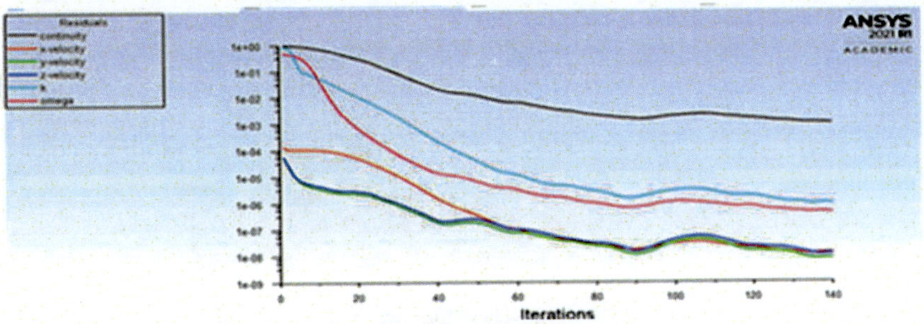

Fig. 1.9 Scaled residual plot, where x-axis is number of iterations and y-axis is equal to residuals

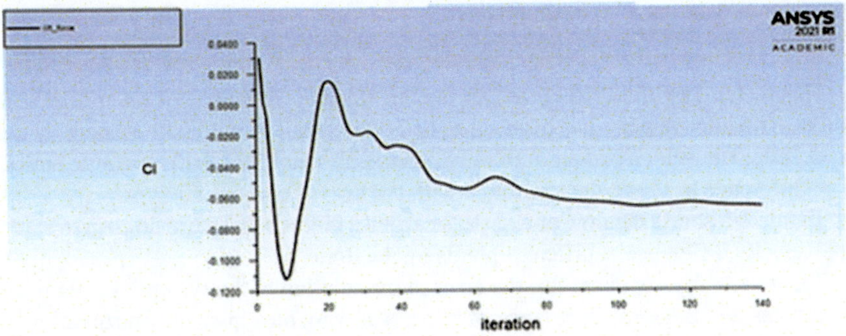

Fig. 1.10 Lift force plot, where x-axis is number of iteration and y-axis is equal to Coefficient of lift

Figure 1.6: In CFD-POST, the figure shows the pressure magnitude by the means of contour lines dawn on a physical body. The pressure values have been maintained at constant atmospheric pressure. The difference in contour lines and its color clearly shows the pressure difference at inlet and outlet.

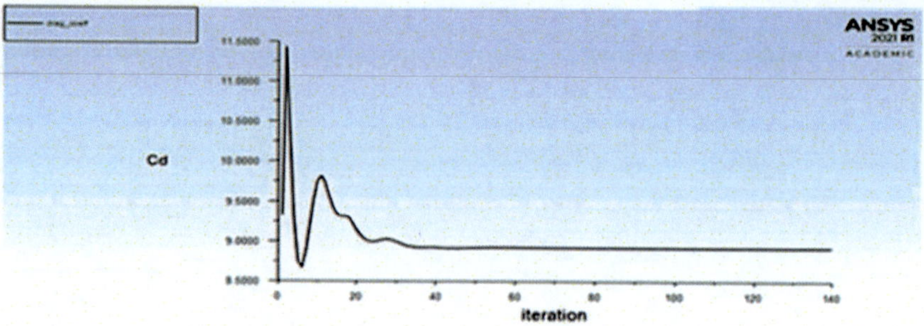

Fig. 1.11 Drag coefficient plot, where x-axis is number of iteration and y-axis is equal to coefficient of drag

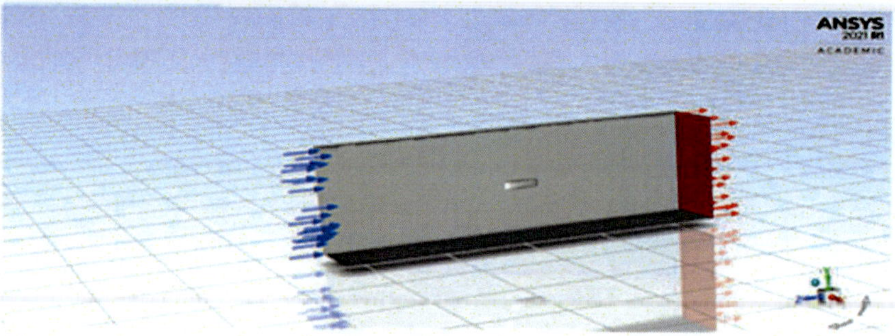

Fig. 1.12 Air inlet and outlet inside enclosure

Figure 1.7: In CFD-POST, we can see the velocity difference of air at inlet and outlet. The inlet velocity has been set to be 30 m/s while at outlet the velocity is set to default. The velocity of air at inlet will be much more than the velocity at outlet. The difference has been demonstrated with the help of contour lines.

Figure 1.8 shows the flow of massless particles through the entire domain to show the velocity difference at different points on the domain.

Figure 1.9 directly quantifies the error in the solution of the system of equations, as it measures the local imbalance of conserved variables in each control volume. The graph represents the residual value of every cell solved equation.

Figure 1.10 is the lift force graph, as the graph stables up after some readings it specifies the lift performance of the new design.

Figure 1.11 is the drag coefficient plot which specifies the air resistance on the surface; the graph gets constant after the initial force, i.e., air gets diverged through first phase of air, increasing the aerodynamic of design. Figure 1.12 is the plot of fluid flow from inlet to outlet, in the figure the blue arrow shows the direction of fluid through inlet section under the enclosure section and red arrow specifies the outlet of fluid.

1.4 Conclusion

From the CFD results, we can say that the newly designed shape of cargo container for drone application is more practical to use as the new design results in less drag due to its curved shape from front and backside which helps in resisting air to a great extent, the new material of construction which is carbon fiber make design durable and helps in decreasing weight.

When the design comes in contact with high wind it becomes aerodynamic and its controls increase, the easier it is for a drone carrying cargo to move, the less energy the system needs making it a valuable part for delivery drones operating at high speed.

References

1. Brar, S., Rabbat, R., Raithatha, V., Runcie, G., Yu, A.: Drones for Deliveries. Berkely university of California (2015)
2. Patchou, M., Sliwa, B., Wietfeld, C.: Unmanned aerial vehicles in logistics: efficiency gains and communication performance of hybrid combinations of ground and aerial vehicles. In: IEEE Vehicular Networking Conference (VNC) (2019)
3. Choi, K.-K., Cho, J.-U.: A convergent investigation on the air flow in driving according to a cargo container and the wind deflector. J. Korea Converg. Soc. (2020). 2713-6353 (eISSN)
4. Cicolani, L.S., Cone, A., Theron, J.N., Robinson, D., Lusardi, J., Tischler, M.B., Rosen, A., Raz, R.: Flight test and simulation of a cargo container slung load in forward flight. J. Am. Helicopter Soc. **54**(3), 32006–3200618 (2009)
5. Ackerman, E., Strickland, E.: Medical delivery drones take flight in east Africa. IEEE Spectr. **55**(1), 34–35 (2018)
6. Campbell, J.F., Sweeney II, D.C., Zhang, J.: Strategic design for delivery with trucks and drones. College of Business Administration, University of Missouri St. Louis, MO, USA (2017)
7. Kovanis, Skaperdas, Ekaterinaris: Design and analysis of a light cargo UAV prototype. AIAA 2010–4957 (2010)
8. Raffler, Wang, Holzapfel: Path Generation and Control for Unmanned Multicolor Vehicles Using Nonlinear Dynamic Inversion and Pseudo Control Hedging. Institute of Flight System Dynamics, Technische Universität München, Germany (2013)
9. Yıldırım, Çabuk, Bakırcıoglu: Design and trajectory control of universal drone system. Measurement **147**, 106834 (2019)
10. Östh, Krajnović: A study of the aerodynamics of a generic container freight wagon using large-eddy simulation. J. Fluids Struct. **44**, 31–51 (2014)
11. Yan, Rakheja, Siddiqui: Baffle Design Analysis for a Road Tanker: Transient Fluid Slosh Approach. Concordia University. 2008-01-2670 (2008)

Chapter 2
The Opportunities and Challenges of Implementing Green Internet of Things (IoT) Towards Energy Saving Practices in the Current Environment

Sonali Vyas, A. Narasima Venkatesh, Purnendu Bikash Acharjee, Satish G. Jangali, Charanjeet Singh, and Zarrarahmed Z. Khan

Abstract The application of internet of things is considered as one of the key cornerstones in the new fifth generation network, it is estimated that more than 40 billion IoT devices will be in place around the work by 2025. The current focus of government and non-profit organisations revolves around implementing energy saving practices, emission of carbon and managing the ecology for sustainable future. The basic objective of Green IoT is to reduce the emission of harmful greenhouse gases and other toxic pollutants, support the government and other stakeholders in conserving the environment and reduce the power consumption. The adoption of the various techniques through Green IoT will lead to creating better and sustainable environment. The Green IoT enables in applying efficient procedures which enables in facilitating the reduction of the harmful substances and to enhance sustainable ecology for the future generation. The smart environment is the one which is focused in leveraging the IoT services, enable in deploying the resources effectively and use the resources so that the energy can be used effectively. This paper attempts to provide the enhanced

S. Vyas
School of Computer Science, UPES Dehradun, Dehradun, India

A. Narasima Venkatesh (✉)
Professor and Head of the Department - Department of HRM and General Management, ISBR Business School, Bangalore, Karnataka, India
e-mail: dr.a.narasimavenkatesh@gmail.com

P. B. Acharjee
Computer Science, The Assam Kaziranga University, Rowmarikhuti, India

S. G. Jangali
Department of Industrial and Production Engineering, U.B.D.T. College of Engineering, Davangere, Karnataka, India

C. Singh
Electronics and Communication Department, Deenbandhu Chhotu Ram University of Science and Technology, Murthal, Sonipat, Harayana, India

Z. Z. Khan
Anjuman I Islam Kalsekar Technical Campus, Mumbai University, Mumbai, Maharashtra, India

S. Yadav et al. (eds.), *Proceedings of Second International Conference in Mechanical and Energy Technology*, Smart Innovation, Systems and Technologies 290,
https://doi.org/10.1007/978-981-19-0108-9_2

opportunities and critical challenges in implementing the Green IoT towards energy saving practices. The Green IoT enables in indicating the application of critical practices for green environment and also supports in saving more cost and energy. This enables in offering systematic solution which will create sustainable growth for the community and also enable in addressing the societal challenges effectively.

2.1 Introduction

The current environment is evolving rapidly due to rapid changes made in technology and Internet. The world is now seeing a faster change in the broadband connections as more individuals' businesses and others are adopting to install such devices due to cheaper cost, faster connectivity and perform business through online. It is noted that there is an exponential increase in the usage of devices which are being connected through Internet which has evolved in creating Internet of Things (IoT). The IoT is mainly driven through the expansion of usage of Internet and connecting the devices so as to provide smarter services to the community, business and individuals [1].

The IoT tends to state the various technologies which are supporting in enhanced connectivity around the world over the physical objects. The IoT can support in sensing the environment, collect data and information from various sources, collaborate with other systems so as to provide the needed services to the stakeholders. The IoT has now become more powerful as there is a growing usage of various systems, understand the environment and respond promptly so that the individuals and business tend to achieve their goals.

The Green IoT focuses in enhancing the efficiencies related to energy usage through the application of different technologies and tools related to IoT. This concept is defined as the application of effective tools which either enable in reducing the greenhouse effects based on existing application or to eradicate them using the resources. Many organisations across the industries and government are focusing in implementing Green IoT for enabling in reducing the greenhouse gases. Various technological solution was stated which covers usage of sensors, software-based tools, cloud computing, application of integrated circuits, smart metering devices etc. which will enable in analysing the usage of electricity, compute the emission of greenhouse gases and suggest measures which will enable in reducing the energy usage efficiently.

The IoT has emerged as a comprehensive tool in the recent years which supports in enhancing the overall quality of creating products and services, enable in enhancing the quality of human life and creating better environment through reducing the emission of harmful gases in the atmosphere [2]. The application of IoT focuses in creating a platform which enables in collaboration of various systems, appliances, application and systems so as to enhance communication, enable in interconnectivity among different devices which supports in performing jobs with more accuracy by communicating and sharing information over the network. IoT has been recently focused in creating better homes, smart cities, enhancing operational activities in the manufacturing companies, reducing the usage of power in factories, warehouses, quicker transportation so that fuel can be saved and in other areas.

The team IoT focuses in stating the various tools and technologies which has been applied for enhancing global communication and connectivity through physical products and objects. The projects used by IoT tend to analyse, sense, collect and transmit the information among each other which enable in taking quicker decision making for the management and individuals. In creating a cleaner and greener environment, Green IoT is being applied in order to understand the usage of power and electricity, enable in reducing the emission of greenhouse gases and support the environment [3]. The major issues facing globally is the climate changes, many governments, organisations and non-profit enterprises are focusing in using various tools and technologies in order to reduce the carbon footprint, enable in creating better and cleaner environment. The IoT can support in analysing the daily usage of power and fuels, create better intelligence which enable in connectivity with the physical world and support in reducing the usage of fuels and power in an efficient manner, which results in lessening the emission, minimise cost and stop the climate change in an efficient manner (Fig. 2.1).

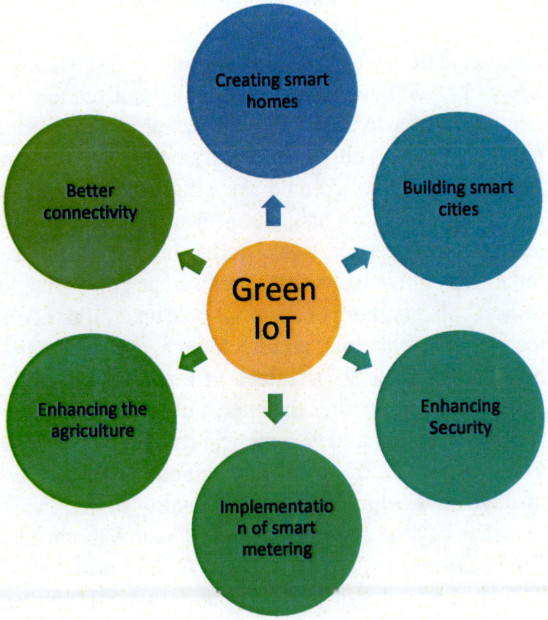

Fig. 2.1 Application of Green IoT. *Source* Albreem et al. [2]

2.2 Review of Literature

Jiet al. [3] discussed various techniques and techniques that can be used to achieve IoT energy efficiency, but did not discuss system models specifically designed for green IoT [4]. Gadreet [5] compared the energy consumption of cloud computers and computers in different scenarios and concluded that choosing the right models for the situation would be the best choice. In addition, their models do not take into account quality of service (QoS), which in some cases can further increase energy consumption. Green technology for IoT implementation has been developed in different sectors, while maintaining the quality of services [6–9], with a special focus on Green IoT solutions. They did not discuss data centres and cloud-based computers, as well as their green solutions that form the backbone of the IoT network. In the ideal load balancing framework [4], it evaluates the workload of different servers in different locations with renewable energy producers, taking into account server resource consumption, energy costs and so on. For green, scalable IoT, an optimization model and a low-power algorithm enable the model to work energy efficiently. The results show that good energy efficiency levels can be achieved with separate network environments. However, the experiments were performed at only 15–20 knots. Therefore, they must be distributed over a large network to achieve valuable energy efficiency results. IoT is used in various industries. Due to the robustness and scalability of the IoT, the medical industry has also decided to store patients in real time [10, 11]. Another strategy that can be adopted to achieve energy efficiency and reduce carbon dioxide emissions is the adoption of basic habits through which we can reduce energy consumption in our daily activities. Although this is a small-scale measure, the addition of small properties around the world can make a big difference [12–14]. One way is to monitor the energy consumption habits of offices, homes and industries with automatic systems and then reduce energy loss during our daily routine tasks. However, it cannot rely too much on this technology, but we can still save a lot of energy [12]. The TREND project collects energy consumption data, evaluates energy saving opportunities from technology, protocols, architectures and experiments with new approaches. It also contains training programmes to promote the green network, i.e. GreenNet. The EARTH project studies the energy efficiency of wireless communication systems. It focuses on theory and the practical constraints of today's energy efficiency to develop a new generation of energy efficient equipment, development strategies and network management solutions to ensure quality of service (QoS). The IEEE Communications Society has also established a Green Subcommittee on Communication and the Green Computer (TSCGCC). TSCGCC works with the development and standardization of energy-efficient computers and communication. It offers opportunities to interact and exchange technical ideas, identify R&D challenges and collaborate on solutions for the development of green communication technology, efficient in terms of energy and resources and environmentally friendly computers [2].

The Internet Academy of Things (IOTA) is London's leading company for improving air quality. IOTA is experimenting with sensors and other technologies to improve air quality in London. The solution proposed by IOTA is the BuggyAir project. IOTA wants sensors in prams that measure pollution at street level and register data. The car's GPS gives an exact indication of the pollution level [1]. Using IoT technology can make air pollution control less complicated and help you better understand the environment.

2.3 Research Methodology

The main purpose of the study is to understand the opportunities and challenges in implementing the Green IoT for implementing energy saving practices in the environment. The implementation of IoT is fast changing and many organisations, government and individuals have started to realise the potential benefits of implementing IoT in safeguarding the planet for the next generation [6]. This study is more conceptual in basis as it focuses in analysing the opportunities faced by the individuals and companies in implementing Green IoT for energy saving practices, the study applies qualitative study so as to understand the merits and challenges faced in implementing the technology. The researchers collate the information from various journals, online publications covering EBSCO, Google Scholar, published thesis, government reports in order to prepare the study [7].

2.4 Critical Discussion

The data centres are considered as the core aspect in order to implement Green IoT in the organisation and residential areas, the data centres enable in analysing the power consumption through analysing the external environment. The servers tend to analyse the temperature in the external environment and continuously sends data to the servers which are processed quickly and the temperature inside the premises are adjusted so as to prevent the usage of electricity efficiently. This architecture tends to require orchestration agent so that the servers receive reliable information, process them and adjust the temperature so that appropriate temperature can be maintained thereby supporting in energy usage.

The sensors are another effective tool which tend to conserve unnecessary usage of energy, the sensors tend to analyse the number of individuals available in the room or building premises and provides the signal in controlling the temperature so that the energy can be saved. With the implementation of embedded server there is a need to ensure optimal data collection and analysis, which supports in taking quick and informed decisions [3] (Table 2.1).

The wireless smart network is one of the effective IoT tool which enable in creating energy efficient measures and leads to creation of Green IoT in the modern business environment, the researchers has applied access control with WSSN which enhances the communication between the system and thereby decreasing the consumption

Table 2.1 Critical elements of Green IoT

S. No.	Elements	Description
1	Documentation	Documentation of gathering the critical information from the point of activity
2	Sensors	Detection of different categories of data and moving them to data warehouse for further processing
3	Enhanced communicating	Various communication tools are implemented for better collaboration
4	Hardware processing	Collaboration of various hardware tools for processing the data
5	Semantics	Information aggregation, extracting historical information so as to predict the future requirement

Source Gadre [5]

of energy, which leads to lowering the cost and protecting the environment effectively. The access control protocols need to be evaluated so as to analyse the energy consumption targets of the business and can provide ways to achieve them, hence they are highly flexible in nature [14].

The application of Green IoT possess various opportunities and challenges in business and home, they are presented as follows.

2.5 Opportunities

There is an increased awareness among the individuals, business leaders and others in protecting the environment for the next generation, enhance the efficient usage of energy and other resources. It is stated that instead of focusing on fossil resources or other related energy needs, the future energy needs are mainly based on renewables resources as it offers cleaner and green energy, protect in emission of harmful gases and support in stopping the adverse effects on climate change. The future electric grid is focused to be more flexible as it can support and balance the power fluctuation, enable in controlling the energy resources, reducing the cost of consumption and protect the environment [13]. Furthermore, it is noted that the implementation of Green IoT offers more opportunities like increasing the quality of life, lowering the cost of energy usage, reducing the latency and power fluctuation, creating greener environment, manufacturing of smart grids etc. Green IoT enable in collating the consumption of energy, analyse the future energy needs and plan wisely so that the requirement is met in a sustainable manner, the implementation of Green IoT can collaborate and communicate with different ICT (Information and Communication technologies) for meeting the energy protocols efficiently (Table 2.2).

The Green IoT is used in waste management as it is a vital issue related to environment, there is an increased cost in recycling the waste without creating adverse impact on the environment, hence through RFID reader, sensor etc. companies tend

Table 2.2 Different Green IoT approaches

S. No.	Major technology	Main purpose
1	Data centre	Analysing the workload, distribute them efficiently for better consumption
2	Sensors	Water metering, minimising the processing etc.
3	Smart buildings	Creating smart meters for better interaction with the grids and utility companies
4	Integrated circuits	Enabling in reducing the network traffic, enable in reducing the energy fluctuations etc.
5	Cloud computing	Predicting energy consumption, tracking various types of energy consumption

Source Arshad et al. [1]

to apply Green IoT in analysing the nature of waste being sourced in the society and take appropriate steps in managing them effectively.

Furthermore, smart water metering is installed so as to measure the quality of municipal water being provided to the citizens, enable in measuring the quantity of water being consumed and position them appropriately so as to maximise the benefits. Smart metering and other IoT tools are used which will provide necessary alarm on the emission of greenhouse gases in the manufacturing facilities, warehouses, supply chain management system etc. This enables in applying novel methods to reduce the emission and focus in generating more carbon credits [5]. The organisation is now focusing in adding more carbon credits as part of sustainable goals, hence Green IoT supports the management in analysing the overall energy usage, identify the emission in different process and supports the organisation in identifying ways to enhance the carbon credit.

The opportunities of Green IoT in the residential sectors are also immense, the property owners can implement IoT tools in order to track the overall energy consumption, the equipments which are consuming more energy and the measures which can be taken in order to save the energy and money [2]. The IoT tends to collaborate with solar panels, batteries etc. which enables in using the appropriate energy sources so that the energy consumption can be optimised. Moreover, the individuals can also implement water sensors in their bathrooms and kitchens which enables in using lesser water resources, maximise the rain water harvesting and check if any of the taps are left open so as to save water [9].

Manu households are now looking to install batteries, solar panels apart from the conventional energy, hence smart metering can be installed in the household for managing the energy sage. Smart metering support in two ways communicating between the meter and the utility companies, enable in measuring the total consumption through traditional meters, support in providing specific information related to the energy usage and also take steps in maximising the usage of energy in efficient manner, this will support in reducing the energy bills and carbon emission.

2.6 Challenges

The IoT is a cutting-edge tool and can support in revolutionising the business and household in enhancing the energy usage, however the technologies are faced with certain challenges. One of the main challenges in the awareness among the individuals in using the system, there are various IoT tools and each of them tend to support for each need and requirements, hence choosing the appropriate IoT is a challenge [10].

There is a constant communication between the sensors and systems placed in controlling the temperature which may lead to increased energy usage rather than controlling them. Moreover, the interconnectivity of the systems needs to be maintained in order to Green IoT to perform efficiently [12]. Though RFID needs lesser unit of energy to pirate based on the needs and requirements, however RFID needs to be active so as to support in implementing energy efficient methods, but when millions of businesses and individuals tend to use such devices, the energy consumption will increase and thereby impacting the environment. Hence, the data centres need to focus in creating unique and low energy consumption-based RFID for easier usage.

2.7 Conclusion

Today's environment is changing rapidly due to rapid changes in technology and the Internet. People today are experiencing a faster change of broadband as more and more companies and others commit to distributing these devices due to lower costs, greater connectivity and faster Internet activity. It is important to point out that there was an exponential increase in the use of devices connected to the Internet, which developed into the creation of the Internet of Things (IoT). IoT is primarily driven by the use of the Internet and expanded device connections to deliver smarter services to society, companies and individuals. The main objectives of the Green IoT are to reduce greenhouse gases and other toxic pollutants, support the government and other stakeholders in preserving the environment and reducing energy consumption. Applying different technologies through Green IoT will lead to a better and more sustainable environment. Green IoT enables the implementation of efficient processes that help reduce harmful substances and improve sustainable ecology for the next generation. The smart environment focuses on the operation of IoT services, enables efficient resource development and uses resources to use energy efficiently.

The potential for Green IoT in the home sector is also enormous, with homeowners being able to use IoT devices to track total energy consumption, which devices consume the most energy and what can be done to save. Energy and money. IoT is usually powered by solar panels, batteries and so on. It works by allowing you to use the right energy sources to optimise energy consumption.

References

1. Arshad, R., Zahoor, S., Shah, M.A.: Green IoT: an investigation on energy saving practices for 2020 and beyond. Special Section On Future Networks: Architectures, Protocols, And Applications (2017)
2. Albreem, M., Alsharif, M.H., Yasin, M.N.M.: Green Internet of Things (GIoT): applications, practices, awareness, and challenges. IEEE Access (2021)
3. Ji, O.A., Tolba, A.: Artificial intelligence-empowered edge of vehicles: Architecture, enabling technologies, and applications. IEEE Access **8**, 61020–61034 (2020)
4. Oliveira, M.C., Pinto, S., Gomes, T.: The future of low-end motes in the Internet of Things: a prospective paper. Electronics **9**(1), 111 (2020)
5. Gadre, M.: Green Internet of Things (IoT): go green with IoT. Int. J. Eng. Res. Technol. (IJERT). ICIOT—2016 **4**(29)
6. Jain, A., Yadav, A.K., Shrivastava, Y.: Modelling and optimization of different quality characteristics in electric discharge drilling of titanium alloy sheet. Mater. Today Proc. **21**, 1680–1684 (2019)
7. Jain, A., Pandey, A.K.: Modeling and optimizing of different quality characteristics in electrical discharge drilling of titanium alloy (grade-5) sheet. Mater. Today Proc. **18**, 182–191 (2019)
8. Gupta, M.K., Magdum, M., Baldawa, Y., Patil, S.: Smart water management in housing societies using IoT. In: Proceedings of 2nd International Conference on. Inventive Communication and Computational Technologies (ICICCT), Apr. 2018, pp. 1609–1613
9. Ebo, O.J.F., Taiwo, O., Olumuyiwa, B.A.: An enhanced secured IOT model for enterprise architecture. In: Proceedings of International Conference in Mathematics, Computer Engineering and Computer Science (ICMCECS), Mar. 2020, pp. 1–6
10. Xu, W.L., Huang, J., Yang, C., Lu, J., Tan, H.: Artificial intelligence for securing IoT services in edge computing: a survey. Secur. Commun. Netw. **2020**, 1–13 (2020)
11. Jain, A., Pandey, A.K.: Multiple quality optimizations in electrical discharge drilling of mild steel sheet. Mater. Today Proc. **8**, 7252–7261 (2019)
12. Panwar, V., Sharma, D.K., Kumar, K.V.P., Jain, A., Thakar, C.: Experimental investigations and optimization of surface roughness in turning of EN 36 alloy steel using response surface methodology and genetic algorithm. Mater. Today Proc.
13. Kunal, A.S., Amin, R.: An overview of cloud-fog computing: architectures, applications with security challenges. Secur. Privacy **2**(4), e72 (2019)
14. Rahbari, Nickray, M.: Low-latency and energy-efficient scheduling in fog-based IoT applications. Turk. J. Electr. Eng. Comput. Sci., 1406–1427 (2019)

Chapter 3
Monitoring of Infrastructure and Development for Smart Cities Supported by IoT Method

Karri Rama Sai Reddy, Chundru Satwika, G. Jaffino, and Mahesh K. Singh

Abstract This manuscript presents a smart monitoring system dependent on a wireless system. This method used Zigbee to collect real-time information about an urban environment. Here using the IoT-based communication for the long-term evaluation, it can be used to monitor from any place. It is consisting of monitoring networks and receiving terminals. Monitoring networks connected to streetlights as routes and taxis nodes and give the address to each node. It is used for the identity of the traffic on network. The system is designed to collect and send the data through the designated terminal in the form of a communication to their monitoring substation. The system was implemented on real-time data for monitoring. The picture and data by routing nodes. The sensor organized by Zigbee throughout the wireless system might be inspired by the smart city infrastructure with the facilitation of the network. It is will be smarter and more secure for society. This paper present how to work the Zigbee and how creation and development of the Zigbee for smart cities.

3.1 Introduction

IoT network brings smart for the city as well as different areas for monitoring. The monitoring of smart cities can be prepared by connecting sensors, lights, meters, etc. The data to be collected from the sensors and it will be stored. The cities then use the stored data to develop the infrastructure, public utilities, services, etc. An urban area can be developed by IoT technology [1]. These are used insight collected from that information to handle belongings, resources, and services professionally and effectively. This contained information gathering from civilians, assets, as well as device which are developed and examined to monitor and managed the different things such as the traffic, hospitals, libraries, schools, colleges, information systems,

K. R. S. Reddy · C. Satwika · G. Jaffino
Department of ECE, Aditya College of Engineering, Surampalem, India

M. K. Singh (✉)
Department of ECE, Aditya Engineering College, Surampalem, India
e-mail: mahesh.singh@accendere.co.in

S. Yadav et al. (eds.), *Proceedings of Second International Conference in Mechanical and Energy Technology*, Smart Innovation, Systems and Technologies 290,
https://doi.org/10.1007/978-981-19-0108-9_3

crime detection, waste management, water supply networks, utilities, power plant, transportation system, etc. [2].

This integrates the information and communication technology (ICT) and different types of substantial devices are associated toward the IoT system. To optimize the effectiveness of smart city operations and their services effectively. It allows the city official to interact openly through equal communities and city infrastructure to scrutinize. To improve the feature and presentation of the city services are used to decrease the cost by using ICT [3]. This method enhanced the contact among populace and government. The applications are urbanized to manage the city flows and allow for real-time response [4–15]. It may be well organized to counter the challenge by a simple "transactional" connection through their citizens. The expression itself ruins indistinguishable to its particulars and consequently, open to a lot of presentation. It provides the capability to slightly monitor supervises and managed the devices, as well as to create novel insight and actionable information commencing huge stream of real-time information [16]. The major characteristics of a smart city contain an important amount of data tools with combination of an exhaustive purpose of data resources [15, 17].

There are some ways in which, this type of technology can be used worldwide today, including following:

(a) Sensor's detection shows the effect of seismic forces and these sensors build a bridge to sense all seismic information [14].
(b) Sensors detected the foundation subsidence, wind forces, government offices [13].
(c) Sensors also detected the persons inside the place to stay and manage lights, heating, and air [12].
(d) Sensors are used to security with buildings government offices and more.

3.2 Literature Review

The tremendous changes in IoT base smart cities in 2014, in this era about the problem of small cities example Ningbo construction; it gives various suggestions about management and measure of main concern to promotion. It solves by using some methods they are efficient investigation and explore with the investigational analysis and synthesis. That is why so many countries take Ningbo as an example to develop our countries [3]. The broad range of interrelated actuators and sensors are available on different portable campaigns and smartphones and jointly distinct as the IoT [4]. A row-based relational database management system is on the whole option for information storehouse and it is used for online transactions. It is also evaluated column and row performances; the purpose of this research is to analyze the line of performance and support storage space scheme are different below the data warehouse kind of DML queries [5]. The growth of the urban area in that turns off the federated simulations model. If it is hand offered to turn then it will develop the urban area. That time presented that one can computerize the formation of code, therefore accelerating the enlargement procedure for creating federation [6].

It represents that if the smart city a large of data will appear that we need large data storage and it will focus on education, medical and transportation so on in the construction of smart city [7]. In that smart transportation brings the form the effect of carbon emissions that can be possible by following these three methods they are (a) clarify the roles. (b) To choose the model of the platform. (c) By using an iterative development model to design the product. In that way, we can reduce the effect of carbon emission [8]. In the smart city, it is using the three integration which one suggested above model for better improvement. The model of smart city forms three perspectives that are, information perception, intelligent application, and data infrastructure constructions. It is required to the smart city governance as traffic congestion, etc. [9, 18].

The flight control system is called a fly by sensors control system, the system associated with them in the flight operation such as their starting, landing, and voice communication, etc. Zigbee wireless device network is aimed at any air vehicle by the Zigbee wireless sensor to regularly monitor on control system [10]. It is a different network topology control problem that could be avoided by the smart city concept in that use the tree standard, algorithms coverage, etc. [11]. The wireless sensor network is used to create the small cities by creating this network. It is measured by various parameters for improved city administration as well as control of urban traffic by using IoT eliminates [12].

3.3 Methodology

The IoT contains smart sensors and uses other devices for their operation. It offers new technical-related issues for urban areas to use IoT systems to manage the traffic, environment pollution, better use of infrastructure and citizens safety, etc. Example: IoT-based smartphones: The smartphone can easily carry anywhere it uses maximum people all day. Smartphones use many tasks daily to connect with other smartphones. Interaction with IoT in smartphone simply because anyone easily can use a smartphone daily. So it is easy to understand how to use IoT in a smartphone. Here problem is not quite right with IoT-enabled devices, for example, your car got check engine light problem we need professional mechanic but the diagnostic tool reads the smartphone it is cheaper than professional mechanic to diagnose the problem. The IoT-based smart city management system is shown in Fig. 3.1.

To bring the smart city we should have the following parameters: ubiquitous connectivity and security and privacy.

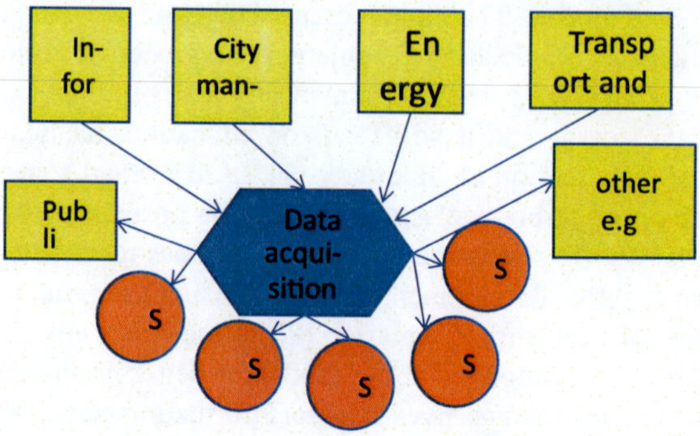

Fig. 3.1 IoT-based smart city management system

Fig. 3.2 Ubiquitous computing technique

3.3.1 Ubiquitous Computing Technique

By using small internet-connected and low-cost computers to help everyday functions in an automated way. Challenges presented by ubiquitous computing across the IoT in systems and it helps to design, in system model, user crossing point design, modern devices, digital audio-players, interactive whiteboards, radiofrequency recognition tags, etc. Ubiquitous computing technique is shown in Fig. 3.2.

3.3.2 Security and Privacy

These are becoming a big challenge in the smart city-related applications cyber security: cyber-attacks are two types of passive attack and active attack both. A passive attack is used for dissimilar information of classification with some changes

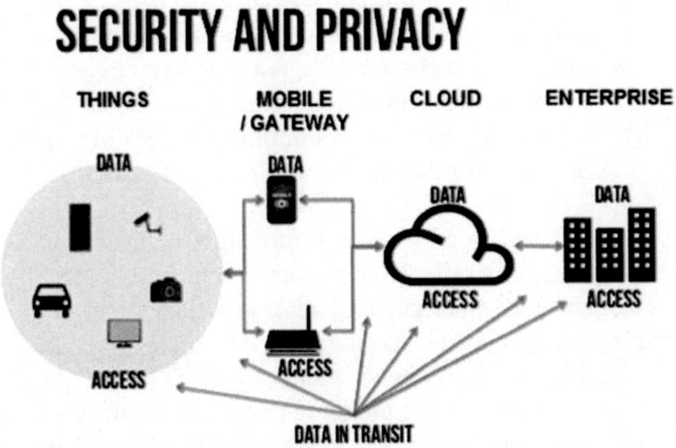

Fig. 3.3 Security and privacy based IoT system

in the possessions of the system. The security and privacy access through IoT is shown in Fig. 3.3.

The major cyber-attacks that could happen in different smart city purpose with briefly describe in this section as follow: traffic analysis, message modification, false information, masquerading, eavesdropping, etc.

3.4 Result and Discussion

The main concept of the paper is to monitor a smart city by using IoT system. To achieve this Zigbee wireless network plays a crucial role in building the smart city. Before the Zigbee, used the "**stack**" but further some drawbacks in the "stack" to implementation, so used the Zigbee concept.

In Zigbee so many advantages The Zigbee has a bending network structure. It has high battery life and also low power consumption it is easy to install and implement. It can support a large number of nodes approximately 6500. It has the low cost compared with the above advantages but here in the system, some have drawbacks also. Zigbee used in official private information is high risk and its low transmission rate. The replacement with Zigbee can be costly and it cannot be used in the outdoor wireless system due to short covering. It is not secure with compare to wi-fi.

1. Zigbee consists of basic remote reception terminal and monitoring system. The essential monitoring structure consists of a variety of terminals and wireless sensor network, Zigbee, 3-G, radio service is used in wireless sensors. For example, the 3-G and GPRS are speared speedily and it is the low financial cost but it faced the issues of instability wi-fi is connected very speed and low cost of maintaining but reaching the wi-fi signal is limited. Zigbee is conversely could

transmit in the long-distance but installation cost high large capacity. But we can use Zigbee it is given more advantages with comparing with wi-fi they are easy protocol, low-energy of the network, safe, reliable, and the capability of Zigbee very strong. In this self-organization and self-healing to fit the Zigbee node into the street lights to organize the basic monitoring network of smart city and the remote receiving the environmental information Zigbee-based smart city shown in Fig. 3.4.

2. Zigbee system is used to collect the information of the humidity, temperature, and succession of terminals of the street light system (see Fig. 3.5). The data can be operated by the mobile monitor and control the applications. The CC2530 sensor work dependent on the z-stack protocol is the fundamental unit of constructing monitoring system of smart city on the street light. The carrying of the basic unit then the sensor circuit is deployed on the street light. A total smart city monitoring system is created which is the Zigbee wireless system to convey data and receive the information from TCP protocol.
3. In this paper discussed the active taxis and road lights in the city. Zigbee knob arranged, routes, top of taxis as well as co-ordinator install in the road lights then solve the two problems, they are limited space and supply. The functions of the

Fig. 3.4 Zigbee enabled based smart city

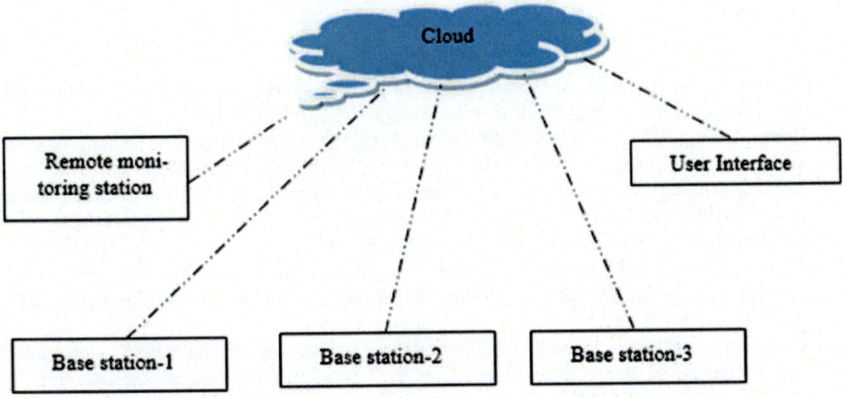

Fig. 3.5 Cloud base Zigbee monitoring of smart city

system: first gathering the illumination data through terminals nodes and show how to work the street lights easily then informing the street light maintenance team of defaults of street light and help them to solve problem, this Zigbee light system is also used for telling the road traffic flow and local information could offer the reliable and real data for the metrological management and traffic management to help them arrange their assigned work shown.

The following important steps are required to implement the smart city:

1. Building the single-hop network.
2. Building the multichip network.
3. Verifying the transferred data.
4. Adding the GPRS/GPS modules to the system for mobile.
5. Developing the remote receiving server.
6. Provide the accessing of backup and data inquiry.

3.5 Conclusion

This paper represented building a smart city by using the IoT. It introduced smart city locations, IoT-based taxis, and street lights. It also adds some more sensors for multifunctional smart city IoT like real-time traffic and weather monitoring. The taxis and streetlight network cover wide range and are systematic scattered and easy to managed. Zigbee wireless sensor system is novel initiative for the construction of smart city infrastructure. By using this system we can control and monitor devices in a real-time manner from any place. So the Zigbee-based IoT system monitoring the smart city is the tremendous change in this field. It is easy to handle remotely with different applications at one time.

References

1. Gyrard, A., Zimmermann, A., Sheth, A.: Building IoT-based applications for smart cities: how can ontology catalogs help? IEEE Internet Things J. **5**(5), 3978–3990 (2018)
2. Chen, S., Yang, C., Li, J., Yu, F.R.: Full lifecycle infrastructure management system for smart cities: a narrow band IoT-based platform. IEEE Internet Things J. **6**(5), 8818–8825 (2019)
3. Zhao, L.: On the policies of mechanism promotion of Ningbo intelligent city construction based on researched analysis. In: 2014 IEEE Workshop on Advanced Research and Technology in Industry Applications (WARTIA), pp. 583–586. IEEE (2014)
4. Shen, C., Choi, H., Chakraborty, S., Srivastava, M.: Towards a rich sensing stack for IoT devices. In: 2014 IEEE/ACM International Conference on Computer-Aided Design (ICCAD), pp. 424–427. IEEE (2014)
5. Kamal, A., Gupta, S.C.: Query based performance analysis of row and column storage data warehouse. In: 2014 9th International Conference on Industrial and Information Systems (ICIIS), pp. 1–6. IEEE (2014)
6. Balaji, V.N., Srinivas, P.B., Singh, M.K.: Neuromorphic advancements architecture design and its implementations technique. Mater. Today Proc. (2021)
7. Punyavathi, G., Neeladri, M., Singh, M.K.: Vehicle tracking and detection techniques using IoT. Mater. Today Proc. (2021)
8. Kanchana, V., Nath, S., Singh, M.K.: A study of internet of things oriented smart medical systems. Mater. Today Proc. (2021)
9. Singh, M., Nandan, D., Kumar, S.: Statistical analysis of lower and raised pitch voice signal and its efficiency calculation. Traitement du Signal **36**(5), 455–461 (2019)
10. Prasanna, G.S., Pavani, K., Singh, M.K.: Spliced images detection by using Viola-Jones algorithms method. Mater. Today Proc. (2021)
11. Singh, M.K., Singh, A.K., Singh, N.: Acoustic comparison of electronics disguised voice using different semitones. Int. J. Eng. Technol. (UAE) **7**(2), 98 (2018)
12. Padma, U., Jagadish, S., Singh, M.K.: Recognition of plant's leaf infection by image processing approach. Mater. Today Proc. (2021)
13. Singh, M.K., Singh, A.K., Singh, N.: Multimedia analysis for disguised voice and classification efficiency. Multimed. Tools Appl. **78**(20), 29395–29411 (2019)
14. Singh, M.K., Singh, N., Singh, A.K.: Speaker's voice characteristics and similarity measurement using Euclidean distances. In: 2019 International Conference on Signal Processing and Communication (ICSC), pp. 317–322. IEEE (2019)
15. Veerendra, G., Swaroop, R., Dattu, D.S., Jyothi, C.A., Singh, M.K.: Detecting plant diseases, quantifying and classifying digital image processing techniques. Mater. Today Proc. (2021)
16. Huanan, Z., Shijun, L., Hong, J.: Guangzhou smart city construction and big data research. In: 2015 International Conference on Behavioral, Economic and Socio-cultural Computing (BESC), pp. 143–149. IEEE (2015)
17. Singh, M.K., Singh, A.K., Singh, N.: Disguised voice with fast and slow speech and its acoustic analysis. Int. J. Pure Appl. Math. **118**(14), 241–246 (2018)
18. Lingli, J.: Smart city, smart transportation: recommendations of the logistics platform construction. In: 2015 International Conference on Intelligent Transportation, Big Data and Smart City, pp. 729–732. IEEE (2015)

Chapter 4
A Review on Various Aerodynamics Control Techniques

Shashank Malik and Faisal Shameem

Abstract The hiking in the fuel price is a great matter of concern. It is very necessary to solve this problem, in a research it is found that 50% of fuel energy consumption is due to aerodynamic drag. From the studies, it is observed that good aerodynamic technique helps in the reduction of drag. In this paper, we review the researches already done on reduction of drag by controlling the flow of wake at rear of the vehicle. We take the studies done in last 10 years. In this paper, we take three techniques for review i.e. passive (splitter plates, non-smooth surface, vortex generators), active (blowing micro jets, suction) and coupled (the combination of active methods). It is observed that all three models help in drag reduction with a broad variation and passive control method found to be the best in all techniques because it requires no extra energy expenditure, not required any input from user and very low in cost than active control techniques.

4.1 Introduction

The rising fuel prices and the norms to control the greenhouse gases emission from vehicles to stop global warming gives a lot of pressure on the engineers to improve the designs of automobiles by taking aerodynamics concepts for increasing efficiency of automotive. Aerodynamic drag is the field which consumes approximate half energy of vehicle as a fuel. Thus, aerodynamic drag reduction is an interesting field to overcome this scenario.

According to *Hucho* [1], the drag of an automotive road vehicle consumes large part of fuel consumption and responsible for 50% loss of fuel energy during highway speeds. Aerodynamics drag have mainly two components—friction drag on skin and pressure drag. Among them pressure drag is responsible for 80% of total drag which

S. Malik · F. Shameem (✉)
Department of Mechanical Engineering, Galgotias University, Greater Noida, Uttar Pradesh 203201, India
e-mail: faisal.shameem@galgotiasuniversity.edu.in

S. Yadav et al. (eds.), *Proceedings of Second International Conference in Mechanical and Energy Technology*, Smart Innovation, Systems and Technologies 290,
https://doi.org/10.1007/978-981-19-0108-9_4

mainly dependent on the geometry or design of the vehicle, because a separation region made behind the vehicle called wake which determines the drag value.

4.2 Techniques of Drag Reduction

For the reduction in drag behind the vehicle, there are mainly two techniques used-

(a) **Passive technique**—This technique is the cheaper technique among them because in this the changing and modification done on the body design of the vehicle at different locations like deflectors, flaps, spoilers etc.
(b) **Active technique**—In this technique a separate device is attached which requires energy to work like Jets, fluid oscillators etc.

There is also another technique which is made from the combination of active and passive technique called **coupled technique**.

4.3 Literature Review

4.3.1 Review Based on the Passive Techniques

(a) **Splitter plates**—The author Gillieron and Kourta, has done an investigation on the use of splitter plates attached at rear side on the Ahmed body at 0.75 scales. This investigation was done in a closed wind tunnel of Prandtl type. They take Reynolds number range $Re = 1.0 \times 10^6$ to 1.6×10^6. They found 28% direct reduction on vertical splitter plates at angle 0° [2–4].

    They vertical splitter plate is shown in Fig. 4.1.
(b) **Vortex generators**—The tiny dispositive used to change forms of vortex is called vortex generators. These are broadly used in higher speeds to control boundary layers. Various studies have been done on vortex generators, Kim and Chen, researches on a minivan vehicle of low mass and achieved 2.8% reduction

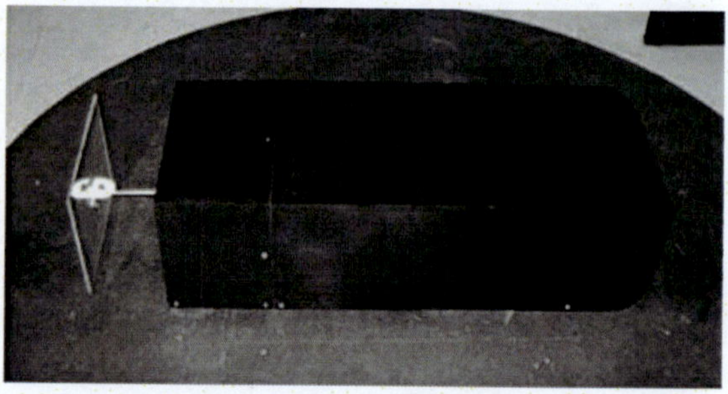

Fig. 4.1 Splitter plate on Ahmed body [4]

Fig. 4.2 Vortex generators installed on car [6]

in drag [5]. Rohatgi et al., investigated VGs on general motors small model and got 1.24% reduction in drag [6]. Arya et al., done a research on passenger car model and analyze done by Ansys CFX, they achieved decrement in drag and lift of 8.7% and 12.8% respectively [7] (Fig. 4.2).

(c) **Deflectors**—Fourrie et al., have done an experiment on Ahmed body with deflector installed on 25° slant rear angle. They did this experiment in a wind tunnel with velocity between 16 and 40 m/s and Reynolds no. Re = 3.4×10^5 and 7.7×10^5 range. They achieved 9% drag reduction and the separation region increases at rear window [8].

Hanfeng et al., studied about the deflectors installed on Ahmed body of 25° slant angle. The experiment was done in wind tunnel of low speed. They take Reynolds number Re = 8.7×10^5 and the velocity of inlet is 25 m/s. Deflectors installed on side and leading edges at slant part and achieved 9.3%, 10.9% and 10.7% with 1%, 3% and 2% width of length [9]. Raina et al., were investigated deflectors at angles—25° to 60° with velocity 16 m/s and 40 m/s on Ahmed body by CFD GAMBIT and fluent, they got 7% reduction of drag [10] (Fig. 4.3).

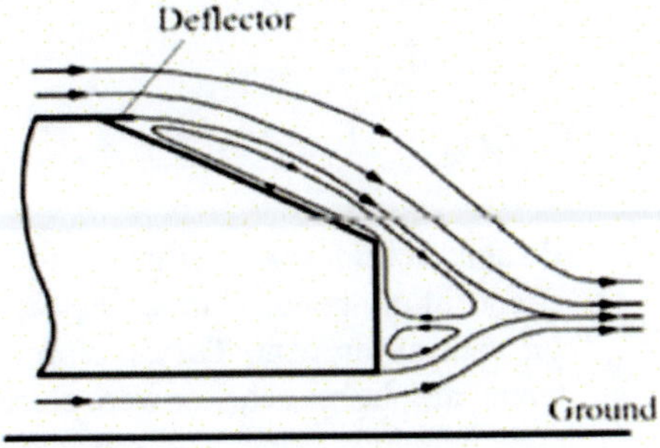

Fig. 4.3 Deflector on Ahmed body [9]

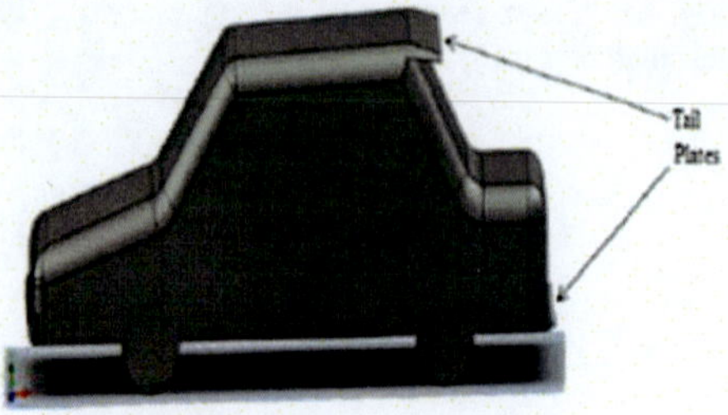

Fig. 4.4 Tail plates model [11]

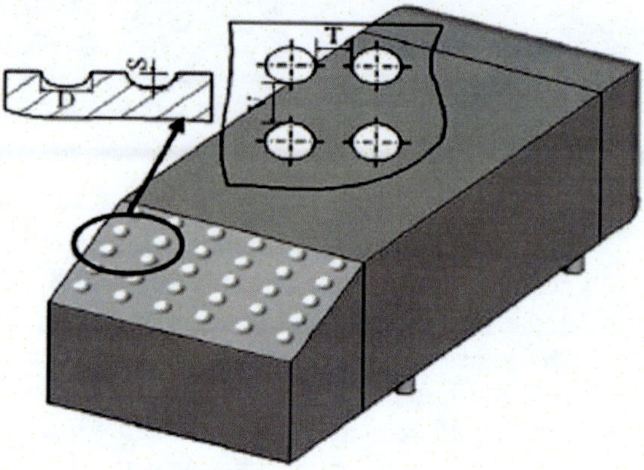

Fig. 4.5 Non-smooth surface Ahmed body [12]

(d) **Tail-plates**—Sharma and Bansal, has done a research on a passenger car model by attached tail plates and simulation was done in Ansys fluent with model K-epsilon of turbulence. The tail plates installed at roof on back side and back bumper with 120° angle. They achieved 3.87% drag reduction and 16.62% coefficient of lift [11] (Fig. 4.4).

(e) **Non-smooth surface**—In this method a shape of golf's ball applied on the rear slant of Ahmed body. Yiping et al., have done a research on the Ahmed body and find the effect of dimples on the back slant surface of Ahmed body. They got the decrement of 5.20% drag coefficient by use of Kriging surrogate model [12] (Fig. 4.5).

4.3.2 Review Based on the Active Control Techniques

(a) **Plasma actuators**—Shadmani et al., have done an experiment to study the effects of plasma actuators. They choose Ahmed body model with 25° slant for their experiment. They perform the experiment in an open circuit wind tunnel the plasma actuator were placed in the center of slant surface of Ahmed body with inlet velocity 10 m/s and Reynolds number $Re = 4.5 \times 10^5$. They

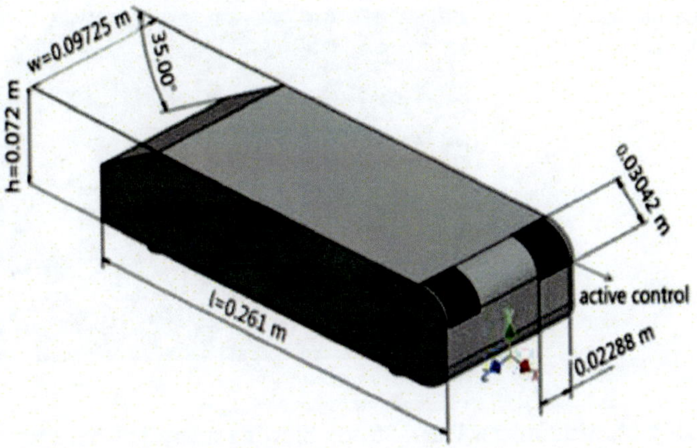

Fig. 4.6 Ahmed body with suction [14]

found that the flow stick back to the surface by actuators and got 3.65% reduction in drag [13].

(b) **Suction**—Harinaldi et al., did a study on flow over the Ahmed body of scale 1:4 with 35° slant by using suction method on front side shown in Fig. 4.6. They used both testing i.e. experimental and numerical. The suction and upstream velocity were 1 m/s and 13.9 m/s respectively. By introducing suction, they obtained decrease of 12.35% in turbulence intensity and increase of pressure coefficient value 26.17%. In the numerical approach, drag reduction of 13.86% obtained by means of suction while in experimental approach 16. 32% drag reduction was obtained [14].

Moussa et al., come with an approach to find the parameters for maximum reduction of drag. This approach was the combination of suction slit with automatic modeling, orthogonal arrays and the computational fluid dynamics. They choose SUV car model for fitting this suction slit with adequate location of opening. From this experiment they found 19% decrement in aerodynamic drag [15].

(c) **Fluid oscillators**—A simple device that contains no moving parts and gives a sweeping jet output from steady flow input called fluid oscillator. Metka and Gregory, conducted a study on Ahmed body 25° slant with blowing of quasi-steady on the roof-slant surface and span wise array type fluid oscillators used. The main goal of their study was to decrease the drag by cut-off the separation region on rear slant. They did this experiment in a wind tunnel of 3 × 5 feet subsonic type State Aerospace Research Center in Ohio. Pressure taps and particle image velocimetry (PIV) were used to define the changes in structure of flow behind the model at Reynolds number close to Re = 1.4 × 106. They used oil flow visualization to understand the working behind oscillator effectiveness. 7% of aerodynamic drag reduction was obtained to control the separation on the rear slant surface [16] (Fig. 4.7).

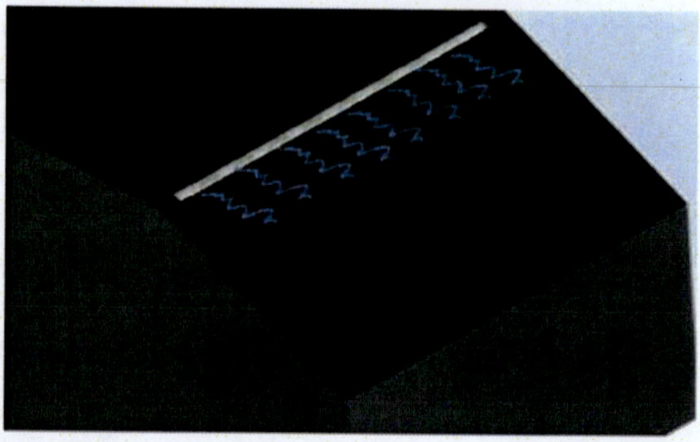

Fig. 4.7 Oscillators on Ahmed body model [16]

(d) **Steady blowing micro jets**—Steady blowing micro jets have jet orifices of array type and their diameter are below the model or wake length. Aubrun et al., investigated the importance of blowing micro jets methods for drag reduction in ground vehicle. They used Ahmed body model of 25° slant as their base model. The micro jets were placed at the line between the slant part and the roof with an array of 6 mm. The investigation was done in a wind tunnel (Lucien Malavard). Wall pressure, PIV and skin friction visuals load measurements of aerodynamics were used to understand changes in flow with velocity 40 m/s and Re $= 1.95 \times 10^6$. A reduction of 9–14% drag and 42% coefficient of lift were observed [17].

(e) **Pulsed jets**—Bideaux et al., experimentally studied the separation of flow control on the rear side on an Ahmed body model with 35° slant and scale of 0.7. This study was conducted in PRISME Institute of Orleans by the help of wind tunnel (Malavard subsonic). A strip of pulsed jets attached on the roof line of model for flow control with velocity 30 m/s. They obtained reduction in drag coefficient about 20% at 500 Hz pulse frequency and coefficient of momentum $C\mu = 2.75 \times 10^{-3}$. This result confirms the importance of pulsed jets in drag reduction and decreasing emission pollutions [18].

4.3.3 Review Based on Coupled Techniques

Bruneau et al. used various active and passive control techniques for drag reduction of the square back Ahmed body. The main goal of this research is to couple passive and active control possibility for drag reduction. They used some active control methods—closed loop jets, pulsed jets and steady jets as well as passive control methods by using porous layers on the geometry and then find the drag reduction results. They find steady jet expensive, the pulsed jet difficult to handle and the closed loop jet more relevant among these. The coupling of closed loop jet with porous layers with good position of sensors and actuator, results as 31% drag reduction, and this is very promising result [19].

Li et al., the objective of their study was to investigate the drag reduction effect of riblets i.e. passive riblets placed in the stream wise direction. Experiment done on riblet on aluminum plate inserted and electromagnetic actuator used as an active control device. They found drag reduction of 4.7% at zero pressure gradient while it increases 6–7% in realistic conditions [20].

Most of the studies were done only on single phase simulation on a car, Mishra, Sujit et al. did investigation on the multiphase mixture K-epsilon turbulence scheme model on a simplified car to obtain drag coefficient. The simulation was successfully conducted on the 2D model of a simplified car and the results found to be more accurate for aerodynamic parameters [21].

This is a review paper of the author Mukut et al. in which they review various researches on the active, passive and combined flow control method for drag reduction. This paper gives the information about various researches on active, passive and combined flow control methods like- movable under body diffuser, steady blowing, synthetic jets, pulsed jets, steady suction, plasma actuator, vortex generators at different location, spoilers, flaps and body modification. They used them individually and combine. It is concluded that the drag reduction occurs 20%, 21.2% and 30% in active, passive and combined flow control methods respectively [22].

4.4 Conclusion

Engineers are trying to solve aerodynamic drag problem, in a research they found that 50% of fuel energy consumes due to aerodynamic drag and they work on reducing it. From the study of researches done on aerodynamics, it is observed that good aerodynamic technique helps in the reduction of drag.

In this review paper, we have studied about various researches already done on the aerodynamic drag reduction techniques and it is found that all techniques help in reducing the drag if they installed properly. Researchers have done various researches by experimentally and numerically and both give satisfactory results and I studied various techniques research papers as mentioned in above section.

If we compare all three techniques—active, passive and coupled then we found that the passive technique is the most economical which gives more output in less input because in this method we did not require any separate operating device while in active control methods we require an add-on device which operates mechanically that is not possible to install on every vehicle. The coupled technique is also good in reducing drag and we can use it on high performance requirement vehicle but not on every vehicle and it is the most complex technique. In the end, it is concluded that the passive techniques have more advantages than any other technique because it requires no extra energy expenditure, not required any input from user and very low in cost than active control techniques.

References

1. Blocken, B., Toparlar, Y.: A following car influences cyclist drag: CFD simulations and wind tunnel measurements. J. Wind Eng. Ind. Aerodyn. **145**, 178–186 (2015)
2. Hucho, W.H., Sovran, G.: Aerodynamics of road vehicles. Annu. Rev. Fluid Mech. **25**(1), 485–537 (1993)
3. Sudin, M.N.: Review of research on vehicles aerodynamic drag reduction methods. IJMME **14** (2014)
4. Gilliéron, P., Kourta, A.: Aerodynamic drag reduction by vertical splitter plates. Exp. Fluids **48**(1), 1–16 (2010)
5. Kim, I., Chen, H.: Reduction of aerodynamic forces on a minivan by a pair of vortex generators of a pocket type. Int. J. Veh. Des. **53**(4), 300–316 (2010)
6. Rohatgi, U.S.: Methods of reducing vehicle aerodynamic drag. In: ASME 2012 Summer Heat Transfer Conference (2012)
7. Arya, S., Goud, P., Mathur, S., Sharma, V., Tripathi, S., Shukla, V.: A review on "Reduction of drag force using ADD-ON devices." Int. Res. J. Eng. Technol. (IRJET) **3**(11), 985–994 (2017)
8. Fourrié, G., et al.: Bluff-body drag reduction using a deflector. Exp. Fluids **50**(2), 385–395 (2011)
9. McNally, J., et al.: Drag reduction on a flat-back ground vehicle with active flow control. J. Wind Eng. Ind. Aerodyn. **145**, 292–303 (2015)
10. Raina, A., Harmain, G.A., Ul Haq, M.I.: Numerical investigation of flow around a 3D bluff body using deflector plate. Int. J. Mech. Sci. **131**, 701–711 (2017)
11. Sharma, R.B., Bansal, R.: CFD simulation for flow over passenger car using tail plates for aerodynamic drag reduction. IOSR J. Mech. Civil Eng. (IOSR-JMCE) **7**(5), 28–35 (2013)
12. Wang, Y, et al.: Reduction in the aerodynamic drag around a generic vehicle by using a non-smooth surface. Proc. Inst. Mech. Eng. Part D J. Autom. Eng. **231**(1), 130–144 (2017)a
13. Li, R., et al.: Drag reduction of a car model by linear genetic programming control. Exp. Fluids **58**(8), 1–20 (2017)
14. Harinaldi, H., et al.: Modification of flow structure over a van model by suction flow control to reduce aerodynamics drag. Makara J. Technol. **16**(1), 15–21 (2012)
15. Moussa, A.A., Yadav, R., Fischer, J.: Aerodynamic drag reduction for a generic sport utility vehicle using rear suction. J. Eng. Res. Appl. **4**(8), 101–107 (2014)
16. Metka, M., Gregory, J.W.: Drag reduction on the 25-deg Ahmed model using fluidic oscillators. J. Fluids Eng. **137**(5) (2015)
17. Aubrun, S., et al.: Separation flow control on a generic ground vehicle using steady microjet arrays. Exp. Fluids **51**(5), 1177–1187 (2011)
18. Bideaux, E., et al.: Drag reduction by pulsed jets on strongly unstructured wake: towards the square back control. Int. J. Aerodyn. **1**(3–4), 282–298 (2011)
19. Bruneau, C.-H., et al.: Coupling active and passive techniques to control the flow past the square back Ahmed body. Comput. Fluids **39**(10), 1875–1892 (2010)
20. Li, W., Olivier, H., Schröder, W.: Experimental investigation of passive and active friction drag reduction. No. RWTH-2017-10471. Lehrstuhl für Strömungslehre und Aerodynamisches Institut (2017)
21. Mishra, S., et al.: Multiphase mixture model simulation over a simplified car. Adv. Appl. Fluid Mech. **23**(1), 17–24 (2019)
22. Mukut, A.N.M.,M.I., Abedin, Z.: Review on aerodynamic drag reduction of vehicles. Int. J. Eng. Mater. Manuf. **4**(1), 1–1 (2019)

Chapter 5
Air Pollution Index Prediction: A Machine Learning Approach

Praveen Kumar Maduri, Preeti Dhiman, Chinmay Chaturvedi, and Abhishek Rai

Abstract Nowadays air pollution has become a major problem for our present generation. Many research and innovations are done in order to deliver fresh and clean air and the first and the most important step in this process is the prediction and monitoring of air pollution index. For accurate prediction of air pollution in surroundings many advanced models are developed taking from different machine learning models like logistic regression model, auto-regression model and many others to the several deep learning models. In this paper, we have proposed three different machine learning models LightGBM, GBM and random forest to compare them on the basis of 21 different physical parameters like humidity, temperature, traffic volume on roads for calculating Air quality index value and their dependency on determining the AQI value. The developed model was found to be more accurate as compared to previously developed models in predicting the level of contamination in surrounding air.

5.1 Introduction

With the increasing industrialization and urbanization, the area of forest-covered land is decreasing which is directly affecting our environmental condition and creating an imbalance in nature. Surrounding air is one of these influenced natural resources which is badly contaminated by different human activities. According to the world health organization, every nine out of ten people living on earth breathe contaminated air which directly or indirectly harms their body organs. The first and the foremost step in solving this issue is the accurate prediction of Air quality index of surrounding air and also determining the parameters affecting it most.

The proposed paper focuses on determining the Air Quality Index (AQI) value using three different machine learning models. For training the model a data set is imported which comprises 21 different parameters like traffic volume, temperature,

P. K. Maduri · P. Dhiman · C. Chaturvedi (✉) · A. Rai
Galgotias College of Engineering and Technology (AKTU), Greater Noida, India
e-mail: chinmaychaturvedi76@gmail.com

S. Yadav et al. (eds.), *Proceedings of Second International Conference in Mechanical and Energy Technology*, Smart Innovation, Systems and Technologies 290,
https://doi.org/10.1007/978-981-19-0108-9_5

wind direction, humidity, clouds, visibility, snow, month, year, etc. Based on the values from the available data set the three machine learning models are trained and compared by analyzing the error generated between actual value and predicted air quality index values. Out of these three models, one model is selected which produces least error. Once the models are compared and selection is done then the effect of the parameters on influencing the AQI value is calculated and plotted. After all these compilation processes the AQI values are predicted for every month in a year. The contribution of the given paper is given below:

- The proposed paper aims to determine the best decision tree-based model out of three different models which are Light GBM, GBM and Random Forest for value prediction purposes.
- The model also aims to predict the air quality index for surrounding air based on different physical parameters.

5.2 Literature Review

In the major developing countries, the monitoring and prediction of air pollution levels are becoming a difficult task and in order to solve this problem, many smart and advanced technologies are being developed and adopted by researchers like implementation of IoT (Internet of things) based smart real-time monitoring system and Machine learning or Artificial intelligence-based system for accurate and efficient prediction of air quality index.

Till now many machine learning models are being developed and deployed in different parts of the world for efficient forecasting of pollution levels in our surroundings. Particulate matter, which is the major component of smog, needs to be accurately predicted for which a combo of 1-D (one dimensional) CNN and multi-dimensional bi-directional LSTM model were developed for guessing the PM 2.5 level [1] in the air. Other than this a machine learning model based on regression and auto-regression models is developed for determining the level of small pollutants in air [2]. Adaptive neuro-fuzzy interface is another model which is deployed for predicting the PM particle level in the environment [3]. Other than particulate matter there are many other harmful gases that are responsible for contaminating surrounding air. Sulfur dioxide is a harmful gas that is generally emitted from power stations and by burning fossil fuels in large quantities. In order to monitor the Sulfur dioxide level, an orthogonal decomposition and machine learning model was proposed which is called parameterized non-intrusive reduced order model for reducing the pollutant transport equation [4]. Along with oxides of Sulfur, there are many other gases that are present in impure air like ground-level ozone, oxides of nitrogen, oxides of carbon such as carbon monoxide, carbon dioxide and others. For predicting these harmful components of polluted air, a classification algorithm was developed in which outputs from different sources were combined like chemical components, meteorological forecast output from different sources for training the model and predicting the air impurity level [5]. Real-time pollution monitoring is another big step in determining air contamination levels in the environment. For such different combinations of hardware and software were set up for efficient prediction of impurity. Image processing is one of the most efficient ways to predict pollution levels by analyzing the images and videos of gases emitting from chimneys of industries [6] and then using different classification programs for guessing the contamination level of air. Implementation of Internet of things with a machine learning model is another potential way of real-time monitoring the degree of impurity in air. Use of different microcontroller-based sensing devices like humidity and temperature sensor along with air pollution sensing modules for recording the real-time data of different polluting components found in air and analyzing it using a proper machine learning model [7] is another way of pollution monitoring of air. Another technique is using previously recorded data along with sensor data simultaneously and a machine learning model to depict the air quality value [8]. But after all these developments there is a need to deploy a proper framework that accurately predicts the AQI (Air quality index) for a large dataset and with a proper machine learning model. The proposed paper has implemented three different machine learning models to predict the AQI value and also has suggested the best model out of three models.

5.3 Flow Chart

See Fig. 5.1.

5.3.1 Methodology

In this paper, we have proposed a machine learning model for predicting the Air Quality index using three different frameworks which are Light GBM, GBM and Random Forest. For value prediction, a proper data set is imported consisting of 21 different parameters and based on these data the AQI value is determined. While after analyzing the data set few data variables are found to be of no use which is rain ph., snow ph. and weekend and thus are removed from the data set. The used dataset has different physical parameters which are recorded from the year 2012 to 2017 as shown in Fig. 5.2 for training the model.

The data set used in the paper is extracted and divided into two parts:

- Train dataset: The train data set compressed 80% of data from overall data set.
- Test dataset: the test data set consists of 20% of the total available data.
- The used data set consists of a balanced set of data which can be visualized from the graph shown in Fig. 5.3.

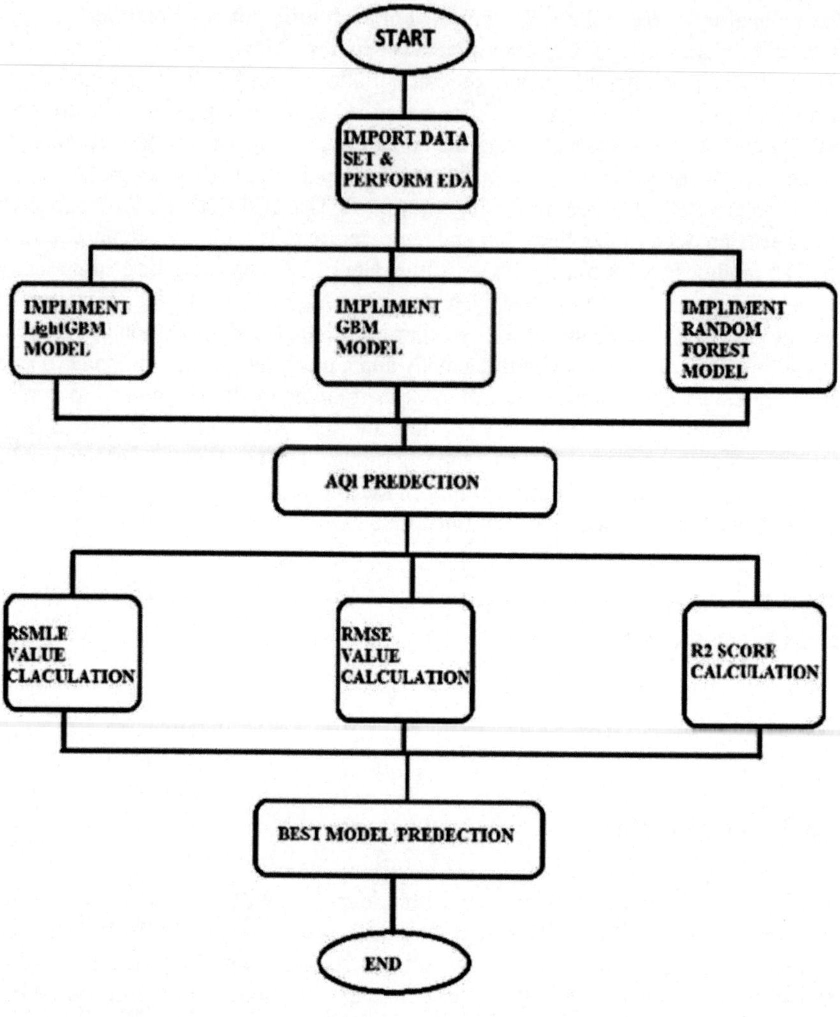

Fig. 5.1 Flow chart

5.3.2 Data Set

The used data set consists of total of 22 parameters out of which 21 are independent parameters which are date-time, holiday, humidity, wind speed, wind direction, visibility in miles, dew point, temperature, rain, snow, clouds, weather type, traffic volume, humidity and temperature ratio, year, day, week of the year, month, day of week, weekend and hour and one is dependent parameter which is Air quality index. The parameters in used data set consist of both integer and string values. The parameters like holiday consist array of different holidays falling in a year such as

index	is_holiday	humidity	wind_speed	wind_direction	visibility_in_miles	dew_point	temperature
0	7	89	2	329	1	1	15.13
1	7	67	3	330	1	1	16.21
2	7	66	3	329	2	2	16.43
3	7	66	3	329	5	5	16.98
4	7	65	3	329	7	7	17.99

clouds_all	weather_type	air_pollution_index	traffic_volume	hum_ratio_temp	year
40	1	121	5545	3.239101	2012
75	1	178	4516	4.318806	2012
90	1	113	4767	4.387576	2012
90	1	20	5026	4.395909	2012
75	1	281	4918	4.479077	2012

day	weekofyear	month	dayofweek	hour
2	40	10	1	9
2	40	10	1	10
2	40	10	1	11
2	40	10	1	12
2	40	10	1	13

Fig. 5.2 Used data set

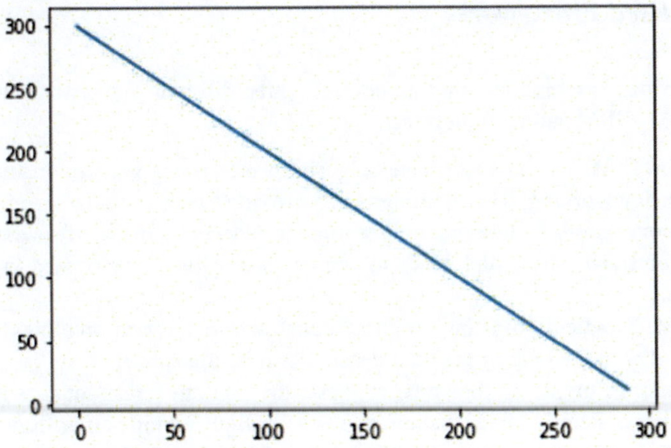

Fig. 5.3 Balanced dataset representation

Christmas Day, Independence Day, labor day, veteran's day, etc. while parameter weather type consists of cloudy, clear, rain, drizzle, fog, etc. Other than these two other parameters contain integer values depending on their type.

For more precise understanding of parameters in dataset different analysis is performed before implementing the models. Being a multivariate regression problem where one variable is depending on multiple independent variables a multivariate analysis is performed to understand how the variables are related to each and to get a view of how they behave with respect to each other.

$$Y = \beta_0 x_0 + \beta_1 x_1 + \beta_2 x_2 + \beta_3 x_3 + \cdots + \beta_{17} x_{17} \tag{5.1}$$

In Eq. (5.1) the dependency between independent and dependent variables is shown where Y represents the dependent variable which is Air Quality Index and β_0, $\beta_1 \ldots, \beta_{17}$ represent the coefficient of dependency while $x_0, x_1 \ldots, x_{17}$ are independent parameters like humidity, temperature, etc.

To get insight into general distribution of dataset variables histogram distributional features plot is used to understand the data distribution of each used variable. The correlation between the variables is shown using heat map plot where the positive values represent that on changing the value of one feature the other feature will tend to follow the same order, i.e., on increasing the value of one variable the value of other will increase and on decreasing the value of one variable the value of other will also decrease. The negative value indicates the reverse of value, i.e., on increasing the value of one variable the value of other variables will increase and vice-versa.

5.3.3 Used Frameworks

The described machine learning model uses three different frameworks which are LightGBM, GBM and random forest.

1. **LightGBM**: This is a tree-based algorithm that uses gradient boosting framework for working. This is designed to provide faster training speed with low memory usage. Unlike other algorithms in place of growing trees depth-wise, it grows leaf wise which tends to achieve lesser loss as compared to previous ones.
2. **GBM**: Gradient Boosting machine constructs a forward level-wise additive model with the help of gradient descent in function space.
3. **Random Forest**: As the name suggests the random forest consists of a large number of decision trees that operate as a multiple learning algorithms to obtain better predictive performance (Figs. 5.4, 5.5 and 5.6).

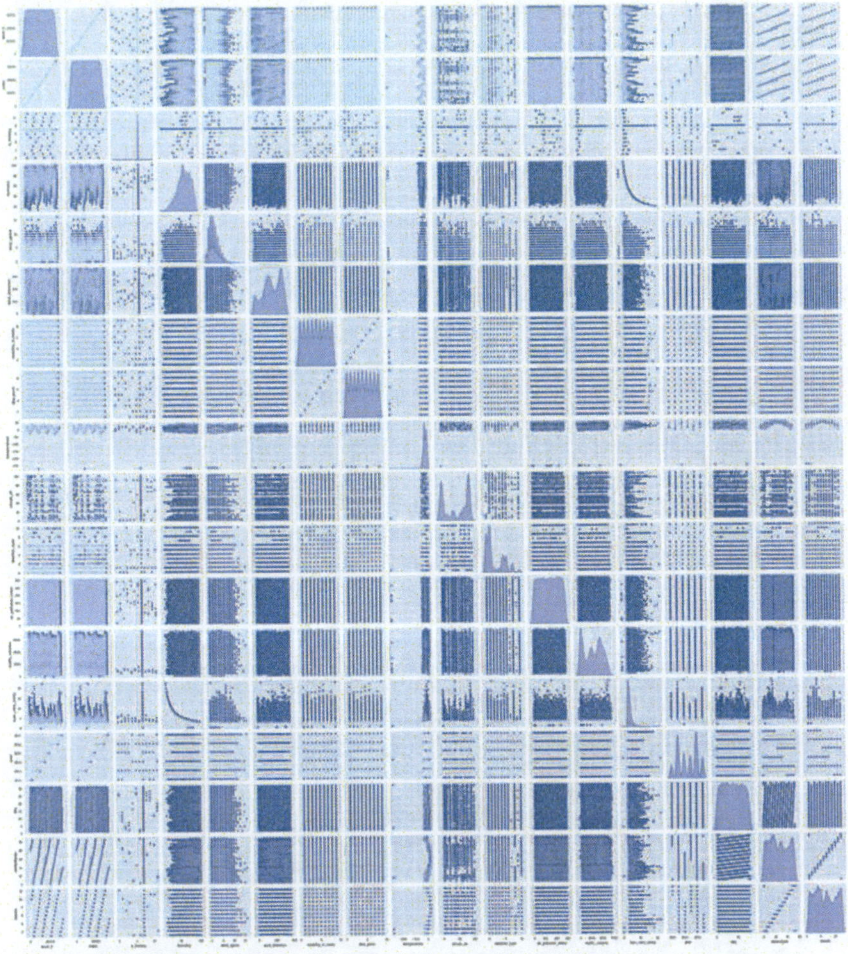

Fig. 5.4 Multivariate analysis chart

5.4 Working

The overall programming of the model is divided into two sections

a. Prediction of AQI value
b. Error calculation for actual value and predicted value of each model.

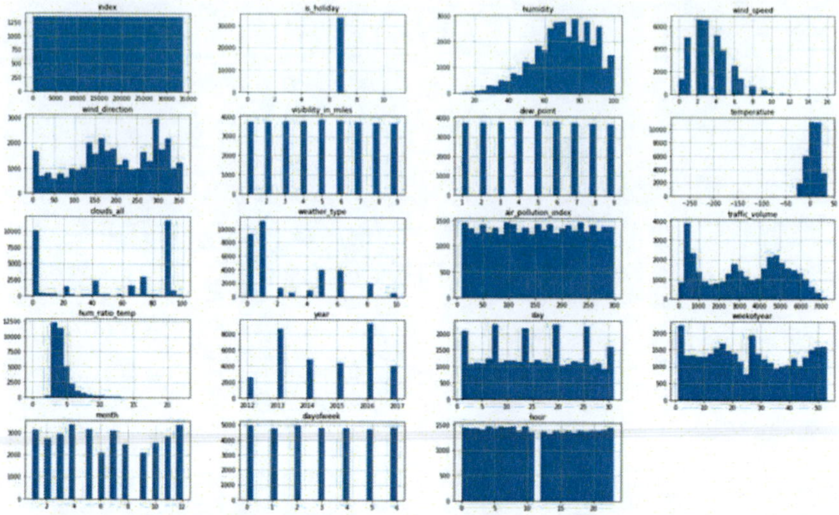

Fig. 5.5 Dataset variable distribution chart

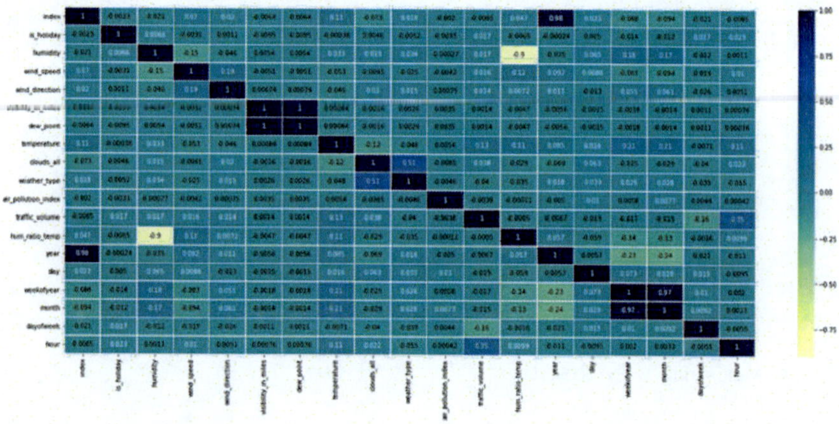

Fig. 5.6 Correlation heat map

5.4.1 Prediction of AQI Value

The model proposed here is a logistic regression problem that is solved using three different decision tree-based frameworks namely LightGBM, Gradient Boosting Machine (GBM) and random forest. The model is trained using 80% of the total data set. At the starting, the dataset is imported and Null values are calculated for data cleaning process and then after the string values present in the dataset are converted into integer values. Once the data processing is done the three frameworks

```
import seaborn as sns
feature_imp = pd.DataFrame(sorted(zip(lgbm.feature_importance(), X.columns), reverse=True)[:50],
                           columns=['Value','Feature'])
plt.figure(figsize=(12, 10))
sns.barplot(x="Value", y="Feature", data=feature_imp.sort_values(by="Value", ascending=False))
plt.title('LightGBM Features')
plt.tight_layout()
plt.show()
```

Fig. 5.7 Implementation of seaborn library for parameter dependency in Light GBM

are imported. Once the framework is imported then the feature dependency is calculated using seaborn library as shown in Figs. 5.7 and 5.9 and respective graph is also plotted as in Fig. 5.8 and fir10 for determining which parameter is largely responsible for determining the AQI value. After compiling all these steps then by using a test dataset the values of Air quality are predicted (Fig. 5.10).

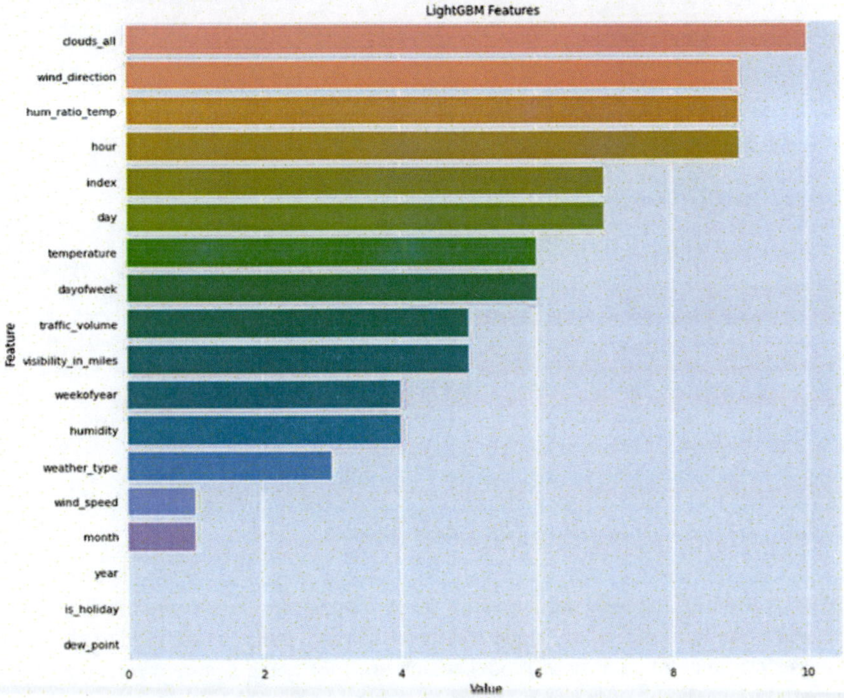

Fig. 5.8 Dependency of parameter for Light GBM model

```
feature_imp = pd.DataFrame(sorted(zip(gb.feature_importances_, X.columns), reverse=True)[:60], columns=['Value','Feature'])
plt.figure(figsize=(12,10))
sns.barplot(x="Value", y="Feature", data=feature_imp.sort_values(by="Value", ascending=False))
plt.title('Gradient Boosting Features')
plt.tight_layout()
plt.show()
```

Fig. 5.9 Implementation of seaborn library for parameter dependency in GBM model

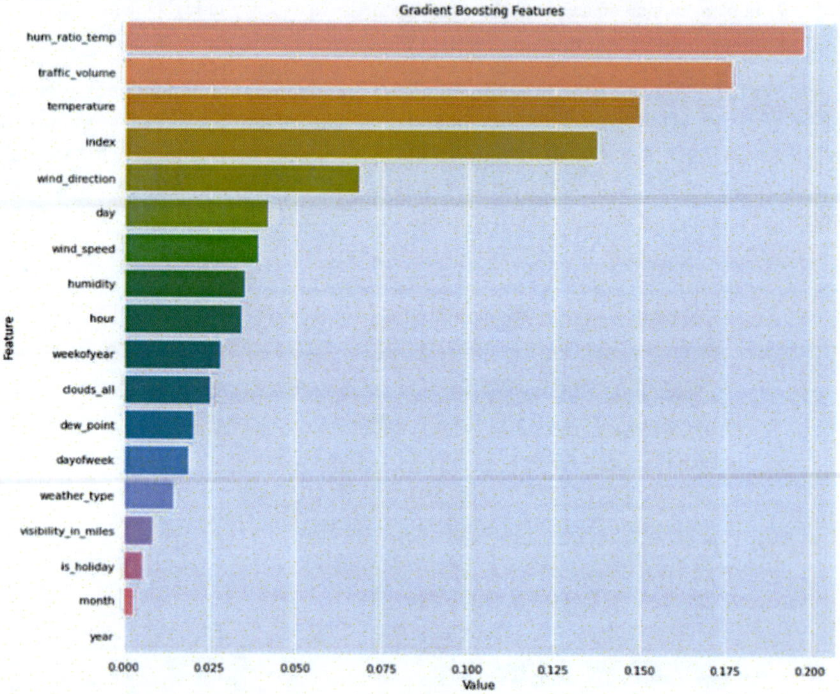

Fig. 5.10 Dependency of parameters for GBM model

5.4.2 *Error Calculation for Each Model*

After predicting the AQI values using three models the data comparison is done between predicted and actual value and three different matrices are used to calculate the performance of each used model which are root mean square error, root mean squared logarithmic error and R2 score.

$$\text{RSMLE} = \sqrt{\frac{1}{n}\sum_{i=1}^{n}(\log(P_i + 1) - \log(A_i + 1))^2} \tag{5.2}$$

The formula written in Eq. 5.2 is used to calculate RMSLE where P_i is predicted value, A_i is actual value and n is total observation in dataset and log(y) is the natural log of y.

$$\text{RMSE} = \sqrt{\frac{1}{n}\sum_{i=1}^{n}(P_i - A_i)^2} \tag{5.3}$$

The formula written in Eq. 5.3 is used to calculate root mean square error (RMSE) where P_i is predicted value, A_i is actual value and n is the total number of observations.

$$R2\ \text{score} = 1 - \frac{\text{Unexplained Variation}}{\text{total variation}} \tag{5.4}$$

The $R2$ score shown in Eq. 5.4 is used to calculate the measure of how well the observations are reproduced by the model.

For each model, the three matrices are calculated and out of all three models, the one which produced lower value is considered to be best model for AQI prediction.

5.5 Results

The proposed model is able to produce the promising results as the available dataset and is able to give the nearly accurate values accordingly. The proposed model is able to give the following results.

1. Prediction of Accurate AQI value.
2. RMSLE, RMSE, R 2 score calculation for each of three models.

5.5.1 Prediction of AQI Values

See Fig. 5.11.

5.5.2 Error Calculation

Once the predicted data and actual data is compared and matric error values are calculated using Eqs. 5.2, 5.3 and 5.4 and mean of error generated from each matric equation is calculated and plotted separately using bar chart as shown in Figs. 5.12, 5.13 and 5.14 and respective values are shown in Table 5.1. After plotting of graph of mean error value for each of three matrices the model which is able to generate

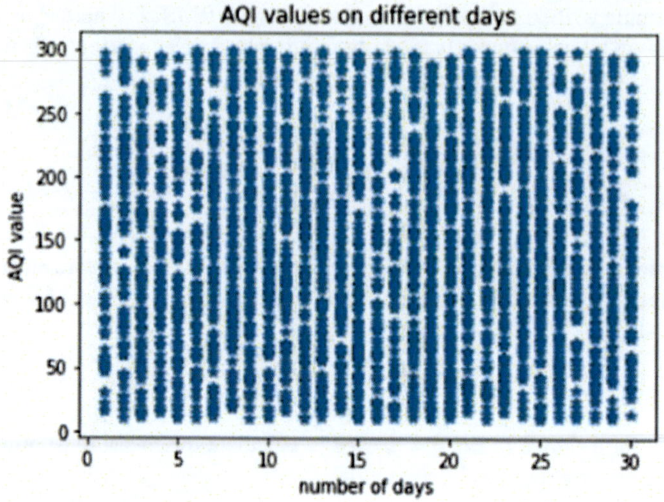

Fig. 5.11 AQI values for different days of a month

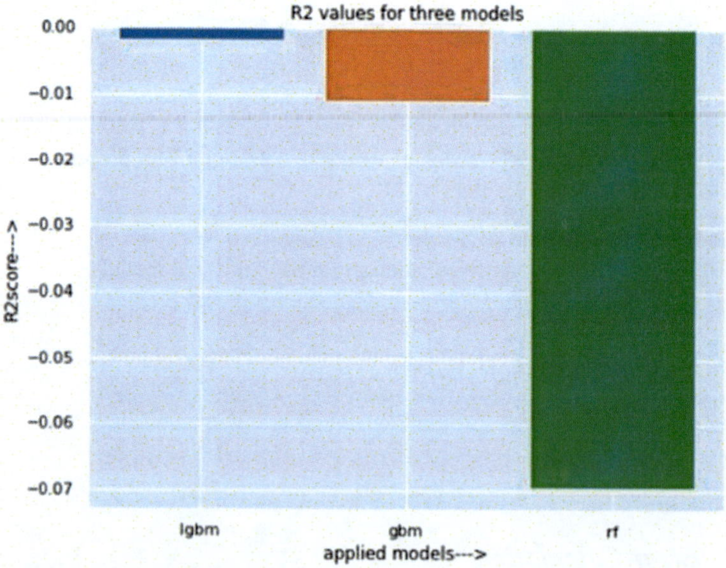

Fig. 5.12 Mean $R2$ score value representation

lesser error value is found to be the best model for the AQI prediction from the used data set.

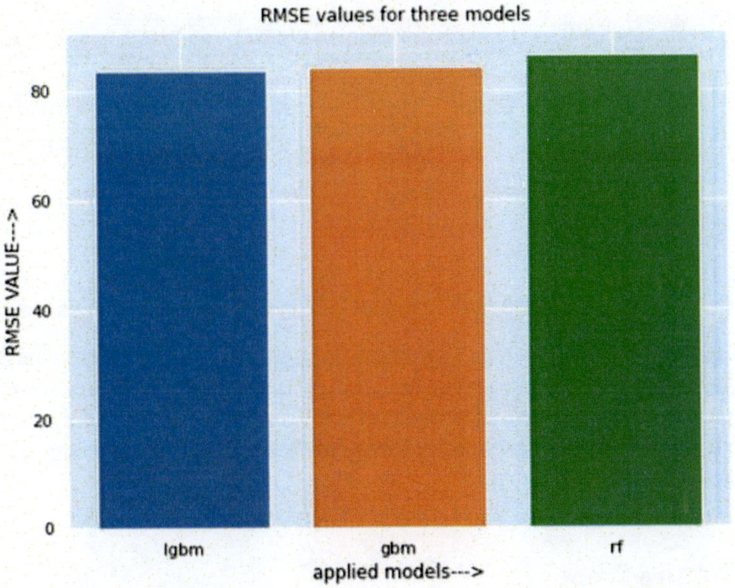

Fig. 5.13 Mean RMSE value representation

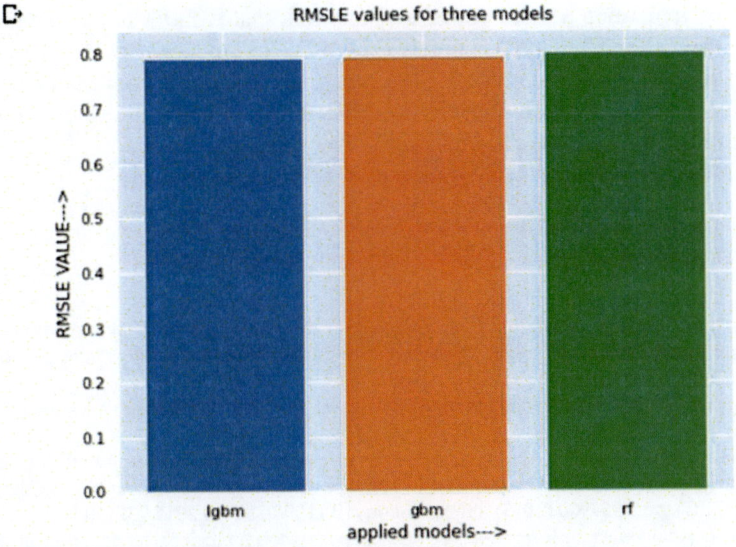

Fig. 5.14 Mean RMSE value for each applied model

Table 5.1 Model and their respective error metric value

Model/matrices	RMSLE	RMSE	*R*2 Score
Light GBM	0.7939	83.79	−0.0071
GBM	0.7953	84.19	−0.0110
Random Forest	0.8045	86.59	−0.6951

5.6 Conclusion

From the proposed machine learning model following conclusions can be drawn out.

1. The proposed model is able to predict the accurate air quality index value (AQI).
2. The best-suited framework for the used dataset is LightGBM model which in comparison to other two models GBM and Random Forest produces lesser error and is found to be most accurate framework for AQI value prediction for the used data set.

5.7 Future Scope

In order to develop more accurate machine learning models for AQI prediction, there is a need to develop a more powerful framework that is faster in the compilation process and also able to predict nearly exact value or produce very less error while undergoing the prediction process. Moreover, there is a need to implement the most accurate model in real-life scenarios which can predict the future values and alert the surrounding about the coming hazard of air pollution.

References

1. Du, S., Li, T., Yang, Y., Horng, S.-J.: Deep air quality forecasting using hybrid deep learning framework. IEEE Trans. Knowl. Data Eng. **33**(6), 2412–2424 (2021). https://doi.org/10.1109/TKDE.2019.2954510
2. Aditya, C.R., et al.: Detection and prediction of air pollution using machine learning models. Int. J. Eng. Trends Technol. (IJETT) **59**(4) (2018)
3. Mihalache, S.F., Popescu, M., Oprea, M.: Particulate matter prediction using ANFIS modelling techniques. In: 2015 19th International Conference on System Theory, Control and Computing (ICSTCC), pp. 895–900 (2015). https://doi.org/10.1109/ICSTCC.2015.7321408
4. Xiao, D., et al.: Machine learning-based rapid response tools for regional air pollution modelling. Atmos. Environ. **199**, 463–473 (2019)
5. Xi, X., et al.: A comprehensive evaluation of air pollution prediction improvement by a machine learning method. In: 2015 IEEE International Conference on Service Operations and Logistics, and Informatics (SOLI). IEEE (2015)
6. Gacquer, D., et al.: Comparative study of supervised classification algorithms for the detection of atmospheric pollution. Eng. Appl. Artif. Intell. **24**(6), 1070–1083 (2011)

7. Ayele, T.W., Mehta, R.: Air pollution monitoring and prediction using IoT. In: 2018 Second International Conference on Inventive Communication and Computational Technologies (ICICCT). IEEE (2018)
8. Srivastava, C., Singh, S., Singh, A.P.: Estimation of air pollution in Delhi using machine learning techniques. In: 2018 International Conference on Computing, Power and Communication Technologies (GUCON). IEEE (2018)

Chapter 6
A Comparative Study of Different IOT Sensors

V. Madhava Sai Teja, B. Sai, G. Veerapandu, and Mahesh K. Singh

Abstract The Internet of Things (IoT) is accepted concepts that have been functioned to many conventional areas for example business processes, medicine and smart cities etc. One of the major problems of an IoT method is the quantity of information; it has to monitor and to manage. This information turns up as procedures that require selecting and processed in instantaneous consecutively to build an accurate decision. The sensors used in IoT produce an enormous quantity of diverse information that requirements to be switch by the suitable techniques. IoT is a revolutionary technology in smart living standard. It's revolutionization, the world through billions of inclination on sensors by using it to create a resourceful environment in the region of us. Sensors are recognized as be a forthcoming field, which has been demonstrated in scientific research in any IoT application, IoT sensors are used effectively to build a smart world. In this manuscript discussed different type of IoT sensors on their applications based. In addition, it is given the data in which different type of sensors are used in many purposes. The IoT is previously approaching to existence particularly in smart environment applications and healthcare by addition of a huge quantity of sensors. The actuators used in IoT are to develope the way of life and initiate innovative services to the area.

6.1 Introduction

The IoT connects all living things that bring about change. Things are linked through a variety of medium to communicate and make it easier. IoT encompasses a broad range of fields includes in cloud, mobile sensors, virtualizes environment, artificial intelligence and radio frequency identification system (RFID). Such IoT are used in a separate application. The technology based on IoT has lead individuals to ubiquitous

V. Madhava Sai Teja · B. Sai · G. Veerapandu
Department of ECE, Aditya College of Engineering, Surampalem, India

M. K. Singh (✉)
Department of ECE, Aditya Engineering College, Surampalem, India
e-mail: mahesh.singh@accendere.co.in

S. Yadav et al. (eds.), *Proceedings of Second International Conference in Mechanical and Energy Technology*, Smart Innovation, Systems and Technologies 290,
https://doi.org/10.1007/978-981-19-0108-9_6

communications and elegant services. Computers can exchange information and provide efficient services through IoT. Alexia and many other IoT enabled products, such as lamps, stream heaters, washing machines, air conditioners etc. [1].

These are apps-based systems available, no screen market without a mobile. Only audio services are provided and a monitor and also provided with screen devices. Sensors are very important in smart applications. IoT permitted this connectivity. It enchants the collection and storing of information in real-time with the cloud-based platform. In the field of IoT, different resource limitation devices, used to communicate over the network using RFID. A fast-growing technology that enables automatic identification of RFID tag objects [2–7]. The different types of sensors are discussed here.

Proximity sensor: The proximity sensor makes it possible to detect the position of any nearby object without any physical touch. By emitting electromagnetic radiation, the presence of an object is detected simply by looking for any difference in the return signal. Various types of proximity sensors are available, such as inductive, capacitive, ultrasonic, magnetic, etc. This particular type of sensor is used primarily in these particular applications [1].

Chemical sensor: Sensors are called chemical sensors by detecting any chemical reaction, material or group of chemicals [1, 5, 8, 9].

Temperature sensor: Temperature sensors are useful in calculating heat energy to detect a physical change in one's body. Temperature sensors are used for monitoring of ambient environmental conditions. They gather data and then use Wi-Fi to send it to the cloud for review [2, 5, 8, 9].

Gyroscope sensor: Gyroscope sensor measuring angular velocity detects any tilt angle movement in the body. This sensor commonly used in 3D mouse games for sports player instruction, industrial automation and much more [6].

Velocity sensor: A sensor to facilitate the measures of the speed of transform in steady point standards at recognized interval. The speed sensor distinguished the motion of an object along a straight line, where angular velocity sensors monitor how quickly the system rotates. It is used for intelligent vehicle tracking in smart cities.

6.2 Related Work

IoT is recently introduced technology. In this world, there are trillions of actuators and sensors using by them for constructing a modern environment. In this paper it is explaining about the various sensors that are used in IoT applications [1]. The IoT is a technology that actives the computing which can be physical and virtual devices. These are connected to the Internet. In our automated home, it can be controlled everything in our home with the help of a mobile. For this, it is using a generic framework that can automatically control all the devices in the automated home [2]. IoT actives an innovative generation of inventive services. Smart devices can be controlled automatically because they are connected to the Internet. IoT devices are

Table 6.1 Billions of units installed based on IoT (2018–2020)

Segments	2020	2019	2018
Transportation	0.08	0.07	0.06
Information	0.37	0.37	0.37
Retail and wholesale Trade	0.44	0.36	0.29
Healthcare providers	0.36	0.28	0.21
Automotive	0.47	0.36	0.21
Manufacturing and natural resources	0.49	0.40	0.33
Physical security	1.09	0.95	0.83
Building automation	0.44	0.31	0.23
Government	0.70	0.53	0.40
Utilities	1.37	1.17	0.98

Source Gartner (August 2020)

transmitted through the untrusting network. So that, it needs security and privacy for that reason which are using attribute-based encryption (ABE) and it is very helpful. By using ABE, there are many advantages like, it is in bearable cost. It is used in many other IoT applications [3].

Nowadays, a smart home, smart meter based on IoT, are using widely which replaces the Analog based meters. The data is communicated in wireless modes, so that the manual work is reduced. The communication of the smart home network is reduced by the energy theft. This attack cannot be completely identified by using present techniques. So in this manuscript expanded an energy recognition method called Smart Energy Theft Systems (SETS). The SETS is planned for energy theft identification. Its results have three stages. Stage first has an energy theft accurateness result of 57.40%, after adding Stage two it increased to 77.80%, after the presence of stage three finally get 98.52% [4]. As the population increases worldwide, there is a need that arises to make available appropriate healthcare services by using modern technology like the IoT based healthcare observing system. By using IoT sensors, these devices store information and it has fixed bandwidth on the device that can communicate in the particular channel. For the health monitoring system, it has to work properly by maintaining the sample rate and stoppage constraint of each sensor used for monitoring system is important [5]. Billions of units installed based on IoT, endpoint market segment worldwide (2018–2020) shown in Table 6.1.

6.3 Methodology and IoT Based Sensors Equipment's

The IoT is one of the latest and fastest developing technology in India and worldwide also. In India, many people depend on agriculture. So, this has been replaced by the agriculture traditional technique with modern IoT based methods in farming. Here we are using Wi-Fi, it is used in particular areas. These devices perform when they are connected to the Internet which can be controlled by the remote in anyplace, these devices cost is very low. By using of sensor it collects data and takes individual decisions and autonomous decisions [8, 9]. Here it wants safe vehicular traffic across the busy traffic way. We require real-time monitoring systems. In intelligent traffic monitoring systems, we are using RFID sensors for real-time tracking of vehicles. The RFID sensor can detect the velocity of the vehicle. It is very useful for the traffic department. It alerts the drivers when they exceed the speed limit. It is designed at a low cost [6]. To give the safe supply of drinking water there is a need for quality of monitoring, it can be done easily by using IoT sensors, it can measure physical and chemical parameters of water. The collected information can be seen on the Internet. The parameters such as temperature, pH and dissolved oxygen of the water can be measured [7]. The sensor element used in IoT is shown in Fig. 6.1.

Sensor component: The elementary method of transduction (e.g., an object), which transforms individual structure of power into another. A number of sensors can include additional than one sensor component (for example a complex sensor). Sensor features are counting its internal wrapping and peripheral (e.g., optical or electrical) connections.

Interface Electronics: Interfacing is the procedure of linking one device, in exacting a processor or micro controller with another. Its permit to propose or transform the two electronic devices amount produced and contribution configurations. So it can function simultaneously.

Sensing Processing: Sensor is a piece of equipment, element, appliance, or subsystem, whose principle is to distinguish proceedings or change in its atmosphere and transmit the data to previous electronics, regularly with a processor.

Sensors participate a significant role in automating any request by measuring and analyzing the information obtained to detect the changes in the physical material. Whenever any physical environment in which a sensor is rendered changes, it produces an assessable response. Figure 6.2 shows the related electrical signals and sensing modules. A variety of sensors can vary from simple to multifaceted. The detection of the sensors may be dependent on their description, their mechanism of procedure. The type of objects used the substantial occurrence of the senses, the properties of its method and their area of function.

Fig. 6.1 Consequent electrical signal for sensing elements

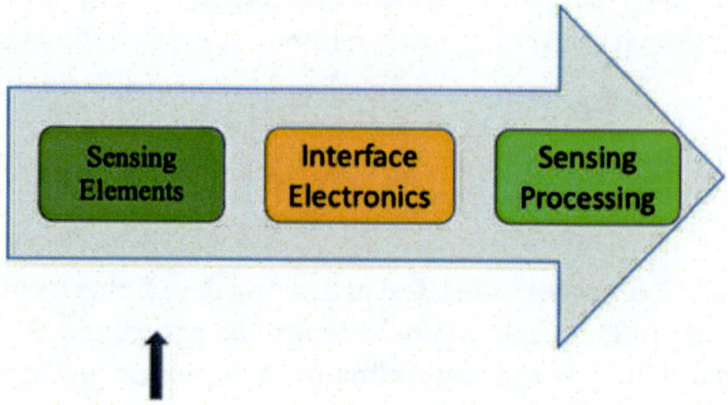

Fig. 6.2 Gyroscope sensor

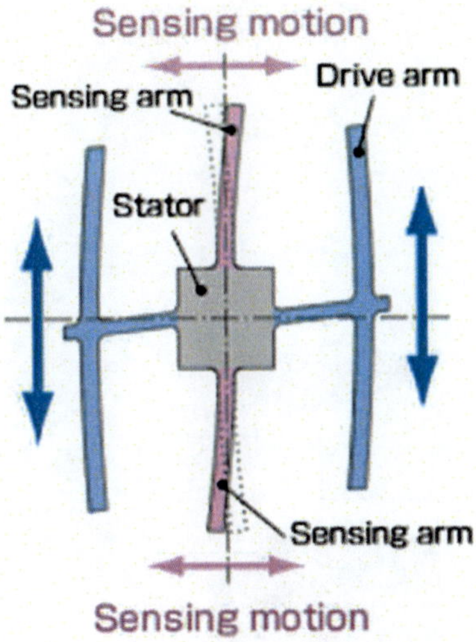

6.3.1 Gyroscope Sensor Equipment

Vibration gyroscope sensors identify angular speed from the energy functional to a shacked body by the Coriolis. Here it is illustrated that how this works, with a double T-crystal dimension of structure of Epson as an example. The Coriolis power act ahead to the constrained artillery at what time the gyroscope is rotating, causing the perpendicular vibration.

6.3.2 Speed Sensor

A speed sensor is a sensor that response to speed rather than absolute location. Movement causes the coil to move concerning the magnet, which in turn creates a voltage proportional to the speed of that movement (Fig. 6.3).

The speed sensors proposed to observe the velocity by induction motor of the three phases. It is used by the joint of MCU regulator. It is also reduced by the switching and harmonic losses of the circuit. The Node MCU transmitted the speed signal throughout Wi-Fi and observes the velocity of the motor [10–21].

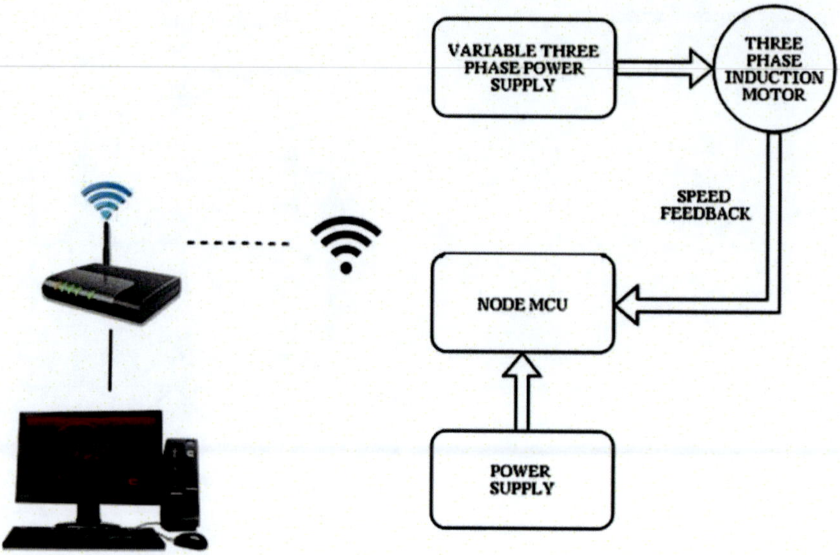

Fig. 6.3 IoT based speed monitoring sensors

6.4 Result Analysis Based on IoT Applications

6.4.1 Application in Smart Water Measurement

Water feature monitoring, ground stage monitoring, pollution level and some types of storage place seepage can put away water. Any such intellectual sensors know how to be used for correct stream superiority application. Water leakage: This water pressure on the outer tanks and pipes can be assessed in order to prevent timely leakage. Regulation of the level of contamination of the sea: the introduction of dangerous leaks or waste to the sea may be controlled by sensors to avoid seawater. Smart swimming pool: the swimming pool conditions can be guarded by the use of sensors and timely preservation can be approved out in this way. The monitoring of water levels and their fluctuations in rivers, dams and reservoirs from time to time is an important task.

6.4.2 Smart Security

Sensors preserve to be installed in different areas to certify the environmental security. Some of those applications everywhere sensors participate a key position to ensure safety is discussed in Smart perimeter access control: Sensors can be used to make an elegant outskirts approximately specific areas, and alarms can be introduced to warn unauthorized persons to enter restricted areas. Devices can be used intelligently to detect unsafe levels of gas, radiation levels and leakage. Sensors can be connected to walls in factories, mines and industries where readily available is a possibility of escape. The use of sensors can be done smartly to identify dangerous gas level, emission level and leakage. Smart Homes: When building or a smart house, sensors linked to different items at house like smart switching on/off machines, distinguishing intruders, and detecting the amount of ingredients used in cooking. This is completed by transferring out warnings, which lead to a healthier lifestyle. Some of the smart home implementations are as follows:

1. Remote piece of equipment: Sharply switch on/off devices makes existence simple and secure by preventing accident and as well save the energy.
2. Intrusion detection system: sensors can detect intruders at the entrance, and well-timed warning is capable of being sent to the authorities worried. It makes existence safer and easier as a person can watch his or her house from anywhere. This can be enforced if entry is only permitted for designated persons after proper approval and the door is opened. In [6], the authors introduced an intrusion detection method that senses indoor motion and intimates owners through a message. It is a sensor that measures the rate of change in constant position values at given intervals [6] in Table 6.2, it is mentioned the types of IoT sensors and their application.

Table 6.2 IoT sensors and their application

IoT applications	Types of sensors
Smart agriculture	Chemical, proximity, water quality, temperature, humidity
Smart security	Temperature, gyroscope, infrared, chemical, light
Smart transport	Infrared, pressure, gyroscope, temperature, pressure, chemical
Smart home	Position, light, accelerometer, proximity, chemical
Smart health	Pressure, accelerometer, light, magneto, bio-sensors
Smart building	Magneto, chemical, light, gyroscope, chemical
Smart water	Water quality, occupancy, humidity, temperature
Smart environment	Optical, chemical, bio-sensors, temperature, light,
Smart city	Infrared, pressure, humidity, proximity, velocity, light

With the aid of sensors, advice can be given on controlling water and electricity consumption. Sensors are using in all IoT applications. In any small application, we use more than one sensor. After studying different sensor types and smart applications in IoT. In this, we suggest that different sensor required in IoT applications to create a smart world.

6.5 Conclusion

The IoT based different applications are changing our humankind by constructing contemporary, well turned-out surroundings around us. In many IoT based elegant application, the sensor plays an explanation responsibility in the computerization of relevance procedure by assembling the devices smarter and more receptive exclusive of individual involvement. Reduces manual labour and saves time. In this manuscript, many types of sensors are presented in IoT enabled smart environments, IoT sensors are used effectively for agriculture, home appliances, water, health, etc. The sensors can communicate wisely and remotely with each other. Better crop improvement is a major challenge for well developed countries, such as India, to take on new smart technologies as part of the agricultural stream that leads to a green population region.

References

1. Sehrawat, D., Gill, N.S.: Smart sensors: analysis of different types of IOT sensors. In: 2019 3rd International Conference on Trends in Electronics and Informatics (ICOEI), pp. 523–528. IEEE (2019)
2. Gunge, V.S., Yanagi, P.S.: Smart home automation: a literature review. Int. J. Comput. Appl. **975**, 8887 (2016)
3. Girgenti, B., Perazzo, P., Vallati, C., Righetti, F., Dini, G., Anastasi, G.: On the feasibility of attribute-based encryption on constrained IOT devices for smart systems. In: 2019 IEEE International Conference on Smart Computing (SMARTCOMP), pp. 225–232. IEEE (2019)
4. Li, W., Logenthiran, T., Phan, V.-T., Woo, W.L.: A novel smart energy theft system (SETS) for IOT-based smart home. IEEE Internet Things J. **6**(3), 5531–5539 (2019)
5. Dhar, S.K., Bhunia, S.S., Mukherjee, N.: Interference aware scheduling of sensors in IOT enabled health-care monitoring system. In: 2014 Fourth International Conference of Emerging Applications of Information Technology, pp. 152–157. IEEE (2014)
6. Prasanna, G.S., Pavani, K., Singh, M.K.: Spliced images detection by using Viola-Jones algorithms method. Mater. Today Proc. (2021)
7. Kanchana, V., Nath, S., Singh, M.K.: A study of internet of things oriented smart medical systems. Mater. Today Proc. (2021)
8. Padma, U., Jagadish, S., Singh, M.K.: Recognition of plant's leaf infection by image processing approach. Mater. Today Proc. (2021)
9. Lu, D., Liu, T.: The application of IOT in medical system. In: 2011 IEEE International Symposium on IT in Medicine and Education, vol. 1 (pp. 272–275). IEEE (2011)
10. Singh, M., Nandan, D., Kumar, S.: Statistical analysis of lower and raised pitch voice signal and its efficiency calculation. Traitement du Sig. **36**(5), 455–461 (2019)

11. Duan, R., Chen, X., Xing, T.: A QoS architecture for IOT. In: 2011 International Conference on Internet of Things and 4th International Conference on Cyber, Physical and Social Computing (pp. 717–720). IEEE (2011)
12. Singh, M.K., Singh, A.K., Singh, N.: Multimedia analysis for disguised voice and classification efficiency. Multimedia Tools Appl. **78**(20), 29395–29411 (2019)
13. Zhang, Z.K., Cho, M.C.Y., Wang, C.W., Hsu, C.W., Chen, C.K., Shieh, S.: IoT security: ongoing challenges and research opportunities. In: 2014 IEEE 7th International Conference on Service-Oriented Computing and Applications (pp. 230–234). IEEE (2014)
14. Singh, M.K., Singh, A.K., Singh, N.: Acoustic comparison of electronics disguised voice using different semitones. Int. J. Eng. Technol. (UAE) **7**(2), 98 (2018)
15. Veerendra, G., Swaroop, R., Dattu, D.S., Jyothi, C.A., Singh, M.K.: Detecting plant Diseases, quantifying and classifying digital image processing techniques. Mater. Today Proc. (2021)
16. Hassan, W.H.: Current research on Internet of Things (IoT) security: a survey. Comput. Netw. **148**, 283–294 (2019)
17. Singh, M.K., Singh, A.K., Singh, N.: Disguised voice with fast and slow speech and its acoustic analysis. Int. J. Pure Appl. Math. **118**(14), 241–246 (2018)
18. Balaji, V.N., Srinivas, P.B., Singh, M.K.: Neuromorphic advancements architecture design and its implementations technique. Mater. Today Proc. (2021)
19. Punyavathi, G., Neeladri, M., Singh, M.K.: Vehicle tracking and detection techniques using IoT. Mater. Today Proc. (2021)
20. Singh, M.K., Singh, A.K., Singh, N.: Multimedia utilization of non-computerized disguised voice and acoustic similarity measurement. Multimedia Tools Appl., 1–16 (2019)
21. Hassija, V., Chamola, V., Saxena, V., Jain, D., Goyal, P., Sikdar, B.: A survey on IoT security: application areas, security threats, and solution architectures. IEEE Access **7**, 82721–82743 (2019)
22. Singh, M.K., Singh, N., Singh, A.K.: Speaker's voice characteristics and similarity measurement using euclidean distances. In: 2019 International Conference on Signal Processing and Communication (ICSC) (pp. 317–322). IEEE (2019)

Chapter 7
Evaluating the Performance of Deep Learning Methods and Its Impact on Digital Marketing

Ms. Gazala Masood, C. Indhumathi, Pacha. Malyadri, Krishna Mayi, B. K. Sumana, and Khongdet Phasinam

Abstract This article maps and explains how two distinct areas of the marketing sciences are now and potentially related to computer science. It analyses the interplay of DL in the academy, and at the same time proposes a machine research framework that might be appreciated in many respects of the scientific field of digital marketing. In the fields of deep learning there are many research papers (DL). This quantity remains modest, however, with regard to digital marketing elements. Marketing intelligence may benefit in many ways from scientific studies on deep learning (DL). Today only a tiny proportion is linked to particular digital marketing techniques by scientific study on Digital Marketing and Deep Learning (DL). Generic aspects such as e-business, consumer behavior, e-commerce strategies, social media advertising, search engines and consumer prevision modelling are mostly discussed, and are not more closely dependent on specific marketing problems that are more aware of in business, such as social media consumption, target commercials, social media marketing, and transformation optimization. In spite of the extensive field of study and a lot of publications, it seems that scholarly papers on digital marketing and deep learning particularly lack (DL). Nevertheless, some highly comprehensive research

Ms. Gazala Masood (✉)
Faculty of Commerce and Management, Rama University, Kanpur, India
e-mail: ghazala1999@yahoo.com

C. Indhumathi
Department of Commerce, PSG College of Arts and Science, Coimbatore, Tamil Nadu, India

Pacha. Malyadri
Centre for Economic and Social Studies, Begumpet, Hyderabad, Telangana State, India

K. Mayi
Radhekrishna Women's College, Charkaman Osmania University, Hyderabad, India

B. K. Sumana
PG Department of Business Administration, Maharanis Womens Commerce and Management College, Mysore, Karnataka, India

K. Phasinam
School of Agricultural and Food Engineering, Faculty of Food and Agricultural Technology, Pibulsongkram Rajabhat University, Phitsanulok, Thailand

S. Yadav et al. (eds.), *Proceedings of Second International Conference in Mechanical and Energy Technology*, Smart Innovation, Systems and Technologies 290,
https://doi.org/10.1007/978-981-19-0108-9_7

efforts are quite promising in certain areas of digital marketing and deep learning. This article is by mapping applications in the field of digital marketing in the present state of deep learning (DL). It emphasizes the foundational element writings, identifies regions of absence or lack of their existence, and offers a learning engine that may fit into numeric marketing opportunities.

7.1 Introduction

The advancement of technology has provided businesses with the ability to provide consumers with massive quantities of goods in the modern day. Consumers may now purchase goods and services via digital marketing platforms such as the Internet, which allows businesses to promote and sell their products to a wider audience [1]. Digital marketing encompasses all of the techniques that may have a significant effect on individuals at a specific moment, in a specific location, and via a specific channel [2]. The combination of data analytics and academic scientific study on intelligent systems has resulted in significant development in the field of digital marketing. This article examines and explains digital marketing techniques from the viewpoint of deep learning (DL) research, which is presented in depth. Meanwhile, the number of research writings continues to be at an intermediate level, despite the fact that the commercial sector seems to be making strides ahead. Using deep learning (DL) approaches to improve the effectiveness of digital marketing strategies, this article focuses on the technical components of digital marketing strategies in scientific research to help researchers better understand them. Despite the wide study area and a significant number of publications, it seems that there is a paucity of scientific papers, particularly in the areas of digital marketing and deep learning, given the increasing number of journals in the field (DL) [3]. In spite of this, there have been some very extensive research attempts on specific digital marketing fields such as search engine optimization, search engines ranking factors, consumer behavior, web development, and targeted advertisements, which provide encouragement for the future of deep learning (DL) impact on digital marketing data analysis [4].

7.2 Digital Marketing

Digital marketing encompasses any marketing strategies and methods that make use of an electrical gadget or the web to demonstrate, promote, or sell goods or services, as well as industries that make use of online channels that will assist them in their efforts to succeed. Websites, social media pages, targeted advertising, and email are all effective ways to keep consumers up to date and attract more new ones. Consumers, resellers, competitors, suppliers, promoters, the economy overall, placement, segmentation, enlargement, development, goods, branded products, marketing, market share, price, promotional efforts, number of resellers, churn, customer value, and other factors are some of the most significant factors that influence decision making in business. Certainly, judgment is a complex process that involves a variety of factors based on the findings, experiences and intuition. In order to identify the critical function of deep learning (DL) in digital marketing research, we must first map the present state of digital marketing scientific research and relate it to the business community. Deep learning (DL) is a kind of machine learning that can learn from data. Then we will be able to determine to what extent digital marketing in academics is falling behind the advancements in the commercial sector [5].

7.3 Deep Learning Methods and Its Effect on Digital Marketing Include

We examine briefly the most significant commercial digital marketing techniques. Optimization Search engine (SEO), social media marketing (SMM), content marketing, pay per click (PPC), affiliates marketing, native ads, online advertising, chatbots (semantic search), ad targeting, and protectionary marketing. This is how it is having an effect on digital work in general [6]. Table 7.1 shows that different methods for deep learning.

7.4 Deep Learning an Understanding

Deep Learning is a technique that simulates human brain by using "neural networks" to learn by completing a task repeatedly and somewhat differently each time in order to improve the result. Figure 7.1 shows that Deep learning Methods as a result, the computer "thinks" in the same manner that a person does, based on past experiences and knowledge [7]. However, the distinction is that the computers is capable of processing vast quantities of data and doing activities at much quicker rates than a person would be able to, which allows them to solve difficult issues and acquire new abilities in a considerably shorter period of time. Handwriting identification as an illustration of how it works. Traditional computer algorithms need the computer to be taught a set of rules in order for it to identify each individual character in a document [8]. Even though this seems like an impossible job when you consider the number of differences in handwriting, we humans are able to interpret many various types of handwriting without any difficulty so because neural network in our brains is performing the work for us. An artificial intelligence system may be trained to identify personal notes in the same manner that a human brain does so by exposing the computer to a large number of sample characters and understanding how to recognize each character from these instances. Figure 7.2 shows that deep learning steps. The greater the number of samples you provide the computer, the more effective it will get at handwriting identification. In order to perform this kind

Table 7.1 Classification of deep learning methods

SEO (Search engine optimization)	It concerns processes to optimize website traffic, blogs or info graphics which rank well in the results of search engines
Marketing through social media	It relates to methods which optimize and support social media brands to boost organic traffic, enhance brand recognition and create business leads
Marketing through content	This means creating a blog post, e-book, infographic and online brochures for brand recognition, increasing the volume of visitors, generating leads, and consumers. This includes strategic and human-centered content
(PPC) Pay-per-click	It is a way of bringing the visitors into a website when a link is clicked or an activity takes place, by paying for a publication Internet service
Marketing through affiliates	This is a recommendation technique to promote a company through a person or a community of enthusiasts who get a commission either in the form of video or a link on their website
Advertising on native lands	It pertains to publicities which are shown on a digital platform and which follow the natural nature of customer experience accompanying unpaid material
Automated marketing	It includes software that automated fundamental business requirements on a daily basis, such as e-mail, social media, etc.
Marketing through email	It concerns ways to create e-mail promotions in order for the clients to stay in touch, tell them about reductions, new things and happenings
Chatbots	See programmable and autonomous interactive conversation and order completion apps with website users and social media visitors when the latest visit is made
Search using semantics	It pertains to an intelligent data search technique that enables consumers to quicker discover the result they are looking for. Based on machine learning, the question is interpreted and the significance of the search term is understood through data correlations between words. It anticipates what data the search history of users might require
Creation of content	It pertains to the content production approach that uses a deeper learning (DL) methodology dubbed natural language generation, which collects, organized, converts the raw information relevant for the search into understandable sports, financial statements etc.

(continued)

Table 7.1 (continued)

Advertisement targeting	It relates to advanced methods for generating Internet advertising that are important to branding. It produces or optimists ads based on user history and behavioral variables importing such as geographical location, sex, age etc. to better target the user in order to increase sponsors' return on investments
Marketing that is predictive	It concerns a deep learning (DL) method which collects and analyzes users' behavioral data, and finds potential connections of data, including preferences and requirements, via data mining. With this data, the neural network model is supplied and improved predictions are produced which assist policy makers in the business
Search by voice	It applies to the methods of deep learning (DL) in voice searches. Voice search will restore the impression of result speed and relevance and will framework presents recovery an extraordinarily simple operation based on speech recognition and text mining methods
Trial of A/B	The test refers to an Internet marketing strategy which would generate greater conversions or leads in version of the same website. Driven by deep learning (DL), conversions improvements are carried out into variants of many editions of the very same website, as well as the most likely user involvement is determined
Score of lead	It involves a process of evaluating and classifying client leads based on user behavior, interest rate and buying history in order to prevent loss of money and time

of job, neural networks must be built by a software engineer and must include many distinct layers of neurons, or decision-making units, which is why the term deep in deep learning refers to numerous different layers of neurons [9].

7.5 Deep Learning for Personalization

When it comes to marketing, one of the most intriguing uses of deep learning is the ability to provide hyper personalization [10]. Personally tailored marketing campaigns are already becoming popular, and they are proving to be popular with customers as well—transaction rates for customized emails are six times greater than those for generic emails. Privacy issues, as well as a shortage of funds to gather adequate data, may be tripping obstacles on the path to achieving effective customization [11, 12].

Because of the development of the Internet of Things, we are already beginning to gather enormous quantities of data about people from their computers, wearable gadgets, mobile phones, smart TVs, and even household appliances, heating and lighting systems, and other electronic equipment [13, 14]. In order to effectively

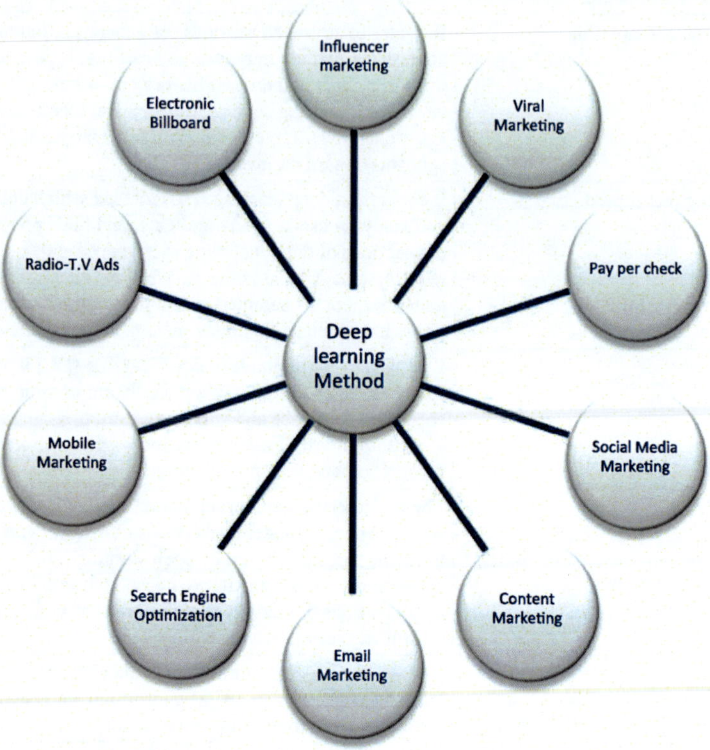

Fig. 7.1 Deep learning methods

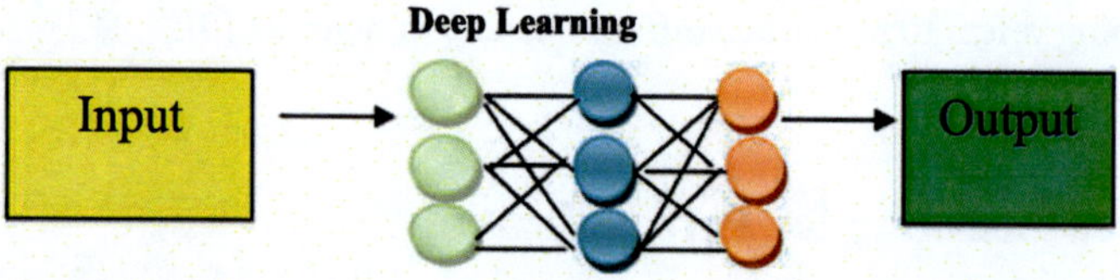

Fig. 7.2 Deep learning steps

interact with consumers on a highly customized level, marketers will be exposed to an increasing amount of data, which will need the development of advanced analytical skills [15].

Consumers are immediately scanned and recognized as they pass by in the movie Minority Report, and they are greeted by name, asked about their previous purchases, and shown a constantly changing array of personalized advertisements customized to their individual needs and preferences, according to the film [16]. Even casually perusing Facebook has an unnerving similarity to the scenario from the film in many

Table 7.2 The following are some instances of deep learning in action

1	Real-time bidding (RTB) software is used to purchase advertising space
2	Artificial intelligence (AI) chatbots that converse with consumers in a "human" manner are being developed
3	Using automatic translation, you may generate several language versions of your webpage in a matter of minutes
4	Copy that has been produced autonomously depending on the style of another line of writing
5	Automatically generated picture captions
6	Using speech recognition to do a voice search

respects. The future of individualized marketing is already here, and it's called Facebook [17]. Customers' personalized experiences would become the most essential consideration of your marketing strategy as your company's access to personal data grows. Care must be taken to strike a balance between offering exactly what the customer wants before they ask for it and crossing a line of that private information.

7.6 Deep Learning Will Play an Increasingly Important Role in Your Digital Marketing Approach in the Future

When it comes to getting begun with the deep learning process for your digital marketing strategy, online businesses may choose from a variety of different starting points. Deliberate learning methods, for example, may aid in the solution of a number of difficult issues, such as analyzing huge quantities of data and developing customized information flows for consumers.

While developing meaningful, customized connections with participating users, DL technologies and chatbots allow future-oriented market research that is far quicker than a human could possibly accomplish [18].

Table 7.2 shows that the following are some instances of deep learning in action: For the contemporary marketing department, data mining methods allow for the generation of foresighted information. Your team may use this data analytics capabilities (predictive analytics) to interact with targeted prospects at numerous points of contact throughout the sales funnel, allowing them to maximize their revenue.

7.7 Conclusion

Based on our findings, the quantity of scientific research that has been conducted and that refers to the use of deep learning (DL) in digital marketing methods is still in its early stages, with a few notable exceptions. Customization, targeting, high conversion rates, high returns on investment, and other features are just a few of the things we may expect to see as a result of the rapid technical advancements in marketing and computer science. Marketers, businesses, and judgment have an amazing opportunity right now to seize the moment and achieve outstanding outcomes. Deep learning (DL) will unquestionably set new norms in digital marketing, both in academics and in the corporate community. The industrial sector is just a few steps ahead of the point at where the academic world is now situated. Despite the development of the deep learning (DL) model used by businesses, academics will ultimately be able to bridge the gap between the two. Marketing science is always evolving, not only in the business sector, but also at institutions, where new knowledge is created, and companies rely on this latter source of information.

References

1. Cameron, G., Cameron, D., Megaw, G., Bond, R., Mulvenna, M., O'Neill, S., McTear, M.: Towards a chatbot for digital counselling. In: Proceedings of the 31st International BCS Human Computer Interaction Conference (HCI 2017) 31 (pp. 1–7) (2017)
2. Eshak, M.I., Ahmad, R., Sarlan, A.: A preliminary study on hybrid sentiment model for customer purchase intention analysis in social commerce. In: 2017 IEEE Conference on Big Data and Analytics (ICBDA) (pp. 61–66). IEEE (2017)
3. Isah, H., Trundle, P., Neagu, D.: Social media analysis for product safety using text mining and sentiment analysis. In: 2014 14th UK Workshop on Computational Intelligence (UKCI) (pp. 1–7). IEEE (2014)
4. Jain, A., Pandey, A.K.: Multiple quality optimizations in electrical discharge drilling of mild steel sheet. Mater. Today Proc. **4**(8), 7252–7261 (2017)
5. Jain, A., Pandey, A.K.: Modeling and optimizing of different quality characteristics in electrical discharge drilling of titanium alloy (grade-5) sheet. Mater. Today Proc. **18**, 182–191 (2019)
6. Jain, A., Yadav, A.K., Shrivastava, Y.: Modelling and optimization of different quality characteristics in electric discharge drilling of titanium alloy sheet. Mater. Today Proc. **21**, 1680–1684 (2020)
7. Krasotkina, O., Mottl, V.: A Bayesian approach to sparse learning-to-rank for search engine optimization. In: International Workshop on Machine Learning and Data Mining in Pattern Recognition (pp. 382–394). Springer, Cham (2015)
8. Li, C., Lu, Y., Mei, Q., Wang, D., Pandey, S.: Click-through prediction for advertising in twitter timeline. In: Proceedings of the 21th ACM SIGKDD International Conference on Knowledge Discovery and Data Mining (pp. 1959–1968) (2015)
9. Mahdian, M., Tomak, K.: Pay-per-action model for online advertising. In: Proceedings of the 1st International Workshop on Data Mining and Audience Intelligence for Advertising (pp. 1–6) (2007)
10. Nie, D., Bai, S., Hao, B., Zhu, T.: Personality prediction for microblog users with active learning method (2015)
11. Orriols-Puig, A., Casillas, J., Martínez-López, F.J.:Automatic discovery of potential causal structures in marketing databases based on fuzzy association rules. In: Marketing Intelligent Systems Using Soft Computing (pp. 181–206). Springer, Berlin, Heidelberg (2010)
12. Panwar, V., Sharma, D.K., Kumar, K.P., Jain, A., Thakar, C.: Experimental investigations and optimization of surface roughness in turning of en 36 alloy steel using response surface methodology and genetic algorithm. Mater. Today Proc. (2021)
13. Rao, S., Srivatsala, V., Suneetha, V.: Optimizing technical ecosystem of digital marketing. In: Artificial Intelligence and Evolutionary Computations in Engineering Systems (pp. 691–703). Springer, New Delhi (2016)

14. Singh, R.K.: E-message the effective tool of online marketing in present scenario. Int. J. **2**(1) (2017)
15. Spentzouris, P., Koutsopoulos, I., Madsen, K.G., Hansen, T.V.: Advertiser bidding prediction and optimization in online advertising. In: IFIP International Conference on Artificial Intelligence Applications and Innovations (pp. 413–424). Springer, Cham (2018).
16. Wierenga, B.: Marketing and artificial intelligence: Great opportunities, reluctant partners. In: Marketing Intelligent Systems Using Soft Computing (pp. 1–8). Springer, Berlin, Heidelberg (2010)
17. Zhang, W., Yuan, S., Wang, J.: Optimal real-time bidding for display advertising. In: Proceedings of the 20th ACM SIGKDD International Conference on Knowledge Discovery and Data Mining (pp. 1077–1086) (2014)
18. Zhu, C., Wu, G.: Research and analysis of search engine optimization factors based on reverse engineering. In: 2011 Third International Conference on Multimedia Information Networking and Security (pp. 225–228). IEEE (2011)

Chapter 8
Effectiveness of Machine Learning Technology in Detecting Patterns of Certain Diseases Within Patient Electronic Healthcare Records

Dilip Kumar Sharma, Dhruva Sreenivasa Chakravarthi, Raja Sarath Kumar Boddu, Abhishek Madduri, Maruthi Rohit Ayyagari, and Md. Khaja Mohiddin

Abstract The research article sheds light on the effects of artificial intelligence and machine learning technologies to detect several diseases within patient electronic healthcare records. Apart from that, the usage and importance of machine learning technologies in detecting certain diseases is another major description in this research article. Thus, the purpose of this research article is to investigate the significance of machine learning to develop the healthcare system. In this research article, the researcher has adopted the positivism research philosophy, inductive research approach and the descriptive research design to make the research article presentable. It is identified that machine learning technologies are very important for the development of healthcare organizations and healthcare systems. On the other hand, machine learning technologies are impactful in detecting patterns of certain diseases. Therefore, machine learning technologies are largely impactful on healthcare organizations. In conclusion, it can be said that artificial intelligence and machine learning technologies are essential in the healthcare system as it helps to detect certain diseases quickly. Therefore, quick detection of diseases is helpful for

D. K. Sharma (✉)
Department of Mathematics, Jaypee University of Engineering and Technology, Guna, Madhya Pradesh, India
e-mail: dilipsharmajiet@gmail.com

D. S. Chakravarthi
Prashanth Hospital & Research Scholar, KL University, Vaddeswaram, Andhra Pradesh, India

R. S. K. Boddu
Department of CSE, Lenora College of Engineering, Rampachodavaram, Andhra Pradesh, India

A. Madduri
Engineering Management, Duke University, Durham, USA

M. R. Ayyagari
College of Business, University of Dallas, Irving, USA

Md. Khaja Mohiddin
Bhilai Institute of Technology, Raipur, Chhattisgarh, India

S. Yadav et al. (eds.), *Proceedings of Second International Conference in Mechanical and Energy Technology*, Smart Innovation, Systems and Technologies 290,
https://doi.org/10.1007/978-981-19-0108-9_8

protecting the health of the people. On the other hand, machine learning technologies help in the development of the healthcare system.

8.1 Introduction

Artificial intelligence is an important system nowadays and there are a lot of benefits of using artificial intelligence in machines. Therefore, machine learning is one of the important applications of artificial intelligence as it is helpful to improve several types of systems and the usage of this technology is increasing day by day. Thereafter, nearly 65% of the companies are planning to use machine learning globally. Furthermore, artificial technologies and machine learning have impacted the healthcare system as well and it helps to develop the healthcare system. Therefore, the rate of using artificial intelligence or machine learning in the healthcare system is 40% in the year 2021 [1]. Thus, it can be said that machine learning is important for not only the businesses but also healthcare. There are certain diseases that can be detected by the usage of artificial intelligence or machine learning (Fig. 8.1).

On the other hand, it was not easy to detect those diseases without artificial intelligence previously. Therefore, the usage of artificial intelligence and machine learning is not only important for the healthcare system but also important for the health of the people [2]. Thus, in this research article the usage of machine learning and artificial intelligence in medical diagnosis is described. Therefore, the effects of machine learning technologies on the healthcare system is described in this research article. Furthermore, the importance of the usage of machine learning technologies or

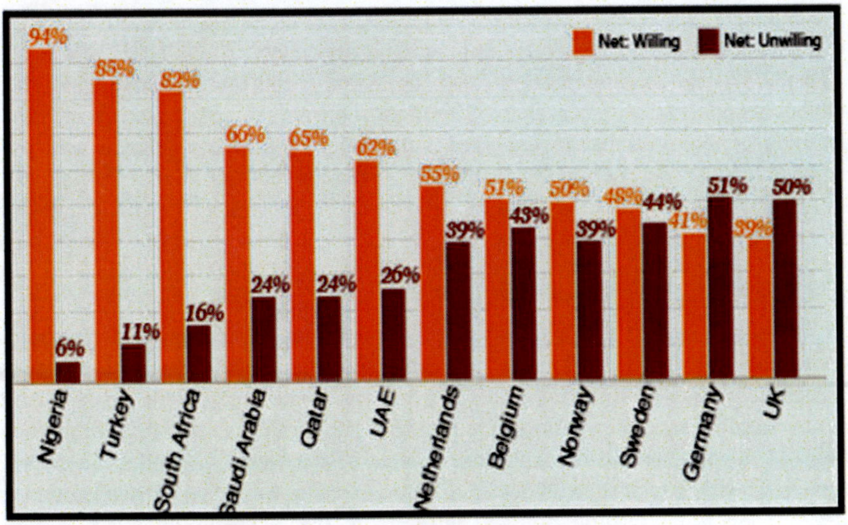

Fig. 8.1 Increment of artificial intelligence in healthcare [1]

artificial intelligence to detect certain major diseases is another important description in this research article. Thereafter, the usage of theory for understanding the concept of machine learning technologies is analyzed in this research article. Apart from that, the research methods that are accepted by the researcher to present a better research study are described in this research article briefly.

8.1.1 Aim and Objective

The aim of this research article is to analyze the effects of machine learning technology in detecting the patterns of certain diseases within patient electronic healthcare records.

The objectives of this research article are

- To investigate the importance of machine learning technology in development of the healthcare system
- To analyze the effect of machine learning technology in detecting patterns of certain diseases
- To understand the usage of machine learning technologies or artificial intelligence.

8.2 Literature Review

8.2.1 Machine Learning Technologies

Machine learning is one of the most important artificial intelligence and the usage of machines in several platforms is increasing nowadays. Thus, there are a lot of different types of technologies in machine learning and those technologies have their own impact on several systems. Therefore, there are six most important machine learning technologies that are analyzed in the table. The first one is Keras, which is an open software source and this technology is very popular for its user friendliness. Furthermore, this technology can be used in CPUs as well as in GPUs and this technology is helpful for fast prototyping. In addition to that, the second one is Torch which was launched in 2002 and is an old technology [3]. Thus, this technology has several algorithms that help to increase the speed in machine learning (Table 8.1).

Therefore, there is Caffe that is a recently released technology and that helps to increase the speed, modularity and expressiveness as well. Moreover, there is TensorFlow that was released in 2015 and it was created by Google firstly then it is now used by several platforms. On the other hand, machine learning technology helps in flowgraphs and helps in development of the neural networks. Therefore, there is Theano which is an advanced machine learning technology that helps in the development of the fashion industry. Furthermore, the next technology is Microsoft

Table 8.1 Six most important machine learning technologies [3]

Machine learning technology	Usage
Keras	It helps to simplify the making of models of deep learning
Torch	It has several algorithms that helps in deep learning
Caffe	It helps to innovate the speed and expressiveness of deep learning
TensorFlow	This technology helps to improve the neural networks
Theano	It helps in processing in defining and optimizing
Microsoft Cognitive Toolkit	This technology helps to make a next level machine learning project

Cognitive Toolkit which is also a new released technology and this helps to make a presentable and next level project.

8.2.2 *Usage of Machine Learning Technologies*

The usage of this technology is increasing day by day and along with that, there are a lot of benefits of using machine learning technology such as it helps in predictive maintenance. Apart from that, machine learning technologies help in product recommendation as a good product is very important for evaluating projects [4]. Along with that, these technologies are helpful to detect spam and help in several medical diagnoses. Thus, these are the benefits of machine learning technologies and the effect of machine learning technologies in detecting certain diseases is analyzed below in this research article (Fig. 8.2).

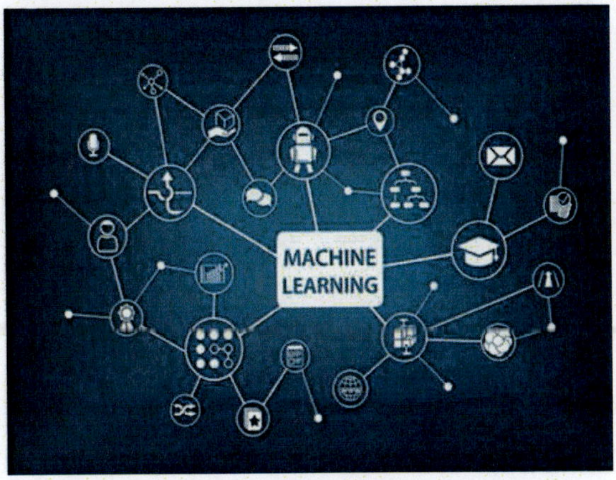

Fig. 8.2 Machine learning technology [4]

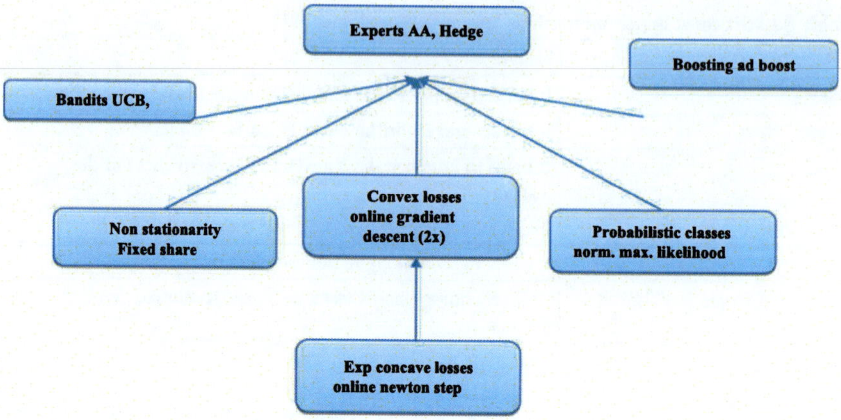

Fig. 8.3 Machine learning theory

8.2.3 *Theoretical Framework*

There are a lot of theories that are helpful to gain knowledge about machine learning technologies. Thus, the researcher has adopted the ***machine learning theory*** in this research article to understand and analyze the concept of machine learning technologies or artificial intelligence [5]. In addition to that, the machine learning theory was helpful for the researcher to formalize the problems of learning in a statistical way. Thereafter, this theory helps to enhance the efficiency of the researcher to analyze the algorithms of learning. Apart from that, there are a lot of benefits of machine learning theory. Thus, it can be said that the machine learning theory was helpful for the researcher to collect information about machine learning technologies (Fig. 8.3).

8.2.4 *Literature Gap*

The specific machine learning technologies that are used for detecting several diseases in medical diagnosis are described in this research study briefly. Thus, in the previous literature this is not analyzed significantly. Therefore, the usage and importance of artificial intelligence or machine learning technologies in medical diagnosis is another brief description in this literature review. Thus, this is not described properly in the previous literature review.

8.3 Research Methods

There are several types of research philosophies that are used for gathering essential information about the research topics. Therefore, the researcher has adopted the ***positivism research philosophy*** for this research article. The positivism research philosophy is helpful to find reliable, representable and generalizable data and this is the reason for choosing this research philosophy by the researcher for this research article. Apart from that, the positivism research philosophy is one of the important research philosophies as it helps to understand the social world in an objective way [6]. Thus, these are the reasons for adopting the positivism research philosophy for this research article. Furthermore, the researcher has adopted the ***inductive research approach*** for this particular research article. Therefore, there are a lot of advantages of the usage of this inductive research approach in research studies. First of all, by using the inductive research approach the researcher can focus on the usage of proper language in several projects.

On the other hand, there are a gamut of probabilities in an inductive research approach and this is one of the largest benefits of the usage of this inductive research approach. Thus, the inductive research approach is very important among all the research approaches [7]. Hence, these are the causes for choosing the inductive research approach in this research article. Apart from that, the researcher has adopted the ***descriptive research design*** in this research article among all the research designs. Thus, descriptive research design helps to make a remarkable and presentable project [8]. Along with that, the research design helps in data collection as well and these are the reasons for choosing this descriptive research design for this research article. Thus, it can be said that the research paradigm such as positivism research philosophy, inductive research approach and descriptive research design are very helpful to create better research study and was quite helpful for this research article as well (Fig. 8.4).

Method of data collection is the process of gathering accurate and proper data for research studies. There are two types of methods of data collection such as primary data collection method and secondary data collection method. Therefore, in this research article the researcher has adopted the ***secondary data collection method*** as there are a lot of benefits of using the secondary data collection method. Apart

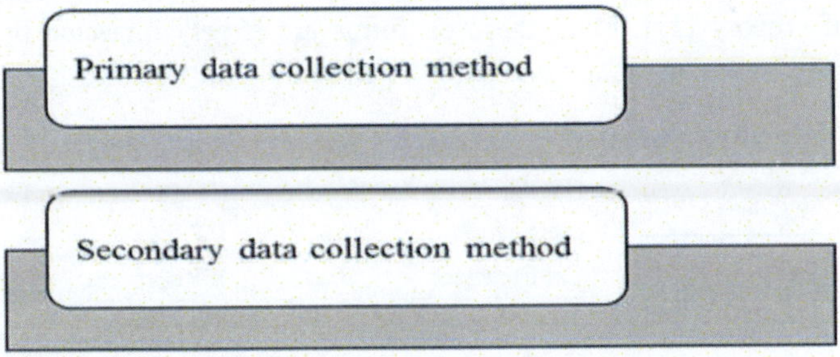

Fig. 8.4 Methods of data collection [9]

from that, there are a lot of sources of secondary data collection methods such as books, magazines, newspapers, government records, public records, journals and other published sources [9]. Thereafter, the secondary data collection method is easy to access, time saving and not expensive and these are the reasons for choosing this data collection method. Furthermore, the secondary data collection method was helpful to gain more data about machine learning technologies and its usage in the healthcare system.

8.4 Result and Discussion

Theme 1: Effect of machine learning technology in detecting patterns of certain diseases

There is a large impact of machine learning technology on medical diagnosis and that effects are discussed in this part of this research article. Thus, there are some diseases that need artificial intelligence and machine learning technologies to recognize such as lung cancer, skin lesions, cardiac attack, heart diseases, diabetic retinopathy and others [10]. Thereafter, machine learning technologies are easy as well as quick and that help to detect the disease more fatly. Furthermore, there are several machine learning methods or technologies that are used in disease detection such as SVM or support vector machines, K-nearest neighbours, random forest and others. On the other hand, there are several technologies for different types of diseases such as CT scan technology that is used to detect lung cancer. Moreover, MRI technology is used to detect cardiac attack and electrocardiogram technology is used to detect several heart diseases [11]. Furthermore, there is an x-ray technology machine that helps to take images of several body parts and that helps to detect disease (Table 8.2).

Theme 2: Several platforms that need the usage of machine learning technologies

The usage of machine learning technologies is increasing day by day among several platforms. Therefore, there are several platforms that need the usage of machine learning the most such as, psychology, artificial intelligence, control theory, neuroscience, information theory, philosophy, Bayesian method and computational complexity theory [11]. Thus, these platforms are largely impacted by machine

Table 8.2 Five diseases that can be detected by machine learning technologies [10]

Diseases detected by machine learning technologies	Name of the machine learning technology
Lung cancer	CT scan
Cardiac attack	MRI process
Skin lesions	Skin images
Heart diseases	Electrocardiograms
Diabetic retinopathy	Eye images

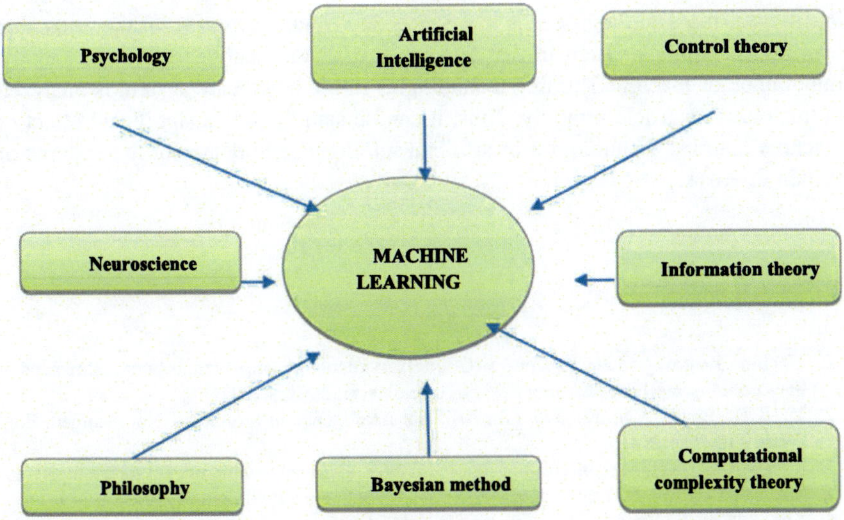

Fig. 8.5 Usage of machine learning technologies

learning technologies and artificial intelligence. Along with that, machine learning theory is used in several ways for several platforms. In addition to that, the usage of artificial intelligence and machine learning technologies in healthcare organizations and medical diagnosis is increasing day by day globally (Fig. 8.5).

8.5 Future Work

Machine learning technologies should be more aware about the social risks and should take more essential steps for protecting the people from those risks. Therefore, they should give proper training to the new artificial society to understand this technology more significantly. Apart from that, transparency is a very important part of artificial intelligence and they should enhance and boost this transparency for improving the technology. Furthermore, they should increase their communication skills so that they can make better teamwork. Thus, these are the future work or recommendations for artificial intelligence to improve the technology.

8.6 Conclusion

In conclusion, it can be said that there is a major impact of machine learning technologies on medical diagnosis. Therefore, there are some diseases that need artificial intelligence for recognition, thus, the usage of machine learning technologies in different medical diagnoses is increasing day by day. Apart from that, machine learning technologies help in the development of the healthcare system and the importance of machine learning technologies in the healthcare system is analyzed in this research article properly. Thus, it can be said there is a significant effect of machine learning technologies or artificial intelligence in detecting the patterns of certain diseases.

References

1. Machine Learning.: Latest Findings (Online). Available at: https://cio.economictimes.indiatimes.com/tag/machine+learning (2021). Accessed on: 8th June 2021
2. Yu, K.H., Beam, A.L., Kohane, I.S.: Artificial intelligence in healthcare. Nat. Biomed. Eng. **2**(10), 719–731 (2018)
3. Kashyap, P.: Machine learning algorithms and their relationship with modern technologies. In: Machine Learning for Decision Makers (pp. 91–136). Apress, Berkeley (2017)
4. Ahn, J., Cho, S.: Dead-band vs. machine-learning control systems: analysis of control benefits and energy efficiency. J. Build. Eng. **12**, 17–25 (2017)
5. Mello, R.F., Ponti, M.A.: Machine Learning: a Practical Approach on the Statistical Learning Theory. Springer (2018)
6. Marsonet, M.: Philosophy and logical positivism. Academicus Int. Sci. J. **10**(19), 32–36 (2019)
7. Ditlmann, R.K., Kopf-Beck, J.: The meaning of being German: an inductive approach to national identity. J. Soc. Polit. Psychol. **7**(1), 423–447 (2019)
8. Atmowardoyo, H.: Research methods in TEFL studies: descriptive research, case study, error analysis, and R & D. J. Lang. Teach. Res. **9**(1), 197–204 (2018)
9. Garcia, N.M., Mayorga, O.J.: The threat of unexamined secondary data: a critical race transformative convergent mixed methods. Race Ethn. Educ. **21**(2), 231–252 (2018)
10. Peiffer-Smadja, N., Rawson, T.M., Ahmad, R., Buchard, A., Georgiou, P., Lescure, F.X., Birgand, G., Holmes, A.H.: Machine learning for clinical decision support in infectious diseases: a narrative review of current applications. Clin. Microbiol. Infect. **26**(5), 584–595 (2020)
11. Schnack, H.G.: Improving individual predictions: machine learning approaches for detecting and attacking heterogeneity in schizophrenia (and other psychiatric diseases). Schizophr. Res. **214**, 34–42 (2019)
12. Cuperlovic-Culf, M.: Machine learning methods for analysis of metabolic data and metabolic pathway modeling. Metabolites **8**(1), 4 (2018)

Chapter 9
Sparse Function Learning for Alzheimer's Disease Detection Dependent on Magnetic Characteristics Imaging with Mark Information

Y. Suma, G. Jaffino, and Mahesh K. Singh

Abstract To detect and classify Alzheimer's disease accurately with moderate cognitive destruction, neuroimaging approaches had been used. The proposed role of the classification technique is further approved out in the SVM model. The investigational results based on the disease of Alzheimer neuroimaging scheme show the efficiency of the proposed technique for the diagnosis of mild cognitive destruction and Alzheimer disease. In the current study, depends on testing the effectiveness of a system using brain shape knowledge to identify healthy subjects and Alzheimer's disease patients. Compared with traditional volume ratio, the current finding indicates that shape information may be more useful in diagnosing. While Alzheimer's disease is not healed, near the beginning analysis is highly significant for both social care and patient. This will be suited even additional important formerly disease modifying representatives are accessible to prevent, treat, or even postponed disease development. One of the most involved fields of medical research in recent years has been the classification of AD by means of deep learning techniques.

9.1 Introduction

Disease of Alzheimer is an irretrievable, progressive brain situation to facilitate progressively, affects the facility to achieve the basic functions of remembrance and thought [1]. These signs occur first in most people with these diseases in their mid-60s. Till 2050 85% of people will be affected by Alzheimer's, around 50 million people have dementia. It will be affected by not only one with dementia, but also on their careers and families. Biomedical signal processing methods, such as magnetic resonance imaging, have been commonly used for Alzheimer's disease research [2]. In such applications, machine learning methods will be used. In the past, Alzheimer's disease applications have been already achieved [3].

Y. Suma · G. Jaffino · M. K. Singh (✉)
Department of ECE, Aditya College of Engineering, Surampalem, India
e-mail: mahesh.singh@accendere.co.in

S. Yadav et al. (eds.), *Proceedings of Second International Conference in Mechanical and Energy Technology*, Smart Innovation, Systems and Technologies 290,
https://doi.org/10.1007/978-981-19-0108-9_9

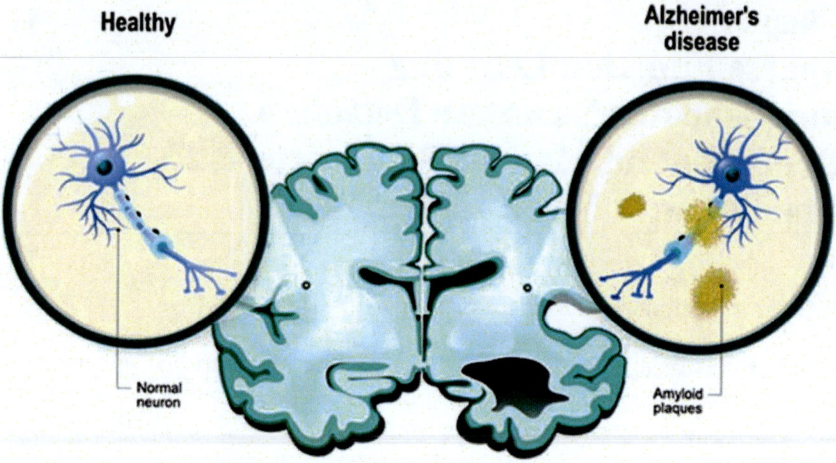

Fig. 9.1 Alzheimer's disease diagnosis in human brain

It is related to brain communication network dysfunction and is known as connection path or communication disorder. It is the most prevalent type of dementia that leads to death and has no cure. The disorder has tremendous significance in reducing the symptoms. In support of the automated analysis of Alzheimer's disease, there are several different forms of classification approaches [4, 5]. It is very difficult to diagnose AD, since the elderly will be affected by this disease. The contraction of the intelligence that occurs outstanding to the progressive loss of nerve cells is one of the main brain changes associated with Alzheimer's disease. A diagnosis of AD effects on brain is shown in Fig. 9.1.

A lot of handcrafted machine learning methods for premature analysis of AD have been established during recent days [6]. With age, the risk of developing AD rises dramatically and after 65, the risk doubles every 5 years, some of the early signs of the disease are memory loss, impaired decision-making, changes in mood and personality, and this is called Mild AD level. MCI induces a drop in cognitive skills such as memory and reasoning ability [7]. MRI is a powerful method for AD diagnosis, it is expensive and simple or convenient to do always not possible. Detection systems for AD could be introduced in the future [7].

9.2 Literature Review

These images may be helpful in creating a powerful AD classifier in 2011, while AD or HC may not belong to their cognitive status. The investigational result on ADNI information shows, that our proposed technique preserve significantly improve the output between AD and HC [8]. This manuscript introduced an image-based detection system and applied it to the detection of brain MRI scan of community with soft cognitive impairment (MCI) [9]. They conclude that this enhanced performance was due to multiple learning methods and semi-supervised classification representing all paired image interactions [10]. Computer Integrated Diagnosis approach for diagnosing the Alzheimer's disease is being proposed. In this result shown the efficiency gets 91% accuracy by using Nonnegative Matrix Factorization-Support vector machine process [11].

Alzheimer's disease is progressive mind disarray which gradually leads to failure of memory confusion, disorientation and communicable inability. The voxel intensity method is the one with the lowest accuracy values in their classification process [8, 12]. In 2014 this article, we are designing an AD detection replica which will be able to support health professional in forecasting disease condition depending on patient health report [13–19]. This leads to refine the decision tree to improve its accuracy and generates that most events can be predicted in the future [13]. The computer aided diagnosis of Alzheimer's disease based on neuroimaging information has attracted a lot of attention. The current HR-SSCRFS algorithm would check the multimodal neuroimaging data in future work [14]. In diffusion of MRI offers new indications not available using traditional anatomical MRI for disease diagnosis [15]. In 9 of the 14 regions of investigation, characteristics exceeded. 85.6%, area involved in many previous AD studies in the hippocampus coagulum, was the highest accuracy [16]. In 2016, the measurement of grading biomarker is then per MCI subject. In this result, they suggested a biomarker-based on the MCI-to-AD global ranking [17].

In 2016 it is gradually destroying the affecter's ability to reason and memory. Current treatment for treating AD is not available. Yet early diagnosis may reduce disease symptoms and may develop the excellence of life [18]. This study is dependent on the comparison and assessment of the related work done to determine the diagnosis of Alzheimer's disease using different methods including MRI [17]. In 2018 currently, it can find it possible that knowledge shape might be more useful, we used support vector machine to perform classification. The accuracy rate here is 87.1% [18].

Neurodegenerative infection is currently the 6th most important cause of casualty in the US in 2018. 1 in 3 elderly people suffer from Alzheimer's or other dementia. In addition, in order to diagnose other 3D MRI diseases, similar techniques may be used. This paper can also serve as an inspiration for other forms of 3D image processing using deep learning [19]. A non-invasive and repeatable approach for the diagnosis of brain disorders is electroencephalography in 2018. Experimental samples of HC, MCI and AD were detected with greater precision than spectral analysis of the Fast Fourier Transform [2]. In 2019 Multi-model biological, imaging markers showed the best performance in reducing Alzheimer's disease from relatively normal elderly patients [2]. In 2019, as per the World health organization (WHO), the number of populace breathing in the midst of AD is estimated in the United States. They also proposed a multiclass diagnostic system in the test. Efficiency is strong [3].

9.3 Method and Methodology for Detection of AD

This technique, which is called multiple regularized sparse learning features with label data, will be proposed for detection shown in Fig. 9.2. It implements the spare regression model feature selection and then presents our process specifications and the right iterative optimization algorithm.

9.3.1 *Regression Sparse Model*

Without taking into account the local geometric spatial structure between samples, the model selects a single common subset of features and the features selected may be less selective. Where the aesthetic coefficient matrix is a weighting parameter that not only balances the value of the relative between the loose term and the term of regulation, but also regulates the sparsely of the elements in the matrix, the 2:1 standard promotes row sparsely in jointly between samples matching the multi-class mark matrix.

$$\min 1 \div 2|B - AC|_U^2 + \lambda|C|_{2,1} \tag{9.1}$$

9.3.2 *Sparse Feature Extraction*

$$p = \sum_{i,j}^{n} S_{ij}|a_i c - a_j|_2^2 = t_r\left[C^T A^T \mathrm{MAC}\right] \tag{9.2}$$

$$S_{ij} = \begin{cases} 1, a_i \in N_k(a_j) \text{ or } a_j \in N_k(a_i), & i \neq j \\ 0, \text{otherwise} \end{cases} \tag{9.3}$$

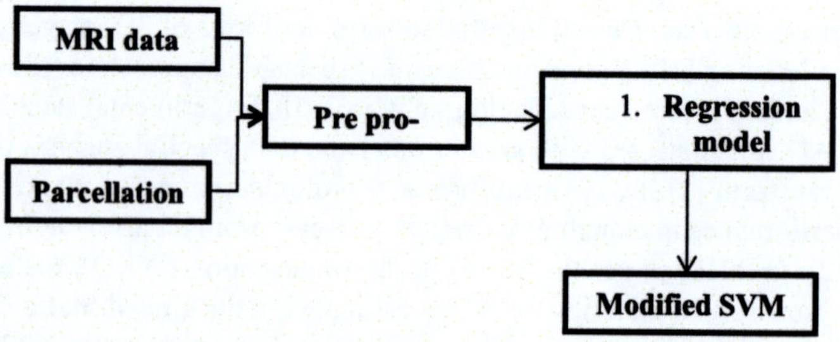

Fig. 9.2 Block diagram the modified SVM based model

where '$N_k(a_i)$' to k-Neighbors denotes subject 'a_i'. How to decide the most favorable k-Neighbors among sample of dissimilar curriculum is the main issue in building the neighborly relationship. To address this problem, we first introduced an adjustable brand in sequence parameter to control the complete Euclidean expanse among sample, and after that the relation detachment will depart as follow.

$$ed_{ij} = \begin{cases} \delta\left|a_i - a_j\right| \text{ if } & a_i \text{ and } a_j \text{ belong to the same class } 0 < \delta \leq 1 \\ & 1 \div \delta\left|a_i - a_j\right| \text{otherwise} \end{cases} \quad (9.4)$$

$$\min 1 \div 2|B - AC|_U^2 + \lambda \sum_{i,j}^{n} S_{ij}\left|a_i c - a_j c\right|_2^2 + \lambda_2 |W|_{2,1} \quad (9.5)$$

when λ_1 and λ_2 parameters are matched. In this proposed approach, the relationship between optimal neighbors should be maintained and more discriminatory features should be chosen, most importantly, to attain better presentation in following classification tasks, the collection of features is immobile assumed to be carried out in the unique room, that is simple to understand or analyze the preferred features.

9.3.3 *Modified SVM Based*

The pattern recognition and classification of the support vector machine is widely used for its ability to identify patterns in a variety of applications in the experimental database. The hyper plane of these data is defined for separation at what time the conservative SVM scholarship method is performed and the resultant SVM classifier is qualified with a series of training observations. Ideally, when re-entered for testing and entry into the SVM, all training data should be correctly classified.

9.4 Result

The efficiency of the future technique with cortical width of MRI information obtains from the ADNI record. For every ROI, they calculate the average cortical depth of the intelligence area, and each subject also has 78 characteristics. EEG multichannel signals were acquired using 19 electrodes that set their position by following the 10–20 foreign systems and using monopoly contacts with the landmark of the earlobe electrode. In resting state and closed eyes, subject brain activity was measured to the electrical potential. In addition, to normalize the sampling regularity, they transformed every signal to 256 Hz.

To repeatedly categorize the samples according to their categories by controlling their relevant functions, they conducted a supervised learning analysis. In dealing with the HC versus MCI, HC versus AD, HC versus CASE grouping of EEG signals

Table 9.1 Classification results for regression sparse and sparse feature extraction analysis

Performance measures	HC versus AD	HC versus MCI	MCI versus AD	HC versus CASE
Regression sparse model				
accuracy	73.1	72.6	81.3	75.6
F-measure	72.3	72.9	81.2	75.6
Sensitivity	73.1	72.6	81.3	75.6
Precision	72.0	77.5	81.3	75.1
specificity	60.1	77.1	77.4	43.6
Sparse feature extraction				
Accuracy	73.1	72.6	81.3	75.6
F-measure	72.3	72.9	81.2	75.6
Sensitivity	73.1	72.6	81.3	75.6
Precision	72.0	77.5	81.3	75.1
specificity	60.1	77.1	77.4	43.6

here intended to derive a human decipherable model unique to every sample category based on a diminutive subset of Fourier analysis features (Table 9.1).

9.5 Conclusion

In this paper, it is proposed a novel range of regularized spare characteristic knowledge technique for AD and MCI detection. The hierarchy dependent on structural attractive character imaging then established a new multiple regularization with class mark information to protect. The association between the optimal neighbors, investigational consequences show that compared to numerous condition of the art method for AD and MCI classification. Computed aided diagnosis technique is associated with Alzheimer's disease with NMF-SVM. This is based on the grouping of non-negative matrix factorization for the selection and decrease of features, and SVM for confidence bound classification. This decrease in function is especially suited to techniques of machine learning such as SVM.

References

1. Padilla, P., López, M., Górriz, J.M., Ramirez, J., Salas-Gonzalez, D., Alvarez, I.: NMF-SVM based CAD tool applied to functional brain images for the diagnosis of Alzheimer's disease. IEEE Trans. Med. Imaging **31**(2), 207–216 (2011)
2. Li, H., Fan, Y.: Early prediction of Alzheimer's disease dementia based on baseline hippocampal MRI and 1-year follow-up cognitive measures using deep recurrent neural networks. In: 2019 IEEE 16th International Symposium on Biomedical Imaging (ISBI 2019), pp. 368–371. IEEE (2019)

3. Jabason, E., Ahmad, M.O., Swamy, M.N.S.: Classification of Alzheimer's disease from MRI data using an ensemble of hybrid deep convolutional neural networks. In: 2019 IEEE 62nd International Midwest Symposium on Circuits and Systems (MWSCAS), pp. 481–484. IEEE (2019)
4. Tong, T., Gao, Q., Guerrero, R., Ledig, C., Chen, L., Rueckert, D., Initiative, A.D.N.: A novel grading biomarker for the prediction of conversion from mild cognitive impairment to Alzheimer's disease. IEEE Trans. Biomed. Eng. **64**(1), 155–165 (2016)
5. Veerendra, G., Swaroop, R., Dattu, D.S., Jyothi, C.A., Singh, M.K.: Detecting plant diseases, quantifying and classifying digital image processing techniques. Mater. Today Proc. (2021)
6. Singh, M.K., Singh, N., Singh, A.K.: Speaker's voice characteristics and similarity measurement using euclidean distances. In: 2019 International Conference on Signal Processing and Communication (ICSC), pp. 317–322. IEEE (2019)
7. Zhang, D., Shen, D.: Semi-supervised multimodal classification of Alzheimer's disease. In: 2011 IEEE International Symposium on Biomedical Imaging: From Nano to Macro, pp. 1628–1631. IEEE (2011)
8. Prasanna, G.S., Pavani, K., Singh, M.K.: Spliced images detection by using Viola-Jones algorithms method. Mater. Today Proc. (2021)
9. Singh, M.K., Singh, A.K., Singh, N.: Multimedia utilization of non-computerized disguised voice and acoustic similarity measurement. Multimedia Tools Appl. 1–16 (2019)
10. Amulya, E.R., Varma, S., Paul, V.: Classification of brain images for Alzheimer's disease detection. In: 2016 IEEE International Conference on Computational Intelligence and Computing Research (ICCIC), pp. 1–4. IEEE (2016)
11. Singh, M.K., Singh, A.K., Singh, N.: Disguised voice with fast and slow speech and its acoustic analysis. Int. J. Pure Appl. Math **118**(14), 241–246 (2018)
12. Ullah, H.T., Onik, Z., Islam, R., Nandi, D.: Alzheimer's disease and dementia detection from 3D brain MRI data using deep convolutional neural networks. In: 2018 3rd International Conference for Convergence in Technology (I2CT), pp. 1–3. IEEE (2018)
13. Singh, M.K., Singh, A.K., Singh, N.: Acoustic comparison of electronics disguised voice using different semitones. Int. J. Eng. Technol. (UAE) **7**(2), 98 (2018)
14. Kanchana, V., Nath, S., Singh, M.K.: A study of internet of things oriented smart medical systems. Mater. Today Proc. (2021)
15. Ye, D.H., Pohl, K.M., Davatzikos, C.: Semi-supervised pattern classification: application to structural MRI of Alzheimer's disease. In: 2011 International Workshop on Pattern Recognition in NeuroImaging, pp. 1–4. IEEE (2011)
16. Balaji, V.N., Srinivas, P.B., Singh, M.K.: Neuromorphic advancements architecture design and its implementations technique. Mater. Today Proc. pp. 1–4 (2021)
17. Singh, M.K., Singh, A.K., Singh, N.: Multimedia analysis for disguised voice and classification efficiency. Multimedia Tools Appl **78**(20), 29395–29411 (2019)
18. Punyavathi, G., Neeladri, M., Singh, M.K.: Vehicle tracking and detection techniques using IoT. Mater. Today Proc. (2021)
19. Singh, M., Nandan, D., Kumar, S.: Statistical analysis of lower and raised pitch voice signal and its efficiency calculation. Traitement du Signal **36**(5), 455–461 (2019)

Chapter 10
A Study on the Relationship Between Cloud Computing and Data Mining in Business Organizations

Dilip Kumar Sharma, A. Dharmaraj, Alim Al Ayub Ahmed, K. Suresh Kumar, Khongdet Phasinam, and Mohd Naved

Abstract In order to enhance information services quality, data extraction is a method through which potentially helpful data may be extracted. Integrated with cloud computing, data mining technology enables customers to obtain valuable information from a data warehouse that lowers infrastructure and storage cost. In recent years the data mining sector and society as a whole have become quite attractive large data availability and urgent need for the transformation of such information into valuable expertise and data Market research, fraud prevention and retention of customers, manufacturing control and scientific research. The natural development of information technology may be seen as a consequence of data mining. Security and privacy of user data is of major issue in the usage of cloud computing in data mining. In the context of discovering new, legitimate, usable, and comprehensible data formats, data mining is regarded to be an essential activity. The integration of data mining techniques with cloud computing offers a flexible and scalable architecture that can be utilized to efficiently extract large quantities of information from

D. K. Sharma
Department of Mathematics, Jaypee University of Engineering and Technology, Guna, Madhya Pradesh, India

A. Dharmaraj
Karpagam Academy of Higher Education, Coimbatore, India

A. Al Ayub Ahmed
School of Accounting, Jiujiang University, 551 Qianjindonglu, Jiujiang, Jiangxi, China

K. Suresh Kumar (✉)
MBA Department, Panimalar Engineering College, Varadarajapuram, Nasarathpettai, Poonamallee, Chennai, India
e-mail: pecmba19@gmail.com

K. Phasinam
School of Agricultural and Food Engineering, Faculty of Food and Agricultural Technology, Pibulsongkram Rajabhat University, Phitsanulok, Thailand

M. Naved
Department of Business Analytics, Jagannath University, Delhi-NCR, India

S. Yadav et al. (eds.), *Proceedings of Second International Conference in Mechanical and Energy Technology*, Smart Innovation, Systems and Technologies 290,
https://doi.org/10.1007/978-981-19-0108-9_10

near-connection data sources in order to provide valuable, decision-making knowledge. This article offers a general explanation of the necessity for data mining to be integrated into cloud computing, to provide its customers effective and secured services and to decrease infrastructure and operating requirements in business.

10.1 Introduction

The Internet has become an essential resource in our everyday activities in recent years, because the quantity of information generated by people who utilize online services is huge. In this data, hidden data may be utilized to make successful choices. Cloud infrastructure is utilized to substantially find valuable information in data mining integration techniques. Cloud computing is designed to alter the conventional computing method by delivering both physical assets and programming. These services are offered via the internet [1]. Its cheap cost, portability, and enormous availability make it more attractive. It offers limitless storage and processing power, resulting in huge numbers of data. For finding knowledge on databases, data mining techniques are employed. It is utilized for the analysis of data from various origins and the provision of relevant information. The Data Mining and the Cloud Computing Relationship Approach offers quick access to technology and an information finding system that consists of several decentralized data analysis services.

We are frequently called the information era in age. We think that information leads to power and success in that information era and that we have gathered enormous quantities of information due to advanced technology such as computers, satellites, and so on [2]. Initially, we began gathering and storing all kinds of data with the introduction of computers and methods for mass data knowledge relying on the computer's ability to assist sort this fusion of information. Sadly, these huge data sets, housed on separate structures, became overpowering very quickly. The early turmoil resulted in organized systems and database management systems being created (DBMS). We have now far more knowledge than we can deal with: from transactions, scientific knowledge, satellite photos, documentary evidence, and military intelligence. Data mining was an efficient technique for analysing data from many perspectives and obtaining valuable data insights. Classifying data, categorizing data and finding connection between datasets data patterns. Many companies are now beginning to use data mining as a technique to cope with the aggressive data analysis environment. The cloud allows you to access your information at all times from everywhere. The cloud eliminates the requirement being in the similar physical place as your gear [3].

10.2 Concept for Cloud Computing

The term "cloud computing" is a catchall term for a new sort of Internet-based computing. It's a unique concept that explains the use of computers as a service, and it's gotten a lot of attention recently. Cloud computing, according to the National NIST is a concept that gives on-demand access to a shared pool of customizable computing assets (e.g. networks, servers, storage, applications, and services) that can be swiftly delivered and discharged with minimum administration effort or service provider engagement.

Cloud computing is an emerging concept change where computers are transferred from particular computers or servers to a "cloud" of computers. Cloud users must only consider the computer service requested, since the fundamental facts on how this is done are concealed. The high-performance computing technique is carried out by bringing together all computer resources and by software instead of a human being [4].

10.3 Cloud Computing Advantages

- Less computer charge: no high performance and affordable computer is needed for the use of web-based cloud computing apps.
- Enhanced Performance: Computers start and operate quicker in a cloud computing environment because they have less memory loaded processes and applications.
- Lower software expenses: you can obtain much of what you need free of charge instead of buying costly software programmes.
- Another benefit of cloud computing is that you are no longer confronted with the choice between outdated software and expensive upgrade fees. Updates occur automatically if the programme is on the Internet
- Cloud computing provides practically infinite storage capacity.
- Greater information reliability: In contrast to desktop computing, where computer accidents in the cloud do not affect data storage if the hard drive fails or loses all important data.

10.4 Cloud Computing Disadvantages

- It does not function well with low-speed connections.
- It needs a continuous internet connection.
- Information or data stored in cloud computing may not be secure.

10.5 Clouds Are Classified into Some Categories

Many firms share the services in a community cloud, and only those companies have access. Companies or a cloud service provider could own and control the technology [5].

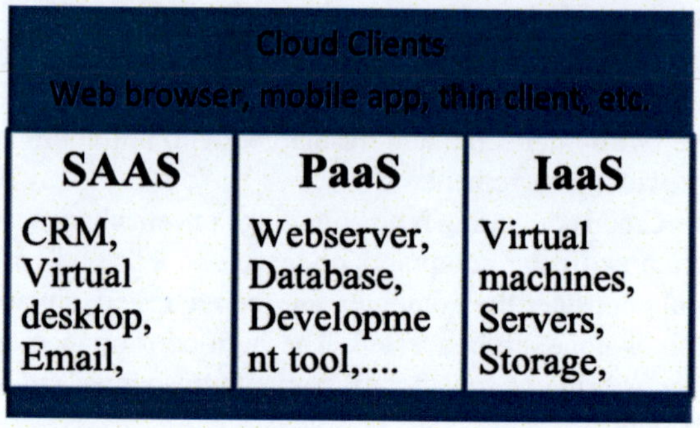

Fig. 10.1 Fundamental models for the provision of cloud services

A cloud provider owns and maintains a public cloud that is accessible through the Internet. Public services such as online photo storage, e-mail services, and social networking sites are examples. The term "hybrid cloud" refers to a system that includes different methods for sharing resources (for example, combining public and community clouds). A private cloud is a cloud infrastructure that is reserved for a single company and is managed by the business or a third party.

10.6 Models for Cloud Computing

(i) **Service as a software (SaaS)** The client is given a whole application as a service over the internet under this approach. A single service instance is running on the cloud and many end users are being served [6]. On the client side, an initial investment in servers or software licensees is not necessary, whilst expenses for providers are reduced since just one application has to be hosted and maintained. Headquarters like Google, Sales force, Microsoft, Zoho, and others are offering nowadays SaaS (Fig. 10.1).

(ii) **Platform as a service (Paas)** Here is wrapped and provided a service that allows other higher levels of support to be built on software layers or development environments. The client may develop its own apps on the infrastructure of the provider.

(iii) **Infrastructure as a services (Iaas)** As a standardized network service, IaaS offers basic storage and computer capacity. The workloads are pooled and made accessible via servers, storage systems, networking systems, network infrastructure space etc.

10.7 Companies in Cloud Top and Key Characteristics

Table 10.1 shows the cloud names with their key components.

Table 10.1 Companies in cloud top and key characteristics

Cloud name	Key component
Sun microsystems sun cloud	More applications are available than on any other open operating system
Amazon EC2	The goal is to make web scale computing more accessible to developers
Microsoft	A discount plan for development accelerators is now available. Discounts ranging from 15 to 30% off consumption costs are available for the first six months
GoGrid cloud computing	Automatic load levelling as well as free, round-the-clock assistance
IBM dynamic infrastructure	Processing power in the context of relationships may assist you in planning, predicting, monitoring, and actively managing the power usage of your Blade Center servers
Google app engine	If you do not surpass the quota given, there is no time restriction on the testing period
AT&T synaptic hosting	You may either use completely on-demand architecture or mix it with specific components to fulfill particular needs

10.8 Overview of Data Mining

Data mining is the process of detecting undiscovered, meaningful patterns and correlations in massive data sets using sophisticated data and technology. These tools include statistical models, mathematical algorithms, and machine learning techniques (algorithms that improve their performance automatically over time, such as neural networks or decision trees) [7]. As a result, data mining comprises more than simply data collection and management; it also includes estimation and forecasting. Data mining is becoming more common in the business and government sectors. Data mining is commonly utilized in industries such as banking, insurance, medicine, and retailing to save money, improve research, and increase sales.

Data mining applications in the public organizations were originally employed to identify fraud and waste and were increasingly being utilized for measurement and improvement of programmed performance. Data Mining's primary purpose is to automatically discover patterns with little human input and effort. Data mining is a significant tool for managing future market trends and judgement. Data mining equipment and methods in many areas in different forms may effectively be used [8].

Data mining, also known as Knowledge Discovery in Databases, is the time-consuming process of extracting implicit, previously unknown, and potentially important information from databases (KDD). In databases, the terms "data extraction" and "knowledge discovery" are commonly interchanged (or KDDs), data mining is an important aspect of the knowledge discovery process. The diagram below depicts data mining as part of an iterative knowledge discovery process.

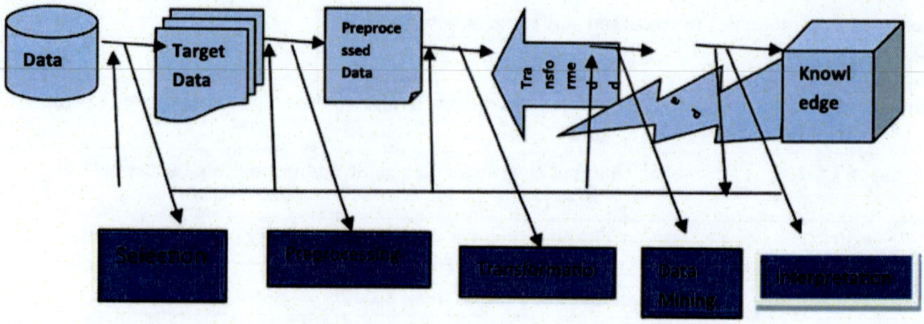

Fig. 10.2 Cloud computing and data mining in business organizations

10.9 Knowledge Discovery Process (KDD)

The several stages in the process of KDD.

- Data integration—Relationship information from various data sources.
- Data Selection and cleaning—Analytical data are collected from the database; noise is eliminated and inconsequential data is deleted.
- Data transformation—This phase includes consolidating and converting data to mining formats e.g. by aggregating data summaries.
- Mining of data—That's the most essential phase and is achieved via smart training datasets.

7Evaluate patterns the most popular method for forecasting a particular result, such as a response, high/medium/low value client, that is susceptible to buying/not buying.

- On-assessment consists on identifying intriguing trends.
- Presentation of knowledge—different visualization and presentation methods are used to show the end user the gathered or mined information (Fig. 10.2).

10.10 Techniques for Data Mining

Table 10.2 discusses the Data Mining Techniques along with their key components.

10.11 Data Mining Applications

The following are the main areas of applications for data mining:

(i) **Fraud** utilized for monitoring fraud by credit cards, monitoring millions of accounts. Fraud detection: They are used to detect financial transactions that may suggest money laundering.

Table 10.2 Techniques for data mining

Data mining	Key component
Removal of feature	It develops new qualities as a continuous combination of current features. This technique's uses include text data, latent semantic analysis, data compression, data decomposition and projection, and pattern recognition
Clustering	This programme may be used to analyse data and find natural groupings of data. Inhabitants of one cluster are more similar to one another than residents of another cluster, which is a positive thing. The discovery of new consumer categories and the discovery of new life sciences are two examples
Identification of anomalies	The way occurrences deviate from the norm is used to describe them as strange or suspicious. Health-care fraud, expense-report fraud, and tax evasion are all instances of common fraud
Association	Criteria linked with commonly encountered products may be discovered and utilized for market basket analysis, cross-sell opportunities and root cause analysis. Product packaging, in-store positioning, and fault analysis are all possible with this tool
Regression	Lifetime value, home worth, and process yield rates are all examples of continuous numerical outcomes that may be predicted using this method
Significance of attribute	The characteristics are ranked in order of their strength of connection with the target attribute. Examples of applications include identifying the variables that are most closely linked with consumers who react to an offer and the factors that are most closely associated with healthy patients
Categories	When forecasting a particular result such as a response or no response, high, medium, or low value client, or whether someone will buy or not will purchase anything. The most frequently utilized method

(ii) **Investment** Many businesses utilize investment data mining, although most do not disclose their methods. Investment: LBS Capital Management is one example. It utilizes portfolio management experts, neural networks, and genetic algorithms.

(iii) **Marketing** Database marketing systems are the main application in marketing, analysing customer databases in order to identify and forecast various consumer groupings.

(iv) **Telecommunications** the TASA provides methods for pruning, grouping, and sorting to improve the outcomes of a simple brute-force rule search. With versatile information extraction methods, interaction and iteration may be explored with several sets of found rules.

10.12 Data Mining and Cloud Computing Relationship in Business

Data mining techniques and applications are crucial in the cloud computing sector. The practice of organizing information from unstructured or semi-structured internet data sources is known as data mining [9]. Cloud computing data mining allows organizations to centralize software and data storage management, assuring trustworthy, efficient, and reliable services for their consumers. The use of data mining technologies such as SaaS, PaaS, and IaaS to collect data in cloud computing is examined. Cloud data mining is used to analyse and extract important data in a variety of industries, including banking, medical, and marketing [10]. With only a few mouse clicks, you can acquire all the information you need about customer behaviour, habits, interests, and location. Smaller businesses can use the cloud to employ a cloud service for efficient analysis of all data in the company that was previously solely reserved for big data [11].

When dealing with large volumes of data, Data Mining is preferred, and related approaches usually require large data sets to achieve classification accuracy. Data mining is used by cloud operators to improve client experience [12]. Customers may extract important information from virtually any data source using cloud computing data mining techniques, which reduces equipment and storage costs. Cloud Computing is a new trend in internet services that focuses on job-processing servers' clouds [13]. Data mining is the technique of obtaining unstructured or semi-structured data from internet data sources in cloud computing. Cloud computing is the delivery of software and hardware as a service via the Internet, and cloud computing data mining technology is no exception [14].

The following are the advantages for data mining and cloud computing.

- The client pays just for the data mining tools he requires.
- The client does not have to keep a physical server since he may use the browser to do data mining.
- Inefficient storage capacity.
- Virtual computers can be launched shortly.
- No relational database queries.
- Communication channel queue.

10.13 Conclusion

This study provides a high-level overview of data mining. Relationships with cloud computing service providers are critical for companies in order to make effective decisions when predicting current performance and patterns. Computing is the component that is given, and data mining is the component that is offered as well. Data mining cannot be completed without the use of a Data Mining software, as well as without the use of cloud computing or cloud computing services. They are all too tasty and effective when combined, just like cake and ice cream are when served separately. Cloud computing is dependent on computer systems that are capable of accepting tasks from a distance. Obtaining organized data from unappropriated or semi-structured Web data sources is accomplished via the process of data mining. Cloud Computing data mining allows businesses to centralize software and data storage management, thus providing customers with more cost-effective and reliable services.

References

1. Ambulkar, B., Borkar, V.: Data mining in cloud computing. In: MPGI National Multi Conference, vol. 2012 (2012)
2. Dillon, T., Wu, C., Chang, E.: Cloud computing: issues and challenges. In: 2010 24th IEEE International Conference on Advanced Information Networking and Applications, pp. 27–33. IEEE (2010)
3. Fayyad, U., Piatetsky-Shapiro, G., Smyth, P.: From data mining to knowledge discovery in databases. AI Mag. **17**(3), 37–37 (1996)
4. Feyyad, U.M.: Data mining and knowledge discovery: making sense out of data. IEEE Expert **11**(5), 20–25 (1996)
5. Jain, A., Pandey, A.K.: Multiple quality optimizations in electrical discharge drilling of mild steel sheet. Mater. Today Proc. **4**(8), 7252–7261 (2017)
6. Jain, A., Pandey, A.K.: Modeling and optimizing of different quality characteristics in electrical discharge drilling of titanium alloy (grade-5) sheet. Mater. Today Proc. **18**, 182–191 (2019)
7. Jain, A., Yadav, A.K., Shrivastava, Y.: Modelling and optimization of different quality characteristics in electric discharge drilling of titanium alloy sheet. Mater. Today Proc. **21**, 1680–1684 (2020)
8. Kamala, B.: A study on integrated approach of data mining and cloud mining. Int. J. Adv. Comput. Sci. Cloud Comput. ISSN 2321-4058 (2013)
9. Mell, P., Grance, T.: The NIST definition of cloud computing (2011)
10. Panwar, V., Sharma, D.K., Kumar, K.P., Jain, A., Thakar, C.: Experimental investigations and optimization of surface roughness in turning of en 36 alloy steel using response surface methodology and genetic algorithm. Mater. Today Proc. (2021)
11. Patil, V., Nikam, V.B.: Study of data mining algorithm in cloud computing using MapReduce framework. J. Eng. Comput. Appl. Sci. (JEC&AS) **2**(7), 65–70 (2013)
12. Petre, R.S.: Data mining in cloud computing. Database Syst. J. **3**(3), 67–71 (2012)
13. Simmon, E.: Evaluation of cloud computing services based on NIST SP 800–145. NIST Spec. Publ. **500**, 322 (2018)
14. Viloria, A., Acuña, G.C., Franco, D.J.A., Hernández-Palma, H., Fuentes, J.P., Rambal, E.P.: Integration of data mining techniques to PostgreSQL database manager system. Procedia Comput. Sci. **155**, 575–580 (2019)

Chapter 11
Study of Different Types of Smart Sensors for IoT Application Sensors

Ch. V. N. S. Mani Kiran, B. Jagadeesh Babu, and Mahesh K. Singh

Abstract IoT appliances form part of the wider home automation definition that is included security systems, television, air conditioning, heating, and lighting. Long-term reimbursement could consist of power savings and making sure the lighting and appliances are shut off automatically. Wireless sensors are important component for building the sensing system for the Internet of Things (IoT). The IoT is a technology in which one or more technologies are replaced by another technology. IoT building a small eco-system around us, with the number of sensors and actuators. These manuscripts show the numerous IoT sensors and in addition explain a variety of sensor dependent IoT application. In addition, subsequent to analyzing is diverging sensor application. These manuscripts enlighten that IoT function requires which category of sensor. In the opportunity, this effort will provide as the source for additional explore the work in the connected area. In this paper mainly we are discussing the scrutiny of dissimilar types of sensors like humidity sensor, temperature sensor, pressure sensor, optical sensor, proximity sensor, position sensor, and velocity sensor. We also discuss sensor applications.

11.1 Introduction

The IoT connect everything, animates and inanimate things which lead to drastic change. Their main objective is to make things more energetic and comfortable. This allows for many devices to behave as intelligent equipment. IoT technologies have inspired smart services for people. IoT permitted smart devices to provide the possessor with many services, such as washing machines, water heaters, lights such as outdoor electronic devices and indoor etc. It is additional obliging for blind and deaf persons with a disability. Such procedures are obtainable on the mobile markets, and with no display, they only provide audio services with a display [1].

Ch. V. N. S. Mani Kiran · B. Jagadeesh Babu
Department of ECE, Aditya College of Engineering, Surampalem, India

M. K. Singh (✉)
Department of ECE, Aditya Engineering College, Surampalem, India
e-mail: mahesh.singh@accendere.co.in

S. Yadav et al. (eds.), *Proceedings of Second International Conference in Mechanical and Energy Technology*, Smart Innovation, Systems and Technologies 290,
https://doi.org/10.1007/978-981-19-0108-9_11

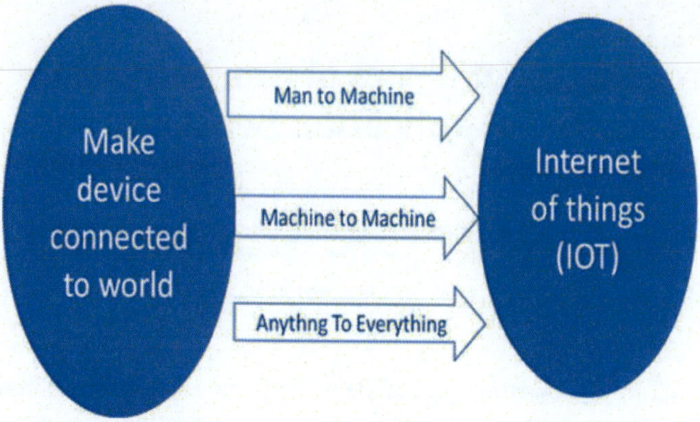

Fig. 11.1 IoT interfacing technique

Sensors are all over the place. They are in our homes and places of work, in our shopping centers and in our hospitals. We are found in smart phones and are an integral part of the IoT [2]. For a long time, sensors have been around. In the late 1880s, the first thermostat was introduced, and since the 1940s infrared sensors have been around. The IoT and its equivalent, the commercial IoT, are bringing a new level of sensor use. Sensors are very critical in the smart application that shown in Fig. 11.1 [1].

Temperature Sensor: Temperature sensors are effective in calculating heat energy to identify the substantial change in one's body. People were using temperature sensors to track ambient environmental conditions. The collected data is subsequently forwarded to the obscure by Wi-Fi for review. All this is completed with the mobile app. A comparable kind of sensor is too used in smart farming and enables farmer to enlarge their generally capitulate and item for consumption superiority by collecting survive information from their land in real-time [2, 3] (Fig. 11.2).

Humidity Sensor: A humidity sensor monitors both air temperature and humidity, which shows moisture in the atmosphere. People use moisture sensor for elegant farming and allow farmers to boost their yield and manufactured goods feature by obtaining survive information of their property in real-time [3, 4].

Pressure Sensor: Pressure sensors can sagacity the power and turn it into signal. This category of sensor can be used for tracking your physical condition [5].

Proximity Sensors: With the proximity sensor, the location of any close entity can be simply detected any substantial make contact with each other. By emitting electromagnetic emission like an inflator, it detects an object's presence through basically look for any difference in the arrival signal. There are various types of propinquity sensors that target different applications, such as magnetic, photoelectric, ultrasonic, capacitive, inductive etc. This exact type of sensor is mainly used in application where safety and effectiveness are required [6, 7].

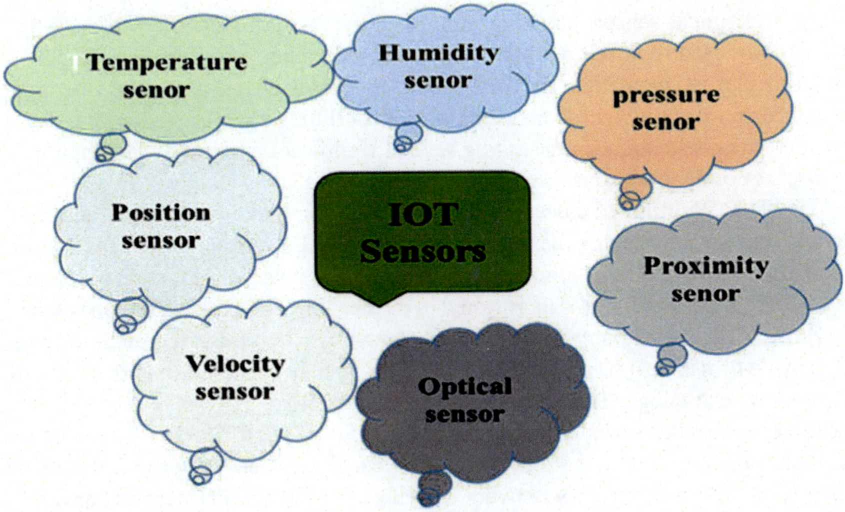

Fig. 11.2 Different type of IoT sensor

Optical Sensors: For the detection of electromagnetic energies like light, optical sensors are useful. They are commonly used in IoT application as well as in digital cameras, being sensitive to all types of electrical interfaces. Optical sensors are good for aerospace, chemical, oil refineries, environment, health, energy and other IoT applications [8].

Velocity Sensors: This type of sensor that measures the rate of transform at defined intervals in the quantity of the constant position and location values. Sensor velocity can be linear or angular. This can be used in advanced vehicle tracking applications in Smart Cities [9, 10].

Position Sensors: By detecting their motion the position sensor distinguishes the occurrence of humans or things in a scrupulous area. It could be used in home safety to allow the landlord to monitor room and appliance doors and window from anywhere [11, 12].

11.2 Related Work

In 1993: The convey of an objective amount beginning one power area to an additional is based on sensors and actuators. The renowned healthy, chemical domains, thermal, magnetic, electrical, mechanical are these. Actuators and sensors can be sub-divided into three stages depending on the dynamic cross effects, static cross effects, and physical principles involved [1]. **In 1995**: Many possibilities are proposed for modern fiber-optic sensors based on two-photon spectroscopic technique. Types include chemical sensor, counting multiple sensors, and dielectric sensors [2]. It is also proposed to develop a lightweight, low-cost fiber optics system for precise time–frequency calculation. **In 2000**: This work provides an overview of fiber optic sensor technology advances and innovations, outlining the main issue foundation current investigate and demonstrating several significant application and input area of successful fiber optics sensor in advancement [3, 13].

In 2012: With the advancement of IoT technology, the use of more and more media sensors for tracking the health of the persons. Each time, tons of IoT sensors recognize the huge sensor information. All this large sensor information is stored in the HDFL and a number of data is stored in the heading a database based on columns [14]. **In 2014**: As the populace is raising universal, an enormous requirement arises to provide proper health care service. The need-gap may be increased with the advent of modern technology. The sensor is one such technology that can be worn to allow health care monitor systems based on the IoT [15]. **In 2016**: The IoT technique has carried rebellion to each common man in area of existence by production of the intelligent. IoT referred to a network of things that create a system optimized by itself. Invention of IoT-based intellectual elegant cultivation systems transforms the features of agricultural production day by day by not simply improving it but also making cost-effective and reducing the desecrate [16].

In 2016: The real-time monitor has turn out to be a critical requirement for today's intellectual interchange monitor system to ensure safe traffic across the expressway. In this paper, presented a low-cost and secure IoT system consisting of a variety of RFID sensor for the real-time tracking. The motor vehicle is during its transportation beginning one end to a different of the high speed expressway [16]. **In 2018**: The IoT is accepted to participate a key part in our daily life during a large extent of sensor systems that surround our world. Such systems are made to track critical objective properties of matter and energy. In IoT, the information can be used to active term acts such as remote control of heat equipment or cool equipment. **In 2018**: Climate change has lead to the increased the significance of monitor the atmosphere [3]. Continuous tracking of the environmental parameter is needed to determine the ambient quality. Given that IoT is the mainly up-and-coming technology. It plays a significant position in the collection of sensing unit information. The IoT helps bring everything together and allows us to connect with our very own things [2].

In 2018: The increasing number of expensive information sources, developments in the big data technologies and IoT as well as the development of a broad choice of appliance knowledge algorithms give the new probable for providing predictive services to people and decision-makers in urban areas [4]. **In 2019**: The IoT is a technology in which one or more technologies are replaced by another technology. IoT building a small eco-system around us, with the number of sensors and actuators. Ubiquitous sensing technologies deliver shared information to create a growing image of operations. The IoT sensors are used to build a smart environment by using the IoT application. This paper shows IoT sensors and different IoT applications based on sensors [5].

11.3 Method and Methodology

The temperature sensor LM35 provides ADC0804 with the analog Temperature Data, which it transforms to Digital Values, and sends to 8051. The 8051 Microcontroller performs a brief measurement after obtaining the digital values, and then shows the temperature on the LCD.

It is as quick to use the automatic fan control to choose the desired temperature range and allow it to do its job. When the temperature rises above the target temperature range, the fans will work in the forward direction automatically and will create a cooling effect shown in Fig. 11.3.

Humidity is defined in terms of the quantity of stream in the nearby air. The stream contented in the soil is a crucial feature in human health. For though the heat is 0 °C with fewer moisture i.e. the atmosphere is fresh, we'll feel comfortable. Moisture sensors are used in products such as incubators, sterilizers, and medication manufacturing devices. Humidity control sensor through IoT is shown in Fig. 11.4.

There are various types of humidity sensors available, which work on different concepts, such as capacitive, semiconductors, and resistive. Humidity Sensor Fan Control senses excess humidity in a room and triggers the ventilation fan automatically to minimize excess condensation. The use of this sensor as an added benefit helps to reduce energy use by automatically running the fan only when required, reducing constant or repetitive usage. An optical sensor is generally part of a larger system. A sensor is defined as an instrument that transforms the physical stimuli into a readable output. A sensor's position in a control and automation system is to detect and quantify any physical effect while supplying the control system with

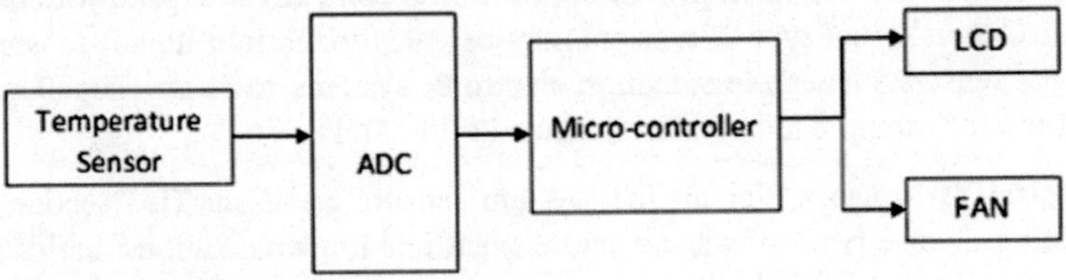

Fig. 11.3 IoT based temperature sensor

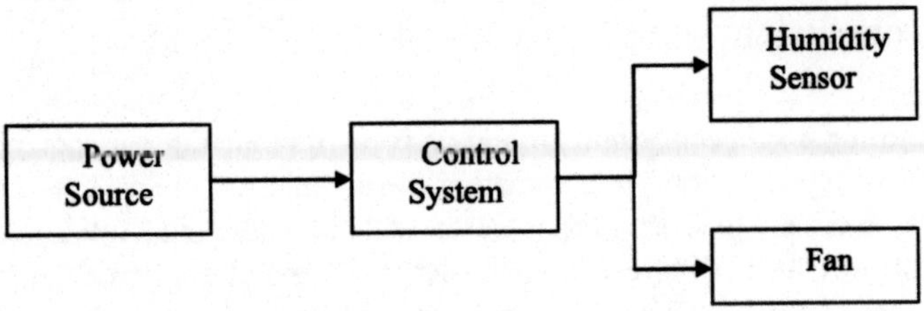

Fig. 11.4 Humidity control sensor through IoT

Table 11.1 IoT application and type of sensors

IoT applications	Types of sensors
Smart cities	Pressure, Humidity, Proximity, Temperature, Position, Light accelerometer, Velocity
Smart environment	Optical, Light temperature, Accelerometer, Biosensors, Gyroscope, Chemical, Humidity, Chemicals
Smart transport	Motion, Gyroscope, Temperature, Chemicals, Magneto, Pressure, Accelerometer
Smart agriculture	Position, Proximity, Chemical, Water quality, Humidity, Temperature

this information. By controlling the sum of movements and strokes, the shipping conditions can be increased efficiently.

Air quality monitoring, field-level monitoring, emission level, and some sort of bottle leakage will save water. Many other intelligent devices can be used for applications for proper air control.

11.4 Results and Discussion

Based on different types and based on IoT sensors application like temperature, humidity, pressure, proximity, optical, velocity, positional sensors the outcome of the IoT is enhancing performance. Data is the key to making IoT systems effective in increasing profitability and operational efficiency. A majority of the executives surveyed say IoT is a significant contributor to their company to expand your outlook on IoT. Generally, all type of sensors may be categorized into digital sensors and analogue sensors. Nonetheless, in most electronic systems, there are a small number of sensors for example temperature sensors (Table 11.1).

Analysis IOT Sensors: For all IoT, system sensors are used. This section, after examining diverse types of sensors and elegant IoT implementations, indicates the kind of sensors needed in building a smart environment in an IoT framework.

11.5 Conclusion

IoT by building a safe ecosystem around us is revolutionizing our society. Sensors participate an input position in automating the submission in any IoT-based smart device. By creation it is smarter to react with no individual interference. This manuscript introduces different kinds of sensor in the elegant world activated by IoT. IoT sensor can be efficiently warned for physical condition, transport, water, environment, agriculture. This paper analyzes various IoT sensors and IoT applications based on sensors and illuminates what sensor is used in many IoT applications.

The Real-time vehicle speed regulation is crucial for preventing fatal accidents on the expressways. During times that vehicles surpass the thresholds, warnings can be issued for passengers.

References

1. Sehrawat, D., Gill, N.S.: Smart sensors: analysis of different types of IoT sensors. In: 2019 3rd International Conference on Trends in Electronics and Informatics (ICOEI), pp. 523–528. IEEE (2019)
2. Zafar, S., Miraj, G., Baloch, R., Murtaza, D., Arshad, K.: An IoT based real-time environmental monitoring system using Arduino and cloud service. Eng. Technol. Appl. Sci. Res. **8**(4), 3238–3242 (2018)
3. Nayyar, A., Puri, V.: Smart farming: IoT based smart sensors agriculture stick for live temperature and moisture monitoring using Arduino, cloud computing & solar technology. In: Proceedings of the International Conference on Communication and Computing Systems (ICCCS-2016), 9781315364094-121 (2016)
4. Padma, U., Jagadish, S., Singh, M.K.: Recognition of plant's leaf infection by image processing approach. Mater. Today Proc. (2021)
5. Singh, M., Nandan, D., Kumar, S.: Statistical analysis of lower and raised pitch voice signal and its efficiency calculation. Traitement Signal **36**(5), 455–461 (2019)
6. Balaji, V.N., Srinivas, P.B., Singh, M.K.: Neuromorphic advancements architecture design and its implementations technique. Mater. Today Proc. (2021)
7. Singh, M.K., Singh, A.K., Singh, N.: Multimedia utilization of non-computerized disguised voice and acoustic similarity measurement. Multimedia Tools Appl. 1–16 (2019)
8. Punyavathi, G., Neeladri, M., Singh, M.K.: Vehicle tracking and detection techniques using IoT. Mater. Today Proc. (2021)
9. Singh, M.K., Singh, A.K., Singh, N.: Multimedia analysis for disguised voice and classification efficiency. Multimedia Tools Appl. **78**(20), 29395–29411 (2019)
10. Kanchana, V., Nath, S., Singh, M.K.: A study of internet of things oriented smart medical systems. Mater. Today Proc. (2021)
11. Singh, M.K., Singh, A.K., Singh, N.: Disguised voice with fast and slow speech and its acoustic analysis. Int. J. Pure Appl. Math. **118**(14), 241–246 (2018)
12. Prasanna, G.S., Pavani, K., Singh, M.K.: Spliced images detection by using Viola-Jones algorithms method. Mater. Today Proc. (2021)
13. Singh, M.K., Singh, A.K., Singh, N.: Acoustic comparison of electronics disguised voice using different semitones. Int. J. Eng. Technol. (UAE) **7**(2), 98 (2018)
14. Jamliya, S., Mahajan, A., Choubey, A., Nandan, D.: An efficient VLSI architecture of multiplier-less 1-D DWT using CSD technique. In: IET International Conference on Signal Processing (ICSP), SATI, Vidisha, 11–13 Nov 2016
15. Veerendra, G., Swaroop, R., Dattu, D.S., Jyothi, C.A., Singh, M.K.: Detecting plant diseases, quantifying and classifying digital image processing techniques. Mater. Today Proc. (2021)
16. Kumar, S., Kumar, R., Singh, M.K., Nandan, D.: A multiple band-notched monopole antenna with incorporated GSM and UWB for wireless applications. Int. J. Adv. Sci. Technol. **28**(16), 362–378 (2019)

Chapter 12
Experimental Study and Investigation of Mechanical Properties of Material SS304/SS316 Using Rotary Friction Welding Technique

Vaibhav V. Kulkarni and Prafulla C. Kulkarni

Abstract Rotary Friction Welding (RFW) is the progressive welding process, widely used in the manufacturing industry as appropriate for the joining of symmetrical geometry and shape of parts. The process offers several advantages like no fumes extraction, no release of harmful gases and no pollution, no melting at solid welding joints; high accuracy and precision of operation even at high speed; and less economical nature. Apart from the above advantages, another significant benefit is accompanied by the weld joint interface region, which produces the partial smelted state of the welded parts. This phenomenon is predominantly substantial for the welding of two unrelated material combinations and precisely where the slender heat-affected zone is encouraged. In this research work, the study has been conducted using the rotary friction welding process machine. The material selected for the study and experimentation for RFW process is austenitic stainless-steel SS304 (specimen dia. ϕ 12 and 14 mm) and SS316 (specimen dia. ϕ 10 and 14 mm). Several parameters have been studied for the RFW process like rpm, temperature, and pressure. A study of different mechanical properties like tensile strength, hardness, and microstructure analysis has also been carried out for the same welded specimen.

12.1 Introduction

The Rotary friction weld technology was developed at the beginning of the 1950s, and introduced during the Second World War, as the most popular process of all friction welding, and named as rotary friction soldering technique. It is used broadly for the joining of structural materials with symmetrical rotational geometry by utilizing conventional lathe. However, they were found unworkable due to incorrect techniques

V. V. Kulkarni (✉)
Dr. Vishwanath Karad MITWPU, Kothrud, Pune, MITCOE-Savitribai Phule Pune University (SPPU, Pune.) (M.S)., India
e-mail: vaibhav.kulkarni@mitwpu.edu.in

P. C. Kulkarni
Principal, R. H. Sapat COE, MS Research, SPPU, College Road, Nashik, India

S. Yadav et al. (eds.), *Proceedings of Second International Conference in Mechanical and Energy Technology*, Smart Innovation, Systems and Technologies 290,
https://doi.org/10.1007/978-981-19-0108-9_12

and required several changes for the correct set up on the lathe. This phenomenon is further divided into two processes based on rotational energy conversion to frictional heat. The first process demonstrates the process of welding direct or continuous friction, and the second is the welding process for inertia friction. Welding of similar and dissimilar metals by using friction is the most common and raised its application in metal joining processes in all over the world's manufacturing industries. The rotary friction welding process develops the excellent quality of the welding joint between the two metals. The joint is developed by rotating one work piece over another, which seems solid without melting, leading to heat release. It utilizes simultaneous frictional force produced by rotating metal pieces once elevated to a suitable temperature. The rotating part rests at the nominal temperature state, and the stationary component continues to coalesce under increased pressure and defines metal contact, as illustrated in Fig. 12.1. Friction welding serves as the most advanced welding process and extensively employed in the manufacturing industry. Its solicitations include no fumes extraction and no harmful gases released in the environment. In this process, the weld joint interface region generates the partially molten state of the welded parts. This phenomenon is particularly significant for the welding of two different material combinations and favors favorable narrow heat-affected zone.

Manufacturing industries employed several welding and metal joining methods for joining different parts as permanent joint or temporary joint, dependent on specific requirements and applications of the component. The market demand and welding setup differ following the type of metals to be weld. Designers and manufacturing engineers must be aware of all the available methods of metal joining. The primary objective of this research work defines the experimentation of joining specimens of austenitic SS304 and SS316 of the same cross-section using rotary friction welding. It executes comparative study for mechanical properties like the strength of welded joint using tensile test, hardness at different locations in the heat-affected zone, and microstructure analysis of weld joints.

Fig. 12.1 Rotary friction welding process

12.2 Literature Review

The first patent of friction welding process was granted to J. H. Bevington, 1891. He applied friction welding techniques to weld metal pipes. After that, various researchers performed the welding studies for plastic joining materials, and related to this; multiple experiments were carried out during the 1940s in the USA and Germany. A. J. Chdikov, who was a Russian machinist, had conducted experimental scientific studies and suggested the commercial use of rotary friction welding and patented the process in 1956. Whereas, many researchers from American Machine and Foundry Corporation have operated thermal and parametrical analysis of rotary friction welding. Experimental investigation studies of friction welding in England were carried out by The Welding Institute (TWI) in 1960–1961. The Caterpillar Tractor Co. also improved the friction welding process to develop the method of inertia welding during 1961–1962. Seshagiri Rao [1] reviewed the experimental investigation of rotary friction welding parameters using similar and dissimilar material for Al (H-30) and MS (AISI-1040). Hence, the conclusion states that the friction welding process can be successfully utilized for the welding of different ferrous and non-ferrous materials [1, 2].

The experimental study by Handa and Chawla [2] defined the mechanical properties of Friction welded AISI 1021 steel. The experimental installation was developed and contrived in order to achieve the Friction welded joints between the austenite stainless steel and the low-alloy AISI 30 sheets of steel as produced by mechanical joining along with the exploration of axial pressure effects on mechanical properties. The literature review of research on rotary friction welding has been carried out by Bhate and Bhatwadekar [3] and various methods have been investigated. The study on a statistical analysis of rotary Friction of steel with different carbon content in workpieces and subsequent effects of variations in carbon content was performed by Kalsi and Sharma [4]. An experimental setup was intended to achieve the process with equal diameter workpieces. Shubhavardhan and Surendran (2012) carried out a study of friction welding for joining stainless steel and aluminum materials. The study was performed through continuous drive friction welding process, by combining heat produced by friction between two surfaces and plastic deformation, to join dissimilar material (aluminum allocation A AA6082 and stainless steel AISI 304). Different welding process parameters were used for testing. Tensile tests, Vickers micro hardness test, fatigue test, Impact tests, and SEM–EDX (energy-dispersive X-ray) analyses were carried out to determine the phases during welding [5, 6].

The research of the experimental characterization of the properties of 100 Cr6 steel, combined with the rotary welding process, is demonstrated in the study by Mourad (2017). The primary objective of the present work has been to manufacture identical metal joints of these steel using the rotary friction soldering process and experimentally explored the sold joints properties. The joints exhibit different diameters and interface geometries when made from steel rows 100Cr6. A friction welded specimen was performed with optical microscopy for microstructural characteristics.

The Vickers hardness distribution was scrutinized at the weld joint. In the central zone, the welds are highly strong, described by a martensitic structure [7–9].

The main objective of the survey on dissimilar material by Alves et al., 2010 was to monitor the temperature at the bonding interface during rotary friction welding of AA1050 aluminum and AISI 304 steel [10–12]. Research by Kondapalli [13] demonstrated the different welding processes for welding of AISI 304L austenitic Stainless Steel. The literature shows that austenitic AISI 304L Stainless Steel is the best material for intergranular corrosion problems. It is the most recurrently used best material for the manufacture of heat-resistant components. The effect of welding parameters on the burn of length for friction welding of 2 dissimilar metal inconel718 and SS304 [14, 15] was explored after the research by Patel and Patel (2017). The recent advancements and the numerous above literature have revealed that, due to the lower welding temperature and shorter welding period, rotary friction welding is one of the most efficient and conventional methods of welding-related and differentiated metals like copper and aluminum joint.

12.3 Experimentation and Trials

The experimental trials were conducted using rotary friction welding machine having 5-ton capacity, model name SPARTAN-5, as shown in Fig. 12.2. Rotary Friction Welding Machine is available at Friction welding Technology [FWT] Company, Pune. The materials selected for trial were SS304, and SS316, where SS304 specimen have ϕ 12 and 14 mm 05 each and specimen SS316 have ϕ 10 and 14 mm 05 each. All the readings and observations were noted during the conduction of the trial. The length of each specimen ranges from min. 100 mm to max. 104 mm, and total length reaches out as 201 mm to 205 mm, respectively, as shown in Table 12.2. The speed of the rotation was fixed for the entire specimen, i.e., 1600 rpm. During experimental study the following observation were taken while considering follow-ing as an input parameters. The input parameters are total length of Sample size (Spindle size),

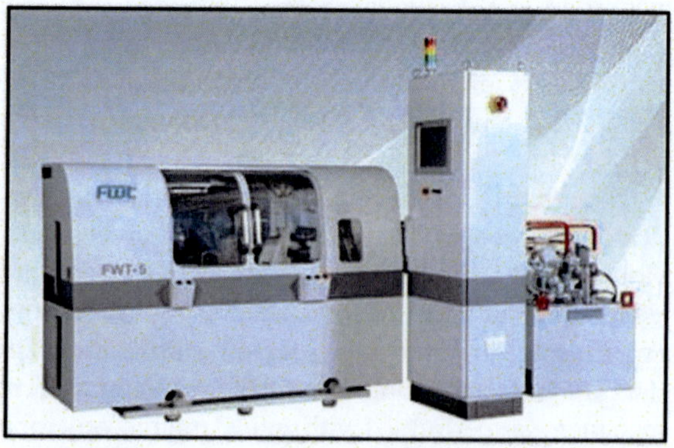

Fig. 12.2 SPARTON-5 Machine Setup

Table 12.1 Chemical composition of material SS304 and SS316 specimen

Grade	C	Si	Mn	P	S	Cr	Ni	Mo
SS304	0.08	1.00	2.00	0.045	0.030	18.0–20.0	8.0–11.0	–
SS316	0.08	1.00	2.00	0.045	0.030	16.0–18.0	10.0–14.0	2.0–3.0

Soft Friction Pressure considered 8.1kg/mm^2.The Friction Pressure is 17kg/mm^2 and upset Pressure is considered 30.5kg/mm^2. Since we have prepared twenty nos. specimens for material SS304 and SS316 as mentioned above. According to the diameter of the speci-men the above parameters has been changed, with keeping the same diameter with same parameters as shown in the table. After the all setting has been done on the machine the five samples of each diameter trials were taken. Where final length of actual weld measured and actual loss is also calculated. Our aim was to conduct the experiment on the friction welding machine and find out the actual loss, temperature during process and also calculate shrink-age of the material after the friction welding trails. (Table 12.1, Graphs 12.1 and 12.2).

Austenitic stainless steels are possibly the most frequently used material among all different stainless steels. The high concentration of chromium and nickel (18–20 and 8–12 wt. % respectively) signifies the corrosive resistant nature of the 300 series materials, which also defines their non-magnetic and non-hardenable property after inducing heat treatment. However, cold treatment can harden them significantly. Austenitic stainless steels are extensively utilized in petrochemical, nuclear, and corrosive chemical environments. Stainless Steel serves as the best material which effectively deals with the problem of intergranular corrosion and is commonly employed for the manufacture of non-heat-treatable components [15].

12.4 Rockwell Hardness Test

The Rockwell hardness test is performed when friction welding is observed at the heat- affected zone area of specimen positioned at 1A 2 mm, 2A 5 mm, and 1B 2 mm, 2B 5 mm locations. Friction welding can be intended by computing the depth of an indent once the specimen material is forced through indenter at a given load. It is evaluated after measuring the depth of an indentation with a diamond carbide point. The Rockwell hardness test has been carried out at 60 kgf. The hardness results are depicted in Table 12.3. The hardness illustrates 51–56 HRA at the point of contact (Figs. 12.3 and 12.4).

It has been observed from the Table 12.3, that the hardness value of SS304 is more as compared to SS316 due to high chromium and nickel percentage in SS304. Minimum hardness and high welding strength are observed at the point of the weld as the material gets soften at the center. The hardness value gradually deviates with each point of indentation from 2 to 5 mm at welding contact. Increasing heat, change

Table 12.2 FWT readings and observations on friction welding machine

Specimen size and quantity		Spindle side length 1	Spindle side length 2	Total (L1 + L2)	Final length (after weld)	Loss (actual)	SHRINK	Weld temperature	Soft friction pressure = 8.1 kg/mm^2	Friction pressure = 17 kg/mm^2	Upset pressure = 30.5 kg/mm^2
SS304 ϕ 12 mm	1	103.3	104.2	207.5	202.6	4.9	6.1	1425	35	62	125
	2	103	104.5	207.5	202.4	5.1	5.9	1413	35	62	125
	3	103.5	102.8	206.3	200.8	5.5	4.1	1432	35	62	125
	4	103.3	100.3	203.6	199	4.6	4.1	1420	35	62	125
	5	105	101	202	197.8	4.2	4.1	1435	35	62	125
SS304 ϕ 14 mm	1	104	100	204	199.5	4.5	5.7	1405	47	84	150
	2	103	101	204	199.2	4.8	5.6	1415	47	84	150
	3	102	101.5	203.5	198.4	5.1	5.6	1400	47	84	150
	4	105.4	102	207.4	202.5	4.9	5.6	1412	47	84	150
	5	103	102	205	200	5	5.6	1393	47	84	150
SS316 ϕ 10 mm	1	103.5	105.5	209	204.2	4.8	5.0	1395	24	43	86
	2	104	109.2	213.2	209	3.8	4.8	1405	24	43	86
	3	104	106	210	205.5	4.5	4.8	1390	24	43	86
	4	104	104.5	208.5	204	3.5	4.7	1402	24	43	86
	5	103	104	207	203	4.5	4.7	1383	24	43	86
SS316 ϕ 14 mm	1	104.2	109	213.3	210.3	3	3.0	1385	47	84	150
	2	112.3	109	221.3	216	5.3	7.4	1392	47	84	150
	3	106	110	216	211.8	4.2	5.9	1380	47	84	150
	4	108	109	217	212.5	4.5	5.8	1395	47	84	150
	5	106	105	210	204.5	5.5	5.6	1373	47	84	150

Table 12.3 Hardness test results

Material	Diameter	Point No.	2A	1A	Point of contact	1B	2B
		Weld center	5 mm	2 mm	0	2 mm	5 mm
SS304	12	HRA	58	57	56	57	59
SS304	14		57	56	53	56	58
SS316	10		54	55	51	55	58
SS316	14		55	57	52	58	57

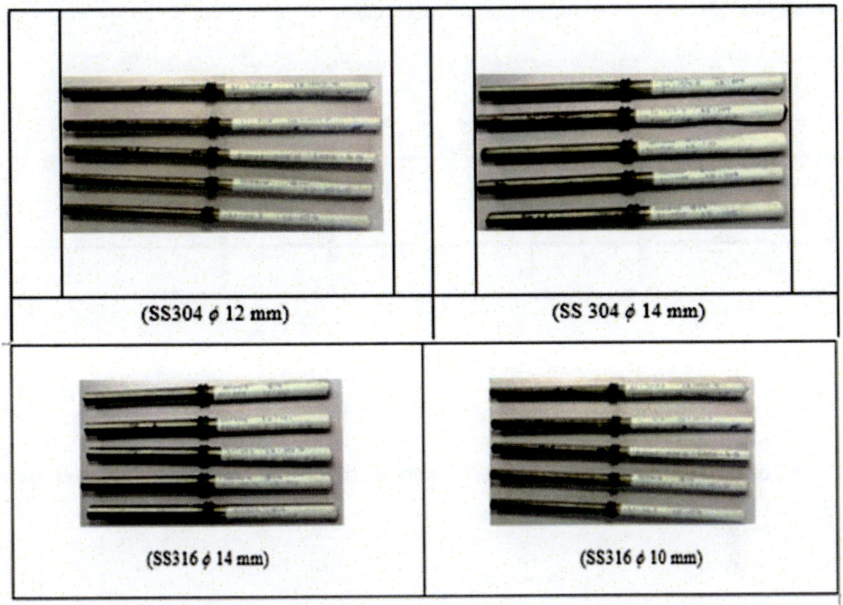

Fig. 12.3 Rotary friction welded specimens

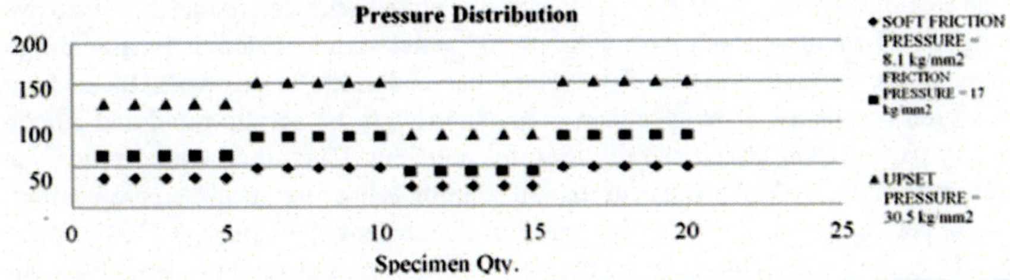

Graph 12.1 Pressure distribution

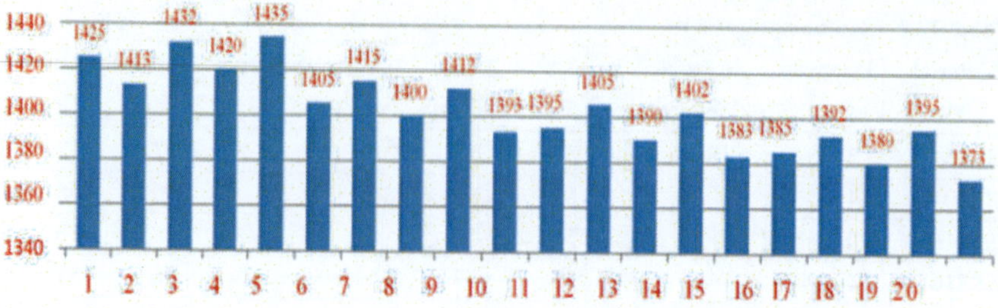

Graph 12.2 Temperature versus samples (X-axis samples and Y-axis temperature)

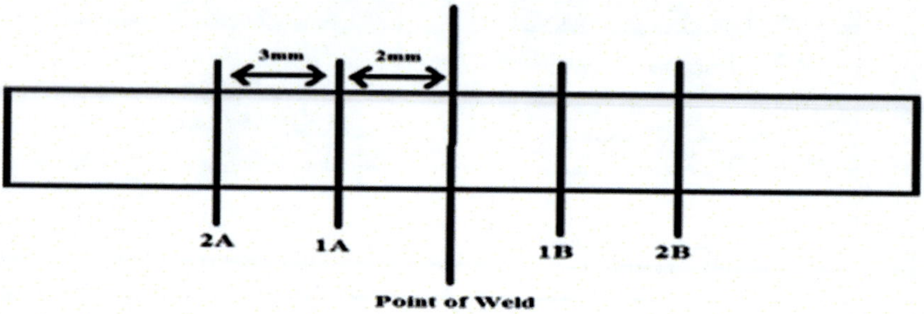

Fig. 12.4 Rockwell hardness indentation

in the grain size at the weld point is observed with a significant increase in hardness value, elongation, and strength.

12.5 Microstructure

The etching of stainless steel grade material is quite tricky due to the anti-corrosive property of stainless steel. Austenitic or 300 series stainless steel is having a high content of chromium and nickel, making it harder to etch. In this work, the etching has been completed after specimen preparation with the acrylic mold and silicon emery paper (1500 grit) generally used for polishing. Then the lapping process is performed after 5–6 ml nitric acid etchant solution is used for specimen preparation.

As per the microstructure observations, the austenitic stainless SS304 and for SS316 grade material super picral means picric acid plus alcohol prepared the solution, and on the surface of the specimen drop, 2–5 ml solution and phases reveals clearly. It was observed on the metallurgical microscope, temperature, grain structure, pressure, and oxidation affects the overall rotary friction welding. Due to temperature differences, the austenitic structure is converted into a martensitic structure.

In microstructure examination, fine grains are observed at some locations, and grains are distorted entirely at the weld zone in the absence of twin boundaries of

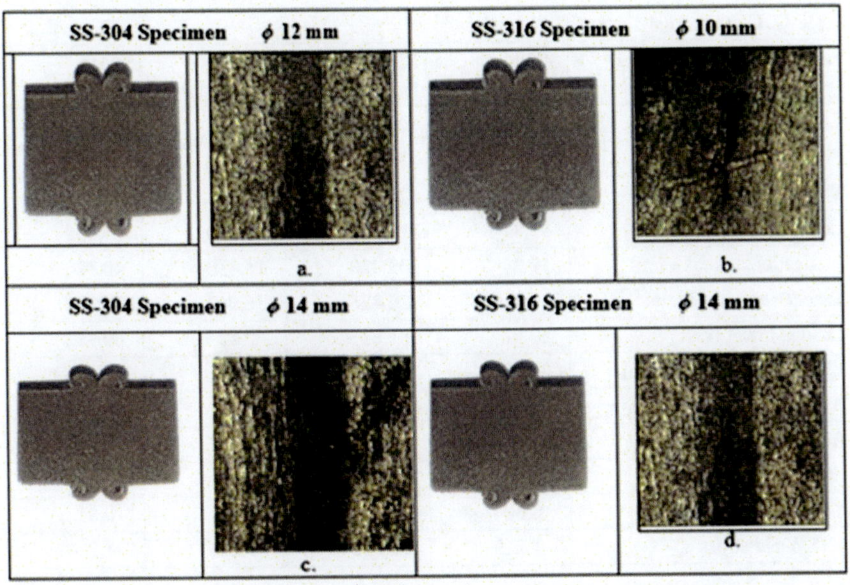

Fig. 12.5 HAZ microstructure images (200 × magnifications) of specimen SS304 and SS316

base metals. Stringing action is found in dynamic crystallization, which may affect the mechanical properties of metals. Higher the percentage of Nickel and Chromium, the higher the percentage of breaking load and elongation. The following Fig. 12.5a to d shows the 200 × magnification microstructure cutting section of heat affected zone [HAZ] of rotary friction welding for all specimen sizes.

12.6 Tensile Test Results

A tensile test of any material states the effectiveness and behavior of a material when a stretching force acts on it and also determines the maximum strength or load which can withstand by the material. The study utilizes a universal testing machine for the measurement of the tensile strength of the rotary Friction welded joint. The load resolution and machine capacity of 20 tons are used with an onboard extensometer facility having a one-micron resolution, 20 mm travel, 20 data set storage (00–19). The one data set contains the 75 results storage, and parameter selection through keyboard non-volatile memory was used for the result as well as for test data storage. The Maximum load for each specimen is 1000 kN, and Maximum elongation is 250 mm. The following Table 12.4 states the tensile test results for SS304 dia. 12 and 14, SS316 dia. 10, and 14 specimens, respectively. In the Fig. 12.6a, b, c, d displays the tensile tested specimens.

Table 12.4 Output tensile test observations

Parameters/Specimen size	SS304 (ϕ 12 mm)	SS304 (ϕ 14 mm)	SS316 (ϕ 10 mm)	SS316 (ϕ 14 mm)
Load at yield (kN)	82.09	109.72	51.95	84.46
Elongation at yield	9.800	12.92	9.600	10.00
Yield stress(n/mm^2)	725.836	620.889	661.783	548.663
Load at peak (kN)	91.87	128.370	65.700	104.250
Elongation at peak (mm)	19.24	33.320	13.430	40.320
Tensile strength (n/mm^2)	812.310	726.426	836.942	677.221
Load at break (kN)	50.92	81.83	60.82	69.98
Elongation at break (mm)	28.35	41.49	14.40	47.48
Breaking strength (n/mm^2)	450.232	463.063	774.777	454.599
Reduction in area (%)	73.65	69.75	37.75	69.75
Elongation (%)	9.11	14.63	6.87	16.28
Yield stress/UTS	0.894	0.855	0.791	0.81
UTS/Yield stress	1.119	1.170	1.265	1.234

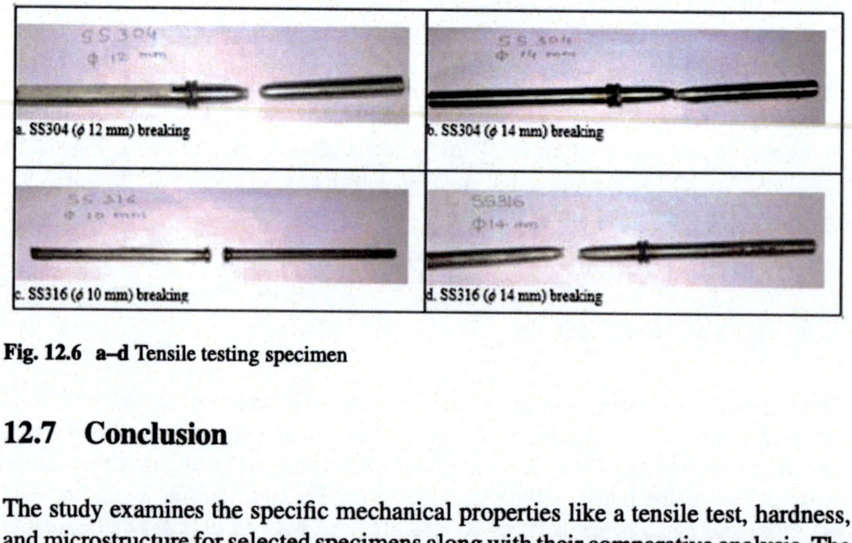

a. SS304 (ϕ 12 mm) breaking

b. SS304 (ϕ 14 mm) breaking

c. SS316 (ϕ 10 mm) breaking

d. SS316 (ϕ 14 mm) breaking

Fig. 12.6 a–d Tensile testing specimen

12.7 Conclusion

The study examines the specific mechanical properties like a tensile test, hardness, and microstructure for selected specimens along with their comparative analysis. The results concluded with the findings that the maximum yield stress attained for the ϕ 12 mm, i.e., 725.836 N/mm^2, whereas, minimum breaking load appears 50.920 kN for SS304 material and maximum yield elongation and load at break calculated as 12.92 mm and 81.830 kN respectively for ϕ 14 mm SS304 material. In comparison with SS304 material, the minimum yields stress appears as 548.663 N/mm^2 for SS316 material. The maximum tensile strength of SS316 comes out as 836.942 N/mm^2 compared with the SS304 material. Rockwell hardness test has been conducted at the HRA scale and demonstrated that the hardness of SS304 is more by 2 to 5 HRA as compared to SS316 material. Minimum hardness is observed at the point of the weld. The hardness value progressively varies with each point of indentation at 0, 2, and 5 mm for welding contact. Following the microstructure testing and related specimen observations at HAZ, the overall rotary friction welding operation for the austenitic SS304 and SS316 grade material is affected by temperature, grain structure, pressure, and oxidation. Significant temperature variations in the material, the austenitic structure, get transformed into the martensitic structure. The microstructure examination defines the first observation as localization of fine grains and a complete distortion of grains in the welded zone with no twin boundaries at base metals. Stringing actions lead to dynamic crystallization, which may affect the mechanical properties of metals.

12.8 Future Scope

In future work, the study of fatigue strength of welded joint, static, and dynamic characteristics of welding, heat flow variations during underwater welding joint, and strength analysis on underwater welds also can be conducted. To continue this research, and comparative study can be performed by welding the same material specimen of SS304 and SS316. Numerical issues are also caused by large strain increments, which can be reduced by introducing a visco-plastic model. The visco-plastic model is used for numerical reasons but, since steel is to be considered visco-plastic at high temperatures, such a model could provide more accurate results. The mathematical model can be developed to predict the shear strength of the rotary friction welding process.

12.9 Acknowledgement

I would like to show my sincere gratitude to everyone who made this research study possible. A special thanks to my research guide without his inspiring words, adequate recommendations, and full-hearted association, this work would not be possible. I would like to thanks to our Principal and Head of the Department for their support and I am grateful for their motivation during this work. I would like to give my sincere thanks to Managing Director and staff of FWT Pvt. Ltd. company for their support, cooperation and valuable guidance during execution of this research study in a well-planned manner.

References

1. Seshagirirao, B., Sivaramakrishna, V., Sai, K.G.: Experimental investigation of rotary friction welding parameters of aluminum (H-30) and mild steel (AISI-1040), India. IJIRSET **4**(5), (2015). ISSN (online): 2319-8753
2. Handa, A., Chawal, V.: Experimental study of mechanical properties of friction welded AISI 1021 steel. Sadhana **38**(Part 6), 1407–1419 (2013)
3. Bhate, S.S., Bhatwadekar, S.G.: Literature review of research on rotary friction. IJITR **4**(1), 2601–2604 (2016)
4. Kalsi, N.S., Sharma, V.S.: A statistical analysis of rotary friction welding of steel with varying carbon in work pieces. IJAMT **57**, 957–967 (2011)
5. Kumar, S., Bhardwaj, D., Sangwan, J.: A research paper on temperature modeling of friction welding of Al and SS-304. IJERSTE **3**(6), 319–327 (2014). ISSN: 2319-7463
6. Shubhavardhan, R.N., Surendran, S.: Friction welding to join stainless steel and aluminium materials. IJMMSE **2**(3), 53–73 (2012). ISSN 2278-2516
7. Mohammad, A.K., Khalil, H.: Effect of frictional welding between different stainless steel materials on their impact properties. IJESRT (2016). ISSN: 2277-9655
8. Samuthiram, G., Kannan, T.T.M., Sureshkumar, M., Natarajan, V.A., Vijayakumar, P.: Evaluation of mechanical properties of friction welded joints of EN-24 steel cylindrical rods. IJMER **3**(4) (2014). ISSN 2278-0149
9. Mourad, D., El Hedj, O., Rachid, L., Ahmed, M.: Experimental characterization of the heat affected zone (HAZ) properties of 100Cr6 steel joined by rotary friction welding method. MMEP **4**(1) (2017). ISSN Online 2369-0747
10. Ali, M.: Study of heat affected zone (HAZ) in friction welding process. J Mech. Eng. **1**(1) (2012)
11. Shete, N., Deokar, S.U.: A review paper on rotary friction welding. ICIIIME **5**(6), 1557–1560 (2017). ISSN. 2321-8169
12. Alves, E.P., Neto, F.P., An, C.Y.: Welding of AA1050 aluminum with AISI 304 stainless steel by rotary friction welding process. JATM **2**(3), 301–306 (2010)
13. Kondapalli, S.P., Chalamalasetti, S.R., Rao, D.N.: A review on welding of AISI 304L austenitic stainless Steel. J. Manuf. Sci. Prod. **14**(1), 1–11 (2014)
14. Khany, S.E., Mehdi, S.N., Sayeed, M.A.: An analytical study of dissimilar materials joint using friction welding and its application. IJSR **5**(2) (2015). Pub. ISSN 2250-3153
15. Patel, A. I., Patel, J.: Effect of welding parameter on burn of length for friction welding of two dissimilar metal inconel718 and SS304. IJAER **4**(5) (2017). ISSN (P): 2348-6406

Chapter 13
Critically Analyzing the Concept of Internet of Things (IOT) and How It Impacts Employee and Organizational Performance

A. Anbazhagan, K. Guru, Gazala Masood, Meeta Mandaviya, Viney Dhiman, and Mohd Naved

Abstract It connects people, organizations, and smart objects, the internet of things (IoT) has the potential to profoundly alter the way we live, work, and interact, and it has the potential to reshape a broad variety of industrial areas. The purpose of this research article is to offer a critical examination of the Internet of Things (IoT) and how much it affects individual and organizational performance in the workplace. We examine the literature on the Internet of Things, including what it is, why it is important, what it has historically been, its benefits, pros and cons, security and privacy issues, and its impact on organizational performance, and we conclude that a better recognizing of how the Internet of Things will impact the ways that organizations and employees conduct their business is required.

13.1 Introduction

In computing, the internet of things, also known as IoT, is a networked system of interconnected computing devices such as system of interrelated computing devices

A. Anbazhagan
PSV College of Arts and Science, Puducherry, India

K. Guru
SRM Valliammai Engineering College, Chennai, Tamil Nadu, India

G. Masood (✉)
Faculty of Commerce and Management, Rama University, Kanpur, India
e-mail: ghazala1999@yahoo.com

M. Mandaviya
Marwadi University, Rajkot, India

V. Dhiman
MSW, Post Graduate Institute of Medical Education and Research, Chandigarh, India

M. Naved
Department of Business Analytics, Jagannath University, Delhi-NCR, India

S. Yadav et al. (eds.), *Proceedings of Second International Conference in Mechanical and Energy Technology*, Smart Innovation, Systems and Technologies 290,
https://doi.org/10.1007/978-981-19-0108-9_13

machineries, items, mammals, and human beings that are all assigned unique identifiers (UIDs) and have the ability to transfer data over a network without any need for direct human or human-to-computer communication. It is possible for a person with a heart monitor implant, a farm animal with a biochip transponder, an automobile with built-in sensors to alert the driver when wheel pressure is low, or any other natural or man-made object that can be allocated an Internet Protocol (IP) address and is able to spread across the globe to be considered "things" in the internet of things. Businesses in a number of sectors are increasingly relying on the Internet of Things (IoT) to improve operational efficiency, better understand consumers in order to provide improved customer care, improve decision-making, and raise the value of their businesses. The Internet of Things (IoT) has been identified as a critical element in computer systems that has an impact on their overall performance independent of how they are used. Data security is the most important study topic in the Internet of Things since the systems are infiltrating further and deeper into our personal lives, where they detect, analyze, and store all kinds of data. This situation presents a number of challenges in terms of security and privacy, particularly for applications running on resource-constrained systems. In this scenario, appraiser selected well-established data protection mechanisms that ensure the confidentiality and integrity of data are maintained. Furthermore, they evaluated the effectiveness of various cryptographic blocks and flow pass codes, encoding algorithms, messaging authentication codes signature methods, and symmetric encryption protocols when implemented on the most sophisticated resource limitation systems available.

13.2 Analysis of Internet of Things (IoT)

13.2.1 In What Ways Does the Internet of Things (IoT) Help? Definitions of the Internet of Things

The Internet of Things is a physical object interconnecting to the Internet and other networks through different Ip addresses, which allow data collection and information to be sent via integrated sensors, electronics, and software. Physical devices are intended for the Internet of Things or they are assets outfitted with data sensing and communications transmission, including live creatures. In the Internet of things, the data collected from linked objects is utilized to characterize, beyond the endpoint dimensions with devices, sensors, actuators, and communications networks [1].

Internet of Things characterizes a variety of applications, protocols, standards, architectures, and data processing and analysis technologies, in which the Internet and other networks, through a single IP address or URI, can be attached to devices and objects (applications, dresses or animals) embedded with sensors and specially designed software or other digital and electronic systems. As can be seen here, data and how it is collected, processed and integrated into value chains of knowledge and advantages are important. Indeed, the real value of the internet of things consists in the manner in which alternative types and data types for completely new business models, insights, forms of commitment, living methods, and social benefits may be used. The Internet of Things is a key phrase, because, as stated, the consumer Internet of Things (CIoT) and the industrial internet of things are frequently distinguished (IIoT). They are frequently used, we discuss both below. CIoT and IIoT, however, also encompass many uses and apps and are thus umbrella words. In addition, crossovers exist between the two.

More from an internet of everything that again forms part of a wider framework, the Internet of things is seen [2]. It is described below what this signifies. There's no internet of things. Data collected, submitted, processed or sent to devices, mostly through Web, fixed routes, via cloud ecosystems or (adapted) wireless connection systems, designed for particular IoT applications (e.g. wireless technologies for the IIoT). The transformation into wisdom and action of digital, physical, and human realms via networks, interconnected activities and data is important in this equation. In recent decades, the emphasis in internet of affairs has moved to this intercontentedness of devices, data, business objectives, people and processes from the very element of connecting devices and collecting data, in IIoT definitely [3].

13.2.2 The Importance of the Internet of Things

The internet of things enables individuals to work and live intelligently and to manage their life completely. IoT is important for companies as well as providing clever gadgets to automate households. IoT offers companies an in-house assessment of how their systems actually function and a view into everything, from machine performance to supply chain and logistics.

IoT allows enterprises to automatically automate operations and decrease labor expenses. It also reduces waste and enhances services, making the production and delivery of products less costly and provides transparency in the transaction of customers. As such, IoT is one of the most essential daily innovations and will decide to lead flow as more companies recognize the potential of linked devices to compete effectively.

13.2.3 History of IoT

The internet of things was originally suggested by Kevin Ashton, a co-founder of the Auto-ID Center at MIT, who first shared his ideas with P&G in a presentation in 1999. Ashton dubbed his presentation the Internet of Things in order to make RFID radio frequency identification (RFID) more popular with P&G's management. When Things Start to Think by MIT professor Neil Gershenfeld was published in 1999 as well. Rather than explicitly utilize the word Internet of Things, it gave a clear picture of where the IoT was going.

The transition from the convergence of wireless technologies, micro electro mechanical systems (MEMSes), micro services, and the internet into the Internet of Things has progressed significantly [4]. Convergence has brought down the silos between OT and IT, and as a result, machine-generated unstructured data is now being examined for insights to enhance performance. The concept of linked devices has existed since the 1970s, when the terms embedded internet and ubiquitous computing were used.

In the early 1980s, the first internet equipment was a Coke machine at Carnegie Mellon University, the place where the original Internet protocol was developed. Programmers might use the web to discover whether or not they would be getting a cool drink if they choose to go to the machine. IoT developed from machine-to-machine (M2M) communication, which is a kind of communication where machines may communicate without human involvement. The gadget must connect to the cloud, be managed, and gather data in order to be considered an M2M. Connecting people, systems, and other apps to gather and exchange data is the next step for M2M. On top of this connection, M2M provides the basis for IoT.

With regards to the internet of things, we are also discussing an expansion of supervisory control and data acquisition (SCADA), a category of software applications that assist with control procedures and data collection in real time from distant places. The hardware and software components of SCADA systems are known as SCADA systems. Data is received by a hardware device, which then feeds it into a computer that comes preinstalled with SCADA software, where it is processed and provided to users in a timely way. SCADA has evolved to the point where the latest generation of SCADA systems has gradually transformed into the first phase of IoT systems. Although the idea of the IoT ecosystem did not truly emerge until the middle of 2010, when the Chinese government announced that it will emphasize IoT in its five-year plan, the term IoT ecosystem gained traction in that time period [5] (Fig. 13.1).

The internet of things may help businesses in a number of ways. Many advantages apply across many sectors, while others are industry-specific. Table 13.1 shows the common benefits of IoT enable businesses.

The rise of the Internet of Things (IoT) is prompting company to think outside the box and offer them the means to build better business strategies. An abundance of sensors and other IoT devices is present in the manufacturing, transportation, and utility sectors, but IoT may also be found in companies across many industries, including agricultural, infrastructure, and cloud computing [6].

Farmers in agriculture may profit from IoT since it makes their work simpler. Farming methods may be automated by using sensors to gather data on rainfall, humidity, temperature, and soil content as well as other variables. IoT can assist organizations in monitoring activities that are involved with infrastructure. Additionally, sensors, for example, may be used to keep tabs on structural buildings, bridges, and other infrastructure, in order to identify change or suspicious activity. It provides many advantages, such as reduced costs, less time, more workflow flexibility, and transparent workflow.

Fig. 13.1 The many ways IoT may help businesses

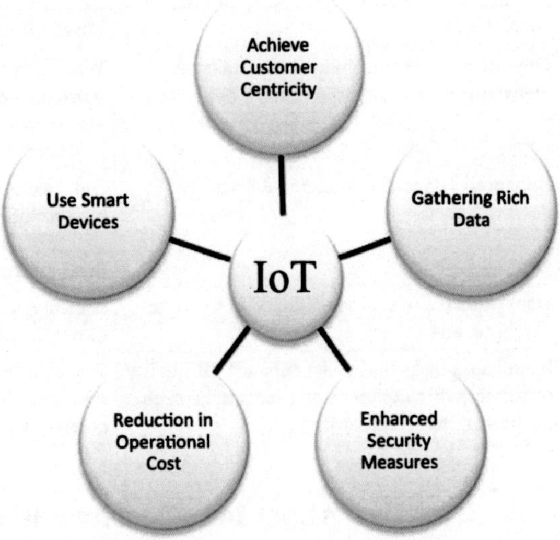

Table 13.1 Common benefits of IoT enable businesses

S. No.	Common benefits of IoT enable businesses to
1	Monitor your business operations
2	Enhance the overall client experience
3	Save both time and money by using our services
4	Increase the productivity of your employees
5	Business models should be integrated and adapted
6	Improve the quality of your business choices
7	Increase the amount of money that you make

To automate building mechanical and electrical systems, home automation businesses may use IoT. Smart cities have the potential to help residents save money and use less energy. Anything is connected to the Internet of Things (IoT). This includes everything from companies inside healthcare, banking, retail, and manufacturing.

13.3 Pros and Cons of IoT

Given the advantages and disadvantages of IoT in Table 13.2.

Table 13.2 Advantages and disadvantages of IoT

Advantages	Disadvantages
Opportunity to control any user data from anywhere	With increasing numbers of linked devices and more data exchanged between devices, there is also increased possibility for a hacker to steal sensitive information
Enhanced connectivity among electrical devices linked	Businesses may ultimately handle a huge number of IoT devices, perhaps millions, and it will be difficult to gather and manage data from all these gadgets
Data packets are sent via a linked network to save time and money	If the system has a flaw, every linked device is likely to be damaged
Automated tasks that assist enhance the quality of services of a company and reduce the need for human involvement	Since there is no worldwide IoT interoperability standards, it is difficult to connect with devices from multiple companies

13.4 Concerns About IoT Security and Privacy

In the internet of things, everything will be connected to the internet, and each of these objects will have an immense amount of data points associated with it. A larger attack surface means more worry for IoT security and privacy. An especially notable DDoS assault was one that targeted domain name server provider Dyn in 2016 and caused internet disruption for many websites over a lengthy period of time. The network was breached via vulnerable Internet of Things (IoT) devices. Hackers may exploit only one weakness in IoT devices and therefore access all the data, making it completely useless. If a gadget is not kept up to date, it becomes more susceptible to hackers. Personal information, such as names, ages, residences, phone numbers, and social media profiles, is an essential part of connecting devices [7] (Fig. 13.2).

Privacy is another significant issue for IoT users, since hackers aren't the only danger to the internet of things. Let's say, for examples, that businesses that manufacture and distribute consumers IoT devices have the devices to gather personal data on customers and resell it. Additionally, the introduction of Internet of Things (IoT) increases the danger of having vital infrastructure, such as the energy, transportation, and banking sectors, fall victim to system vulnerabilities.

How IoT Impacts on Organizational Performance

The internet of things is allowing companies to change as the technology fuels company development and helps firms navigate the commercial environment. This article may teach you case studies of companies that have used IoT solutions in the real world [8].

The Internet of Things (IoT) has gained enormous ground in today's commercial world. One of the advantages of IoT devices like sensors is that they're cost-effective and significant, which allows businesses to collect a lot of data, monitor operations,

Fig. 13.2 IoT security and privacy

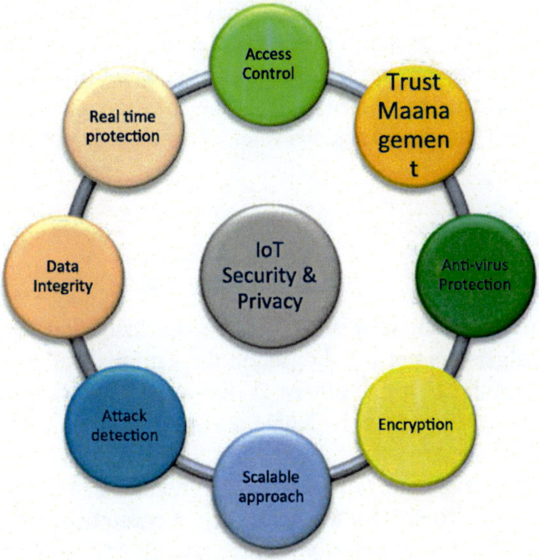

and anticipate equipment failures. The theory of IoT offers a treasure trove of potentially good ideas for products that may address both big and everyday problems in a simple and cost-effective manner. IoT statistics supports the utility of IoT: The number of devices continues to increase year after year across sectors.

The gadgets and smart home markets have seen an approximate compound annual growth rate of approximately 16% between 2017 and 2018. However, the number of IIoT devices (specifically, those with web and app access) has risen from 356.5 million to 440.8 million. On the other hand, how IoT is applied in real life is a separate issue. To demonstrate how technology helps address industry problems, we've selected two IoT scenarios from our IoT solutions range. Additionally, we will demonstrate several IoT solutions to make your IoT navigation simpler [9].

How could the IoT impact the Employees?

Many of the advantages we now take for granted because of the digital revolution that is happening all around us have also been applied to the workplace environment. Computers in the workplace have only accelerated the pace of digitization. It wasn't so long time ago that we'd have to dial-in to network through a modem, emailing customers or using discs to transmit information.

In today's work environment, changes in technology are less about facilitating the office's mobility and are more focused on creating the best possible work environment for employees, where hours spent on administration are kept to a minimum, wellbeing is the top priority, and costs are kept to a minimum. The driving force behind this corporate transformation is implementing smart devices and systems, however the importance of IoT security must be a priority while undertaking this journey.

More security measures that include installing additional devices that utilize significant amounts of data in the workplace must be considered seriously due to data breaches and GDPR penalties at an all-time high. However, it is also important understanding the advantages the IoT may offer to a company, so businesses can make educated decisions whether or not to adopt this technology [10].

13.5 Personal Digital Assistants

IoT devices have seen significant usage in the house for digital assistants. While these gadgets are not widely used in the workplace, they have the potential to become a helpful addition to offices. A digital assistant may be used by office workers to aid with private chores including making phone calls. Nevertheless, they have enormous promise.

For example, office assistants who use digital systems may advise you on which meeting rooms are available, adjust equipment settings in conference rooms, and place orders based on your purchase history. You may be concerned about needing to look out the conference ID and foreign country code number in advance for a meeting that you have on your calendar, but your digital assistant will not. You'll also have your assistant activate the smart TV, lower the lights, and start the screen saver without you needing to touch anything.

13.6 Wellness and Intelligent Desk Items

Occupational Safety and Health Administration (OSHA) estimates that the cost of treating workplace ergonomic problems including poor posture, physical work, and stress burnout is about $1 billion each week. Many organizations now have programmed in place to teach their staff about the IoT, which can also be utilized to assist in helping to minimize the occurrences of these issues by supplying on-the-fly analytics on employee health and well-being.

When you have a smart desk or smart chair, you will be warned if you have been sitting for too long, and your posture will be monitored so that you may benefit from improved posture. Employees may be given health devices to inform them when they need to get up and move, or make little changes to their workstation, such as changing the lighting or drinking water. By using cutting-edge analytics, employers may increase focus levels and optimize the workplace environment to help workers be more productive. According to an expected $57 billion office sector smart product market by 2025, we are just on the very beginnings of a smart product industry for the workplace [11].

To an employee or company, creative IoT (Internet of Things) technologies have the potential to provide more decision-making power in both their day-to-day and long-term tasks. Conventional wisdom: Not only may these gadgets assist improve workplace efficiency, but they can also save company's money by decreasing the overall and financial expenses [12].

13.7 Conclusion

The purpose of this conceptual study is to offer a critical examination of the Internet of Things (IoT) and how it affects employee and organizational performance in a practical setting. We examine the literature on the Internet of Things, including what it is, why it is important, what it has historically been, its benefits, pros and cons, security and privacy issues, and its impact on organizational performance, and we conclude that a better understanding of how the Internet of Things will impact the ways that organizations and employees conduct their business is required. In computing, the internet of things, also known as IoT, is a networked system of interconnected computing devices such as mechanical and digital machineries, objects, animals, and people that are all assigned unique identifiers and have the ability to transfer data over a network without the need for direct human or human-to-computer communication.

References

1. Dezhina, I.G., Nafikova, T.N.: Evolution of internet of things concept and state policy. Mirovaia Ekonomika Mezhdunarodnye Otnosheniia **63**(7), 23–31 (2019)
2. Kuhwah, V.S., Goyal, S.K., Sharma, A.: Applications of Internet of Things in Health Care: A Survey Study
3. Abu, M.A., Nordin, S.F., Suboh, M.Z., Yid, M.S.M., Ramli, A.F.: Design and development of home security systems based on internet of things via favoriot platform. Int. J. Appl. Eng. Res. **13**(2), 1253–1260 (2018)
4. Tzafestas, S.G.: Ethics and law in the internet of things world. Smart Cities **1**(1), 98–120 (2018)
5. Aryal, A., Becerik-Gerber, B., Anselmo, F., Roll, S.C., Lucas, G.M.: Smart desks to promote comfort, health, and productivity in offices: a vision for future workplaces. Front. Built Environ. **5**, 76 (2019)
6. Kar, G., Hedge, A.: Effect of workstation configuration on musculoskeletal discomfort, productivity, postural risks, and perceived fatigue in a sit-stand-walk intervention for computer-based work. Appl. Ergon. **90**, 103211 (2021)
7. Aryal, N., Kvist, T., Ammam, F., Pant, D., Ottosen, L.D.: An overview of microbial biogas enrichment. Biores. Technol. **264**, 359–369 (2018)
8. Jain, A., Yadav, A.K., Shrivastava, Y.: Modelling and optimization of different quality characteristics in electric discharge drilling of titanium alloy sheet. Mater. Today Proc. **21**, 1680–1684 (2020)
9. Jain, A., Pandey, A.K.: Modeling and optimizing of different quality characteristics in electrical discharge drilling of titanium alloy (grade-5) sheet. Mater. Today Proc. **18**, 182–191 (2019)
10. Jain, A., Pandey, A.K.: Multiple quality optimizations in electrical discharge drilling of mild steel sheet. Mater. Today Proc. **4**(8), 7252–7261 (2017)

11. Panwar, V., Sharma, D.K., Kumar, K.P., Jain, A., Thakar, C.: Experimental investigations and optimization of surface roughness in turning of en 36 alloy steel using response surface methodology and genetic algorithm. Mater. Today Proc. (2021)
12. Karlov, A.A.: Cybersecurity of internet of things—risks and opportunities. In: Proceedings of the XXVI International Symposium on Nuclear Electronics & Computing (NEC'2017), pp. 25–29. Budva, Montenegro (2017)

Chapter 14
Based on Internet of Things Tracking of Hand Hygiene and Practices

Sujay Mugaloremutt Jayadeva, Krishanveer Singh, Lokanayaki Karnan, L. Franklin Telfer, Anupama Bharti, and Viney Dhiman

Abstract The study explores the tracking and practices essentiality in hospitals and health facilities, because related infections, such as higher risks and acceleration of illness, can have significant effects on health care. Human observers tracking and Practices have problems of their own, such as loss of observation concentration, accumulation of human mistake and expert deformation. In this investigation, the results of two internet-based techniques for controlling the hand hygiene of medical personnel during patient visits were carried out and evaluated. As base stations and smart phones, we employed ESP modules as mobile nodes and calculated the range to be located in the patient's room using Bluetooth RSSI values. We analyzed the RSSI in the proximity solution from a sensor host measured on different ESP nodes/module and used the premise that the sensor host is nearest to the ESP node which offers the maximum RSSI values. We utilized the RSSI value to measure the location between mobile nodes and every ESP node in the trilateration method and we applied the trilateration technique to identify the mobile node inside the room. Our tests have shown that 20% of the solution based on proximity was erroneous, whilst the 8% of the solution based on trilateration was erroneous.

S. M. Jayadeva
Department of Health System Management Studies, JSS Academy of Higher Education and Research, Mysuru, Karnataka, India

K. Singh (✉)
GLA University, Mathura, India
e-mail: krishanveer.singh@gla.ac.in

L. Karnan
Department of Computer Science, St. Francis de Sales College, Bengaluru, India

L. Franklin Telfer
Electronics and Communication Engineering, Rajalakshmi Institute of Technology, Chennai, India

A. Bharti
Department of Sociology and Social Work, Himachal Pradesh University, Shimla, Himachal Pradesh, India

V. Dhiman
Department of Cardiology, PGIMER, Chandigarh, India

S. Yadav et al. (eds.), *Proceedings of Second International Conference in Mechanical and Energy Technology*, Smart Innovation, Systems and Technologies 290,
https://doi.org/10.1007/978-981-19-0108-9_14

14.1 Introduction

Research study has revealed that health care workers (HCW) are linked to the number of hand hygiene (HH) infections in health-care institutions (HI) (HAIs). Moreover, the larger the conformity to HH is, the less the number of HAIs increases. As the method for how and when based on HI standards is well-defined, conformity may be automatically verified and Tracking. Tracking and practicing is crucial because of the various implications of HAIs. First of all, death increases and secondly morbidity increases. In this approach, patients and their families get unconsidered charges. In addition, in many nations across the world, they have major economic repercussions. Therefore, the HH monitoring and monitoring is of considerable relevance, from both the humanitarian and economic point of view of the corresponding author. Surveillance, tracking, and measurement of HH conformity is conducted through the use of observers. The standard is regarded. However, it also has its own problems. Complications such as loss of observational concentration, accumulation of human mistake, and occupational deformation may be dealt with. Researchers identified several strategies and ways of handling these instances [1].

This includes the computation of consumption to sterilize materials to mark HCWs with electronic equipment according to its location. We propose in this study a technique combining localization tracking with the communication skills of the web (Internet-of-things). Our System can recognize the place of HCW in the patient room with Bluetooth low energy (BLE) beacons and ESP modules. And so, it chooses to complete HH action by comparing the determined location with the handwashing/Sanitation locations. This provides a means of alternate patrolling of humanly controlled HH [2].

14.2 Objective

In this topic we deeply talk about Approach based on nearness, Approach based on fraternization. Three intersecting circles to calculate the location and the trilateration methodology block definition diagram (BDD), Assessment and debate (Tables 14.1 and 14.2), Categorization of hand hygiene occurrence and Categorization of tracking using hand hygiene.

Table 14.1 Research resulted for an approach based on trilateration

Situation	Model assumption	Designs identified	Mistakes
1	p1, p2, p1, c, p2	c, p2, p1, c, p1, p2	1
2	p1, c, p2, p1, c	p1, p2, c, p2, p1,c	0
3	p2, c, p1, c, p2	p2, c, p1, c, p2	1
4	c, p1, c, p1, p2	c, p2, p1, c, p1, p2	0
5	c, p2, p1, c, p2	c, p2, p1, c, p2	1

Table 14.2 The nearness technique experiment results

Situation	Model assumption	Designs identified	Mistakes
1	p1, p2, p1, c, p2	p1, p2, p1, c, p2	1
2	p1, c, p2, p1,c	p1, c, p2, p1, c	2
3	p2, c, p1, c, p2	p2, c, p2, p1, c, p2	0
4	c, p1, c, p1, p2	c, p2, p1, c, p2, p1, p2	1
5	c, p2, p1, c, p2	c, p2, p1, p2, c, p2	2

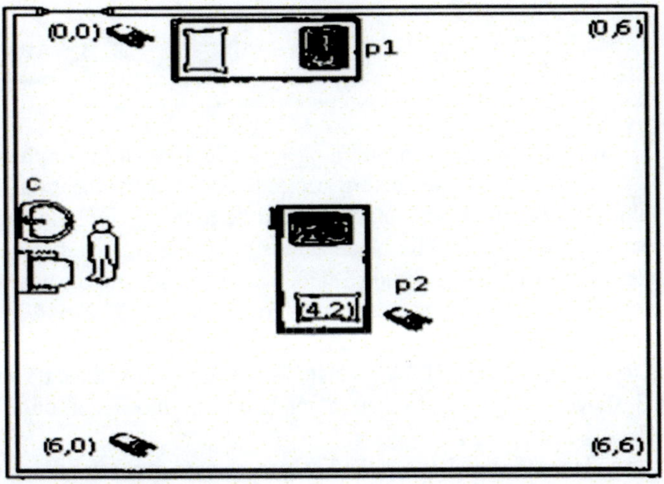

Fig. 14.1 The test bed model in the trial

14.3 Approach Based on Nearness

The ESP node, which gets the greatest RSSI from the mobile node, determines mobile node position under the implemented proximity approach. ESP nodes in the environment are deployed regularly for brief scans and communicate the Id to a server running a Node-Red app, together with its own ID, for Bluetooth devices and received RSSI values from the devices accessible. Figure 14.1 shows that the test bed model in the trial Three ESP nodes has been installed in a room and a smartphone.

14.4 Approach Based on Trilateralization

For trilateral estimation approaches, the three known located access sites shown in Fig. 14.2 are employed. Three ESP32 nodes that scan the region for the mobile

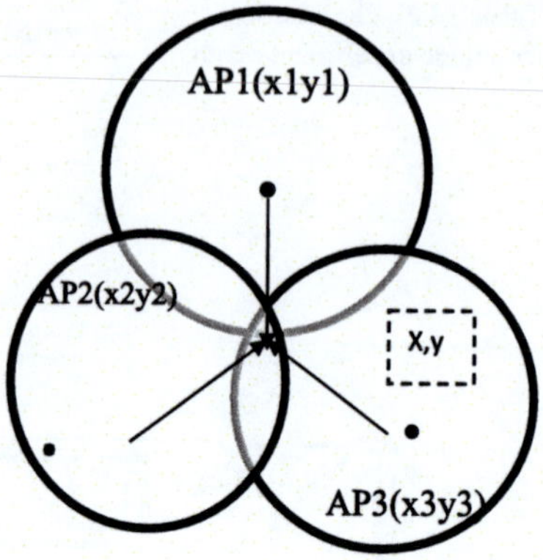

Fig. 14.2 Three intersecting circles to calculate the location [10]

phone RSSI value determine the position of a mobile node in the implementation. Figure 14.2 shows that three intersecting circles to calculate the location [7–9].

RSSI values are the main variables gathered periodically for calculating three distance vectors from each Mobile Node access point. Calculated distance vectors thereby generate three cross-sectional circuits that assist estimate the position of the device node. The Java programme estimates are calculated. BLE and WiFi capabilities are employed for this purpose. ESP32. The data gathered is sent over TCP connection to a server using ESP32. As stated in Section IV-A, the data are parsed. Figure 14.3 shows that The trilateration methodology block definition diagram (BDD) [11].

In order to enhance the precision of the Java programme's estimate, the mean RSSI value is also calculated. Active areas are identified to discover where the doctor is positioned in order to obtain more trustworthy findings. These areas are the beds of the patients. If a doctor is near or near a bed, the estimates are marked by the results. For example, if a physician is in patient 1 and the predicted co-ordinates are inside this field, then the estimated results succeed [12].

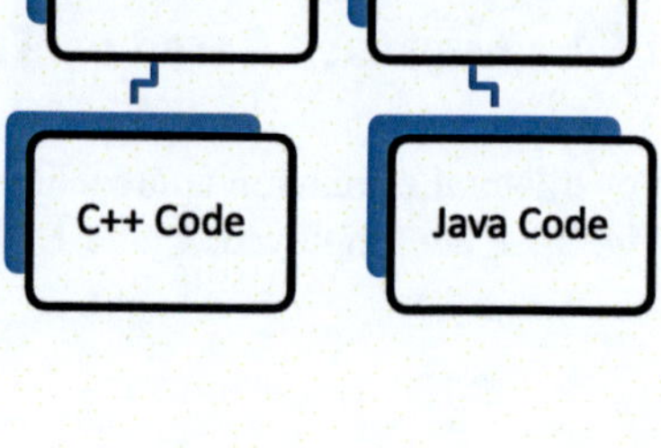

Fig. 14.3 The trilateration methodology block definition diagram (BDD)

Trilateration demands that a mobile node is visible since barriers impact the signal intensity in the surroundings. Based on RSSI, assessment of distances might produce false estimates of weaker signals. The atmospheric constant should be addressed in order to produce a more accurate estimate. The trial and error approach are calculated in this study [13].

14.5 Assessment and Debate

We have developed a test bed in a room with two patient beds (p1 and p2) and a laundry/washing station for evaluation of techniques described in the preceding Section 3 ESP nodes and a mobile nodes (c). An ESP Node was installed on each bed and an ESP Node was also placed next to the cleaning station. For addressing the patient and cleaning station we utilized a Smartphone to mimic an HCW and create 5 distinct scenarios. The examining arrangements and predicted arrangements in the nearness solution are shown in Table 14.1. Scan 1, patient 1, patient 1, cleaning station, 2 following patient visits the HCW (carrying the smart phone) and remains for approximately 30 s at every place [14–16].

Table 14.1 illustrates the pattern of visiting the solution based on trilateration and its estimated patterns. Without any improper detection, the trilateral solution has estimated the right position of the mobile node in Situation 1. In situations 2 and 4, the trilateral solution was recorded in the wrong patient visit 2 and all sites were accurately identified without improper detection in situation 3 and 5. With the 25-location visitors, the trilateration technique reported 2 inaccurate detections, resulting in 8% failure [17].

Table 14.2 indicates that the predicted sites match the expected pattern using the nearness-based method. Likewise, the characteristics for situation 2 that were identified and predicted are the same, which implies the mobile device's nearness approach has assessed its proper position without defect. Situation 3, though, records 1 erroneous visit for patient 2 in the developed application. We have 2 wrong patient visits recorded in situation 4. The application finally logged 1 wrong visit for patient 2 in situation 5. There are 5 distinct places to visit per situation, leading to 25 places to see in total. Five erroneous visits were recorded for 25 visits based on the proximity technique, which gives 20% failure for the full trial [18].

14.6 Categorization of Hand Hygicnc Occurrence

At this point, we recognized all individuals in the hospital unit on their tracks (i.e. position on the global ground plane). Finally, hand hygiene activities are detected and linked to a specific path. In other words, every path must be labelled clean or unclean. When you implement one of the many gel distributors in the hospital, a person gets clean.

14.7 Variability of the Learned Perspective

Real-world sensors are often susceptible to mistakes. Construction teams and services professionals place detectors that change in angle and location, usually deliberately or not. To offer a non-interference-based visual system for tracking the activity of hand hygiene, our model has to be resistant to these variations. By predicting the hand location and deriving a front body outline, we increase this depth picture. The posture network is a twin network and the categorization module employs motion information to make modifications in the backdrop. A binary forecast is the output of the classifier of hand hygiene: whether or not a hand dispensation event has happened.

Conventional counterrevolutionary network designs are not stable in general. This problem is addressed by an implicit transfer learning: a space transformers network that can convert any input characteristic map in space [19].

14.8 Categorization of Tracking Using Hand Hygiene

Our objective is to build a system that understands the hospital in its entirety. A tracking system can't grasp coarse granular motions by itself, and it's insufficient to understand human behaviour over time. The outputs of these devices need to be fused. This is a challenge with spatial time matching, where we have to match detection of hand hygiene with certain recordings. We must match it to a single track for each classification of manual hygiene (i.e. dispenser is utilized). A connection between both the classification and tracking system happens, when a track T meets two requirements:

(a) Track T comprises points $P(x, y)$ that occur at some moment at the very same time as hand hygiene recognition E.

(b) The sensor accountable for the recognition event E is physically close to at least one-point p—step P. The threshold of closeness around the patient's door determines this.

If many paths fulfil these conditions, then we break links by choosing the one closest (x, y) to the door.

14.9 End Result

Our model's ultimate output is a T-list of paths where each track consists of a list (t, x, y, a), with t indicating the time stamp, x, y indicating the 2D ground plane coordination, and an indication of the current action or occurrence labelling. From T, the conformance rate may be calculated or the basis of assessment metrics can be compared.

14.10 Conclusion

To overcome the problem of hand hygiene compliance in hospitals, two alternative indoor location alternatives were introduced. As ground stations and mobile cellphones, ESP modules/nodes were employed, and Bluetooth RSSI measurements were used to calculate ESP Nodes and Mobile Nodes are separated by this distance. We simply looked at the RSSI from the mobile node recorded in several ESP node systems and picked the nearest ESP node by checking for the highest RSSI value of all ESPs in the Nearness solution. Using the trilateration methodology, the RSSI values are employed in the trilateral approach to compute the distance between mobile nodes and each ESP node, as well as to identify the mobile node in the room. Three ESPs and one mobile node were used to build up a test bench (representing a HCW). In a room we have conducted five distinct scenarios to compare the performance of both techniques. In our studies, the closeness solution reported 20% wrong locations visited whereas the closeness solution showed 8% to be more dependable than the closeness solution.

References

1. Alaca, O.F., Tezel, B.T., Challenger, M., Goulão, M., Amaral, V., Kardas, G.: AgentDSM-Eval: A framework for the evaluation of domain-specific modeling languages for multi-agent systems. Comput. Stand. Interfaces **76**, 103513 (2021)
2. Allegranzi, B., Pittet, D.: Role of hand hygiene in healthcare-associated infection prevention. J. Hosp. Infect. **73**(4), 305–315 (2009)
3. Bal, M., Abrishambaf, R.: A system for monitoring hand hygiene compliance based-on internet-of-things. In: 2017 IEEE International Conference on Industrial Technology (ICIT), pp. 1348–1353. IEEE (2017)
4. Cawthorne, K.R., Cooke, R.P.D.: A survey of commercially available electronic hand hygiene monitoring systems and their impact on reducing healthcare-associated infections. J. Hosp. Infect. **111**, 40–46 (2021)
5. Demirkol, S., Challenger, M., Getir, S., Kosar, T., Kardas, G., Mernik, M.: SEA_L: a domain-specific language for semantic web enabled multi-agent systems. In: 2012 Federated Conference on Computer Science and Information Systems (FedCSIS), pp. 1373–1380. IEEE (2012)
6. Ferguson, A.G.: The internet of things and the fourth amendment of effects. Calif. L. Rev. **104**, 805 (2016)
7. Guest, J.F., Keating, T., Gould, D., Wigglesworth, N.: Modelling the costs and consequences of reducing healthcare-associated infections by improving hand hygiene in an average hospital in England. BMJ Open **9**(10), e029971 (2019)
8. Han, G., Choi, D., Lim, W.: A novel reference node selection algorithm based on trilateration for indoor sensor networks. In: 7th IEEE International Conference on Computer and Information Technology (CIT 2007), pp. 1003–1008. IEEE (2007)
9. Han, G., Choi, D., Lim, W.: Reference node selection algorithm based on trilateration and performance analysis in indoor sensor networks. In: 2007 IEEE International Conference on Intelligent Computer Communication and Processing, pp. 177–184. IEEE (2007)
10. Jiang, W., Majumder, S., Subramaniam, S., Li, X., Khedri, R., Monday, T., ... Deen, M. J.: A wearable tele-health system towards monitoring COVID-19 and chronic diseases. IEEE Rev. Biomed. Eng. (2021)

11. Jumaa, P.A.: Hand hygiene: simple and complex. Int. J. Infect. Dis. **9**(1), 3–14 (2005)
12. Karaduman, B., Challenger, M., Eslampanah, R.: ContikiOS based library fire detection system. In 2018 5th International Conference on Electrical and Electronic Engineering (ICEEE) (2018)
13. Mathur, P.: Hand hygiene: back to the basics of infection control. Indian J. Med. Res. **134**(5), 611 (2011)
14. Ojanperä, H., Kanste, O.I., Syrjala, H.: Hand-hygiene compliance by hospital staff and incidence of health-care-associated infections, Finland. Bull. World Health Organ. **98**(7), 475 (2020)
15. Pineles, L., Robinson, G., Morgan, D.J.: New strategies to monitor healthcare workers' hand hygiene compliance. Curr. Treat. Options Infect. Dis. **9**(1), 11–17 (2017)
16. Schweizer, M.L., Reisinger, H.S., Ohl, M., Formanek, M.B., Blevins, A., Ward, M.A., Perencevich, E.N.: Searching for an optimal hand hygiene bundle: a meta-analysis. Clin. Infect. Dis. **58**(2), 248–259 (2014)
17. Situation This, S.B.A.R. NHS Improvement: SBAR: Automated/Electronic Systems for Hand Hygiene Monitoring NHS Improvement IPC Fellow (2018)
18. Sow, D., Turaga, D.S., Schmidt, M.: Mining of sensor data in healthcare: a survey. In: Managing and Mining Sensor Data, pp. 459–504 (2013)
19. Ward, M.A., Schweizer, M.L., Polgreen, P.M., Gupta, K., Reisinger, H.S., Perencevich, E.N.: Automated and electronically assisted hand hygiene monitoring systems: a systematic review. Am. J. Infect. Control **42**(5), 472–478 (2014)

Chapter 15
A Detailed Exploration of Artificial Intelligence and Digital Education and Its Sustainable Impact on the Youth of Society

Korakod Tongkachok, Baig Muntajeeb Ali, Madhurima Ganguly, Sonu Kumar, M. Malathi, and Muthukumar Subramanian

Abstract Artificial intelligence is a software-based modern technology that helps to improve the communication system. Therefore, by using artificial intelligence some digital applications are launched for education digitally. Thus, the use of artificial intelligence and digital applications for education has increased during the COVID-19 pandemic situation. As all the institutions, colleges, and schools were closed for a certain time due to the pandemic situation; the stoppage of studies was impacting the productivity of the students. Therefore, the digital applications and artificial intelligence help the students to continue their education through those applications. Apart from that, these applications and systems for digital education impact the youth of the society largely. The students have to stay in their homes and they can use these digital applications to continue their education and this impacts their communication skills. Along with that, there are a lot of beneficial sites of artificial intelligence and sustainable impacts on the youth of the society. On the other hand, the purpose of this particular research study is to analyse the sustainable impacts

K. Tongkachok
Faculty of Law, Thaksin University, Khao Rup Chang, Thailand

B. M. Ali
College of Teacher Education Darbhanga, Maulana Azad National Urdu University, Hyderabad, India

M. Ganguly (✉)
Heritage Business School Kolkata, Maulana Abul Kalam Azad University of Technology, Kolkata, West Bengal, India
e-mail: mganguly10@gmail.com

S. Kumar
National Level Coordinator, Ignite, Bhumi, India

M. Malathi
Electronics and Communication Engineering, Vivekananda College of Engineering for Women (Autonomous), Elayampalayam, Namakkal, Tamil Nadu, India

M. Subramanian
Nagarjuna College of Engineering and Technology, Bangalore, India

S. Yadav et al. (eds.), *Proceedings of Second International Conference in Mechanical and Energy Technology*, Smart Innovation, Systems and Technologies 290,
https://doi.org/10.1007/978-981-19-0108-9_15

of artificial intelligence and digital applications of education on the youth of the society. Artificial intelligence has a great impact on the education system. Therefore, the advantages and disadvantages of the uses of artificial intelligence in education are analysed in this particular research study. Furthermore, the researcher has used several types of methods and techniques for collecting and analysing more data and information about the particular research topic. Thereafter the researcher has used the secondary methods and sources for collecting data about artificial intelligence and its impacts on youth of the society. Apart from that, the research has used the qualitative techniques for analysing all the collected data in this particular research study in a proper way.

15.1 Introduction

In contemporary times, Artificial intelligence (AI technology) as a software has helped in fighting and enhancing the field of education. In both business and education AI currently remains high in demand and has the opportunity to sustain for the long term. Due to the constant development of the world, modifying technology is always acceptable and beneficial in increasing knowledge as well as skill. Mainly in the educational field, the emergence of every new technology has proved vital for supporting society and the economy. The digital education action plan is an inclusive, high-quality, and accessible digital platform of learning. It is considered as an important approach in developing learning areas for an individual student [1].

AI technology has proved its impact on social media and online games. Recent works have concentrated on the development of digital technology frameworks for educational organizations along with educators. Due to the enhancement of higher education more effectively the majority of educational sectors are involved in the adoption of Artificial Intelligence in their education process. It has been mentioned by several researchers that developing an effective future technological framework can help in modifying both personal and social development. In this study, the impact of Artificial Intelligence and digital education on the youth of the society will be discussed to understand its value to establish a better future opportunity in this competitive world [2].

15.2 Literature Review

15.2.1 Concept of Artificial Intelligence (AI) and Digital Education

Artificial intelligence is a technological machine that is capable of understanding different languages, thinking individually, and solving problems. Basically, it is a computer system that properly organizes intelligent beings to provide better outcomes. It is managed by following a particular scientific discipline and it has three hierarchically connected levels regarding human behaviour. As stated by Yigitcanlar et al. [12], the ability to change society can help in bringing revolution and provide a better structure. As learning is a highly significant element in this current provision, AI technology can provide a great support in this area [3–9].

UNESCO, which is the International advisory Board, also supports AI software in leading the educational development [11]. UNESCO has declared near about five areas where AI can be implemented in the educational field. It includes development of value, empowers teaching, educational management, learning assessment, and learning opportunity. As argued by Pham et al. [10], getting high quality and accessible knowledge regarding education, digital education is recognized as most supportive. Accompanied with multimedia along with mobile phones, the education process can be sustained properly whilst staying long distances [10].

15.2.2 Impact of Artificial Intelligence (AI) and Digital Education on the Youth of the Society

In the context of education of this current decade, education has become more challenging and competitive. At this moment, AI technology adoption in the educational field has been considered than any other approach. As per the view of Mhlanga [9], AI has shown immense potential in all walks of life even in education by supporting both teachers and students. Accompanied with development data, educational quality, the entire educational system can be updated with the help of AI technology [11] (Fig. 15.1).

Based on *The New Education Policy 2020,* AI adoption in the education process has made **6.6%** development in India [2]. Personalized learning and predictive quality is capable of maintaining due to with AI technology in the education system. Utilizing digital platforms within AI software has become more flexible in this recent decade. As argued by Yu [13], due to this pandemic situation, delivering long distance education through digital platforms has proved easy and profitable within AI involved in the education program. The youth of this generation can get a better experience of digital platforms to learn with artificial intelligence technology [12] (Fig. 15.2).

15.3 Methodology

Due to conducting a particular research approach, it is important to select an appropriate method. Similarly, to evaluate the ideology of the entire relevant collected data, the researcher has identified a ***positivism research philosophy*** will be a better approach. As utilizing a positivism research philosophy can support a research topic

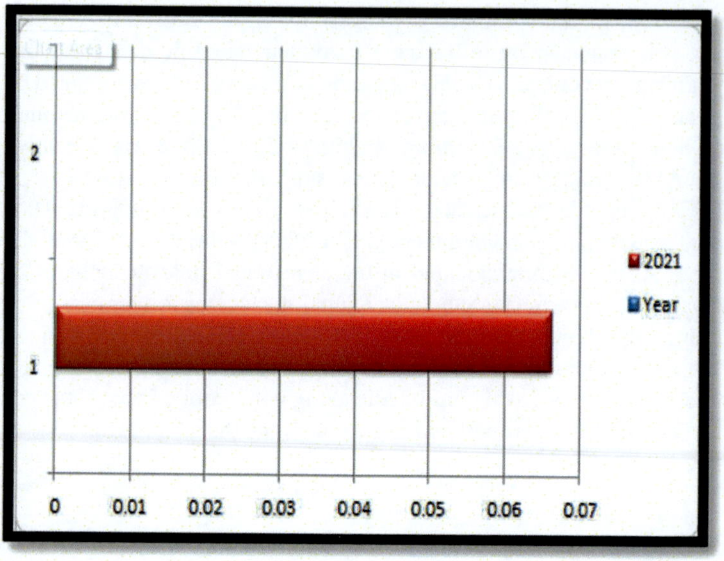

Fig. 15.1 Educational development by adopting AI. *Source* Influenced by Mhlanga [9]

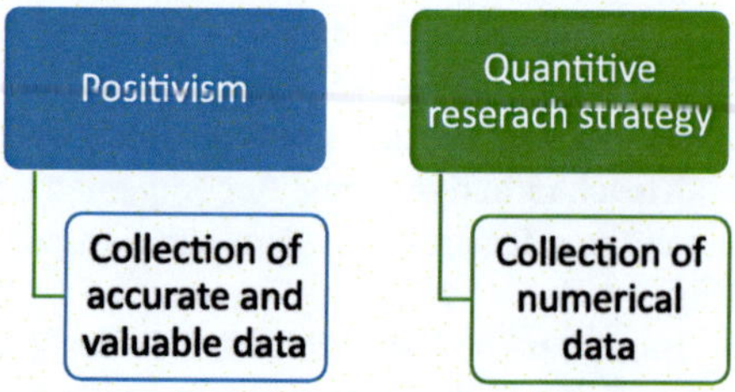

Fig. 15.2 Research methodology. *Source* Self-created

to develop its quality and reliability this philosophy will assist the researcher in a correct way. As proposed by Fuchs [5], this philosophy can help in gaining a better knowledge and understanding regarding the topic of artificial intelligence and digital platforms.

In addition to this, in order to legitimize the research objectives, the research study has used a ***descriptive research design*** to explain the research goals and objectives. After justifying the research objectives, the possibility of getting appropriate research outcomes has been increasing a lot. As stated by Akyol and Ustaoğlu [1], accompanied with justified research objectives the research will be capable of adopting a descriptive research design to make it approachable and understand the appropriate relationship between different study variables [13].

Apart from this, the researcher has selected a ***quantitative research strategy*** to conduct the research further forward. According to Li et al. [8], it is capable of delivering the aspect to recognize the appropriate variables amongst various criteria

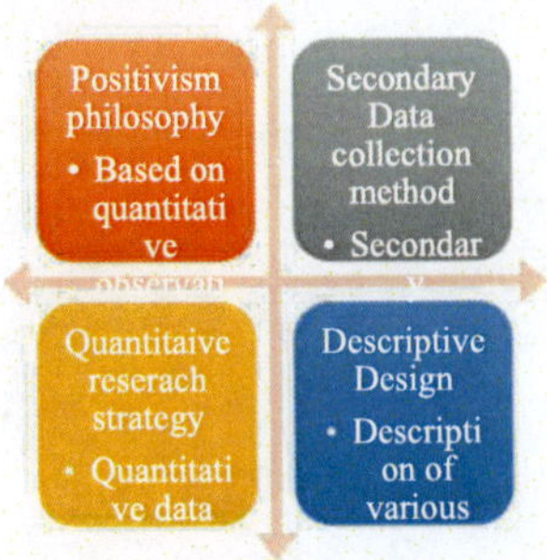

Fig. 15.3 Justification of selecting methodology. *Source* Self-created

to make the selected study more suitable and authenticate. In this scenario, to lead with artificial intelligence in the digital education system, this strategy can provide a potential shape to figure out the accurate data and information.

Therefore making a study valuable, it is important to select an effective data collection method that will help the study to be developed and perfect. In this scenario, the researcher has chosen a ***secondary quantitative data collection method*** to collect valid and relevant data to get better outcomes. As observed by Hsu [7], this method is helpful in gathering relevant information and knowledge to mitigate the existing problems. The researcher has collected data and information related to AI technology and digital education systems depending on books, journals, newspapers, articles, and websites. In addition, as it is capable of reducing time and cost, the researcher has found this data collection method as helpful (Fig. 15.3).

15.4 Result and Discussion

15.4.1 Online Learning is a Part of Digital Education

Accompanied with currently available technologies, it has been understood simply that a digital platform is the only option that can help in the forward-thinking process and develop the education system. The digital platform signifies the system that is included in the electronic systems, tools, and devices. Utilizing technology when the education quality can be improved and modified, it is called the digital education system. As opposed by Blayone et al. [4], it helps to broaden and increase across all curriculum learning areas. Due to the COVID-19 pandemic situation, the demand for digital education has become highly significant. It is the way to make forward to gain knowledge and learning within a flexible way and help to solve the issues that occur in the education system. Thus, this AI technology remains expensive for many institutes; it can help in establishing a strong support by providing a better online experience.

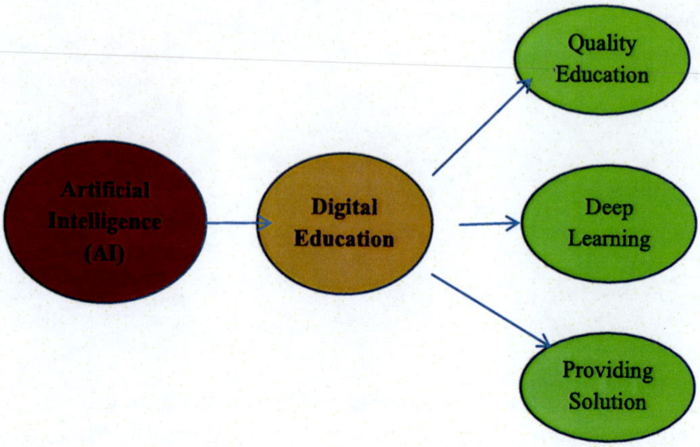

Fig. 15.4 Impact of AI technology on the youth of society. *Source* Self-created

15.4.2 *Future Opportunities with Artificial Intelligence*

The majority of sensible educational sectors have partnered with artificial intelligence in order to recognize the lessons successfully. As the students have to stay in their homes due to the COVID-19 situation, they can use these digital applications to continue their education and this impacts their communication skills. As observed by Bachtiger et al. [3], it has a lot of advantages in the criteria of artificial intelligence and it impacts on the youth of the society. The utilization of artificial intelligence and digital applications has increased in a wide range during the pandemic situation. As almost all schools and colleges are closed for a certain time, this has a great opportunity to sustain for a long-term issue. Along with this, it is available for 24 × 7 which is identified as another positive element for its sustainability (Fig. 15.4).

15.5 Recommendations and Conclusion

15.5.1 *Recommendations*

Depending on the entire study, it can be recommended that by adopting artificial intelligence software, the quality and value of education can be developed. Recognizing the educational risk, the implementation of AI technology can be beneficial by providing a potential solution. As opined by Goyal et al. [6], by boosting the transparency in the education system the digital platform can play a supportive role in providing education. In the activity of implementing AI technology in the digital platform, it needs to adopt training regarding the new artificial intelligence society.

In addition, building bridges and increasing communication, the development of artificial intelligence, and the digital education system can be executed.

15.5.2 Conclusion

Based on the entire study, it can be stated that during COVID-19 pandemic situation, the digital education system has occupied a wide space in the education system. This approach will be beneficial within the collaboration of Artificial Intelligence (AI technology), as it is an effective software that is capable of handling several issues individually. Moreover, this study will be helpful for the researchers to understand both the advantages and disadvantages of Artificial intelligence in the digital education system to balance educational satisfaction.

References

1. Akyol, E.M., Ustaoğlu, E.T.: A descriptive research on the evaluation of knowledge management Themed publications. Int. J. Contemp. Econ. Adm. Sci. **11**(1), 160–177 (2021)
2. Analyticsindiamag.com: State of artificial intelligence in India—2020 [Online]. Available at: https://analyticsindiamag.com/report-state-of-artificial-intelligence-in-india-2020/ (2021). Accessed on 31 Aug 2021
3. Bachtiger, P., Plymen, C.M., Pabari, P.A., Howard, J.P., Whinnett, Z.I., Opoku, F., Janering, S., Faisal, A.A., Francis, D.P., Peters, N.S.: Artificial intelligence, data sensors and interconnectivity: future opportunities for heart failure. Card. Fail. Rev. **6** (2020)
4. Blayone, T.J., Barber, W., DiGiuseppe, M., Childs, E.: Democratizing digital learning: theorizing the fully online learning community model. Int. J. Educ. Technol. High. Educ. **14**(1), 1–16 (2017)
5. Fuchs, C.: From digital positivism and administrative big data analytics towards critical digital and social media research! Eur. J. Commun. **32**(1), 37–49 (2017)
6. Goyal, M., Knackstedt, T., Yan, S., Hassanpour, S.: Artificial intelligence-based image classification for diagnosis of skin cancer: challenges and opportunities. Comput. Biol. Med. 104065 (2020)
7. Hsu, G.C.: Using math-physical medicine and artificial intelligence technology to manage lifestyle and control metabolic conditions of T2D. Int. J. Diabetes Complications **2**(3), 1–7 (2018)
8. Li, X.H., Shi, Y., Li, H., Bai, W., Cao, C.C., Chen, L.: An experimental study of quantitative evaluations on saliency methods. In: Proceedings of the 27th ACM SIGKDD Conference on Knowledge Discovery & Data Mining, pp. 3200–3208 (2021)
9. Mhlanga, D.: Industry 4.0 in finance: the impact of artificial intelligence (AI) on digital financial inclusion. Int. J. Finan. Stud. **8**(3), 45 (2020)
10. Pham, Q.V., Nguyen, D.C., Huynh-The, T., Hwang, W.J., Pathirana, P.N.: Artificial intelligence (AI) and big data for coronavirus (COVID-19) pandemic: a survey on the state-of-the-arts. arXiv preprint arXiv:2107.14040 (2021)
11. Unesco.org: Artificial intelligence in education [Online]. Available at: https://en.unesco.org/artificial-intelligence/education (2021). Accessed on 31 Aug 2021

12. Yigitcanlar, T., Desouza, K.C., Butler, L., Roozkhosh, F.: Contributions and risks of artificial intelligence (AI) in building smarter cities: Insights from a systematic review of the literature. Energies **13**(6), 1473 (2020)
13. Yu, Z.: Visualizing artificial intelligence used in education over two decades. J. Inf. Technol. Res. (JITR) **13**(4), 32–46 (2020)

Chapter 16
Review Paper: Combined Study of Oceanography and Indigenous Method for Effective Fishing

Zalak Thakrar and Atul Gonsai

Abstract There is a complex question for fishermen these days due to various factors such as shortage of fish in the sea. Due to this scarcity of fish in the Pacific Ocean, our fishermen are forced to ship in the neighboring countries for fishing. They are caught and taken hostage during fishing in the maritime boundary of other countries because of lacking sufficient information. As a result, they have to spend a long period in imprisonment of other countries, leaving behind their family in a pathetic condition. In order to avoid all these problems, we can use the machine learning technique to inform the fishermen about the places to go for fishing based on the scientific parameters derived from Ocean Color website of the satellite (MODIS-Aqua and MODIS-Terra Sensor) give you (Ocean Surface Temperature (SST), chlorophyll content) and many others. We also use Indigenous technique (Survey, Questioner, data collection from government agency/Private Fisheries agency) based on that two methodology out proposed system will analyze the data in machine learning technique using Ensemble method based on python, also our android application provides an intelligent boundary alert (IBAL) to fishermen and helps to navigate within fishing zone with the help of Google map using android smart phone.

16.1 Introduction

In India, Gujarat has largest coastline length from all other state it has around 992 miles (1596 km) long, along with from all other port Porbandar has only the first all-weather 24 × 7 port which is located at (21.6417° N, 69.6293° E) [1] and Gujarat's coastline is known for its fishermen. Many of the people there depend on fishing for their livelihood, but the problem of fishermen is increasing day by day due to the

Z. Thakrar (✉)
Shri V. J. Modha College of Information Technology, Porbandar, Gujarat, India
e-mail: zalak.thakrar@gmail.com

A. Gonsai
Department of Computer Science, Saurashtra University, Rajkot, Gujarat, India
e-mail: amg@sauuni.ac.in

S. Yadav et al. (eds.), *Proceedings of Second International Conference in Mechanical and Energy Technology*, Smart Innovation, Systems and Technologies 290,
https://doi.org/10.1007/978-981-19-0108-9_16

shortage of fish in the seas which is a major cause of increasing sea pollution, such as the chemical water of the companies. Other reasons such as small size of fishermen's nets due to premature catching of small fishes lead to shortage of fishermen. This makes fishing a difficult problem for fishermen. Due to which fishermen have to spend maximum time in the sea. So that related expenses like diesel, rations, time etc. are wasted and fishermen have to go farther to catch fish, which is why fishermen often reach the boundary of other countries to fish unknowingly. In earlier Indigenous methods are used when technology was not used, fishermen went fishing keeping in view the water type, wind direction, sun, moon, and wind direction. In which high tides and low tides in water were considered as the main factor. We will try to reduce the problem by using modern technology in place of tradition technology. For this we will be able to give him potential fishing zone information about the places where fishing can be done using Ensemble method (Indigenous + scientific method) based on both technique applied using machine learning algorithm [2] tolled used in python. We use python library to provide a PFZ location on Map, also the research alert to fishermen boundary line using his smart phone GPS, furthermore the application shows the border distance.

16.2 Materials and Methods

16.2.1 Study Area

The study is conducted in the Arabian Sea water of Porbandar at Gujarat State, India Country (see Fig. 16.1). This area is situated of the Gujarat Coast fisheries management area and is a major fishing ground for fishermen on the west coast of India. It is an area with significant fishing potential in India. With overall productivity of Gujarat's second largest catching area and available fish resources off Mumbai in the 50–100 m water depth, and Location of potential thread fin bream resources (Rani fish) between Maharashtra and Gujarat coast in the depth range of 100–200 m have also been explored during the survey. Horse mackerel supplies occur in significant amounts beyond the traditional operating limit of industrial boats. Other deep sea tools such as Priacanthidae (Bulls eye), Drift fish, and Scad can be found in large amounts between 50 and 100 m deep. Cuttlefish supplies are found in south of Maharashtra and off the coast of Gujarat. Off the coast of Gujarat, there are ribbon fish stocks, India in the depth range of 30–70 m, Deep Sea Marine Resources on the Gujarat Seaboard Coastal Plain and Slope [3].

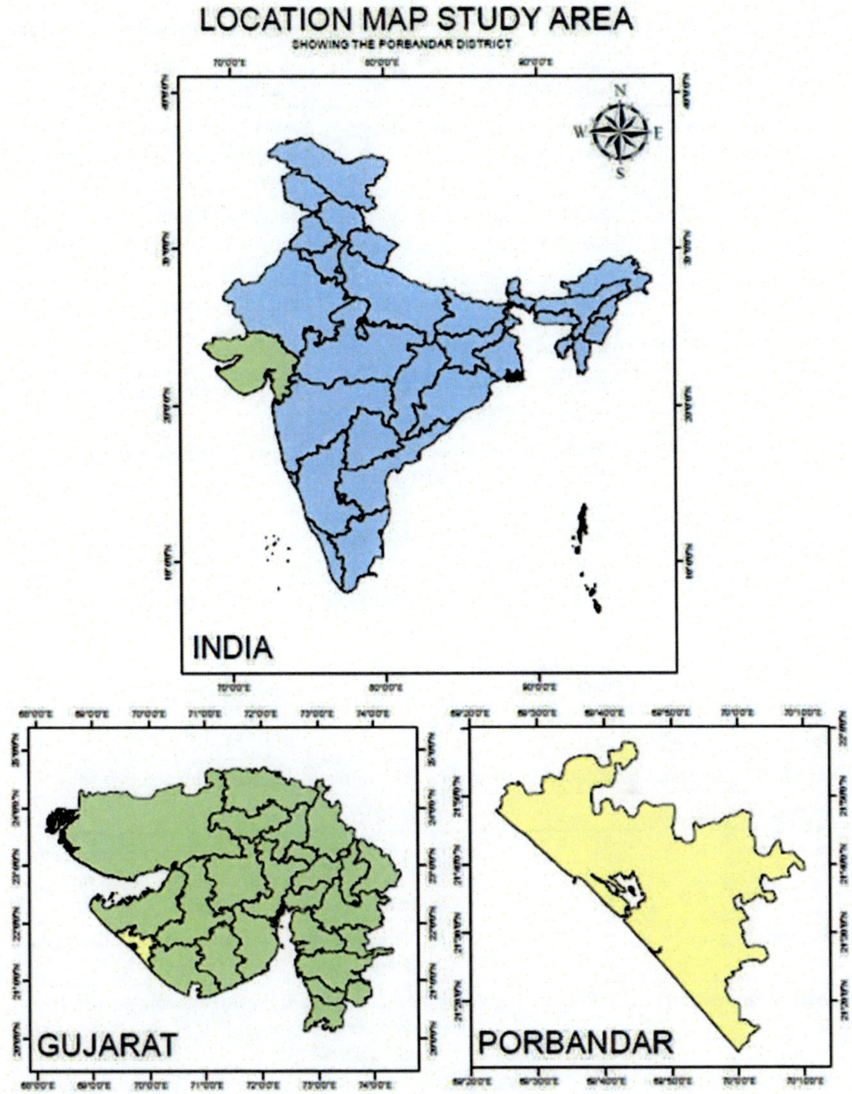

Fig. 16.1 Study area in the Arabian Sea water of Porbandar Gujarat State India (Self Source)

16.2.2 Types of Methods

In this study data should be analyzed with the help of daily fisheries collected data and also examine these data using two type of method (1) Scientific method (see Table 16.1). (2) Indigenous method (see Table 16.2).

Table 16.1 Scientific parameter used in research with help of different satellite website and weather forecast website [4–7]

Attribute/Parameter	Description	Satellite/Sensor/Website
SST—(Sea Surface Temperature)	Average, highest, and lowest temperature (Monthly)	AVHRR-NOAA-18 (https://oceancolor.gsfc.nasa.gov/)
Chl—Chlorophyll	Average, highest, and lowest chlorophyll (Monthly)	MODIS-TERRA (https://oceancolor.gsfc.nasa.gov/)
Humidity	Average pressure, average humidity, and lowest relative humidity (Monthly)	https://www.esrl.noaa.gov/
Wind speed	Average, maximum, and minimum wind speed	https://developer.nrel.gov
Wind direction	Most frequent direction and max direction speed of wind speed (daily)	https://developer.nrel.gov
Sunshine/Sunrise	Daily season wise sunshine and sunrise timing	https://www.timeanddate.com/
Tide	Low tide and high tide data (daily)	https://incois.gov.in/
Current	Daily mean sea surface current data	NOAA-14/AVHRR-2 https://podaac.jpl.nasa.gov/

Table 16.2 Indigenous technique to collect data from different research methodology [8]

Indigenous technique	Description
Field survey	In field survey we collect the tradition method used by fishermen to catch fish based on Sample-based fishery survey
Questioner survey	We will prepare fishing survey to collect data for train model
Government agency/Private Fisheries agency	Collected data from Government agency for month wise report for fish catches by fishermen and exported fish

16.2.2.1 Scientific Method

With the help of scientific method data are collected from satellite based on sensor and other parameter used (see Table 16.1).

SST and Chl—a level 3(4 km) standard features extract data were taken and collected from multiple cloud services from July 2010 to 2021 (see Table 16.1).

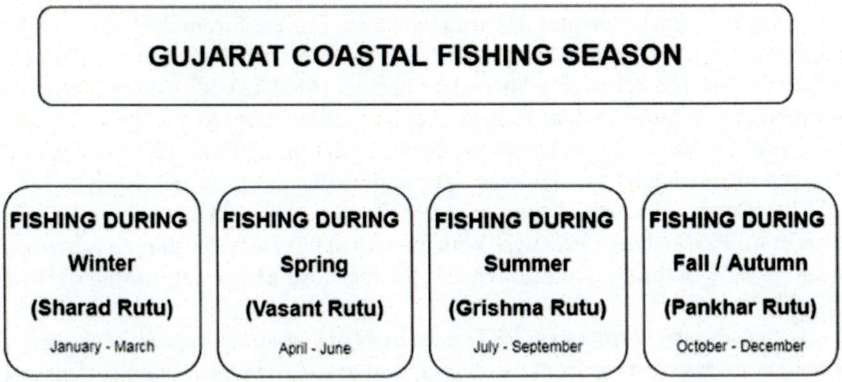

Fig. 16.2 Gujarat Coastline fishing season time line [9]

16.2.2.2 Indigenous Method

With the help of indigenous method data are collected with different terminology (see Table 16.2).

Both data will be gathered around the Arabian Sea based on the fishing season, as (see Fig. 16.2) of the Gujarat coastline between 2010 and 2020. The catching weight (kg), fishing location (latitude and longitude), and used gears have all been collected.

16.3 Literature Survey

In this paper, the researcher can try to get a powerful tool for potential fishing zone which is basically based on the aspects like the chlorophyll-a (CHL) and the temperature of the sea surface (SST). This paper also tries to reduce the time to find the potential fishing zone in ocean area. We may also use the automatic technique which has been developed to calibrate and validate the development of frequent maps of planned good fishing grounds in near-real time for the FEDERPESCA fleet. The same technique could be developed for other seas as well [10].

Psychological fish trap scaling was proposed as a way to implement the experience among fishery workers into an automated FCP. According to questionnaire reports on classification error tolerance, fisherman continues to assess errors for only small catches. The Weber–Fechner law was also enhanced to calculate the psychological fish capture. GBDTs are being equipped in a psychological error minimization lane change. When the fish catches were small, this method was successful in reducing prediction errors (e.g., for the end period of fishing). As a consequence of this discovery, the proposed psychological measure may be useful in developing forecasts that are suitable for fishermen [11].

As per the research studies, the data which are retrieved from the images which are provided by the satellite such as Land sat Enhanced Thematic Mapped 8 (ETM+), Moderate Resolution Imaging Spectra radiometer (MODIS), MODIS-AQUA. This study shows the research done from the Vembar and extended to the Tuticorin coastline. The retrieved pre-processed and then the data on SST and CC are retrieved. The pre-processing of data is done using various feature in R and using Arc-GIS 10.2.1 software. The rank is given to the retrieved data to get the result of the High PFZ/Low PFZ/Medium PFZ/High. With the help of this study, fisher man can reduce their 30–70% of time and it also reduces the fuel cost, manpower, fishing duration [12].

It concludes the Searching of PFZ on the basis of the various features of the ocean. The data are also derived from the satellite and get analyzed. Net Ekman transport, measured in the forward and backward components of the wind in reference to the frontal angle, is helpful for PFZ area monitoring and potential shifts. The retrieved data got analysis with the help of the proposed algorithm and bio-physical model. The gradient approach is used for the identifying the thermal power of the ocean. The eddies in the oceans indicate the sign of the PFZ. The advantage of the respective model which is used in this study is that, it is available continuously and provides the PFZ [13].

In the proposed system's architecture the AIS, VHF collection, micro controller, voice module, loud speaker, and ultrasonic sensor are the main components. This tool was provided to help fishermen in staying within their nation's borders. There will be two modes of operation for the AIS system. AIS acts as a transceiver in active mode and as a receiver in passive mode. The AIS is set to passive mode by coast guards, but it is triggered by fishermen. This paper proposes an integrated novel strategy for fishermen's survival based on GPS and AIS [14].

It concludes that the proposed android application helps to save the lives of the fisherman and stop them to cross the maritime boundary. The latitude and Longitude are set in the application which gives alert to the fisherman when they're crossing the Line so that they can stop over there and save their life [15].

This paper has researched the potential and no potential fishing areas based on the regular fishing are using the various models and techniques and algorithm. Proposed Model is used to get the result of the data and respective proposed framework of it. 87.11% result has been derived from the proposed model on the basis of the oceanographic characteristic such as SST and SSC. The result shows that heuristic rule model has been outperformed on the proposed data-mining framework. In Future, with the help of oceanographic Characteristic they'll try to provide more prediction method to enhance the prediction performance [16].

In previous days fishermen could not easily find out the borders and had no proper information about climatic condition. With the help of this paper, we can easily identify the border and continuous monitoring of climatic condition using LoRa wireless technology and wind speed sensor. As a consequence, fishermen can easily recognize national sea barriers, stopping them from entering their territories. As an effect, their lives have been saved, and good relations with neighboring nations were established. Consequently, vessel theft can be safely charged [17].

As a conclusion the crossing of the boundary is prevented. This developed the system with the various technologies such as GPS, and also use IOT devices to prevent maritime boundaries alert. A revered mechanism is used to stop the engine of the boat if it crosses the boundary. The United Nations Convention on the Law of the Sea (UNCLOS) introduced and established the maritime border structure, and the nations are divided by the International Maritime Boundary Line (IMBL) boundaries. If the boat crosses the threshold, a warning is sent to the boat via the LCD and alert system, as well as to the nearest coastguard. The bouncy castle was designed with the aid of the machine. Up to 74% of cases are avoided [18].

The app developed in the Android is useful to save the life of the fisherman and as it helps to track their position. It also gives an alert to the fisherman if he is crossing the border. The position of the phone is identified based on the mobile's s International Mobile Equipment Identity (IMEI) number which will be sent along with the lat. and lon. coordinates. It also provides the alarm to ring if any issue or problem occurs in the sea [19].

16.4 Discussion

The goal of the study is to provide a fishing location based on SST and CHL analysis with addition to use Indigenous method approach used by traditional fisherman. We will implement and predict the fishing location using machine learning algorithm and provide a location with the help of Google map in smart phone. This research also provides the boundary alert system to fisherman. In the current paper we give introduction about the problem faced by fisherman during fishing and what are the parameter we have to take care for good fishing for fishermen to catch fish from ocean bases of remote sensing in the field of oceanography. We have collected the different parameters based on two types of methodology (Scientific + Indigenous Method). In Scientific method we provide the major Four Parameter that are (satellite time-series derived parameter from Level 3 browser with MODIS aqua and tera sensor on the basis of periodically daily data with resolution of 4 km dataset of resolution, Wind direction, Low/High tide data, Ocean Current, etc.) and in indigenous method we have also used field survey, questionnaire survey, government/private agency data along with dissuasion about and intelligent boundary alert for fisherman.

16.5 Conclusion

In the future, researcher can research how the machine learning algorithm will be applied in Oceanography parameter as well as on indigenous parameter for predictions fishing spots.

References

1. Gujarat Coastline: Google Search. https://www.google.com/search?q=gujarat+cost+line&rlz=1C1RXQR_enIN932IN932&oq=gujarat+cost+line&aqs=chrome..69i57j35i39l2j46i433j0i433j69i60l3.4882j0j4&sourceid=chrome&ie=UTF-8. Accessed 7 Mar 2021
2. Mohammed, M., Badruddin Khan, M., Bashier, E., Bashier, M.: Machine Learning: Algorithms and Applications. Google Books. https://books.google.co.in/books?hl=en&lr=&id=X8LBDAAAQBAJ&oi=fnd&pg=PP1&dq=machine+learning+algorithms&ots=qQzvyoE6yE&sig=yvjyCu3fno4Xk3xuXH9dw2iP6HU#v=onepage&q=machinelearningalgorithms&f=false. Accessed 7 Mar 2021
3. FISHERY SURVEY OF INDIA Porbandar Base of Phone. http://www.fsi.gov.in/ob2.htm. Accessed 6 Mar 2021
4. NASA Ocean Color. https://oceancolor.gsfc.nasa.gov/. Accessed 7 Mar 2021
5. ESSO-INCOIS-Indian National Centre for Ocean Information Services. https://incois.gov.in/portal/osf/tide.jsp. Accessed 7 Mar 2021
6. India Wind Data Download API | NREL: Developer Network. https://developer.nrel.gov/docs/wind/wind-toolkit/india-wind-download/. Accessed 7 Mar 2021
7. Indian Meteorological Department—Meteorological Data. https://www.imdtvm.gov.in/index.php?option=com_content&task=view&id=29&Itemid=43. Accessed 7 Mar 2021
8. Kothari, C.R.: Research Methodology: Methods and Techniques. Google Books. https://books.google.co.in/books?hl=en&lr=&id=hZ9wSHysQDYC&oi=fnd&pg=PA2&dq=research+methodology&ots=1tUcoAbYF2&sig=fYqIuPT7beoxd0Ej5ikWeo0fSx4#v=onepage&q=researchmethodology&f=false. Accessed 7 Mar 2021
9. Wang, W., Zhou, C., Shao, Q., Mulla, D.J.: Remote sensing of sea surface temperature and chlorophyll-a: implications for squid fisheries in the north-west Pacific Ocean. Int. J. Remote Sens. **31**(17), 4515–4530 (2010). https://doi.org/10.1080/01431161.2010.485139
10. Tijani, K., Chiaradia, M.T., Morea, A., Nutricato, R., Guerriero, L., Pasquariello, G.: Fishing forecasting system in Adriatic sea—A model approach based on a normalized scalar product of the SST gradient and CHL gradient vectors. In: International Geoscience and Remote Sensing Symposium (IGARSS), vol. 2015, pp. 2257–2260 (2015). https://doi.org/10.1109/IGARSS.2015.7326256
11. Kokaki, Y., Kobayashi, T., Ogawa, T.: Psychological measure on fish catches and its application to optimization criterion for machine-learning-based predictors. In: OCEANS 2019—Marseille, pp. 1–5 (2019). https://doi.org/10.1109/oceanse.2019.8867405
12. Karuppasamy, S., Ashitha, T.P., Padmanaban, R., Shamsudeen, M., Silva, J.M.N.: A remote sensing approach to monitor potential fishing zone associated with sea surface temperature and chlorophyll concentration. Indian J. Geo-Marine Sci. **49**(6), 1025–1030 (2020)
13. Jishad, M., Sarangi, R.K., Ratheesh, S., Ali, S.M., Sharma, R.: Tracking fishing ground parameters in cloudy region using ocean colour and satellite-derived surface flow estimates: a study in the Bay of Bengal. J. Oper. Oceanogr. **14**(1), 59–70 (2021). https://doi.org/10.1080/1755876X.2019.1658566
14. Saravanan, K., Aswini, S., Kumar, R., Son, L.H.: How to prevent maritime border collision for fisheries?-A design of real-time automatic identification system. Earth Sci. Inf. **12**(2), 241–252 (2019). https://doi.org/10.1007/s12145-018-0371-5
15. Parkhande, R.H., Pawar, S.B., Nalawade, S.U., Yadav, P.S.N.: Marine boundary alert system for fishermen. Int. J. Innov. Sci. Res. Technol. **3**(2), 722–724 (2018)
16. Fitrianah, D., Hidayanto, A.N., Gaol, J.L., Fahmi, H., Arymurthy, A.M.: A spatio-temporal data-mining approach for identification of potential fishing zones based on oceanographic characteristics in the eastern Indian Ocean. IEEE J. Sel. Top. Appl. Earth Observations Remote Sens. **9**(8), 3720–3728 (2016). https://doi.org/10.1109/JSTARS.2015.2492982
17. Shreyas, S., Sridhar, S.M., Safulla, S., Kumar, D.C.P.: Monitoring with Self-Helping System for Fishermen, no. 8 (2020)

18. Nair, L.R., Subramaniam, K., Prasannavenkatesan, G.K.D.: A Review on Multiple Approaches to Medical Image Retrieval System, vol. 1125 (2020)
19. Dinesh Kumar, R., Aldo, M.S., Joseph, J.C.F.: Alert system for fishermen crossing border using Android. In: International Conference on Electrical, Electronics, and Optimization Techniques, ICEEOT 2016, pp. 4791–4795 (2016). https://doi.org/10.1109/ICEEOT.2016.7755630

Chapter 17
The Dynamic Role of Big Data Analytics in Learning and Development and Its Impact on Risk Analysis in Stock Market

P. A. Mary Auxilia, Joseph Alvarado-Tolentino, Tatiana Gonzales-Yanac, Antonio Huaman-Osorio, S. Durga, and Mohammed Faez Hasan

Abstract Big data analytics is making a paradigm shift in all industries, including banking, financial institutions and stock broking companies. The financial big data analytics (FBDA) is considered as one of the key areas of enhancing the management, governance, performing critical analysis and support the management, regulators and managers to make profound decisions. It is noted that the big data is transforming the finance sector mainly stock market domain in assisting the risk managers and investors to measure the risk in an effective manner so that they can invest in better assets like stocks, bonds, mutual funds, exchange traded funds (ETF) etc. for enhancing the return on investments. Every day there are millions of financial transactions which occur in the financial community around the world, the financial practitioners and risk management analyst need to upgrade their knowledge through continuous learning and focus in analysing the risk efficiently so as to enhance their investment plan and realise the financial goals. The risk management is considered as the systemic modelling which enables in the estimation of the interrelationship amongst the financial institutions, big data applies the major characteristic which is comprised of 4 Vs—Volume, Velocity, Variety and Veracity this enables in analysing the large volume of data available in stock prices, forecast the volatility and manage the risk in an appropriate manner. This paper enables in presenting a conceptual understanding on the dynamic role of big data analytics in learning and development and its impact of risk analysis in stock market, this paper enables in presenting sophisticated methods for risk control based on big data analytics in creating new finance aspects by analysing the overall feelings based on massive textual information from

P. A. Mary Auxilia
Loyola Institute of Business Administration, Chennai, India

J. Alvarado-Tolentino · T. Gonzales-Yanac · A. Huaman-Osorio
Universidad Nacional Santiago Antunez de Mayolo, Huaraz, Peru

S. Durga (✉)
K L Business School, KLEF, K L University, Guntur, India
e-mail: sdurga1234@gmail.com

M. F. Hasan
Department of Finance and Banking Sciences, Kerbala University, Karbala, Iraq

S. Yadav et al. (eds.), *Proceedings of Second International Conference in Mechanical and Energy Technology*, Smart Innovation, Systems and Technologies 290,
https://doi.org/10.1007/978-981-19-0108-9_17

the social networks, apply various technologies in creating ways of estimating risk and enhance their return on investments (ROI).

17.1 Introduction

In the age of dynamic business environment, the technological innovation is driving competitive edge for the business. Companies are focusing in using the technologies for upgrading the skills and abilities of their employees which will enable in increasing their performance, support the organisation to enhancing their service offerings, engage the customers, create strategies which will out beat the competition effectively. Through the use of big data analytics. The management can analyse and forecast the precise learning and development needs which will enable them to support in various functions. Human resource managers along with operational heads can use the power of big data analytics in analysing the training needs in the organisation, measure the skills required by the individuals to enhance their performance and provide better services for realising the mission and vision of the organisation [1].

Stock market is considered as one of the key fields in the world, which involves in forecasting the price movement of the assets involving stocks, bonds, commodities, foreign exchange etc. The analyst, researchers and employees in the stock market tend to apply various learnings which cover fundamental analysis, technical analysis, analysing the global economy, company specific analysis etc. hence, the knowledge and skill is highly important for enhance the prediction level and thereby support the organisation and other stakeholders to realise their financial objectives. The employees need to analyse different data sets related to historical information and current aspects in order to forecast the price of the security, all of these aspects tend to add more data each data and hence business analytics enable in analysing the requirements of the individuals. On the other hand, big data analytics enables in supporting the management in creating a new means of creating personalised learning requirements, responsive to assessment and actively engage the learners so that they can access the learning anytime based on their convenience. Hence, the employees, analyst and risk managers can unleash the role of business analytics to equip themselves and enable in analysing the risk of securities in the stock market. Hence, this paper is focused in analysing through conceptual aspects the overall role played by the business data analytics in learning new concepts, skills and knowledge in stock market and how it supports the intended users in predicting the risk and other key aspects in their profession.

17.2 Review of Literature

Anware et al. [2] have mentioned that big data intends to refer to the large volume of data which is stored, processed for making effective analysis. Big data uses different tools and methods which enable in extracting patterns from the large volume of data which are available in the hands of the organisation. The greatest resource in the financial services sector like stock broking, risk management, insurance companies is stated to be the human capital, hence organisations tend to provide necessary training and enhance their skill so as to gain competitive edge. On the other hand, Sun et al. [3] have stated the big data analytics has enabled in addressing the training needs of the risk analyst so as to perform their function effectively. In stock market, the risk is mainly categorised into systematic risk and unsystematic risk, furthermore other risks like interest rate risk, currency risk, inflation risk, business risk etc. tend to impact the performance of the stock in the market. Hence, appropriate training needs to be provided in order to measure the risk and identify strategies which will enable in managing the stated risk so as to enhance the return on investments.

Furthermore, the impact of data analytics enables the organisation to measure the performance of the individuals and devise unique training plans so that they can perform effectively in their profession [4]. The application of big data tends to consider various inputs and thereby perform detailed analysis so that the overall training needs can be assessed. The career progression in the modern business environment is dynamic, the individuals need to be prepared and equipped with different skills so as to meet the requirements of the organisation, hence using data analytics the current skill set required to perform the job can be assessed and employees can enrol for the program accordingly (Table 17.1).

Table 17.1 Opportunities of big data analytics in learning and development

Categories	Possible outcomes
Measuring training needs	Business data analytics can apply tools and techniques in order to identify the pattern and can implement flexible training solutions
Methodologies	Business data analytics tend to create new methodologies, create better engagement to support the training needs
Optimal balance	The data analytics focuses in creating optimal balance between the job responsibilities and upgrading the skill set of the employees
Sustained collaboration	There is a sustained partnership between the management and application of data analytics in meeting the training and development goals

Source Rabhi et al. [4]

17.3 Research Methodology

The main purpose of the study is to apprehend various academic research which has been carried out in the area of big data, machine learning and stock market. The researcher will use descriptive research study so as to understand the critical nature of big data analytics in financial services mainly in the stock market domain and how the managers and analyst utilise them in order to enhance their skill set for forecasting the risk and managing them efficiently so that they can enhance the profits for the investors, clients and other stakeholders.

The researcher uses secondary source for collecting the past literature reviews, major electronic database used is Scopus and Google Scholar, the major words used are big data, risk management, learning and development in organisation, stock market, big data in financial services etc. [5]. The researcher focused mainly on the peer reviewed and published journals, however certain dissertations and other related reports were also considered in order to perform the research. The study uses qualitative data design as it enables in collecting the academic journals in the area related to big data analytics and its overall impact in providing training to the individuals for forecasting the risk and managing effectively in the stock market. The researcher does not apply any primary data collection for performing the study.

17.4 Critical Discussion

The purpose of implementing business data in business is that it can extract meaningful patterns and analysis from the available data sourced internally and externally for efficient decision-making, also business analytics supports the organisation in forecasting and interpreting the data in a more meaningful manner [6]. Organisations tend to possess large volume of data which are historical and current in nature, big data supports in applying different tools like machine learning, deep learning, application of statistical tools in order to convert the raw data into meaningful information [7]. The human resources managers and training heads use the data available related to employees to measure their performance and also to devise unique training plan which will enable them to enhance their skills and abilities, make them perform better and realise their personal and organisational goals.

Big data analytics supports the management in the financial services industry mainly in stock broking to understand the overall performance of the employees and assign them the necessary training so that they can enhance the performance. The following are some of the key attributes on the dynamic role of big data analytics in learning and development and its critical influence on risk analysis.

17.5 Identification of Critical Training Needs

In the stock market profession, the individuals need to enhance their knowledge and skills on a constraint basis. The big data analytics supports the organisation and individuals in analysing their current nature of work, performance level and desired career path, provides critical analysis on the nature of training program which can be enrolled for achieving the specific task [8]. Business enterprises are adopting new and sophisticated tools so as to understand the impact of data analytics driven training needs towards employees and organisational performance. The application of such tools has brought in various benefits to the individuals and management through better performance, foster organisational change, delivering enhanced services to the customers and enable in creating better career progression for the individuals (Table 17.2).

In a recent study it is noted that one of the critical aspects of implementation data analytics is to support the management and supervisors to equip the skill set of the employees so that they can perform better, since estimating risk is always dynamic in stock market, the individuals need to possess the necessary knowledge, and data analytics can determine the specific training which the individuals need to undergo at their level.

17.5.1 Focused Approach

The human resource managers and training heads tend to apply business analytics so as to define the specific strength and weakness of the human capital and make the necessary adjustments. Specific training program can be devised and implemented which will enable in creating better and skilled workforce in the organisation. The risk managers need to understand the potential of the individuals and create training strategies based on the experience and skill set of the individuals [9]. The major

Table 17.2 Risk management framework through business analytics

Nature of risk management	Impact of business analytics
Identification	Business analytics enable in identifying the specific set of training needs in the area of risk management for the employees. Enable in identifying the future trend and suggest the best programs for supporting employees and management
Assessment	Analyse the underlying data, estimate the patterns through application of various tools
Management and control	Manage the training programs and provide the necessary modes like online, classroom training etc. based on the employee skill set
Reporting to management	Provide real-time performance measurement of the employees before and after training, calculation of risk and enhance better customer delivery

aspect of big data analytics in risk management learning is that it can integrate the trainee knowledge with the larger dataset and determine which mode of training will be effective for the individuals. Also, the big data support in continuous monitoring and evaluation of the individual progress, measure against the training with that of the actual performance in their job responsibilities and support them in growing in their career path.

17.6 Constant Improvement

The application of big data analytics in learning and development supports the HR managers and senior leadership team to monitor the progress of the employees on a continuous basis and implement the strategies for continuous improvement. Individuals need to enhance their skill and capabilities so as to meet the dynamic market environment, hence to harness such abilities it is necessary for the management to look at different training programs for constant improvements [10]. The cloud-based tools support in getting the real-time updates on the training progression and its impact on their job responsibilities. The clear interconnection amongst the risk management and influence of training has led to increase in implementing the necessary technologies to support the employees in a dynamic manner.

The critical framework in the risk management activities tend to analyse the overall risk and provide the investors, clients and others to invest in securities which generate better ROI and also manage the risk effectively.

17.7 Imminent Outlooks

The application of business analytics and other related tools are highly significant for the organisation to create competitive edge. In the service industries, the employees are considered as critical assets as they tend to be in close relationship with the customers, therefore it is highly necessary to provide the trainings and enhance their skill set so that they can offer better services Through business data analytics, risk analyst, managers, advisors and employees can estimate the risk in stock market in an effective manner and suggest their clients to invest and manage the investments effectively [11]. Hence, the impact of business analytics on learning and development is profound in different industries including the financial services sector. The organisation is using big data analytics and sustainable progression as key trends in order to enhance business performance, make their team adequately trained and implement strategies to achieve the stated mission and vision of the organisation. It can be further stated that the implementation big data analytics enables in forecasting the future trends and hence provide the training to the employees in the current context to meet the expectations.

Developments in process-oriented learning techniques include conversational teachers who "read" short written answers with latent semantic analysis and automatically generate clues based on historical data using advanced methods, Machine learning. In all of these scenarios, the data size is derived from many data points as the student progresses. These data points can be made semantically readable for the student and the teacher in the form of immediate feedback. Over time and for many students (one class, all software users, one demographic), the data size increases proportionally. The financial markets are always looking for technical innovation for various activities, namely technical innovations that are always positively accepted and that have a great impact on the financial markets and that have a real transformative effect.

17.8 Conclusion

The stock market is considered to be one of the most important sectors in the world, which includes forecasts for asset price movements, including stocks, bonds, commodities, currencies and so on. Stock market analysts, researchers and employees tend to apply a variety of knowledge, including basic analysis, technical analysis, world economic analysis, company-specific analysis and so on. Therefore, knowledge and skills are very important to improve the forecast level and thus help the organisation and other stakeholders to achieve their financial goals. In addition, the effect of data analysis enables the organisation to measure individuals' performance and design individual training programs so that they can practice their profession effectively. Big data applications normally take into account different inputs and therefore perform detailed analysis to assess the overall training needs. Comprehensive data analysis helps management in financial services, especially in real estate, to understand employees' overall performance and provide them with the training they need to improve their performance. Below are some key features of the dynamic role of big data analytics in learning and development and its critical impact on risk analytics. Personnel managers and trainers often use business analysis to identify strengths and weaknesses of human capital and make necessary adjustments. A special training program can be designed and implemented to create a better and more specialised workforce in the organisation. Risk managers must understand the potential of individuals and develop training strategies based on individuals' experience and skills. The most important aspect of big data analysis in learning risk management is that you can integrate student knowledge into a larger amount of data and determine which training method is most effective for individuals.

References

1. Cope, B., Kalantzis, M.: Big data comes to school: implications for learning, assessment, and research. Rev. Res. Educ. (2016). https://doi.org/10.1177/2332858416641907
2. Anwar, S., Bascou, N.A., Menekse, M., Kardgar, A.: A systematic review of studies on educational robotics. J. Pre-College Eng. Educ. Res. (J-PEER) **9**, 19–42 (2019). https://doi.org/10.7771/2157-9288.1223
3. Sun, Y., Shi, Y., Zhang, Z.: Finance big data: management, analysis, and applications. Int. J. Electron. Commer. **23**(1), 9–11 (2019)
4. Rabhi, L., Falih, N., Afraites, A., Bouikhalene, B.: Big data approach and its applications in various fields: review. Proc. Comput. Sci. **155**(2018), 599–605 (2019)
5. Jain, A., Yadav, A.K., Shrivastava, Y.: Modelling and optimization of different quality characteristics in electric discharge drilling of titanium alloy sheet. Mater. Today Proc. **21**, 1680–1684 (2019)
6. Jain, A., Pandey, A.K.: Multiple quality optimizations in electrical discharge drilling of mild steel sheet. Mater. Today Proc. **8**, 7252–7261 (2019)
7. Panwar, V., Sharma, D.K., Kumar, K.V.P., Jain, A., Thakar, C.: Experimental investigations and optimization of surface roughness in turning of EN 36 alloy steel using response surface methodology and genetic algorithm. Mater. Today Proc. (2021). https://doi.org/10.1016/j.matpr.2021.03.642
8. Dicuonzo, G.: Risk management 4.0: the role of big data analytics in the bank sector. Int. J. Econ. Financ. Issues **9**(6), 40–47 (2019)
9. Saggi, M.K., Jain, F.: A survey towards an integration of big data analytics to big insights for value-creation. Inf. Process. Manage. **54**, 758–790 (2018)
10. Lackovic, D.I., Kovska, V., Lakovic, V.Z.: Framework for big data usage. Risk management process in banking institutions. In: Central European Conference on Information and Intelligent System, pp. 49–54 (2016)
11. Hossein, H., Xu, H., Emmanuel, S.S.: Digitalisation and big data mining in banking. Big Data Cogn. Comput. **2**(3), 18 (2018)

Chapter 18
Roles of Cloud Computing and Internet of Things in Marketing Management: A Critical Review and Future Trends

Sujay Mugaloremutt Jayadeva, Alim Al Ayub Ahmed, Reena Malik, Asmat Ara Shaikh, Md. Nur-E-Alam Siddique, and Mohd Naved

Abstract Our lives are now comprised of our two very distinct notions of cloud computing and the Internet. Their increasing popularity and use will continue to grow and become essential features in the future Internet. A new age in which cloud and IoT merge should interrupt many application scenarios and facilitate them. This paper focuses on Cloud and IoT integration, which we call the Cloud and IoT idea. Several literary studies have independently examined the Cloud and IoT: its fundamental qualities, technology and future research. Sadly, the notion of cloud and IoT is not thoroughly analyzed to our comprehension. We start analyzing and discuss the need for them to be integrated, the issues which occur and how the literature addresses these issues. We then explain various uses, whether it be in private or open-source efforts, documented in the literature and platforms for the realization of the Cloud and IoT idea. In this research, we will analyze several areas of cloud computing that dominate IoT and highlight open challenges and possible answers for the future Internet in cloud computing.

S. M. Jayadeva (✉)
Department of Health System Management Studies, JSS Academy of Higher Education and Research, Mysuru, Karnataka, India
e-mail: sujay15@gmail.com

A. Al Ayub Ahmed
School of Accounting, Jiujiang University, 551 Qianjindonglu, Jiujiang, Jiangxi, China

R. Malik
Chitkara Business School, Chitkara University, Punjab, India

A. A. Shaikh
Bharati Vidyapeeth's Institute of Management Studies and Research, Navi Mumbai, India

Md. N.-E.-A. Siddique
Faculty of Economics and Management, University Kebangsaan Malaysia (UKM), 43600 Bangi, Selangor, Malaysia

M. Naved
Department of Business Analytics, Jagannath University, Delhi-NCR, India

S. Yadav et al. (eds.), *Proceedings of Second International Conference in Mechanical and Energy Technology*, Smart Innovation, Systems and Technologies 290,
https://doi.org/10.1007/978-981-19-0108-9_18

18.1 Introduction

Intelligent and self-configured devices in a rapidly evolving global network architecture are at work in the Internet of Things (IoT) paradigm. It is one of the most perturbing technology that allows all-embracing and all-round computer situations. IoT is typically characterized by real-life and tiny processing and storage capacities and concomitant dependability, efficiency, security and privacy problems. Cloud computing, but at the other hand, offers almost infinite storage and processing capacity, is a much maturer technology and at least has largely resolved most IoT concerns. Therefore, both existing and future Internet is predicted to be disturbed by a new IT paradigm in which the Cloud and IoT complementing capabilities are fused simultaneously. This is our new Cloud IoT paradigm. This study examines the Cloud and IoT integration literature for academics and industry alike. Integrate on-demand cloud-based computing, data storage, apps and IT assets. The conventional IT system and its infrastructure has infiltrated cloud computing. Maintaining engineering tools with the aid of cloud technology, many technological companies like Amazon, Facebook, Google and Apple create enterprises globally with a variety of approaches [1] (Fig. 18.1).

Futures Internet refers to a group of data transmission networking techniques that will be available in the future. The Internet of Things (IoT) is the most essential idea of the Future Internet for combining seamless networks and networked things

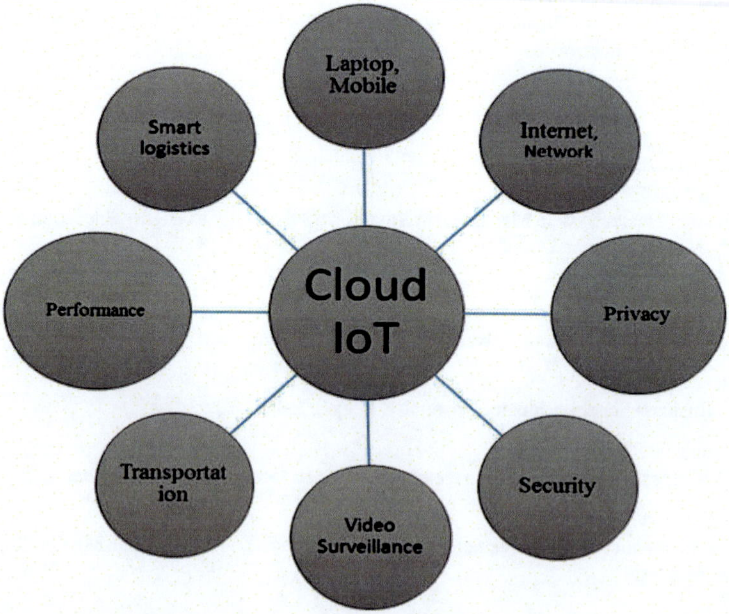

Fig. 18.1 Future internet of things (IoT) and cloud computing

into a single global IT platform. People will be linked to the Internet at anytime, anywhere, with anything and anybody, and will be able to use any network and any business correctly in the future. In other terms, the Internet of Things (IoT) is concerned with the integration, information, collecting, computing, communications, and connectivity of people and things. Cloud Computing, but at the other hand, is seen as a backbone answer for quantity of data streams and calculations while dealing with the difficulties of the future, when everything will be connected by smooth networks. For discussion computing and online service to allow IoT, cloud technologies can provide a virtual, accessible, efficient, and adaptable data center [2].

The Internet of Things (IoT) and Cloud Computing are both Future Internet developments. Nevertheless, IoT technology is evolving at a rapid pace and is not yet compatible. As a result, service providers and administrators are left with no clear guidelines to follow. Cloud computing services, on the other hand, are reliant on network operators. Many multinational organizations are working on defining their standards for a uniform security and network architecture. As a result, we believe that the IP Multimedia Subsystem (IMS) is the best option for meeting the criteria. Yet, the network and software structure combining IoT and Cloud still poses numerous problems to IMS. Throughout this study, the open problems are discussed and the answers for the future Internet are proposed. Lastly, we provide an early IoT bootstrap platform to debate the open issues and answers for IoT deployment on the Internet in the future [3] (Table 18.1).

Government: Governments are attempting to create IoT solutions for smart cities. IoT improves the solutions and technologies of military forces. Cheap and greater gadgets give improved safety beyond boundaries. IoT enables government departments monitor and enhance their services, such as medical, transport and education, in real time (Fig. 18.2).

Advertising Your Products: By providing relevant material and solutions, companies may better evaluate and respond to client preferences using IoT. In live time, it helps to improve company tactics. Now that we are aware of the strong Iot technology in many fields, let us go deeply and study Raspberry Pi, a widely used solution for

Table 18.1 Component of internet of things and cloud computing

Component	Internet of things	Cloud computing
Data analysis	Acts as a big data source	Acts as a technique or a way of managing huge data
Reachability	Very narrow	Large expanded, wide
Storage	Limited or almost none	Large, virtually never ending
Role of Internet	Acts as a convergence point	Acts as a way to provide services
Computing abilities	Restricted	Almost limitless
Parts	Runs on hardware components	Run on virtual machines that simulate components of hardware

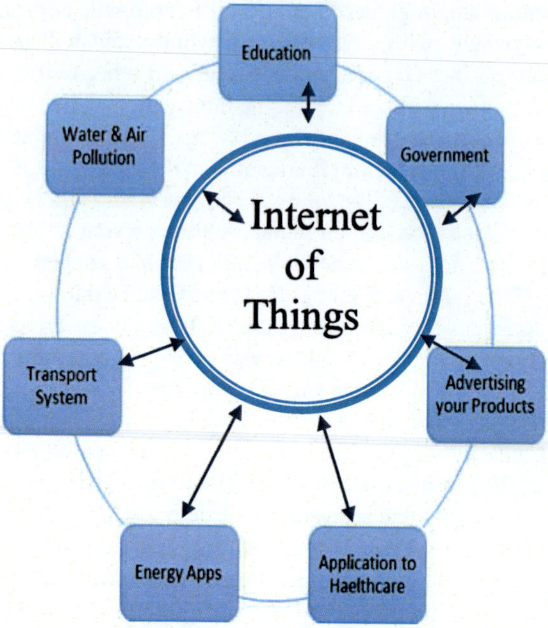

Fig. 18.2 IoT in a range of systems [4]

the preparation of IoT. We will create an IoT application when we have understood Raspberry Pi.

Application to Healthcare: Intelligent watches and exercise equipment have modified the health monitoring routine. Persons can regularly monitor their own health. Not only that, but now if a patient comes to the hospital for treatment, his health report is evaluated by physicians and the therapy starts rapidly as he or she enters the hospital. Information from multiple healthcare apps are currently being acquired and used for the analysis and treatment of various diseases.

Energy Apps: The pace of energy has increased to an instinct. Individuals and businesses both look for strategies for reducing usage and to regulate it. IoT offers a method of monitoring not just power consumption at the device level, but also at housing, grid level or at the distribution system. For the monitoring of energy usage, Smart Meters and Smart Grid are utilized. It also identifies performance of the system and stability risks, which safeguard devices against inactivity and harm.

Transport System: The IoT industry has transformed. Here we have self-driving cars with sensors, stoplights that autonomously feel and change, car parking aid, free parking spot location etc. Moreover, several sensors in your car show you the state of your automobile, so you don't have to encounter any problems throughout your journey.

Water and Air Pollution: We can identify water and air contamination by regular testing using different sensors. This assists to avert major pollution and associated catastrophes. IoT permits operations in the field of analysis and surveillance to minimize human interaction. Systems can detect agricultural, soil, climate, and more changes automatically.

Education: IoT offers training aid to meet the shortages in the educational sector. Education: It not only enhances educational standards, and also optimizes costs and improves administration by taking student reaction and performance into mind.

18.2 What Are the Problems that Cloud and IoT Present?

Transferring particular IoT app data to the cloud offers considerable convenience, including cost savings and complexity in the direct handling of hardware. As a result, the complicated Cloud IoT scenario contains numerous factors linked to various diverse issues, each of which poses problems when specialized skills are required. To provide trustworthy and efficient services, for example, the following characteristics are needed: protection, privacy, dependability, availability, adaptability, and vital link. As important IoT apps go to the cloud, additional issues develop owing to a lack of certain important components: confidence in the network operator, understanding of service level agreements (SLAs), and understanding of the physical location of the storing data [5].

Data Analytics: With the anticipated amount of 50 billion devices to be networked by 2020, the vast amounts of data that they create need to be specifically transported, stored, accessed and processed. IoT will become one of the primary Big Data sources with current technology developments, which Cloud can keep for a long period of time and do complicated analysis. The accessibility and the prevalence of mobile devices and sensors really require scaled computer platforms. Efficient management of this data is an essential problem since the entire performance of the application depends heavily on the features of the data management services.

Tracking: Tracking is a vital activity in the cloud environment for capacity management, resources management, SLAs, performance and security as well as for troubleshooting, as is widely discussed in the literature. As a result, Cloud IoT has the same tracking needs for Cloud, while volume, diversity and vastly improved of IoT significantly impact the linked issues.

Sensor Networks: Sensor Networks were characterized as a key IoT facilitator, one of the five technologies that would alter the world, enabling assessment methods from sensitive ecosystems and environmental assets to urban settings to be measured, inferred and understood. Recent technical developments have made miniature systems available in big, distant sensing applications for efficient, low-cost and low-performance utilization. Smartphones include several sensors—Location, speedometer, motion sensor, microphone and camera—which allow a variety of mobile applications in various IoT fields but they are restricted by battery consumption and durability. Cloud offers new options for the collection and use of sensor data to increase the coverage and relevance, while affecting security and privacy. Cloud offers new chances.

Networking and Communication Protocols: Communicates across numerous smart objects using multiple protocols, depending on the individual application situation. The IoT cloud comprises the machine-to-machine (M2M) connectivity. It is hard to deal with the variability to uniformly handle things while delivering necessary performance. Most applications do not include movement: IoT uses IEEE 803.11.4/6LoWPAN solution in fixed settings commonly. Other situations, like as vehicle networks, on the other hand, usually use IEEE 802.15p standards. Meanwhile, because WiFi and Bluetooth are the most commonly used radio technology for personal networks, their acceptance for Internet of things is growing: they are a less expensive solution, most mobile devices already support them, and both protocols are becoming increasingly low energy.

Service Provision: IoT services are usually offered as a vertical isolation solution, with all parts of the system closely associated with the particular context of the application. Providers must scan target application environments, analyze their needs, choose hardware, connect diverse subsystems, build applications, offer computer infrastructure and provide service support for each application service. Cloud IoT, but at the other hand, enables efficient, scalable and easy-to-expand IoT services by utilizing Cloud delivery methods [6] (Table 18.2).

Sensors for Energy Efficiency: WSN comprises mostly of low-cost, power restrained and reduced detectors. The network energy is consumed in each activity, computation and inter-communication. Also, sustainable energy is a big concern, and energy-efficient management is more than desired if particular sensors are replaced by smart devices. Several approaches for energy conservation may be embraced: for example, compressive sensing permits lower signal readings without having an influence on precise channel modeling and uses continuous communication to minimize the transmitted signal of each sensors. Across the other hand, Cloud IoT is not influenced by power problems with the Cloud concept In fact, on-demand storage and computing resources may be leased from an energy point of view via Cloud in an optimum way. In order to conserve energy in processing resources and devices, local calculations can also be uploaded to the cloud [7].

Computing in the Fog: Fog Computing is an increase to the network edge for classical cloud computing (as fog is a cloud close to the ground). IoT applications defined by delay limitations and mobile and location requirements have been developed to support it. Although both the cloud and the fog have a computer, storage and networking resource, these have particular features: edge location and location awareness of low latency; geographical distribution and a very wide range of cloud-centralized nodes; mobility (wireless access) support and real-time ability to interact (rather than batch proceedings) assistance [8].

Sensors Collaborative: In person-centered sensing, the number of missing samples is crucial. Based on data from consumers, in fact, the capacity to create relevant data

Table 18.2 Future Cloud IoT address

	Capacity	Durability	Adaptability	Endurance	Security	Privacy	Maintainability	Availability	Mobility	Energy saving	Efficiency of cost	Wide-ranging
Data analytics	Yes	Yes	No	No	No	Yes	No	Yes	Yes	No	Yes	Yes
Tracking	Yes	Yes	Yes	Yes	Yes	Yes	No	No	No	No	No	No
Sensor networks	Yes	No	No	Yes	No	No	No	No	Yes	Yes	Yes	Yes
Networking and Communication protocols	Yes	No	No	No	No	No	No	No	Yes	Yes	No	No
Service provision	Yes	Yes	Yes	No	No	No	Yes	No	No	No	Yes	Yes

is severely restricted. Furthermore, problems of data management and data protection must be resolved and suitable rewards for involvement established in order to create meaningful interaction with end-users [9].

Clouds Forms: Security is a key challenge for Cloud IoT, which involves every networked environment. I am, in fact, exposed to several assaults from both its IoT side and cloud. In IoT, data security, authenticity and integrity are guaranteed through protection. But insider assaults are not addressed and computing-restricted gadgets are difficult to install. As no higher level of intelligence can be activated there, RFID is the most susceptible element. Cloud-related security concerns also demand care because Cloud manages economy, data and technologies. Cloud services need to be upgraded by offering field-specific software programs or environments and smooth application forms as well as the use of numerous dynamically varying resources, to support the QoS demands of various users, to enable quick application development. In order to offer services reliably, cloud scheduling algorithms must handle job duplication in the event of failure management.

Prevailing Opinion Requirement: While the scientist has given a variety of contributions to IoT and cloud concepts, a clear need is required for common protocols, architectures and APIs to make it easier for diverse intelligent devices to link together and to provide better services that carry out the Cloud IoT concept. Mobile-to-mobile in specific (M2M) is a concept, but it is not standardized very much. This makes use of ordinary Internet, cellphone and Internet applications for the various current options. Furthermore, state-of the art infrastructures are lacking for quick management and quality implementation of Cloud IoT services. In fact, at the early stage of IoT, many designs are either from the point of view of the WSN or depending on the cloud in the center [10].

18.3 Conclusion

The study explores that IoT and Cloud Computing are the tendencies of Future Internet. But IoT technology advancements are diverse and not compatible. This does not lead to any standard for providers and operators. From the other hand, the solutions for Cloud Computing rely on providers. Because many international organizations are committed to developing their specs for a prevalent service providers and software design. New apps from this combining known as the Cloud IoT bring up new business and research opportunities. Hopefully, this combination will reveal a new concept for the growth of multi-networking and an open user services platform.

References

1. Aceto, G., Botta, A., De Donato, W., Pescapè, A.: Cloud monitoring: a survey. Comput. Netw. **57**(9), 2093–2115 (2013)
2. Akyildiz, I.F., Su, W., Sankarasubramaniam, Y., Cayirci, E.: Wireless sensor networks: a survey. Comput. Netw. **38**(4), 393–422 (2002)
3. Ballon, P., Glidden, J., Kranas, P., Menychtas, A., Ruston, S., Van Der Graaf, S.: Is there a need for a cloud platform for European smart cities? In: eChallenges e-2011 Conference Proceedings, IIMC International Information Management Corporation, pp 1–7 (2011)
4. Botta, A., De Donato, W., Persico, V., Pescapé, A.: Integration of cloud computing and internet of things: a survey. Future Gener. Comput. Syst. **56**, 684–700 (2016)
5. Chen, T., Barbarossa, S., Wang, X., Giannakis, G.B., Zhang, Z.L.: Learning and management for Internet of Things: accounting for adaptivity and scalability. Proc. IEEE **107**(4), 778–796 (2019)
6. Nguyen, B., Simkin, L.: The Internet of Things (IoT) and marketing: the state of play, future trends and the implications for marketing (2017)
7. Parwekar, P.: From internet of things towards cloud of things. In: 2011 2nd International Conference on Computer and Communication Technology (ICCCT-2011), pp. 329–333. IEEE (2011)
8. Pereira, P.P., Eliasson, J., Kyusakov, R., Delsing, J., Raayatinezhad, A., Johansson, M.: Enabling cloud connectivity for mobile internet of things applications. In: 2013 IEEE Seventh International Symposium on Service-Oriented System Engineering, pp. 518–526. IEEE (2013)
9. Zaslavsky, A., Perera, C., Georgakopoulos, D.: Sensing as a service and big data. arXiv preprint arXiv:1301.0159 (2013)
10. Zeinab, K.A.M., Elmustafa, S.A.A.: Internet of things applications, challenges and related future technologies. World Sci. News **2**(67), 126–148 (2017)

Chapter 19
The Internet of Things and Its Aspect for Digital Marketing During COVID-19 Pandemic

Amgad S. D. Khaled, Khaled M. K. Ismail Alshaketheep, Kiran Kumar Agrawal, Dilip Kumar Sharma, Ankit Jain, and Viney Dhiman

Abstract This paper is prepared to identify the influence of the Internet of things and its aspects in digital marketing during the COVID-19 pandemic. The research examines the various aspects of digital marketing before and after the pandemic. The paper also evaluates the tools of IoT for digital marketing. The COVID-19 pandemic has inevitably resulted in digital technologies due to social distancing customs and nationwide lockdowns. It has been found that the COVID-19 outbreak has had an impact on marketing in general and digital marketing in particular. IoT and digital technologies have become essential enablers of connectivity, allowing people to maintain their everyday lives while connecting in digital ways to which they have never been connected before. In this research, the investigator uses secondary data to obtain reliable results. This research methodology employs a descriptive qualitative approach to analyze diverse digital marketing and IoT. In addition, this study proposes a more straightforward method for predicting customer tastes and preferences on digital platforms.

A. S. D. Khaled
Al-Rowad University, Riyadh, UAE

K. M. K. Ismail Alshaketheep
Amman Arab University, Amman, Jordan

K. K. Agrawal
Marwadi University, Rajkot, Gujarat, India

D. K. Sharma
Department of Mathematics, Jaypee University of Engineering and Technology, Guna, Madhya Pradesh, India

A. Jain (✉)
Amity University, Jaipur, Rajasthan, India
e-mail: mittalankit.amity@gmail.com

V. Dhiman
Postgraduate Institute of Medical Education and Research, Chandigarh, India

S. Yadav et al. (eds.), *Proceedings of Second International Conference in Mechanical and Energy Technology*, Smart Innovation, Systems and Technologies 290,
https://doi.org/10.1007/978-981-19-0108-9_19

19.1 Introduction

Due to social distancing conventions and statewide lockdowns, the COVID-19 pandemic has inevitably resulted in an increase in the usage of digital technologies. People and organizations around the world had to adapt to new work and lifestyles. Businesses and individuals grow more digitally savvy as they embrace technology while working from home. It has been seen that most people have turned to the Internet and internet-based services to communicate, connect, and continue their jobs from home. When compared to pre-lockdown levels, utilization of Internet services has increased from 40 to 100%.

As a result of increased digitalization, more businesses and educational institutions adopt work-from-home policies (WFH). Block chain technology will become increasingly essential, necessitating design and regulatory studies. With a rise in digital presence, workplace monitoring and technostress issues will become more prevalent. Most organizations, whether in industry, society, or government, have experienced these shifts. Now employers' priorities have shifted. For example, organizations are now monitoring employees' health and wellness. COVID-19 cases are on the upswing, and IoT software solutions are giving a substantial hit. The COVID-19 epidemic has had an impact, especially on digital marketing. Now, even big-name brands collaborating with a digital marketing agency to develop more digital content than it's ever had to or a shift in the digital experience required by customers. The years 2020 and 2021 will be exciting in the world of digital marketing [1].

Digital marketing refers to marketing strategies that use the most up-to-date online media or communication-related technology breakthroughs to communicate with clients, particularly in the virtual world of the Internet. The disruption caused by COVID-19 has a significant influence which slows the business industry's growth. Hence, in this context, the need for IoT solutions has increased dramatically in digital marketing. Digital agencies will become even more crucial after the pandemic. It has been observed that digital transformation technologies such as Cloud, Internet-of-Things (IoT), Blockchain (BC), Artificial Intelligence (AI), and Machine Learning (ML) have been significantly introduced into businesses as part of their transformation efforts [2].

This research is being carried out in order to have a better understanding of concepts. The objective of the report is to recognize the importance of IoT and its aspects to understand the deeper insights. The research evaluates the role of IoT in digital marketing. Along with this, the paper also recognized digital marketing before and after COVID-19.

19.2 Literature Review

The Importance of Digital Marketing during and after COVID-19 [Internet]: Triton Commerce (2021) states, the pandemic has particularly affected marketing processes in both business-to-consumer (B2C) and business-to-business (B2B) markets due to relevant issues emerging in the management of physical sales channels and interactions with customers. Due to limited interaction with existing and potential clients, COVID-19 has significantly impacted the firm's competitiveness. As a result of the epidemic, businesses are turning to IoT to reform company operations, customer interactions, and marketing methods. In such a scenario, the online channel becomes the most critical place for customers to communicate with the company and conduct business [3].

As per the Marketing I, Grover V. Five Ways IoT is Changing Digital Marketing [Internet]. Martechadvisor.com (2021), Data-processing technologies like Big Data, Artificial Intelligence (AI), and the Internet of Things gave rise to a new wave of marketing technologies like a chatbox. All together opened up new options for value development for everyone engaged. The company offers in terms of marketing new products and services based on such revolutionary technologies [4].

Marketing et al. (2021), states IoT is a formidable technology with tremendous marketing possibilities. It will change the way customers interact with business and boost customer happiness. IoT technology is transforming how people interact with the business because it offers personalization and precise adjustment of technology to customer needs and preferences. The Internet of Things connects a wide range of devices in daily lives, and the amount of data generated by these devices will continue to grow. Marketers can use this information to generate actionable insights, spot patterns in customer interactions, and predict customer behavior and lifestyle. This data can assist marketers in extracting useful information in order to deliver a rich consumer experience. This also aids in visualizing customer preferences and connecting them to buy intent.

According to Ways IoT Is Changing Digital Marketing—IoT Marketing [Internet]. IoT Marketing (2021), Automation of operations is becoming more common, and its good impact on marketing cannot be overstated. The Internet of things deserves credit. Marketers had to spend a lot of time and money obtaining client data before the Internet of Things. Nowadays, marketers can automate acquiring marketing data rapidly with little effort and at a minimal cost. People are accustomed to advertisements throughout the day; however, the Internet of Things (IoT) provides the much-needed assistance that digital marketers have been seeking for years—highly relevant commercials. IoT devices allow marketers to have a more detailed understanding of their target demographic [5].

In accordance with Top 7 Ways IoT is Transforming Digital Marketing in 2020 and Beyond [Internet]. Medium (2021), IoT devices provide a vast amount of data that assist marketers in expanding business prospects while reducing risk. Marketers can better understand their customers' needs via this digital platform, including smart devices. Based on acquired data, various tools can be used to learn about your consumers' everyday lives.

Further, customers are the lifeblood of any business in today's digital environment, and the capacity to contact them at numerous stages is critical. With the help of IoT-connected gadgets, communication between marketers and consumers is improving. Sensors predict problems before they appear. Hence, customer satisfaction can be enhanced because the issue is fixed before it leads to arise damage. The Internet of things aids in product quality improvement and forecasting demand, and planning for new items. Data from IoT devices will assist businesses in making better, faster, and more informed decisions [6].

19.3 Research Question

Q1. What are the impacts of the Internet of Things and its aspects on digital marketing due to COVID-19?
Q2. How digital marketing operated before and after the pandemic?
Q3. What are the tools of IoT which facilitate digital marketing?

19.4 Methodology

The author describes the research methodology, data collecting methods, and research procedure in this section. The researcher in this study collects secondary data to come up with credible conclusions. This study uses a descriptive qualitative technique to examine several aspects of digital marketing and IoT. The goal is to categorize characteristics and build statistical models to explain what's researched in this paper. The researcher reviews the literature on the subject in order to develop research in the areas of digital marketing and IoT technology.

19.5 The Critical Role of IoT in Digital Marketing

The Internet of Things is having a significant effect on the digital marketing industry. By making it easier to listen to customer requirements and respond appropriately, IoT technology assists businesses in making their products and services more appealing to consumers. IoT marketing will have the most impact on three areas, and they are as follows:

E-commerce—IoT quickly targets individuals by advertising with mobile devices and regular Internet access.

Customer Channeling—IoT marketing makes a merchant's or brand's e-commerce services available on any device a customer has, whether at home or work.

Big Data—With loads of customer interaction data, past Internet advertising provided a deluge of big data for marketers, mining consumers from many aspects while taking their interests and routines into account. Hence, the Internet of Things is set to become a critical supporting tool for online marketing, generating more data directly from IoT devices.

Hence, IoT allows firms to collect more data from customers; it helps them promote their products and services more effectively. This information is valuable because it provides insights into consumer behavior, making it easier to promote products and services. Businesses will better understand their customers' current state in the buying process through data obtained by IoT. The Internet of Things allows companies to build entirely new experiences by integrating the digital and physical worlds. In addition, with IoT, digital marketers can target customers with promotional messages and personalized ads based on personal preferences. Digital marketing engages consumers, saves time, and money.

IoT not only benefits brands but the consumer as well. Due to COVID-19, everybody is restricted from going out to buy groceries and other essential products. In such a scenario, IoT helps to find online stores for fulfilling their needs, such as smartphones. This pandemic teaches the value of health and well-being. Hence, people can easily monitor their sugar level, BP, oxygen through IoT devices.

19.6 Digital Marketing Before COVID-19

Brands, particularly those that operate primarily in the physical rather than online world, had to transform their websites into industry knowledge hubs, delivering content that aided the user at every stage of her customer journey, from research to awareness to advocacy. Figure 19.1 shows that the role of IoT in digital marketing.

According to data, customers preferred to buy items physically before the epidemic. Still, that preference had to be set aside when governments throughout the world were obliged to lock down entire countries in many cases to prevent the spread of the coronavirus.

The lockdown of 2020 forced people and markets to stay digitally if they want to connect with the outside world. Digital has evolved into a solution, a way to meet fundamental requirements such as food, as well as interaction with friends and family, entertainment, and the procurement of knowledge, goods, and services that were previously only available in person. COVID-19 is the first significant worldwide issue since the mainstreaming of digital marketing; a situation like this before, with easy access to data on how customer behavior is changing and evolving. Marketers rearrange and reassess their activity and budget allocation thanks to a massive amount of data virtually daily from numerous sources. Before COVID-19, measurable and popular digital marketing trends such as video marketing, influencer marketing, micro-moments, and engagement-based email marketing gained traction [4].

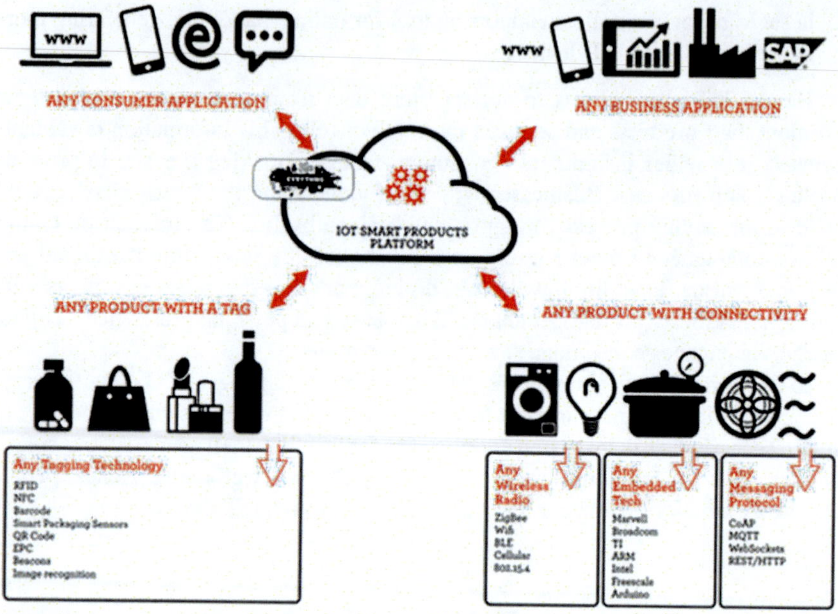

Fig. 19.1 Role of IoT in digital marketing. *Source* The Internet of Things in marketing: the integrated marketing opportunity [Internet]. i-SCOOP. 2021

19.7 The New Era of Digital Marketing

Indeed, people were already using digital media and marketing all day, every day before COVID. However, COVID-19 has only increased the urge to interact with other people and brands online. It has been found that COVID-19 forced businesses to close their physical locations. The days of a customer walking to a store to purchase an item are long gone. Many entrepreneurs are being compelled to reconsider their business tactics, including marketing outlets. Digital marketing became vital in ways they never expected. With the help of IoT technology, purchasing can now be done in a matter of seconds, with just a single click. An essential aspect that IoT brings to digital marketing is direct communication with customers via smart devices [7].

In addition, IoT advertising will allow companies to contact customers after they have purchased a product and "guide" them through its use to guarantee that it is effective. Now, marketers will be able to offer other products and services that are related to previous purchases in the future. Offline advertising is unable to track this degree of specificity, but the Internet of Things can. Brands now can get creative when reaching out to their target audience by developing solutions that stimulate social interaction online. IoT collects data, accumulating a range of information by watching customers, allowing marketing analytics to determine the types of campaigns that customers are most responding to and what conditions and patterns influence a customer's purchase decision. Therefore, IoT data will be used to feed

Table 19.1 Effectiveness of digital channels

S. No.	Digital marketing channels	Its impacts (%)
1	Mobile advertisement	60
2	Social media platforms	70
3	Banners	20
4	TV/Desktop	40
5	Videos	50

Source Self-made

machine learning algorithms, improving predictive analytics capabilities. Table 19.1 shows that the effectiveness of digital channels [8].

In this new era of digital marketing, many businesses are adopting a new strategy with the help of IoT. For example, when the pandemic forced one of the world's largest brands to look for a replacement for their annual Christmas PR live event, which shows products and gift items to the media, they went to a digital solution. Amazon's events team builds an interactive virtual experience that captured the atmosphere of the actual event as well as the spirit of the season.

19.8 IoT Tools for Digital Marketing

Companies that adjust to the new reality the fastest and most adaptable will gain a significant competitive edge and become industry leaders. The Internet of Things is transforming a wide range of digital marketing. IoT technology, in particular, has an impact on network traffic, data volume, and how people handle both. IoT also has a significant effect on everyday life for the customer, as gadgets may communicate data to the web and back to product producers in real-time. Just a few examples of Internet-of-Things (IoT) tools are Smart Watches, Smart Phones, Smart Cars, and more. These are becoming a part of ordinary people's lives [9].

It is a fact. Smartphones are the primary devices that marketers specifically target. This is the way consumers connect with brands and markets. Digital marketers are the main contributor to make changes that will facilitate the brand and industry together. The Internet of Things is a consumer "blueprint" full of data, habits, and preferences that will be utilized to improve marketing strategies. Some consumer demands and aspirations, such as medical access for baby boomers and the aging population, are predictable, whereas other consumer behaviors appear erratic.

Marketers will be able to predict when a customer will shop at a given store, dine at a specific restaurant, attend a cultural event, or visit the gym with the help of IoT. This will be valuable data to assist companies in their marketing efforts. Hence, Smartphone is an effective medium to connect peoples with markets and brands [10]. The primary goal of IoT is to generate real-time data, analyze it, and then use that data to generate various business outcomes.

Fig. 19.2 IoT tools

Figure 19.2 show that the IoT Tools. The purpose of combining IoT and big data is to deliver reliable business insights and analytics to enable digital marketers to target their audience and increase engagement. Companies are incorporating tools such as IoT sensors into their machinery in order to acquire critical operational data. This allows them to take a closer look at how their products are working and anticipate their consumers' future demands. Everyone knows how necessary customization is to the success of campaigns [11].

19.9 Findings/Result

The researcher looked at numerous literature studies to learn more about digital marketing and IoT technology. The findings suggest that understanding purchase patterns and decision-making processes were easily influenced by digital marketing. According to the results, marketers utilize digital marketing to boost brand reputation. Furthermore, Table 19.2 show that the benefits of IoT in digital marketing the survey discovered that marketers use IoT technologies to promote products and services in the market. Below is given how IoT helps digital marketers during COVID-19:

19.10 Conclusion

The objective of the current paper is to identify the influence of IoT in digital marketing in COVID-19. The COVID pandemic has thrown the usual way of life into disarray. This lifetime occurrence has shaken almost every aspect of human life,

Table 19.2 Benefits of IoT in digital marketing

S. No.	Benefits of IoT in digital marketing	Effectiveness (%)
1	Acquire new customer	55
2	Retin customers/Potential leads	54
3	Effective digital marketing and campaigns	40
4	Better advertisement promotions to drive sales	45
5	Maintain engagement with the external and internal audience during lockdown	50
6	Increase brand value	40
7	Staying Top of mind in customers/prospects	30

Source Self-made

from social interactions to corporate finance. It has been discovered that marketing is becoming increasingly data-driven, and the gold mine of Internet of Things data is enticing marketers. Business and technology will continue to evolve. Those in the advertising industry should learn about IoT technology in order to optimize the value of data gathered from web-connected devices. IoT will enrich the marketing industry in the months ahead, according to the trends. Those who can adjust to the change will be successful. Those who refuse to change will eventually lag behind.

Before the Internet of things, analyzing customers' behavior was not an easy task. IoT discloses product consumption patterns, allowing digital marketers to estimate demand accurately. Data from IoT-connected devices can be utilized to gain a better understanding of customers' tastes and preferences. Digital marketers use IoT analytics to determine when customers will need to replace products and what type they prefer, as well as purchase patterns, trends impacting purchasing patterns, and optimal markets where the product would sell the most. Overall, IoT gives digital marketers with more reliable data than any other channel.

As a result, digital marketers are able to effectively promote their products with more clarity and understanding. COVID-19 has exposed gaps in organizations' digital sales readiness. According to the study's findings, consumer preferences and tastes in the digital era are constantly shifting, according to the study's findings as mentioned above. Digital channels are increasing the sales of any brand's products as the globe progresses toward the digital era. As a result, the current study sought to determine the influence of digital marketing on client purchase decisions.

19.11 Future Work

This study examines the role of digital platforms in affecting customer attitude during COVID-19. Several potential research areas are being considered to encourage scholars to explore a wide range of phenomena when doing this study. Many alternative optimizations and experiments have been left for the future due to time restrictions. More investigation is needed, however, on each of these selections: norms and their interrelationships. As such, the customer responds to digital marketing and digital marketing future.

References

1. Jain, A., Pandey, A.K.: Modeling and optimizing of different quality characteristics in electrical discharge drilling of Titanium alloy (Grade-5) sheet. Mater. Today Proc. **18**, 182–191 (2019)
2. Jain, A., Pandey, A.K.: Multiple quality optimizations in electrical discharge drilling of mild steel sheet. Mater. Today Proc. **8**, 7252–7261 (2019)
3. Jain, A., Yadav, A.K., Shrivastava, Y.: Modelling and optimization of different quality characteristics in electric discharge drilling of titanium alloy sheet. Mater. Today Proc. **21**, 1680–1684 (2019)
4. Lata, E.: Digital marketing during COVID-19 times: data-driven insights [Internet]. Search Engine Watch (2021) [cited 29 June 2021]. Available from: https://www.searchenginewatch.com/2020/06/10/digital-marketing-during-covid-19-times-data-driven-insights/
5. Marketing C, Media P, Development W, Outreach P, Marketing E, Social P et al.: The Internet of Things and the Future of Digital Marketing | Power Digital Marketing [Internet]. Powerdigitalmarketing.com (2021) [cited 29 June 2021]. Available from: https://powerdigitalmarketing.com/blog/the-internet-of-things-and-the-future-of-digital-marketing/#gref
6. Marketing me, Grover V. 5 Ways IoT is Changing Digital Marketing [Internet]. Martechadvisor.com (2021) [cited 29 June 2021]. Available from: https://www.martechadvisor.com/articles/iot/5-ways-iot-is-changing-digital-markeing/#:~:text=IoT%20means%20having%20more%20touch,purchase%20decision%20of%20the%20customer
7. The Importance of Digital Marketing during and After COVID-19 [Internet]. Triton Commerce (2021) [cited 29 June 2021]. Available from: https://www.tritoncommerce.com/blog/articleid/190/the-importance-of-digital-marketing-during-and-after-covid-19
8. The Internet of Things in marketing: the integrated marketing opportunity [Internet]. I-SCOOP (2021) [cited 29 June 2021]. Available from: https://www.i-scoop.eu/internet-of-things-guide/internet-things-marketing/
9. Top 7 Ways IoT is Transforming Digital Marketing in 2020 and Beyond [Internet]. Medium (2021) [cited 29 June 2021]. Available from: https://medium.com/@kiran.bhatt270897/top-7-ways-iot-is-transforming-digital-marketing-in-2020-and-beyond-a855101b6607
10. Panwar, V., Sharma, D.K., Kumar, K.V.P., Jain, A., Thakar, C.: Experimental investigations and optimization of surface roughness in turning of EN 36 alloy steel using response surface methodology and genetic algorithm. Mater. Today Proc. (2021). https://doi.org/10.1016/j.matpr.2021.03.642
11. Ways IoT Is Changing Digital Marketing - IoT Marketing [Internet]. IoT Marketing (2021) [cited 29 June 2021]. Available from: https://iotmktg.com/ways-iot-is-changing-digital-marketing/

Chapter 20
Enhancing Data Quality by Detecting and Repairing Inconsistencies in Big Data

Vinaya V. Keskar, Jyoti Yadav, and Ajay Kumar

Abstract In several industries in established countries, a phase of Big Data examination has begun. Big data entails huge amounts of data that are challenging to analyze using standard database and software approaches. When using big data applications, a technological hurdle arises when transferring data across several locations, which is highly expensive and necessitates a huge primary memory for storing data for computation. Big data refers to the transaction and interaction of datasets whose size and complexity transcend the ordinary technical capabilities of gathering, organizing, and processing data in a cloud environment. Expanding data metrics are overflowing into contemporary associations with growing developments in Internet technology. Because of a relentless era of data, data from various devices and channels, such as cellular phones, PCs, government documents, medical reports and web-based media, are increasingly misunderstood. In this article, we explored the anomalies in the banking sector attributable to big data technologies, credit card discrepancies and the manner in which the toolkit is used to assess the incongruity of a specific WAP (Wireless Application Protocol) instrument.

20.1 Introduction

Information are different kinds of data that are usually produced in a particular way. The product is divided into two main categories, namely projects and info. Projects consist of an assortment of data control instructions [1]. In this respect, let us know some fabulous truths, actually after a full comprehension of data and data science.

V. V. Keskar (✉)
ATSS College of Business Studies and Computer Applications, Savitribai Phule Pune University, Pune, India
e-mail: vasanti.keskar@gmail.com

J. Yadav
Department of Computer Science, Savitribai Phule Pune University, Pune 411007, India

A. Kumar
JSPM Jaywant Technical Campus, Pune, India

S. Yadav et al. (eds.), *Proceedings of Second International Conference in Mechanical and Energy Technology*, Smart Innovation, Systems and Technologies 290,
https://doi.org/10.1007/978-981-19-0108-9_20

In the concept of data, the word Big Data is used to describe data in or larger petabyte sizes. The term "big data" applies to the creation and usage of technologies that provide the ideal consumer with the right knowledge from a mass of data which has developed rapidly throughout our general population for some time [2]. In addition to handling the complexity of tracking gradually heterogeneous organizations, Big Data has now become a 5Vs spectrum, number, meaning, truthfulness and tempo, and it is not just the challenge to handle steadily expanding data volumes. In the past, Internet companies have started to spread, Big Data action strategies have evolved and data was viewed as a resource. There are also various benefits of big data such as cost savings, increased performance and enhanced dealing; thus, above and beyond data collection in computer systems, there is growing value of data. As far as data science is concerned, data science is a community consisting of realities [3]. In addition, economical, demographic, welfare and exhibition all have different consequences of details, which eventually provides different responses to results.

20.1.1 5 V's of Big Data

Volume: refers to the volume of data generated inside the frame, processed and run. The increase in the amount of data produced and processed is explained by the in-crease in it, but additionally by the need to use it [4, 5].

Variety: Refers to increasing data types managed by a system for knowledge. This replication allows relations and connections between the data to be multi-faceted. The range often describes the possible applications of a raw data.

Speed: Refers to the recurrence of data produced, processed and transmitted. The data exists by stream and should be continually broken down.

Value: There is one more V that reflects Value! After taking the 4 Vs into account. The mass of data without meaning is just terrible for the enterprise, even if you find it useful. Data itself is of little value or relevance except that it must be translated into something significant in order to distinguish knowledge.

Veracity: It applies to incoherences or vulnerability of data, which ensures data can be accessed often because it is challenging to monitor consistency and accuracy.

20.1.2 Applications of Big Data

With regard to data of the board, each application field has its characteristics and necessities. For the factors underpinning this analysis, further specifics have been selected in the most tested and renowned fields of application: enterprise and manufacturing, medical, legislative and public services, education and a rational study [6].

Data is generated into organized (in the relevant arranging as in databases), unstructured (multimedia, text) and semi-structured (XML archives) systems from various channels such as the social network, e-commerce, healthcare, etc. Utter first step is to resolve these data in order to clean them up.

Continuous Data Analytics: Real-time systems are done with instantaneous knowledge streams which involve accelerated processing in an extremely limited amount of time as judgments are taken when the time line is finished. When large data are converted, the investigator can in a few moments call for their findings from gigantic datasets [7]. Data need to be processed super rapidly in order to react constantly to evolving conditions. Email Analytics: the website text data to Exabyte has been accessed. More than 7 million pages are generated daily. This detail is broken down under the text analysis umbrella. AI, mathematical analysis and machine etymology are the texts of the inquiry. The text overview produces a retrospective that moves core features from individual or different archives. There have been two approaches that have been more evolved in the round: extractive and abstract methodologies [8].

Analytics of Multimedia: multimedia material contains sound, video and images. Multimedia content analysis extracts and acquires interesting knowledge. Multimedia synopsis, multimedia commentaries, multimedia ordinations and retrieval are the key subjects listed for exploring multimedia science.

Spatio-Textual Analytics: Spatio-textual comparability retrieves a sequence of publications in which items in each pair are spatially similar, as are textual measurements. Great spatial textual data, which allow for new strategies for questioning and performing the procedure on this new data form, is created because of the pervasiveness of GPS-enablement devices. The web actually obtains spatial calculation alongside its earlier textual measurements with the expansion of GPS and Internet creativity. Currently, social web applications like Foursquare, Flicker are generating a hundred space textual data and offer the opportunity to use this data in different applications. However, it also demands that processing methods be enhanced, so that the space-text data are tracked and applied through separate activities [9].

20.1.3 Credit Card Related Problem in Banking Sector

The three types of credit-card fraud can be completely categorized into: card-related fraud, trader-related fraud and web-related fraud. The below are different forms of payment card fraud:

Amount Takeover: This kind of misrepresentation arises if a fraudster improperly collects all classified person details. He/she illuminates the bank hanging on his private or company address, which is imitated as the real cardholder. Next he/she announced that his credit card had been misplaced and that his current address was being demanded to mail another card. He/she is issued the card and the suspect will carry over the record easily in that way.

Fake Card Theft (also referred to as Skimming): A fake, cloned and skimmed card has, or has been legally produced and then changed or reported, been printed or embellished, or encoded without approval from the card organization. False misrepresentation also entails skimming, a loop where the desirable strip of the cards is repeated on another card through electronic machines, without the details on the real cardholder. Retail sources—especially bars, cafés and gas stations [10]—can be used to skim.

Postal Fraud: Which arises in situations when a thief catches a replacement card sent by and uses a genuine card holder.

Fraud: Anytime someone unlawfully gains knowledge and frequently uses it for the intent of making new documents or beginning an exchange for the profit of an authentic user. Most burglaries of character occur disconnected like pockets, mail blockage, or the Junk Browsing.

20.2 Literature Review

Sk. Sk. Sk. Credit card fraud was one example of Ref. [11] the banking and financial sectors are experiencing significant problems in the form of cyber fraud. We used a one-class classification technique in big data paradigm to identify credit card fraud. In this article, we also introduced a hybrid architecture for the Spark Spark theoretical paradigm for Particle Swarm Optimization and the Self-Associative Neural Network, as indicated elsewhere.

Fraud has no clear trend [12]. You still alter your behavior; so we have to learn un-attended. Fraudsters learn about emerging tools to enact deception by Internet purchases. Fraudsters. The normal activity of customers is suspected by fraudsters and fraud habits easily shift. Fraud identification mechanisms therefore have to monitor Internet purchases using unattended learning, and certain fraudsters have once perpetrated fraud via online media and then moved to other strategies. This paper aims at (1) designing a model of the deep auto-encoder and a restricted Boltsmann system (RBM). This model is capable of reconstructing regular business transactions to identify irregularities in normal patterns. The proposed auto-encoder-deep learning (AE) is an unattended learning algorithm that applies context propagation by setting the inputs to fit the outputs. The RBM is split into two layers, the input (visible) and the unseen. We use the Google Library from Tensorflow to carry out AE, RBM and H_2O by way of deep learning in this investigation. The findings display the average square error, root mean square error and curve field.

Researcher [13] Credit card use has significantly increased because of rapid credit card growth. As a consequence, credit card theft and defaults have greatly risen on credit card owners and credit card firms. Credit card analysis was commonly used to diagnose irregularities in credit card purchases on the premise that a fraud trend will rely on the previous transaction. Unattended learning should not, though, neglect the likelihood of the fraudsters modifying their tactics depending on consumers' behavior. Three unsupervised approaches, including an auto-encoder, a one-class vector support machine and robust outlier identification, were presented in this review. The data collection used in this analysis is focused on actual credit card purchase data. After training the models, the success of each model was assessed due to the availability of the answer, fraud labels. In the manuscript we examined the success of these three approaches in depth. Standard transaction labels were used for training in one-class SVM and auto-encoders. The benefits of the Mahalanobis system relative to these techniques, however, are that there is no requirement for a training sticker.

There are a broad variety [14] of uses in the use of graphs to collect and present de-tails. These applications could be present in the identification of semanticized and systemic patters and graphs have been increasingly increasing for such applications. In this article we will present one of the most harmful credit card anomalies focused on such a definition. The pace of usage of credit cards has dramatically escalated with advancing banking technologies. In this region, the incidence of fraud has risen, and we are modeled on graphs to resolve such anomalies. The major benefit of the strategy is that while running simulations the device overload rate is decreased in order to detect fraud and thus the detection level is accelerated.

The exponential development [15] in the area of e-commerce and the boom in e-payment has made it extremely necessary for the Fintech sector to recognize online transaction fraud in real time. In order to resolve this problem, we present the TitAnt, a framework for detecting transactions fraud in Ant Financial, which is one of the world's largest fintech firms. The machine will detect fraud in only milliseconds in real-time online transactions. We present the issue description, extraction of functions, detection methods, device installation and deployment and analytical performance. Significant real-world transaction results have been used to illustrate the reliability and efficacy of the proposed method.

The credit card [16] has been potentially the most prevalent way of payment, on daily and on-line orders, and thus there are major rises in fraud connected with those transactions. Fraudulent transacting credit cards per year cost companies and customers significant financial costs, and fraudsters are actively finding innovative technology and ways to participate in fraudulent transactions. The prevention of fraudulent purchases is a big factor in the increased usage of electronic payments. Effective and reliable methods in credit card purchases are also important to identify fraud. This paper suggests a clever solution to fraud prevention utilizing a light gradient boosting machine in credit card transactions (OLightGBM). A Bayesian optimization algorithm is smartly combined with the proposed solution in order to change the parameters of a light gradient boosting machine (LightGBM). Experiments were conducted utilizing two real-world public credit card data sets consisting of fake transactions and legit sets to show that our proposed OLightGBM was successful to identify fraud in credit card transactions. The suggested method, focused on a compare to other methods utilizing the two data sets, outpaced the other techniques and obtained the most exact (98.40%) precision, recepteur-operating curve (AUC) region (92.88%), accuracy (97.34%) and F1 score (56.95%).

Irregularities in data identification [17] is a critical activity and is subject to many high-effect applications in areas including defense, economics, health and law enforcement. In recent years, various methods have been established to spot outliers and anomalies in unstructured arrays of multi-dimensional objects with graph data being omnipresent. Because artifacts are correlated over a long distance, a series of modern technologies for the identification of anomalies in graph data has been created. This survey is aimed at offering a summary of state-of-the-art anomaly detecting methods in figurative data, which is common, systematic and standardized.

A primary problem [18] confronted by major economic entities was the detection of fraud arising from a spike in expenditure for credit cards. This paper proposes a novel approach to the estimation of payment through credit card fraud that relies on isolation forest and local external variables. The approach proposed involves the corresponding phases: data set pre-processing, preparation and sorting, judgment con-verging and test review. In this document, two forms of algorithms are used as a teaching to demonstrate the conduct characteristics of the right and wrong transactions. To date, many scientists have established numerous methods to detecting and increasing these frauds. In this article, we propose an overview of python and its de-tailed experimental performance in isolation forest and local external factor algorithms. When analyzing the dataset, the Local Outlier Factor Algorithm demonstrated strong accuracy of the Isolation Tree.

The collection of characters [19] is considered very necessary to enhance the classification and identification method in order to recognize the credit card danger in large or high-dimensional details. Random Forest Classifier (RFC) is one of the most frequently employed sorting methods for massive data sets. RFC works well and aims to classify the most predictive traits that can substantially boost the efficiency of a classification model in the identifying danger of credit card. In this report, we propose to use the Random Forest Classifier and Vector Machine to detect fraud danger as a tool for enhanced credit card risk ID (CCRI). Our experimental findings indicate that in terms of classification efficiency over a wider data set the proposed algorithm sur-passes the Local Outer Factor, Isolation Forest and Decision Tree.

The method of detection [19] of irregularities is the discovery of accidental artifacts or occurrences in data sets, which are different from the usual. Anomalical identification of unlabeled data is also used compared to traditional classification tasks, taking into consideration only the internal context of the data collection. This challenge is recognized as unmonitored identification of anomalies. It is discussed in many functional areas such as intrusion detection, fraud detection, life science and medical sciences. There have been hundreds of proposed algorithms in this field, but there is still a shortage of universal comparative assessment and common publicly accessible data sets in the research community. In this survey, 19 separate unmonitored anomaly detection algorithms are tested in 10 different data sets from diverse fields of use.

These limitations are resolved. This paper seeks to shape a new, supported foundation for unexamined anomaly detection analysis through the publication of source code and datasets. Furthermore, this assessment demonstrates for the first time the benefits and disadvantages of the multiple methods. Apart from success of anomaly detection, device initiative is highlighted as well as the effect of parameter setting and global/local anomaly detection behavior. Finally, we give guidance on the selection of algorithms for typical activities in the modern world [20].

Financial IoT fraud [21] is a misuse of mobile transactions by robbery and credit card, which is intended as fraudulent money. The usage of the mobile network is illegal. The fast-growing problem of mobile and online transitional resources is financial fraud under IoT. In the real world, the identification of financial fraud under IoT is an extremely accurate method, as financial fraud causes losses. We have therefore re-searched financial fraud approaches, focusing on the advantages and disadvantages of each research, utilizing machine learning and the approach of depth learning, predominantly from 2016 until 2018. In addition, in contrast with artificial neural networks approach to fraud detection and analysis of vast sums of financial data, our approach has been suggested. Our proposed method involves the collection of functions, sampling and implementation of monitored and unattended algorithms to identify financial fraud and to process vast quantities of financial data. The final model has been verified by the latest 2015 financial transaction details in Korea.

20.3 Methodology

In the present research, we suggest a new approach to anomaly detection filtering and refining. We divide anomaly detection phase into two stages: filtering and refinement to deliver extremely reliable results for both performance and stability. First, by deterministic space partition, a limited number of anomaly candidates will be created in sublinear time (DSP). At this point, the algorithm effectively distinguishes normal instances and potential anomalies by removing visible, typically highly unified, normal instances. In the second level, density-based steps are then implemented as refining, which contributes to consistent and reliable final outcomes, but only on the candidates that are confronted with inconsistencies in relative time complexity. This approach often produces attributes that define the outlying degree. There-fore, dividing the phase of anomaly detection into two stages would allow us to profit from various approaches and produce better results with less time complexity (Table 20.1).

Table 20.1 Time–space complexity

Type	Time	Space
Statistical	Various	Low
Density	High	High
iForest	Low	Low
DSP	Low	Low

The second stage time complexity is O(s2), where s is the number of candidates for anomaly. More than 70% of normal instances could be extracted out in the first stage on the basis of our experiments. The total difficulty of time is thus poor. The two-stage method uses time and space as frequently as possible and guarantees the precise outcomes.

When the scripts are configured and the data sets have been downloaded, it's time for the scripts to be performed and the outcomes tracked. Every script's results can be compiled in an Excel table to track the algorithms, data sets, cases, attributes, the processor time, the maximum memory, the prediction rate and the false warning rate. Based on these observations, R may be used to construct linear regressions and diagrams to finish the time–space study that this research needs. For iForest and RF algorithms, the following code can be used to produce these linear models.

```
iForest.model<Im(iForest$CPUTime~iForest$Instances+iForest$Attributes)
RF.model<Im(RF$CPUTime~RF$Instances+RF$Attributes)
```

20.3.1 *Existing Outlier Factor Algorithm*

The CreditCards.csv Dataset is a dataset of Kaggle belonging to Google. This dataset contains 2,84,807 payment card purchase information, 31 separate data sets criteria or attributes. Local Outlier Element is used in the proposed model to measure the forest score of anomaly and isolation algorithm. The "Class," meaning 10,000, in the data collection, means that certain transactions are illegitimate, while "class" is 0.00000 for legitimate transactions, Python Software Language is used to create the models. The "Local Outlier Factor" is an unmonitored algorithm for the identification of outlines:

```
import numpy as np
import matplotlib.pyplot as plt
from sklearn.neighbors import LocalOutlierFactor

print(_doc_) np.random.seed(42)

# Generate train data
X_inliers = 0.3 * np.random.randn(100, 2) X_inliers = np.r_[X_inliers + 2, X_inliers -
2]

# Generate some outliers
X_outliers = np.random.uniform(low=-4, high=4, size=(20, 2)) X = np.r_[X_inliers,
X_outliers]

n_outliers = len(X_outliers)
ground_truth = np.ones(len(X), dtype=int)ground_truth[-n_outliers:] = -1

# fit the model for outlier detection (default)



clf = LocalOutlierFactor(n_neighbors=20, contamination=0.1)
# use fit_predict to compute the predicted labels of the training samples# (when LOF
is used for outlier detection, the estimator has no predict, # decision_function and
score_samples methods).
y_pred = clf.fit_predict(X)
n_errors = (y_pred != ground_truth).sum()X_scores = clf.negative_outlier_factor_

plt.title("Local Outlier Factor (LOF)")
plt.scatter(X[:, 0], X[:, 1], color='k', s=3., label='Data points')
# plot circles with radius proportional to the outlier scores
radius = (X_scores.max() - X_scores) / (X_scores.max() - X_scores.min())
plt.scatter(X[:, 0], X[:, 1], s=1000 * radius, edgecolors='r',
        facecolors='none', label='Outlier scores')plt.axis('tight')
plt.xlim((-5, 5))

plt.ylim((-5, 5))
plt.xlabel("prediction errors: %d" % (n_errors))legend = plt.legend(loc='upper left')
legend.legendHandles[0]._sizes = [10]
legend.legendHandles[1]._sizes = [20]plt.show()
```

20.3.2 Isolation Forest Algorithm

The Forest Isolation "isolates" findings by choosing a function arbitrarily and then automatically splitting the value into the maximum and minimum values of the specified element. The amount of splits needed to separate a sample is equal to the root node path length of the termination node. Recursive dividing will describe a tree. The average duration of this route provides one a metric of normality and the option. This algorithm can be pseudo-coded as:

```
import numpy as np
import matplotlib.pyplot as plt
from sklearn.ensemble import IsolationForest

rng = np.random.RandomState(42)

# Generate train data
X = 0.3 * rng.randn(100, 2) X_train = np.r_[X + 2, X - 2]
# Generate some regular novel observations
X = 0.3 * rng.randn(20, 2) X_test = np.r_[X + 2, X - 2]
# Generate some abnormal novel observations
X_outliers = rng.uniform(low=-4, high=4, size=(20, 2))
#fit the model
clf = IsolationForest(max_samples=100, random_state=rng)clf.fit(X_train)
y_pred_train = clf.predict(X_train) y_pred_test = clf.predict(X_test) y_pred_outliers
= clf.predict(X_outliers)
#plot the line, the samples, and the nearest vectors to the plane xx, yy =
np.meshgrid(np.linspace(-5, 5, 50), np.linspace(-5, 5, 50)) Z =
clf.decision_function(np.c_[xx.ravel(), yy.ravel()])

Z = Z.reshape(xx.shape)

plt.title("IsolationForest")
plt.contourf(xx, yy, Z, cmap=plt.cm.Blues_r)
```

```
b1 = plt.scatter(X_train[:, 0], X_train[:, 1], c='white',s=20, edgecolor='k')
b2 = plt.scatter(X_test[:, 0], X_test[:, 1], c='green',s=20, edgecolor='k')
c = plt.scatter(X_outliers[:, 0], X_outliers[:, 1], c='red',s=20, edgecolor='k')
plt.axis('tight') plt.xlim((-5, 5))
plt.ylim((-5, 5))
plt.legend([b1, b2, c],
["training observations",
 "new regular observations","new abnormal observations"],loc="upper left")
plt.show()
```

20.4 Result Analysis

Local Outlier Component calculates each sample's anomaly score in comparison to its neighbors, and measures the local variance or density of the sample. Isolation Wood-land, on the other hand, relies on the degree to which the data items are separated from the neighborhood by choosing the split value randomly from the dataset. The relation between the test outcomes of the two methods can be seen in Table 20.2. It is found that there is a great deal of skew between processing transactions when re-viewing and comparing the attributes in the dataset.

The non-parametric approach named mutual knowledge can be applied to measure the dependence between two attributes, which can document some kind of statistical reliance between variables. When reciprocal knowledge is 0, this does not mean dependency and a higher value implies stronger dependence between the variables. More training examples are present in the data collection, but shared knowledge is definitely the strongest.

Table 20.2 Comparison of test results of local outlier and isolation forest

Local outlier factor	Accuracy = 99.65417			
		Precision	Recall	Support
	Class 0	1.09	1.08	22,685
	Class 1	0.13	0.15	34
Isolation factor	Accuracy = 99.519			
	Class 0	1.06	1.06	22,685
	Class 1	0.18	0.19	34

20.5 Conclusion

In this article, we find the incoherence of credit cards where we choose the invalid information number based on this aspect. We defined measures to remove it using the WAP toolbox. Big information surveys are being upheld in many banking circles and enable them to move higher management levels inside and outside their customers. The authentication of the credit card number in the web application is both significant and important. These designs are checked and consistency measured on the same credit card data collection. XG Boost is outperforming both of these versions in terms of accuracy, precision and recall efficiency metrics. If the data collection grows more, the issue is that it may contribute to fitting issues. This may be seen as potential work in order to eliminate the challenge of overcoming deception from being identified in real time. Deep learning principles can be implemented better and more reliably in real time and can produce danger ratings. These models may also be used for calculating outcomes such as predictive harm. The usefulness of the danger score is dependent on the model that recognizes trend deviations, distinguishes matches to existing trends and recognizes new patterns.

References

1. Tan, H.: The Competitive Issues in Credit Card Markets (2020)
2. Wewege, L., Lee, J., Thomsett, M.C.: Disruptions and Digital Banking Trends (2020)
3. Rakhman, R.A., Widiastuti, R.Y., Legowo, N., Kaburuan, E.M.: Big data analytics implementation in banking industry—Case study cross selling activity in Indonesia's Commercial Bank. Int. J. Sci. Technol. Res. **8**(9) (2019)
4. Dospinescu, O., Anastasiei, B., Dospinescu, N.: Key Factors Determining the Expected Benefit of Customers When Using Bank Cards: An Analysis on Millennials and Generation Z in Romania. Received 25 Oct 2019; Accepted 21 Nov 2019; Published 25 Nov 2019
5. Meng, C., Zhou, L., Liu, B.: Journal of Physics: Conference Series on "A Case Study in Credit Fraud Detection with SMOTE and XGBoost"
6. Ashrafi, A.Z., Ravasan, P., Trkman, S.: Afshari, The role of business analytics capabilities in bolstering firms' agility and performance. Int. J. Inf. Manage. **47**, 1–15 (2019)
7. Lehrer, C., Wieneke, A., Vom Brocke, J., Jung, R., Seidel, S.: How big data analytics enables service innovation: materiality, affordance, and the individualization of service. J. Manage. Inf. Syst. **35**(2), 424–460 (2018)
8. Côrte-Real, N., Ruivo, P., Oliveira, T.: Leveraging internet of things and big data analytics initiatives in European and American firms: is data quality a way to extract business value? Inf. Manage. (2019)
9. Mikalef, P., Krogstie, J., Pappas, I.O., Pavlou, P.: Exploring the relationship between big data analytics capability and competitive performance: the mediating roles of dynamic and operational capabilities. Inf. Manage. (2019)
10. Gupta, M., George, J.F.: Toward the development of a big data analytics capability. Inf. Manage. **53**(8), 1049–1064 (2016)
11. Vadlamani, R., Kamaruddin, S.K.: Big Data Analytics Enabled Smart Financial Services: Opportunities and Challenges, pp. 15–39 (2017). https://doi.org/10.1007/978-3-319-72413-3_2
12. Pumsirirat, A.: Credit card fraud detection using deep learning based on auto-encoder and restricted Boltzmann machine. Int. J. Adv. Comput. Sci. Appl. (IJACSA) **9**(1) (2018)

13. Mashhadi, R., Mahdi, M.: Anomaly detection using unsupervised methods: credit card fraud case study. Int. J. Adv. Comput. Sci. Appl. **10** (2019). https://doi.org/10.14569/IJACSA.2019.0101101
14. Ramaki, A.A.: Credit card fraud detection based on ontology graph. Int. J. Secur. Privacy Trust Manage. (IJSPTM) **1**(5), (2012)
15. Cao, S., Yang, X., Chen, C., Zhou, J., Li, X., Qi, Y.: TitAnt: online real-time transaction fraud detection in Ant Financial. In: Proceedings of the VLDB Endowment, vol. 12, pp. 2082–2093 (2019). https://doi.org/10.14778/3352063.3352126
16. Taha, A., Malebary, S.: An intelligent approach to credit card fraud detection using an optimized light gradient boosting machine. IEEE Access **8**, 25579–25587 (2020)
17. Akoglu, L.: Graph-based Anomaly Detection and Description: A Survey. arXiv:1404.4679v2. [cs.SI] 28 Apr 2014
18. Veeramani, V., Divya, N., Sarojini, P., Sonika, K.: Isolation Forest and Local Outlier Factor for Credit Card Fraud Detection System (2020). https://doi.org/10.35940/ijeat.D6815.049420
19. Rtayli, N.: Selection features and support vector machine for credit card risk identification. Procedia Manuf. **46**, 941–948 (2020)
20. Goldstein, M., Uchida, S.: A comparative evaluation of unsupervised anomaly detection algorithms for multivariate data. PLoS ONE **11**(4), e0152173 (2016). https://doi.org/10.1371/journal.pone.0152173
21. Choi, D., Lee, K.: An artificial intelligence approach to financial fraud detection under IoT environment: a survey and implementation. Secur. Commun. Netw. **2018**, 1–15 (2018). https://doi.org/10.1155/2018/5483472

Chapter 21
The Integrated Framework of Environmental Internet of Things Based Recruitment Process and the Impact Created by IoT in Inventory Management

A. Apsara Saleth Mary, Sudhir Fulzele, Varsha Rokade, M. Z. M. Nomani, Sushma Jaiswal, and Ghazal Salahuddin

Abstract The internet of things (IoT) is focused in offering new technologies which is changing the business environment by showing the potential in different sectors like retail, transportation, healthcare etc. Through IoT, the managers can gather and process data and information in an effective manner, also they can use them to automate in order to enhance productivity and efficiency. With the usage of such technologies, human resource managers can access information for possible candidates which can be accessed through IoT. The overall behaviour of potential candidate tends to help them in shortlisting and choosing the desirable ones. The application of IoT has emerged in enhancing the recruitment of candidates in search of candidates based on large volume of information, assist in screening the candidates, selecting the most suitable ones. The application of IoT technology can modify the HR roles, enable in changing the company perspective in enhancing recruitment process, source the candidates effectively and select them so as to meet the organisational requirements.

A. A. S. Mary
Faculty, Centre for Tourism and Hotel Management, Madurai Kamaraj University, Tamil Nadu, Madurai, India

S. Fulzele
Dr. Ambedkar Institute of Management Studies and Research Deekshbhoomi, Nagpur 10, India

V. Rokade (✉)
Department of Management Studies, Maulana Azad National Institute of Technology, Bhopal, Madhya Pradesh, India
e-mail: deshpande.varsha@gmail.com

M. Z. M. Nomani
Faculty of Law, Aligarh Muslim University, Aligarh, India

S. Jaiswal
Department of Computer Science & Information Technology (CSIT), Guru Ghasidas Vishwavidyalaya, (A Central University), Koni, Bilaspur, C.G., India

G. Salahuddin
Department of Geography, Kirori Mal College, University of Delhi, Delhi, India

S. Yadav et al. (eds.), *Proceedings of Second International Conference in Mechanical and Energy Technology*, Smart Innovation, Systems and Technologies 290,
https://doi.org/10.1007/978-981-19-0108-9_21

The implementation of IoT driven technologies support in removing the biases in screening the candidates and aids in candidate selection with the support of other technologies like artificial intelligence, predictive analytics etc. Furthermore, IoT is also involved in better inventory management as it supports the organisation in converting the data which are sourced through RFID (Radio Frequency Identification) into more insights related to inventory data, location of the products, tracking the movement of goods until it reaches the hands of the customers. With the application of IoT inventory managers possess better control in managing the inventory through automated ordering request and thereby reduce manual time and cost for the business. This paper is focused in presenting a conceptual understanding on the application of environmental IoT in recruitment process and also the overall impact created by the technology in managing the inventory efficiently.

21.1 Introduction

In a rapidly changing environment, the conventional recruitment system may not be as effective as it were a decade ago. It should be regarded that human resource management is considered as the critical aspect for organisational sustainable development, it is highly important for any organisation across industries to identify, nurture and retain the best human capital for competitive advantage. With the cost of hiring high skilled, qualified employees it is vital for the management to look for different aspects in looking the right candidate for the job who supports in realising the goals of the organisation [1]. On a general parlance, recruitment is regarded as the process of identifying and choosing the right individuals for the job and enable them to channelise their efforts to realise their career and organisational goals.

With the advent of technologies in the HR domain like internet of things (IoT), business analytics, deep learning etc. organisations are using them in order to enhance their hiring strategies and look for best candidate for the job. IoT is considered as the process of collaborating various systems and process through online with the core idea of collecting the data, analyse them and create reports in order to take quick decisions. The increasing complexity of the new technologies is supporting the organisation to meet the rapidly changing business environment and also support in ancillary functions like recruitment, performance management, training etc. In the field of recruitment, IoT is focusing in strengthening the competitive edge by analysing the specific job requirements in the organisation and look for suitable candidates by analysing through different online job portals and social media sites, and short list the candidates based on the job description, this enables in saving more time and cost for the HR department, eliminate any errors or bias in the candidate screening process and enhance productivity for the recruiters.

The adoption of IoT in recruitment process has been implemented since 2018 where the hiring managers focused in sourcing various profiles who will be suited for the open position in the organisation, measure the values and attitude of the individuals, assess them by scanning through the job profiles and past experiences, skill set etc. [2]. Furthermore, the application of IoT technologies support the business in rediscovering the talent through screening the available talent pool and also look for opportunities to cross train the resources so as to use the available capacity, harness the skills and enhance productivity.

Furthermore, IoT is also focused in supporting the management to manage the inventory which is essential for company to hold necessary goods for supporting the manufacturing process and support the company in managing the cost efficiently. The HR managers and logistics heads uses the sophisticated technologies like IoT, Machine learning and other aspects for planning, creating better relationship with stakeholders, frame strategies and deploy the resources in an efficient manner. The IoT powered tools supports in avoiding unconscious biases, support in screening the candidate effectively and support the business to enhance the sustainability and development.

This paper is focused to analyse the integrated framework of environmental IoT in the recruitment process and also the impact in enhancing the inventory management for competitive advantage.

21.2 Review of Literature

IoT can enable in using the sophisticated technologies to perform required actions for supporting the management in various process and activities. Analytical IoT is mainly used in order to analyse the data and screen the information based on the set criteria provided by the organisation to perform certain functions.

In human resource functions, recruiting the right candidate is always the challenging task, due to dynamic nature of job market there are growing need for the company in selecting the right candidate for the job so that they support in realising the organisational goals and objectives [3]. In the modern business environment, the recruiters are investing more time, cost and effort in screening the candidates through different platforms so that they can approach the right candidate for the interview. However, with the advent of modern technologies like artificial intelligence, machine learning, IoT, business analytics etc.

Recruiting is considered as a challenging task in the HR process as the managers need to screen through numerous profiles which matches with the job description and then schedule the interview for further processing. The recruiters tend to post advertisements in the professional network pages, social media, online job portals for inviting suitable application, however this creates huge number of candidates being applied for the specific job. Hence, the managers need to evaluate the application in order to choose the most suitable candidate.

Moreover, there may be bias in choosing the candidates as it involves more human intervention, in order to overcome the specified challenges and make the recruitment process highly efficient through reducing the cost and time for the management [4]. The IoT powered digital tools enable the management to enhance the recruitment system, various tools are being applied which will enable the company to specialise and increase the hiring capabilities more efficiently.

Black et al. [5] has stated that the overall application of IoT tools has enabled in transforming the recruitment process easier, furthermore the interaction with other tools like AI, deep learning, business analytics can support the organisation in screening the right candidates from the available or sourced application forms. This also enables the company in saving more time, lesser human intervention and increase productivity of human resource professionals in the organisation.

Similarly, the application of IoT is supportive in inventory management as it enables in keeping the supply chain system more agile, it enables in analysing the growing demand for specific products, create better supplier engagement and manage the networks for enhancing the movement of goods from production to point of sale. The growing need for IoT tend to influence the management so as to utilise the available space in inventory management, by moving into just in time (JIT) approach, agile supply chain which enables in holding lesser inventory for the company and enable in saving the carrying and ordering cost, reduce real estate space and efficiently manage the supply chain network.

21.3 Research Methodology

This paper is focused in applying the overall framework of internet of things (IoT) in the recruitment process and also to analyse its impact on the inventory management system. The descriptive research design is been applied so as to apprehend the importance of the technologies in supporting the management to support in different domains like human resource, supply chain, production management etc. The authors have focused in sourcing the secondary data from various sources like Google scholar, Ebsco Host and other sources.

The main keywords used were Internet of things, recruitment, human resource analytics, inventory management etc. The authors have intended to use only the peer reviewed, double blinded articles which are published, the study is mainly involved in performing qualitative aspects of the application of IoT in the recruitment process and also in inventory management. More care and focus are taken in selecting the articles and the criticality of its application were analysed and interpreted in the study.

21.4 Discussion and Analysis

This section provides critical discussion and understanding on the integral framework of IoT on the recruitment process in the organisation and the impact which it provides on inventory management.

21.4.1 Connecting the Dots—Application of IoT in Human Resource Management

The increase in usage of IoT has facilitated the business enterprises to connect, analyse and measure the data for taking informed decisions. The application of IoT in recruitment process has enabled in lowering the cost of hire. Since it is widely stated that the data about the prospective candidates is mainly stated in the subjective aspects, also the candidate tend to mention about the skill set, work experience, age and other demographics enables the management to gain access to screen the candidates quickly [6]. These technologies have immensely supported the organisation in tracking and measuring the job profile of the candidates with the open position in the organisation, thereby saving time and cost for the recruiters (Table 21.1).

The IoT enable in creating real time engagement with the candidate, make them be more aware of the job profile, their preliminary interest and other specific aspects. The implementation of chatbot enable the candidate and company to interact freely and support in asking the feedback and required data from the candidate. The data analytics tools enable the HR managers to take better decision making, predict the future organisational requirement in relation to manpower and use the available data for choosing the right candidate. Also, IoT helps in mapping the skill set of the internal candidate, skill level, competency basis and other which may match with the available jobs and support in hiring the internal candidates.

The business managers predicts that IoT is a way to support the recruiters in getting better insights on the candidate profiles, enable in saving more time and effort on manual process, integrate huge volume of data and apply prescriptive analysis for generating key reports to aid in decision making [7]. These tools enable in creating and implementing intuitive solutions to support in screening the application skills, measure their social behaviour, communicative and interaction skills, generate better online interview with the candidates. The IoT solutions also support in accessing the information about the employees and measure them against the job requirements which will enhance their easier mode of choosing the right candidate either internally or externally. The application of IoT not only supports the organisation but also creates a better recruitment industry which enhances competition and decrease cost for the companies while hiring the candidates.

Table 21.1 Connecting the dots in recruitment process

Advanced categories	Role of IoT
Clear communication	Quicker transmission of data and information, web calling facilities and easier connectivity
Application of analytical tools	Support in implementing business analytical tools for quicker and precise analysis
Cloud computing	Supports cloud computing which can enable in storing, retrieval of data at any point in time

Source Geetha and Bhanu Sree [6]

Table 21.2 IoT on inventory management

Key attributes	Impact of IoT
Precise tracking	Supports in precise tracking of movement of goods
Enhanced communication	Enable in communication from the procurement till it reaches the hands of the customers
Touchless data collection	Easier and effective means of data collection through scanning aspects like RFID
Agile warehouse	Focused in creating agile warehouse

21.4.2 *Influence of IoT on Inventory Management*

Real-time data collection with IoT is a relatively new wave generation technology. Early IoT users tend to set goals to drive initiatives with clear revenue and competitive advantages. Manufacturing, consulting, business services and distribution/logistics represent tough and competitive industries where revenue growth is often difficult to achieve. Competitive, revenue-generating IoT initiatives are the fastest and most likely to receive funding and support. Stock defects can be costly [8]. Fixing incorrect transactions and errors requires more work, which means unnecessary costs. IoT devices can reduce manual labour, errors, increase processing speed and increase overall storage performance. Companies install IoT sensors in their warehouses to monitor the movement and use of equipment and other devices on the premises. Companies also use sensors to transmit inventory information in real time to their management system. IoT and intelligent inventory management help eliminate costly and time-consuming errors, ensuring that inventory levels and equipment locations are clearly known and monitored continuously (Table 21.2).

Although it has always been critical, real-time warehousing has never been more important. This is largely due to the increase in the online shop. Today's consumer expects orders to be viewed, clicked on and received in days, not hours. Today, it is important that consumers have accurate, real-time information on product availability and expected delivery dates. This will help reduce the number of products in stock and improve customer service. IoT devices use GPS and other technologies, including radio frequency identification (RFID) chips, smart devices, and mobile sensors to track and authenticate products and transportation [9]. The large amount of data now available across IoT devices, mobile devices and resellers provides real-time visibility in stock. RFID helps improve inventory management by eliminating the need to read barcodes or labels. Instead of tracking numbers, companies can now easily track their inventory and supply chain in real time [10]. Connected IoT devices enable companies to track expiration dates for goods, which greatly helps prevent destruction and waste. IoT devices also help protect against theft and loss of products by enabling companies to track the location of their products.

Intelligent warehouse management uses IoT tools to provide real-time visibility to warehouses, production centres and distribution centres. Strengthening existing inventory management systems with quality and data from sensors and IoT systems helps reduce inventory costs and inventory management errors [11]. When IoT data is integrated into ERP systems, they improve file storage and preventative maintenance, ensuring that the best possible data is available for daily business decisions [12].

21.5 Conclusion

The IoT app is designed to improve candidate recruitment based on a wealth of information, which helps to select the most suitable candidates. The application of IoT technology can change the role of human resources, change a company's perspective by improving the recruitment process, effectively researching candidates and selecting them to meet organisational requirements. Increasing the complexity of new technologies helps an organisation respond to a rapidly changing business environment and also supports additional functions such as recruitment, performance management, training and more. In recruitment, IoT focuses on increasing its competitive advantage by analysing the organisation's specific job requirements and finding suitable candidates through analyses across multiple portals and media sites, spend more time and money with the human resources department, eliminate mistakes or prejudices in the candidate selection process and improve recruitment productivity.

Recruitment is seen as a difficult task in the personnel process because managers need to review many profiles that match the job description and then schedule the interview for further processing. Recruiters usually post ads on corporate networking sites, social media, job portal portals, but this has resulted in a large number of candidates for the position. Therefore, directors must evaluate the candidacy to select the most suitable candidate. By using IoT for the recruitment process, rental costs were reduced. Although it is generally accepted that the potential candidate's data is mostly subjective, the candidate also tends to cite skills, work experience, age and other demographics that enable management to quickly access candidates. These techniques have greatly helped the organisation to track and measure the job profiles of job vacancies in the organisation, saving newcomers time and money.

References

1. Min, J., et al.: A comparative study of potential job candidates perceptions of an AI recruiter and a human recruiter. J. Korea Convergence Soc. **9**(5), 191–202 (2018)
2. Van Esch, Patrick, and J. Stewart Black (2019). "Factors That Influence New Generation Candidates to Engage with and Complete Digital, AI-Enabled Recruiting." *Business Horizons*, vol. 62, no. 6, pp. 729–739
3. Vedapradha, R., et al.: Artificial intelligence: a technological prototype in recruitment. J. Serv. Sci. Manag. **12**(03), 382–390 (2019)
4. Fernández, Carmen, and Alberto Fernández (2019). "AI in Recruiting Multi-Agent Systems Architectu Re for Ethical and Legal Auditing." *The Twenty-Eighth International Joint Conference on Artificial Intelligence (IJCAI-19)*

5. Black, J.S., van Esch, P.: AI-enabled recruiting: what is it and how should a manager use it? Bus. Horiz. **63**(2), 215–226 (2020)
6. Geetha, and Bhanu Sree Reddy D (2018). "Recruitment through artificial intelligence: A conceptual study." *International Journal of Mechanical Engineering and Technology (IJMET)*, vol. 9, no. 7, pp. 63–70
7. Sekhri, Alka, and Dr. Jagvinder Cheema (2019). "The new era of HRM: AI reinventing HRM functions." *International Journal of Scientific Research and Review*, vol. 07, no. 3.
8. Jain, A., Yadav, A.K., Shrivastava, Y.: Modelling and optimization of different quality characteristics in electric discharge drilling of titanium alloy sheet. Material Today Proceedings **21**, 1680–1684 (2019)
9. Jain, A., Pandey, A.K.: Modeling and optimizing of different quality characteristics in electrical discharge drilling of titanium alloy (Grade-5) sheet. Mater. Today Proc. **18**, 182–191 (2019)
10. Jain, A., Pandey, A.K.: Multiple quality optimizations in electrical discharge drilling of mild steel sheet. Mater. Today Proc. **8**, 7252–7261 (2019)
11. V. Panwar, D.K. Sharma, K.V.P.Kumar, A. Jain & C. Thakar, (2021), "Experimental investigations and optimization of surface roughness in turning of EN 36 alloy steel using response surface methodology and genetic algorithm" *Materials Today: Proceedings*,
12. Upadhyay, A.K., Khandelwa, K.: Applying artificial intelligence: implications for recruitment. Strateg. HR Rev. **17**(5), 255–258 (2018)

Chapter 22
An Evaluation of Company Economic Performance Using 5G Network an Integrated Management Framework

M. Sangeetha, Sapna Mathur, K. V. Pradeep Kumar, Yogendra Singh, Sushma Jaiswal, and C. Selvaraj

Abstract The fifth-generation mobile communication, commonly known as 5G network is now available in developed countries and many developing nations are also focusing to implement them so as to boost trade and commerce domestically and internationally. 5G is set to revolutionise technology in various devices covering smartphones, internet connectivity etc. and also capable of using the artificial intelligence applications at ease. This will enable in enhancing the overall performance of many business enterprises as they are highly connected through technology for managing the factors of production. Moreover, it is estimated that the overall investments in the 5G networks is expected to reach $1 trillion by 2025, which represents the comprehensive interests of the global mobile network operators. Furthermore, the 5G network when coupled with other technology solutions like internet of things (IoT), Web 2.0 and big data etc. tend to deliver large-scale economic performance for the organisation and enhance sustainable growth and development. The 5G network uses advanced ultra-wide bans which possess higher band breadth with lower energy costs, helps the organisation to consume less power, furthermore, the 5G network supports the business through easier connection, enhanced communication and perform task quickly and efficiently. The application of 5G can enable the organisation to reduce their cost as more operational benefits can be ascertained, support

M. Sangeetha
Shri Shankarlal Sundarbai Shasun Jain College for Women, Chennai, India

S. Mathur · Y. Singh
Management Studies, JSS Academy of Technical Education, Noida, India

K. V. Pradeep Kumar (✉)
Department of Mechanical Engineering, Ramaiah Institute of Technology, Bengaluru, India
e-mail: pradeepkv@msrit.edu

S. Jaiswal
Department of Computer Science and Information Technology (CSIT), Guru Ghasidas Vishwavidyalaya (A Central University), Koni, Bilaspur, Chhattisgarh, India

C. Selvaraj
Business Administration, Saveetha College of Liberal Arts and Sciences, Saveetha Institute of Medical and Technical Sciences (Deemed University), Chennai, India

S. Yadav et al. (eds.), *Proceedings of Second International Conference in Mechanical and Energy Technology*, Smart Innovation, Systems and Technologies 290,
https://doi.org/10.1007/978-981-19-0108-9_22

in servicing more clients, and manage the human resources in an effective manner for generating better profits. The management can enhance operational effectiveness through the implementation of 5G network, this enables in greater communication specifically in different departments covering logistics, supply chain, information technology and other related areas. It is also stated that the work place can be enhanced through adopting such technologies and thereby achieve the mission and vision of the organisation.

22.1 Introduction

The cellular wireless has been transformed significantly since its introduction in 1981 as 1G (1st Generation), there were constant updates every 5–6 years and in the past 30 years the mobile communication has transformed to 5G technology. The 1st generation was focused in offering mass market mobile communication, whereas the 2G was deployed to enhance interoperability and reliable communication which made SMS possible and easy. 3G was focused in offering high speed data transfer and also download information from the Internet and other sources. 4G was introduced to enhance the data capability, high speed, access online platforms using smart mobile and were easily available for masses. The current 5G technology is highly focused in creating most powerful mobile services with higher data capabilities, increased call volumes and more data broadcast.

The fastest and intelligent connectivity is being enabled by the implementation of 5G technology and is expected to generate nearly $3.6 trillion economic output and 22 million jobs by 2035 in the value chain [1]. This intends to translate around $13 trillion in economic value across different industries covering manufacturing, information and technology, retail, public services, constriction companies etc. In general, 5G is estimated to enhance application of more digitalisation across industries in the economy, offers scalability for business growth through implementation of artificial intelligence (AI), automation, robotics, augmented virtual reality etc. Also, the investments in 5G networks are expected to increase worldwide and estimated to reach nearly $1 trillion by the end of 2025 [2]. The industry experts are contemplating that the global investment cycle in this technology is expected to be longer than the 4G and hence coexistence of the previous version and the new version will be there until 2030.

When the new version of mobile network is coupled with the advanced technology covering internet of things, big data analytics, Web 2.0 etc. Furthermore, 5G tend to hold more potential in enhancing economic performance of the business across industries. For example, in a study conducted by Tech4i2 stated that 5G intends to support nearly 135,000 jobs in Switzerland and create economic value of nearly €42 billion by 2030 [3]. Also, from the European Commission Study, it is stated that the potential economic performance is forecasted at €140 billion and nearly 2 million jobs will be created in the European Union [4].

Table 22.1 Key factors of Different mobile technologies

Key factors	1–2G	3–4G	5G
Major innovation	Connecting call, low speed data	Moderate and above moderate Internet access which has a limit of nearly 10 Mbps	Very high speed is proposed with nearly 100 Mbps
Major applications	Telephony	Sharing online contents, making transactions and smartphone accessibility	Enhanced virtual and augmented reality, internet of things Web 2.0
Economic performance	Enhanced productivity growth through quicker call connectivity which reduced transaction cost Entrepreneurship opportunities	Enhanced innovation by the major companies through applying digitalisation and enhanced interaction between stakeholders Creating better access to markets by reducing the entry barriers in different sectors Governments are moving into digitalisation for faster processing	More innovation is mainly created through digitalisation among the various stakeholders of business Application of robotics, AI and automation

The following Table 22.1 shows the major innovations, applications and economic performance of different mobile technologies.

22.2 Objectives of the Study

The main objectives of the study are stated as follows:

1. To analyse the organisational economic performance through the application of 5G mobile networks
2. To apprehend the estimated benefits which 5G network offers to various industries
3. To provide future directions to the business in adopting the new mobile networks.

22.3 Research Methods

The research methodology enables in constructing the path for the researchers in order to perform the research in an effective way. The research methodology assists the researcher in formulating the problems and objective, apply necessary research design and techniques in order to achieve the stated objectives [5] (see Fig. 22.1).

Fig. 22.1 Research approaches. *Source* Prepared by the authors

In the current study, the researchers are interested in applying the exploratory research design which will enable in analysing the research question in an effective manner. This research is used in the areas where there are very less studies are available for references. 5G network has been implemented only in few countries and other countries are exploring its implementation in the coming years, hence very less data and information are available in relation to the topic, hence exploratory research design will enable in exploring more information on the topic and present in a comprehensive manner [6].

Furthermore, the researchers apply qualitative research approach as it involves in non-numerical investigation of the data. The researchers intend to provide critical information on the application of 5G networks on company economic performance through integrated framework, hence qualitative data is best suited. Moreover, the descriptive approach for performing the detailed study on application of 5G networks towards economic performance.

The research uses secondary data analysis for performing the study and to measure the economic performance through implementation of 5G networks, there are two major sources of collecting the data which are classified as primary data source and secondary data source. In the given study, the researchers' sources information from research journals, business magazines, companywide reports and other sources which are published and available. Moreover, secondary data sources enable in gathering additional knowledge and cost effective for performing the study.

22.4 Critical Discussion

This section presents a critical understanding on the evaluating of company economic performance through the application of 5G networks, the business leaders need to identify the key impact areas of using the 5G network, collect the necessary data and information, map the usage requirements and evaluate the cross-industry perspectives so as to enhance collaboration and unleash the opportunities in the organisation [7]. The following chart (see Fig. 22.2) depicts the potential economic performance of the organisation through 5G networks.

This section provides analysis on the critical evaluation on the economic performance of organisation by implementing 5G networks, the research is focused in analysing the potential impact on selected industries covering automobile, manufacturing and healthcare sectors.

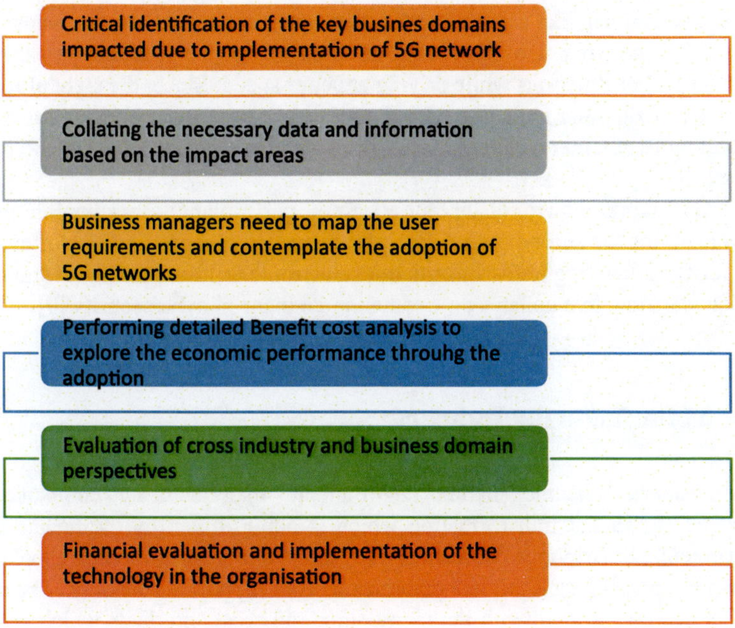

Fig. 22.2 The potential economic performance of the organisation through 5G networks

22.5 Automobile Industry

The application of 5G in automobile industry will be a great advantage for companies like Tesla, Ford, Toyota, General Motors etc. due to the change in the customer mobility behaviour. In the recent years individuals are contemplating in using different modes of transportation to complete their journey, the state of delivery is also considered as another key areas for the individuals to complete their movement from one place to another [8].

The car ownership is increasingly becoming burden due to climate change, higher competitive value and other aspects. Hence, governments and companies are contemplating in using autonomous electric vehicles which will support in realising the stated benefits. This new business model is stated as Transport-as-a-Service (TaaS). Furthermore, automakers are focusing in enhancing the technology on the self-driving cars, implementation of 5G networks in these companies will enable in enhancing the digital transformation in the auto industry. The business leaders of the auto industry have stated that the smart cars consume more bandwidth, enable in quick responses and demand continuous connectivity. Hence, 5G network supports the necessary bandwidth and reduces the latency issues, this supports the usage of smart cars in an effective manner. Moreover, the 5G technologies enhances wireless network capacity and data speed which allows the manufacturers and customers to use these services in an extensive manner, like using driverless cars, quicker connectivity for maps, directions, etc. The network also offers robust Internet connection for communication and performing other functions while driving at better speed. Since, the technology also collaborates with other tools like AI, robotics etc., it supports in making the driving safer and more efficient which reduces risk of accidents, make quick calls to protect the lives.

So, 5G networks tend to enhance the economic performance of automobile companies by enabling the autonomous rides, safer means of transport and also offers additional features covering traffic information, electronic tolling, warning of emissions, collision warning, live updates on weather and enhanced cybersecurity [9].

22.6 Manufacturing Industry

In the modern dynamic market environment, manufacturing companies are witnessing significant transformation which is being driven by the technological advancements. Smartphone companies like Apple, Samsung, Xiaomi and others are focusing in updating their product pipelines so as to support 5G network. The manufacturers need to keep up with the momentum and create products which supports such advanced technological devices. Table 22.2 shows the growth of Industrial revolution. Industry 4.0 is stated as the fourth revolution in the industrial sector which represents new stages of implementing smart machines, which can enable in implementing appropriate control through the use of technology, smart devices needs extensive network coverage and high speed bandwidth in order to function efficiently, also the expansion of Internet of Things (IoT) requires management to use higher spectrum in order to engage customers, receive their feedback and meet their needs and requirements 24 × 7.

Sensors and tags are mainly attached to the parts which enable in tracking their supply chain management, from the area of procurement till it reaches the hands of the customers. This enables to forecast the requirements and learn how the machines can be enhanced for increasing the efficiency and productivity. This will support in reducing the maintenance cost of the companies and also reduce the downtime by

Table 22.2 Growth of the industrial revolution

Industry 1.0	Industry 2.0	Industry 3.0	Industry 4.0
• Mechanisation and manufacturing of engines • Growth of textile industry • Steam factories and machine tools	• Electrification • Enhancing the product line • Production begins	• Growth of internet • Digital manufacturing • Robotics, Digitisation • Networking	• Autonomous machines • Cloud computing and smart technologies • Machine learning and Internet of Things

more than 40%. The 5G network supports the manufacturing companies in providing the necessary network connectivity which are highly essential for the manufacturing sector [10]. The network will support the top management as a chance to build more smart factories, like Lenovo creating better work place structure, deploy advanced tools like AI, Robotics, Augmented reality for analysing the issues. Moreover, the implementation of 5G technology support in digitisation of the business process, enhance connectivity among the functions, reduce latency and increase bandwidth. The new technology also supports the management in overcoming production issues and release mundane tasks.

22.7 Healthcare Industry

The 5G networks and service providers supports in enhancing the health platforms which integrates mobility and life support for patients. 5G technology also supports the doctors and nurses to possess quicker interaction with the patients, many companies like Novartis, Glaxo Smith Kline, Astra Zeneca are exploring ways to get in touch with the patients, provide them the medication and other supports to protect their lives. Also, the 5G enhanced broadband offers faster speed which will assist medical practitioners to have access to client's information for quick monitoring and diagnosis thereby save more time and cost [11].

22.8 Future Directions

Overall, a faster availability of 5G in emerging markets will require a coordinated effort between the public and private sectors, including private companies outside the telecommunications sector. However, these efforts cannot replace current reforms to improve the competitiveness of digital infrastructure. With the disruption caused by 5G, most mobile network operators would have extra revenue, mainly from the introduction of value-added products. In emerging markets, especially in low-income countries, specific product innovations depend on the development challenges facing each country. This was true for mobile financial services with the advent of 3G. For telecom operators, innovations in 4G products have so far been limited. With the advent of 5G, some telecom providers may need to partner with digital service providers to develop more uses and share the economic benefits. 5G will continue to evolve as companies enter the next phase, although it will take time to fully develop and deploy 5G networks. 5G is expected to grow rapidly after its launch in 2020, with coverage reaching just over a third of the world's population in 5 years [12].

The emergence of the autonomous electric fleet has enormous consequences for the transport industry, society and the automotive industry. 5G plays an important role in the implementation of electric vehicles and autonomous passenger transport. 5G enables autonomous vehicle networks to exchange data, communicate with traffic lights, road sensors, drones and more in milliseconds. In addition, freight trains, carriages and even airplanes may soon be on the horizon.

5G wireless also plays a key role in the growing number of consumer electronics technologies and companies and is changing the fundamentals of industries. The 5G wireless network enables companies to go up in the growth wave while investors, customers and employees are satisfied. Thus, the near future will be one of the most exciting periods of the lives in business, full of challenges, opportunities and risks [13].

22.9 Conclusion

5G networks and services will be gradually developed over the next few years to provide a platform for the development of new digital services and business models. 5G will be a turning point in the future of communications and provide high-performance connectivity to billions of devices. This allows machines to communicate in an IoT environment that can control an almost infinite range of services. As more and more devices connect and IoT usage grows exponentially, 5G networks will drive IoT's rapid growth and provide significant benefits to businesses and consumers. 5G networks revolutionize transportation and reliably connect patients and physicians around the world, providing better access to healthcare. As the digital transformation shifts the user experience from text, image and video to iconic virtual reality and augmented reality, 5G cellular technology facilitates this new shift by delivering high speed, excellent reliability, high capacity and low latency.

References

1. IHS Markit.: The 5G economy: how 5G will contribute to the global economy (2019). Retrieved from: https://www.qualcomm.com/media/documents/files/ihs-5g-economic-impact-study-2019.pdf
2. International Telecommunication Union (ITU): Setting the scene for 5G: opportunities and challenges (2018). Retrieved from: https://www.itu.int/dms_pub/itu-d/opb/pref/D-PREF-BB.5G_01-2018-PDF-E.pdf
3. Tech4i2 Ltd: 5G socio-economic impact in Switzerland (2019). Retrieved from: https://asut.ch/asut/media/id/1465/type/document/Study_Tech4i2_5G_socio-economic_impact_switzerland_February_2019.pdf
4. European Commission: Identification and quantification of key socio-economic data to support strategic planning for the introduction of 5G in Europe (2016). Retrieved from: https://connectcentre.ie/wp-content/uploads/2016/10/EC-Study_5G-in-Europe.pdf
5. Hilmola, O.P.: Research methodology and identifying key references used. Supply Chain Cases. Palgrave Pivot (2018)
6. Patel, M.: Exploring research methodology: review article. Int. J. Res. Rev. (2018)
7. Attaran, M.: The impact of 5G on the evolution of intelligent automation and industry digitization. J. Ambient Intell. Human Comput. (2021)

8. Kaur, S., Singh, I.: A survey report on Internet of Things applications. Int. J. Comput. Sci. Trends Technol. **4**, 330–335 (2016)
9. Hu, F.: 5G overview: key technologies. Opportunities in 5G Networks, 1st edn, pp 1–557. CRC Press, Boca Raton (2016)
10. Barreto, A.N., Faria, B., Almeida, E., Rodriguez, I., Lauridsen, M., Amorim, R., Vieira, R.: 5G-wireless communications for 2020. J. Commun. Inf. Syst. **31**, 146–163 (2016)
11. De Matos, W.D., Gondim, P.R.L.M.: Health solutions using 5G networks and M2M communication. IT Professional. **18**, 24–29 (2016)
12. Tariq, U.U., Ali, H., Liu, L., Panneerselvan, J., Zhai, X.: Energy-efficient static task scheduling on VFI-based NoC-HMPSoCs for intelligent edge devices in cyber-physical systems. ACM Trans. Intellectual Syst. Technol. **66**, 22 (2019)
13. Saha, R.K., Saengudomlert, P., Aswakul, C.: Evolution towards 5G mobile networks—a survey on enabling technologies. Eng. J. **20**(1), 87–112 (2016)

Chapter 23
The Future Impact of Technological Developments on Digital Marketing Through Artificial Intelligence

K. Guru, S. Raja, J. Sasiganth, Dilip Kumar Sharma, Mohit Tiwari, and Tripti Tiwari

Abstract Digital marketing includes a combination of marketing activity and implication of computer science to optimise the activity of advertisement and sales generation. In this study a comprehensive discussion is conducted to evaluate the aspect of technological advancement in Artificial intelligence and its implication in digital marketing activities. Different methods of digital marketing are discussed in this regard that facilitate modern companies to attain an effective customer engagement through direct interaction. SEO, PPC, Chatbots, SMM and other methods are found to be used extensively in recent times to optimise the marketing operation in the global market. Implication of AI is significant in this regard to enhance the aforementioned methods and improve the process of online marketing. AI facilitates the business entities in gathering customer's data at the time of searching about products. Brain modelling, time series prediction, image recognition and other techniques are used by AI to offer more relevant data to the business entities regarding customer behaviour and purchase intention. Augmented reality is another important feature in this regard that facilitates customers to see and feel products offered by companies that contribute to increasing the amount of sales volume in the targeted market. Such kind of technological advancement contributes in evaluating customers' perception

K. Guru (✉) · S. Raja
SRM Valliammai Engineering College, Chennai, Tamil Nadu, India
e-mail: guruvpm@gmail.com

J. Sasiganth
PG and Research Department of Management Studies, Sacred Heart College (Autonomous), Tirupattur, Tamil Nadu, India

D. K. Sharma
Department of Mathematics, Jaypee University of Engineering and Technology, Guna, MP, India

M. Tiwari
Department of Computer Science and Engineering, Bharati Vidyapeeth's College of Engineering, 4, Rohtak Road, Paschim Vihar, Delhi, India

T. Tiwari
Department of Management Studies, BVIMR, 4, Rohtak Road, Paschim Vihar, Delhi, India

S. Yadav et al. (eds.), *Proceedings of Second International Conference in Mechanical and Energy Technology*, Smart Innovation, Systems and Technologies 290,
https://doi.org/10.1007/978-981-19-0108-9_23

about the product and services that helps the marketer to implement suitable strategies to meet customer demands. It can be done by optimising the quality of customer services and products by differentiating it from others available in the similar market. Hence, AI influences to enhance the technique of product promotion to improve the sales volume in target market by effective technical advancements.

23.1 Introduction

Technological advancements lead to enhance business activities in a significant way by providing enormous opportunities to reach more customers with diversified products and services. Digital marketing includes utilization of online platforms and software to promote a wide range of products and services to global customers fast. It also facilitates gathering customer's feedback in a massive manner to improve the quality of existing marketing strategies. Artificial intelligence (AI) is an advanced digital tool that integrates cloud computing, network devices, computer and digital content of marketing. In this paper technological advancements in AI and its influence on digital marketing activities is discussed in a comprehensive manner.

23.2 Overview of Digital Marketing

Electronic devices and Internet platforms are key components of digital marketing that facilitate business entities to display products and make advertisements to increase sales volume. Different Internet channels such as websites, social media platforms, email and others are used to keep in touch with customers and increase their numbers [1]. Modern companies utilise AI software to optimize marketing activities, reduce turnaround time and overhead cost and improve the business output. An unprecedented technological growth is evident in this regard that enhances incorporation of robotic components in marketing activities. It contributes in enhancing interaction between business units and customers by means of different online channels. Thus, companies can experience notable profits by increasing their sales volume in the global market.

23.3 Methods of Digital Marketing

In today's commercial activities several digital marketing tools are utilised in an extensive way. Likewise search engine optimization (SEO), social media marketing (SMM), pay per click (PPC), content marketing, chatbots, affiliate marketing and others are the most used digital marketing methods in recent times [2]. SEO facilitates to increase visibility of a business unit on the online platform that optimises

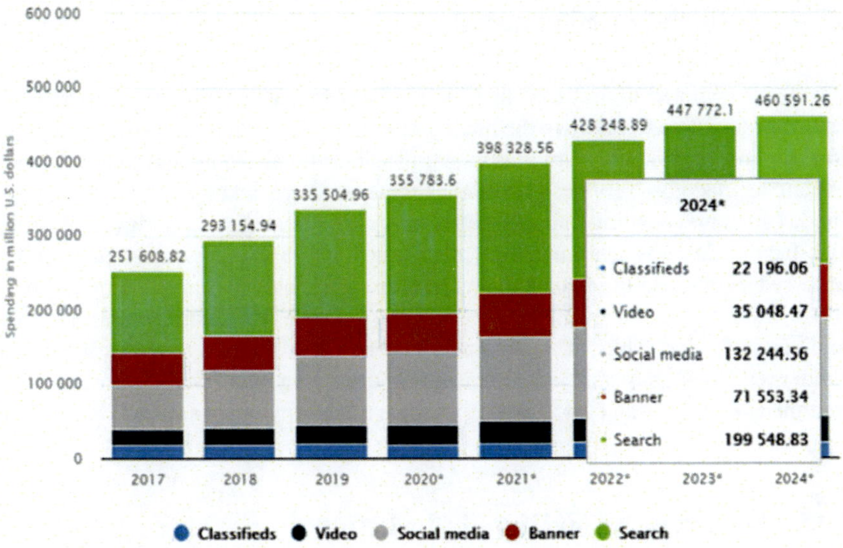

Fig. 23.1 Expenditure increase in video content of digital marketing. *Source* [3]

promotion of a brand among a wide group of customers. Among all other components of digital marketing video content is found to be more effective to attract customers as it depicts products and services in an alluring manner in front of the customers. It is estimated that a total of $ 35.05 billion will be spent on digital video contents to promote products and services globally in 2024 [3]. Chatbots are significant in respect of chatting with customers on a regular basis for ensuring smooth completion of an order made by a customer (see Fig. 23.1).

Semantic search is another important method that facilitates faster attainment of data searched by customers on a commercial website. It is mainly a machine learning technique that predicts information that can be asked by a customer based on the past search history [4]. AI based technology is also used in this regard to create natural language for transforming raw data of customer's query into readable report; that is termed as content creation. Based on the user's search history and additional information such as gender, geographic location and age, suitable advertisement contents are designed on digital platforms to reach targeted customers in an optimum way [5].

Predictive marketing is an AI based technique to collect behavioural data of customers and analyse the same to find any suitable interrelation between customers' preferences and digital advertisement through data mining [6]. Another advanced technique of voice search is also based on AI that facilitates users to search any complement on the commercial websites at a faster rate. Hence, it can be stated that a wide variety of methods are utilised by digital marketers to promote their products among global customers effectively.

23.4 Artificial Intelligence (AI)

AI is an integrated digital tool that incorporates human intelligence in machines such as computers to improve the quality of existing operations. This technology is also applied in marketing activities to optimise the advertising operations along with enhancing customer services [7]. It helps to detect current data trends in the global market for implementing suitable digital marketing strategies to mitigate probable risk factors. Moreover, a huge amount of sales and marketing documents can be analysed through AI to identify the reason for compliance failure [8]. Combination of AI with robotics can leverage future business activities by making it more scalable and cost effective. Thus, it is estimated that future business operations would be influenced by AI in respect of implementing marketing strategies, improving sales volume and optimizing customer services in a transformative manner.

23.5 Impact of AI on Digital Marketing

AI implication includes several techniques such as data mining, pattern resignation, natural language processing and others. Main advantages of this system include consistency in processes and conduction of all the activities by following a rule-based programme that leads to minimising the errors [9]. Machine learning is the main component used in AI based marketing that gathers all the data regarding a customer while someone browses any commercial website. Those data are then manipulated with computer programmes and suitable recommendations are created based on that data. An application based recommendation is sent to customers according to his/her interest and behavioural record. In this way companies utilise AI based machine learning techniques to improve brand positioning in the triggered market. Besides attainment of business goals; AI based digital marketing also facilitates companies to fetch suitable data regarding customers' experiences. However appropriate input data is required to generate suitable programmes, codes and loops to maintain high speed and accuracy of data mining through AI.

On the other hand, AI is also significant in changing the customer relationship management (CRM) activities by incorporating suitable software. For example AI based techniques can mitigate the problems that occur due to huge human intervention required by Zoho, Salesforce and others software to conduct CRM activities [10]. As AI facilitates self updation, auto correction and less human intervention an effective CRM is attained by modern business units optimally. For instance automatic notifications are sent to a customer's mobile prior to finishing the data pack for ensuring quick recharge. Discount offers are also sent to the user to enhance customer retention in a better way along with improving the sales volume. Hence, a better return on investment (ROI) is ensured by modern business units by making customers aware of all the recent offers and facilities offered to them through the digital platform instantaneously.

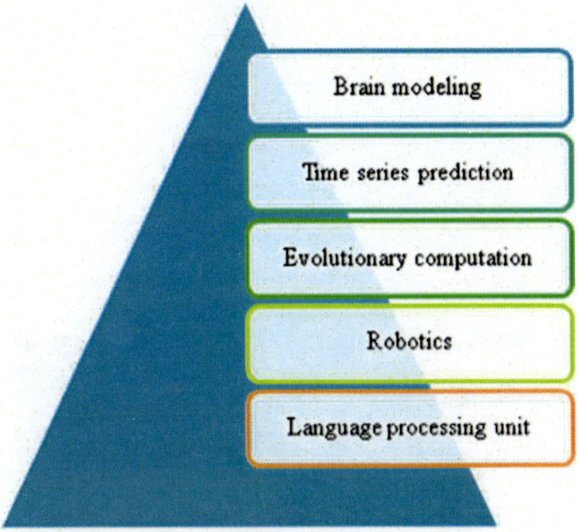

Fig. 23.2 Technological features of AI. *Source* [11]

Different components are used in AI based digital marketing that optimises the entire process of market research, customer evaluation and decision making to resolve problematic issues. Likewise, brain modelling technique is utilised by AI through neural networks that also helps in conducting a time series prediction regarding customer' behaviour [6]. Genetic algorithm is utilised in conducting an evolutionary computation through a genetic programming procedure. Another technology of vision is used to recognise objects and identify images. An intelligent control of all the marketing activities is conducted by robotics that also contributes to autonomous exploration. AI also helps in supporting decision making by means of an expert system (see Fig. 23.2) [11]. Another component of speech processing is used to recognise speeches and machine translation is conducted by a language processing unit. Thus, it is understood that a combination of several components facilitate effective utilisation of AI based platforms in digital marketing activities.

23.6 Changes in Digital Marketing by Implementing AI

Global business activities are associated with continuous innovation to survive in the competitive environment through process improvement. AI plays a significant role in this regard to collect suitable data from the customers and utilise it to product customizstion [12]. This can differentiate products offered by a company with other competitive business groups and leads to achieving the competitive advantage globally. AI is a scientific electronic data processing tool that helps in planning and controlling business data along with reporting it through appropriate management. It is termed as real time marketing that includes development of suitable

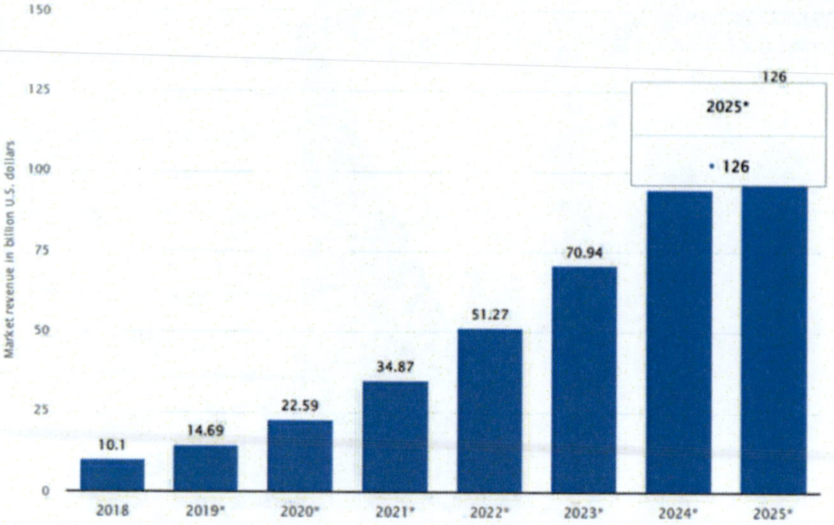

Fig. 23.3 Progression of implication in global marketing activities. *Source* [13]

marketing strategy based on electronically processed data obtained through online communication and digital information retrieval system.

AI offers a wide range of services to detect customers' feedback in the form of voice recognition, image identification, machine learning and semantic searching. Moreover information about latest technological updates is conveyed to customers through AI based platforms. As example, Siri technology of Apple and Google's Deep Mind are based on Natural Language Processing (NLP) for voice interpretation and providing quick response to customers accordingly [13]. Effectiveness of this system includes development of connections with customers rather than relying on predetermined behavioural algorithms. Hence, efficiency of the digital system also gets increased in terms of energy saving and cost effectiveness. Moreover, it also prevents data leakage from a system regarding marketing activities and leads to attain the competitive advantage in a sustainable manner. Therefore, AI based techniques are found to be used in global marketing activities in a progressive way through the recent period (see Fig. 23.3).

23.7 Technological Evolution of AI

A technologically advanced data driven approach is facilitated by AI to improve the process of marketing and decision making. It evaluates a huge set of online information regarding customer's behaviour and detects their digital identity [14]. Automated system is involved to analyse customer' profiles that eases the process of decision making to implement suitable techniques for meeting customers' demands in a more effective manner. SMM is another potential tool to gather customer's information and that data undergoes an analysis by utilising artificial intelligence. AI analysis of data ensures optimisation of information and provides the most relevant data to marketers that lead to an effective marketing activity globally.

23.8 Extant of AI Technology and Future Scopes of Innovation in Digital Marketing

AI is an effective tool to conduct statistical analysis of numeric data through the technology of machine learning. It is applied to optimize the pricing strategy aligned with competitive issues that lowering the price is significant to attract more customers. On the other hand, more profit can be earned by a business unit by increasing the price of products. Companies utilise AI to analyse a huge number of numeric data prior to setting the real time price of products. For example Tesco utilises the techniques of AI to gather customer insights and warehouse data prior to setting the price of products [15]. Non-numeric data can also be controlled by AI for effective lead generation to identify the ideal customer. Potential clients can also be identified by utilising AI techniques based on the gathered information. Hence, a significant amount of time can be saved in business to business (B2B) activities along with staffing and recruitment processes. Thus, marketers can get more time for performing sales calls and pitching to customers to promote their profits in the targeted market (Table 23.1).

Customer characteristics are also affected by adoption of AI techniques that can be ensured by utilising AI based chatbots to interact with customers in a natural language environment. Marketers use this platform in an extensive way in recent times due to the increasing rate of social media traffic occurring in private messaging platforms. AI technology is effective in respect of automated content generation that is utilised by global companies to create logos, advertisement scripts and attractive audio tunes [16]. Hence the maximum portion of advertisement generation is estimated to be done by utilising AI in future.

Marketers are benefited to use AI in PPC campaigns as it helps to identify new channels of advertising to promote products. As these channels may not be used by any of the competitors; it benefits the marketer to gain the competitive advantage effectively. Augmented reality is another main contribution of AI in digital marketing that offers better customer satisfaction [17]. As customers can both see and feel the products offered by a company they can understand features of the product prior to purchasing it. For instance Leskart's 3D Trial provides an opportunity to its customers to feel the frame through webcams. Hence, by providing an optimum level of service

Table 23.1 Evolution in the utilisation of AI

Utilisation of AI	
Controlling numeric data	Controlling non-numeric data
Pricing strategy	Ideal customer identification

to customers AI facilitates markets to promote their products and services in the targeted market along with achieving desired level of profit in the targeted market.

23.9 Conclusion

AI and its technologically advanced features include a set of innovative characteristics to attract a wide number of customers for improving the sales volume. This technology mainly utilises the strategy of developing a potential engagement with customers and making them familiar with global brands through direct and regular interaction. B2B marketing and allied activities are approaching complete automation by laying down the idea of incorporation of AI and robotic components in marketing operations. Certain components such as Big data, Internet of Things and Machine learning techniques are subsidiary processes of AI that contribute to strengthening digital marketing activities in a profitable manner. Thus, markets of the modern era prefer to utilise AI to get detailed customer information and take suitable steps to meet customers' demands in an optimum way for attaining the competitive advantage on the targeted market.

References

1. Murgai, A.: Transforming digital marketing with artificial intelligence. Int. J. Latest Technol. Eng. Manage. Appl. Sci. **7**(4), 259–262 (2018)
2. Schaefer, M., Hetman, O.: Effective tools of digital marketing implementation. Економічний вісник університету **41**, 67–74 (2019)
3. Statista Research Department: Digital advertising spending worldwide from 2017 to 2024, by format (2021). [Online]. Available at: https://www.statista.com/statistics/456679/digital-advertising-revenue-format-digital-market-outlook-worldwide/. Accessed on 9 Jun 2021
4. Liu, X.: Analyzing the impact of user-generated content on B2B Firms' stock performance: big data analysis with machine learning methods. Ind. Mark. Manage. **86**, 30–39 (2020)
5. Behera, R.K., Gunasekaran, A., Gupta, S., Kamboj, S., Bala, P.K.: Personalized digital marketing recommender engine. J. Retail. Consumer Services **53**, 101799 (2020)
6. Thomas, P.S., Theocharous, G., Ghavamzadeh, M., Durugkar, I., Brunskill, E.: Predictive off-policy policy evaluation for nonstationary decision problems, with applications to digital marketing. In AAAI, pp. 4740–4745 (2017)
7. Vishnoi, S.K., Bagga, T.E.E.N.A., Sharma, A.A.R.U.S.H.I., Wani, S.N.: Artificial intelligence enabled marketing solutions: a review. Indian J. Econ. Bus. 167–177
8. Yeğin, T.: The place and future of artificial intelligence in marketing strategies. Ekev Akademi Dergisi **24**(81), 489–506 (2020)
9. Qin, X., Jiang, Z.: The impact of AI on the advertising process: the Chinese experience. J. Advert. **48**(4), 338–346 (2019)
10. Banjac, I., Palić, M.: Analysis of best practice of artificial intelligence implementation in digital marketing activities. CroDiM: Int. J. Market. Sci. **4**(1), 45–56 (2021)
11. Verma, S., Sharma, R., Deb, S., Maitra, D.: Artificial intelligence in marketing: systematic review and future research direction. Int. J. Inf. Manage. Data Insights, p. 100002 (2021)
12. Brock, J.K.U., Von Wangenheim, F.: Demystifying AI: what digital transformation leaders can teach you about realistic artificial intelligence. Calif. Manage. Rev. **61**(4), 110–134 (2019)

13. Mustak, M., Salminen, J., Plé, L., Wirtz, J.: Artificial intelligence in marketing: topic modeling, scientometric analysis, and research agenda. J. Bus. Res. **124**, 389–404 (2021)
14. Gherheș, V., Obrad, C.: Technical and humanities students' perspectives on the development and sustainability of artificial intelligence (AI). Sustainability **10**(9), 3066 (2018)
15. Monti, M.: Automated journalism and freedom of information: ethical and juridical problems related to AI in the press field. Opinio Juris in Comparatione **1**, 2018 (2019)
16. Martınez-Plumed, F., Loe, B.S., Flach, P., O hEigeartaigh, S., Vold, K., Hernández-Orallo, J.: The facets of artificial intelligence: a framework to track the evolution of AI. In International Joint Conferences on Artificial Intelligence, pp. 5180–5187
17. Kane, G.C.: The evolutionary implications of social media for organizational knowledge management. Inf. Organ. **27**(1), 37–46 (2017)

Chapter 24
Order Winning Criteria: A Case of Small Manufacturing Sectors

Brijesh Singh and Manoj Kumar Lohumi

Abstract The small manufacturing enterprises (SMEs) comprise Micro, Small, and Medium manufacturing Enterprises. These have significant contributions to nation's economic and social development. These not only contribute to employment but also contribute significantly toward Gross Domestic Production, export and acts as raw material suppliers for big enterprises. More orders are needed to sustain and grow in this world. The investigations, based on 84 surveyed SMEs, reported that most of the orders are obtained directly from customers; and quality, product durability, and product cost are the influencing criteria. So, attention should be paid to improving these factors to improve the overall performance of manufacturing enterprise.

24.1 Introduction

Small manufacturing enterprises (SMEs) comprise Micro, Small, and Medium manufacturing Enterprises. These are one of the significant contributors in economic growth, innovation, and jobs. The development of SMEs is vital as they provide larger amount of jobs at lower investment compared with larger enterprises, help in development of rural and backward locations, and contribute in more equitable distribution of national income and wealth. In India, the SMEs contribute in a significant way to the growth of the country's economy with a vast network of about 63.38 million enterprises which contribute about 45% to manufacturing output, more than 40% to exports, over 28% to the Gross Domestic Production (GDP) and creating employment for about 111 million people, which in terms of volume stands next to the agricultural sector [1–3].

This is in line with the findings of Tahir et al. [4] who quoted that Malaysian SMEs consist of about 99.2% of total enterprises and contribute 32% of GDP. As per China Banking News, with respect to total enterprises, there are about 90% are

B. Singh (✉) · M. K. Lohumi
Department of Mechanical Engineering, Galgotias College of Engineering and Technology, Greater Noida 201 306, India
e-mail: brijesh.singh@galgotiacollege.edu

S. Yadav et al. (eds.), *Proceedings of Second International Conference in Mechanical and Energy Technology*, Smart Innovation, Systems and Technologies 290,
https://doi.org/10.1007/978-981-19-0108-9_24

SMEs and contribute to employment (about 80%), patents (about 70%), GDP (60%), and tax revenues (50%) [5]. Further, in the UK, at the start of 2018, about 99.9% of the 5.7 million enterprises are SMEs which contributed to 60% of all private-sector jobs. Hence, SMEs are very important to the UK's economy [6].

As elucidated in these reports, it is thus well acknowledged that SMEs are the backbone of a nation's sustainable economic growth. In a growing economy, the growth of SMEs is of utmost desirable. With this, the study is being done to analyze various order winning criteria that affect the growth of SMEs.

24.2 Review of Literature

World Bank states that formal SMEs contribute nearly 60% of total employment and approximately 40% of national income in emerging economies [7]. So, SMEs are significant contributors to economic development, job creation, and poverty alleviation [8]. Majama and Magang [9] added that SMEs contribute vastly to the nation's economy. Internationally, the consumer demands customized lower-cost products with high quality and reliability [10].

Manufacturing enterprises are being established to manufacture a product that is offered for sale to the customers that satisfies their needs. In return for this sale, the enterprise gets money (as return on investment) and arranges all the expenditure to run itself on profit. For maximum profit or earnings, sales must grow. More sales can be achieved by getting more orders (orders to supply manufactured products). Hill [11] introduced the concept of order winners and order qualifiers. As also discussed by [12], the criteria that an enterprise must meet as supplier—categorized under the order qualifiers and the criteria that help in getting the order over the rivals—are grouped under the order winning criteria. In other words, to provide order qualifiers, enterprises need only to be as good as competitors but to provide order winners they must be superior too.

SMEs are facing the problem of low-quality products [13] and thus to grow, manufacturing organizations are to be motivated to focus on improving the products and processes' quality [14]. Enlightening from the available literature and additional researches of [15–17], product durability, product quality, product range, product design, product cost, attractive packaging, promotional schemes, and early development of new products are considered as order winning criteria, to be analyzed in respect of SMEs. In this regard, a null hypothesis is being generated as:

H_N: Various sectors of SMEs do not agree on any of the order winning criteria.

24.3 Methodology

A survey was conducted to get responses through questionnaire that comprises of questions on five-point Likert scale, followed by personal visits. About 57% of participants were the managers while remaining were the owner or partners of SMEs. Descriptive Statistics is being used for analysis of responses and Chronbach's Alpha is being computed for reliability analysis.

Hypothesis has framed referring the available literature, responses, and prevailing situations. This framed hypothesis is being validated by computing the value of Cronbach's Alpha and tested statistically by using one-way ANOVA and independent sample t-test (p-value at 95% confidence level). Further, factor analysis (Varimax rotation method) and correlation analysis (Pearson's correlation coefficient) is being done to find the relationship among few identified variables and performance parameters.

24.4 Demographic Profile of Surveyed SMEs

In this survey, 31 glassware enterprises, 30 shoe/footwear manufacturing enterprises, and 23 lock manufacturing enterprises participated. A total of 17 enterprises reported having employees of less than 10, 29 enterprises were having employees between 10 to 49, 30 enterprises were having employees between 50 to 199, and eight enterprises were having employees between 200 to 250. Export was reported by 26 enterprises. Annual sales turnover up to 50 Lakh was reported by 15 enterprises, between 51 to 100 Lakh was reported by 12 enterprises, between 101–200 Lakh was reported by 19 enterprises, between 201 to 500 lakh was reported by 13 enterprises, and more than 500 Lakh was reported by 25 enterprises. Market share up to 1% was reported by 71 enterprises and 1–5% by 12 enterprises (one enterprise reported for having 11–15% market share). Over the past three years, sales growth up to 10% was reported by 54 enterprises, 11–20% by 28 enterprises, 21–50%, and no increase was reported by one enterprise only.

24.5 Findings of Survey

24.5.1 Receiving of Orders and Possible Changes

The sales of products are being accomplished by receiving orders either from the sales representatives, dealers (in case of dealer network), directly from customers that may be called as "Party" who purchases the products, and from a combination of these modes. The customer may be the end-user or the middle supplier. Many of the enterprises which are not involved in direct export are manufacturing products for those "parties" who are engaged in export or sell in country level market with their brand name. In such a case, the "party" does stress upon quality checks enormously. In this research, all such type of customers was considered under the head of "Directly

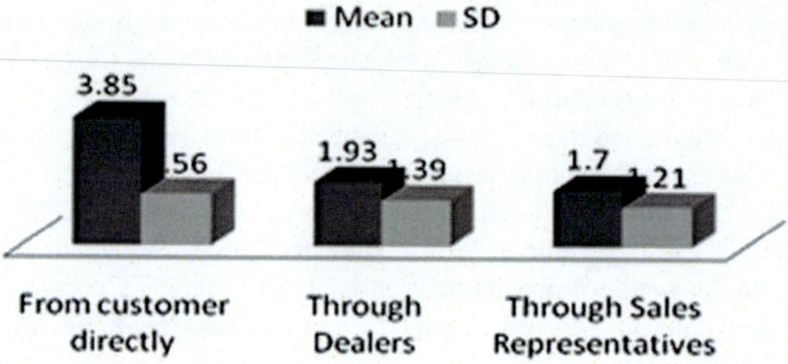

Fig. 24.1 Receiving of orders

through customers", as they provide their own drawings, designs, quality targets, etc. Refer to Fig. 24.1 for mean and SD values.

Referring Fig. 24.1, it is observed that the highest number of orders were received directly from customers (reported enterprises = 87%, mean = 3.85, SD = 1.56), followed by the order received through the dealers (reported enterprises = 41%, mean = 1.93, SD = 1.39) and the least number of orders were received by the sales representatives (reported enterprises = 34.5%, mean = 1.70, SD = 1.21). An enterprise can obtain the orders through more than one type of mode of order receiving. With these outcomes, it might be possible that most of the enterprises were producing products for other brands of repute. For example, many enterprises were manufacturing locks for Harison and Ramson brands. It might also be possible that the party was receiving the products and did export with its name, which is the common practice adopted by shoe/footwear enterprises. Here, the question arises that "when these enterprises are engaged in producing quality products for other brands/parties for exports, why they could not do so with their own name, including the export?"

Accepted orders, sometimes, need to be revised in terms of change in quantity, change in date of supply, change in address, and change of product and/or ultimately, its cancellation. Respondents were enquired about this situation. Regarding the cancellation of orders, about 70% of participants stated that they never entertain the cancellation of accepted orders. About 18% reported for seldom, 6% reported for doing so sometimes, 1% reported for mostly and 5% reported for always. When asked for change in quantity, address, and due date, 10% of respondents reported that they never accepted such changes, while 24% reported for seldom, 37% reported for sometimes, 24% reported for most, and 5% reported for doing so always. In a response to demand for other products, 56% of respondents reported that they never accepted the request for the supply of another product (change of the product), while 12% reported for seldom, 13% reported for sometimes, 13% reported for mostly and 6% reported for doing so always.

Those enterprises, whose products are not customized, producing for mass production, and have faith in their brand/product quality and reliability, can accept the change in the order at any time/stage including even the cancellation of the order, while others cannot. Enterprises having customized products, cannot accept the change in product type and cancellation of order. For example, in case of shoe/footwear enterprises, only change in due date, quantity (increase), and change in address is possible as always customized products are produced. The reason for this is that the design/pattern of stitching/look of upper are always revised to new. Sometimes cancellation of the order is cited because of poor quality, change in design, poor finish, exploiting rates, or any other defect/circumstances including the closure of business. To retain the customer and to remain in business, supply of changed product/cancellation of order should be entertained.

When asked for needed time to accept any change in order, about 16% of respondents reported that they could accept the change in order one day earlier, while about 74% respondents reported for the need of one week time and about 10% reported for the need of two weeks' time to accept any change in order. More time is required in case of shoe/footwear enterprises as procurement of raw material is based on the ordered quantity. The enterprises, whose products are not customized and maintain a large inventory of finished products, can accept the change of orders even before a day as their production is to stock (not for a particular customer). More time (of generally one week) is needed when orders are placed in large quantities. These enterprises are not having large inventories of finished products. Successful adoption of suitable Advance Manufacturing Technology (AMT) offers flexible and customized production with shorter time at a lower cost of production that enables least effects on business owing to cancellation/modification of orders.

24.5.2 *Manufacturing Strategy*

Manufacturing strategy is a long-run strategy of an enterprise for production planning and control. In the manufacturing enterprises, various types of manufacturing strategies included "make to order", "make to stock", "make to assemble", and "assemble to order". Respondents were asked to rate the adopted manufacturing strategies. Responses are presented in Table 24.1.

Referring to Table 24.1, it can be observed that most of the enterprises considered make-to-order (mean = 3.85, SD = 1.71), followed by make-to-stock (mean = 2.46, SD = 1.83) as prominent manufacturing strategy. The reason may be that they manufacture customized products or commence production only after confirmation of the order. Almost nil respondents reported for make to assemble (mean = 1.04,

Table 24.1 Manufacturing strategy

Manufacturing strategy	Mean	SD
Make to order	3.85	1.71
Make to stock	2.46	1.83
Make to assemble	1.04	0.33
Assemble to order	1.04	0.24

SD = 0.33) and assemble to order (mean = 1.04, SD = 0.24). Study by Hoffmann and Orr (2005) also stated that 54.2% of respondents adopted manufacture to order while only 8.3% adopted manufacture to stock. It indicates that most of the enterprises manufacture the products after receiving the order. This study also reflects that most SMEs follow make-to-order manufacturing strategy, followed by make-to-stock.

24.5.3 *Order Winning Criteria*

Order winning criteria are those criteria or factors which help in obtaining the orders over the competitors. With more orders, sales will improve. Hence, these are necessary for the growth of an enterprise. In this respect, participants were asked to rate the potential order winning criteria (conformance to quality, product durability, product cost, product range, product design, new product development, attractive packaging, promotional schemes) of their enterprises. The value of Cronbach's Alpha for these eight order winning criteria appears to be 0.734). Refer to Fig. 24.2 for value of mean and SD.

From Fig. 24.2, the observed foremost three order winning criteria included quality (mean = 4.54, SD = 0.67), product durability (mean = 3.95, SD = 1.11) and product cost (mean = 3.62, SD = 0.90). These are the factors that not only help in improving sales but also help in retaining the customers. Refer to Fig. 24.3 for the frequency distribution of these order winning criteria.

Let, analyze the hypothesis related to order winning criteria for its statistical validation. Computations are represented in Table 24.2.

Refer to Table 24.2. The significant difference of mean among all the three sectors of SMEs was shown for product durability, product range, product design, and new product development. Thus, these could not be considered as potential order winning criteria for all the sectors of SMEs. Further, the significant difference of

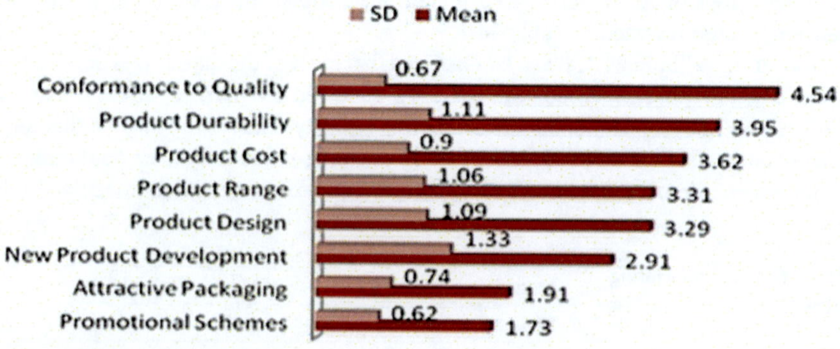

Fig. 24.2 Order winning criteria

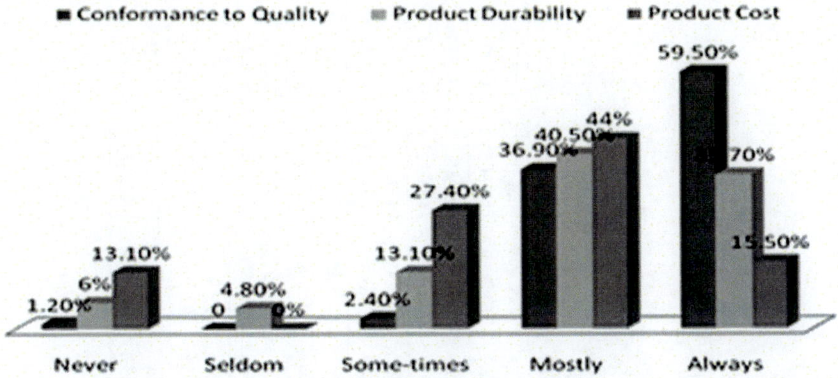

Fig. 24.3 Frequency distribution of foremost three order winning criteria

mean between any two sectors was observed for product quality (between glassware and shoe/footwear enterprises), product cost (between shoe/footwear and lock enterprises), and promotional schemes (between glassware and shoe/footwear enterprises). Attractive packaging had a very low value of overall mean (mean = 1.91 on a scale of 5). Hence, these also could not be considered as potential order winning criteria. In this way, it can be inferred that all the sectors of SMEs do not adhere to any of the order winning criteria.

From these observations, it cannot be rejected that various sectors of SMEs do not agree on any of the common order winning criteria. Hence, the null hypothesis is statistically validated.

24.5.4 Impact of Order Winning Criteria on Performance

Let's examine the impact of these considered order winning criteria on the performance of SMEs. In this respect, Table 24.3 represents the outcome of factor analysis.

Based on the shown data in Table 24.3 of factor analysis of order winning criteria, three components were identified and these were grouped into three factors that comprise as Product Attributes (referred to as Order Win F1 in Table 24.4) that included product durability, product quality, product range, product design, and new product development; Product Offer (referred as Order Win F2 in Table 24.4) that included attractive packaging and promotional schemes and Product Cost (referred as Order Win F3 in Table 24.4) that included itself only. These three factors were considered for further correlation analysis (refer to Table 24.4).

Referring to Table 24.4, it can be observed that product attributes are strongly correlated with performance parameters like annual sales turnover, sales growth,

Table 24.2 Values for order winning criteria

Order winning criteria	Overall		Glassware manufacturing enterprises (1)		Shoe/footwear manufacturing enterprises (3)		Lock manufacturing enterprises (5)		F-value	*t*-value (Sig.)		
	Mean	SD	Mean	SD	Mean	SD	Mean	SD		1–3	3–5	5–1
Product quality	4.54	0.67	4.32	0.79	4.67	0.48	4.65	0.65	2.611 (0.080)	−2.046 (0.045)	0.094 (0.926)	1.632 (0.109)
Product durability	3.95	1.11	3.07	1.24	4.40	0.50	4.57	0.66	25.238 (0.000)	−5.498 (0.000)	−1.037 (0.305)	5.277 (0.000)
Product cost	3.62	0.90	3.58	0.85	3.87	0.90	3.35	0.94	2.255 (0.111)	−1.279 (0.206)	2.046 (0.046)	−0.955 (0.344)
Product range	3.31	1.06	2.77	1.09	3.83	0.79	3.35	1.03	9.043 (0.000)	−4.339 (0.000)	1.945 (0.057)	1.963 (0.055)
Product design	3.29	1.09	2.71	1.13	3.83	0.79	3.35	1.03	9.835 (0.000)	−4.481 (0.000)	1.945 (0.057)	2.130 (0.038)
New product development	2.91	1.33	2.26	1.26	3.57	1.31	2.91	1.04	8.730 (0.000)	−3.979 (0.000)	1.969 (0.054)	2.026 (0.048)
Attractive packaging	1.91	0.74	2.0	0.82	1.70	0.54	2.04	0.83	1.855 (0.163)	1.691 (0.096)	−1.835 (0.072)	0.193 (0.848)
Promotional schemes	1.73	0.62	1.94	0.73	1.57	0.50	1.70	0.56	2.866 (0.063)	2.295 (0.025)	−0.881 (0.383)	−1.318 (0.193)

Table 24.3 Factor analysis of order winning criteria

Order winning criteria	Component-1	Component-2	Component-3
Product durability	0.725	−0.437	−0.064
Product quality	0.686	−0.360	−0.305
Product range	0.845	0.142	0.356
Product design	0.859	0.122	0.380
Product cost	0.113	−0.063	0.888
Attractive packaging	0.043	0.799	−0.199
Promotional schemes	0.035	0.749	0.127
New product development	0.772	0.234	−0.007

Rotation converged in six iterations

Table 24.4 Correlation matrix for order winning criteria and performance variables

	Sales turnover	Market share	Sales growth	Export	Order win F1	Order win F2	Order win F3
Sales turnover	1						
Market share	0.109 (0.324)	1					
Sales growth	0.338** (0.002)	−0.057 (0.604)	1				
Export	0.607** (0.000)	−0.035 (0.749)	0.297** (0.006)	1			
Order win F1	0.425** (0.000)	−0.331** (0.002)	0.381** (0.000)	0.325** (0.003)	1		
Order win F2	0.172 (0.117)	0.238* (0.029)	−0.013 (0.905)	−0.017 (0.881)	0.000 (0.993)	1	
Order win F3	0.136 (0.218)	−0.263* (0.015)	−0.049 (0.655)	0.055 (0.622)	0.213 (0.052)	−0.017 (0.878)	1

**Correlation is significant at the 0.01 level (2-tailed); *Correlation is significant at the 0.05 level (2-tailed)

and export, which shows that by improving the product attributes, overall performance can further be improved. Product attributes can be enhanced by employing suitable AMT successfully as these are concerned with manufacturing, primarily. Thus, adoption of suitable AMT in SMEs is required to improve the state of product attributes. Product offers are correlated to marketing strategy, while product cost is the primary motivating force based on which customers think and decide to buy a product. Product offering is correlated with the market share that can be understood that offers and packaging influence customers for purchasing of product and motivates shopkeepers and dealers to promote sales. Product cost is also correlated with the market share as this helps in improving the sales in the same manner.

24.6 Discussion

SMEs, the backbone of industrial economy, are generating significant amount of jobs and contributing a lot to the nation's GDP. So, care should be taken to improve their functioning and performance. By offering customized, and highly reliable quality products for sale to the customer internationally, the customer satisfaction will be maximized as the offered product will serve extraordinarily without any possibility of failure or even complaint. It sought to retain customers everlastingly and thus, the sales will improve. Increase in sale will turn to improved turnover and maximize the earned profit. Thus, by emphasizing the order winning criteria, an enterprise can maximize its sales.

24.7 Conclusions

The outcome of this survey concluded that the most of the orders are received by the party (or customer) itself. The care should be taken to identify the correctness of the order with authenticity. The flexibility in change of the order is also reported. Quality, product durability, and product cost are the most important order winning criteria. These factors help in increase in sales by retaining the customers. However, by improving the product attributes (product durability, product quality, product range, product design, and new product development), the overall performance of an enterprise can further be improved. These product attributes can be improved by adopting suitable newer technologies that are primarily concerned with the manufacturing.

Further related research will widen the aspects of need, awareness and will help in taking initiatives for the reduction in problems and imparting their growth.

References

1. Report of the expert committee on Micro, Small and Medium Enterprises, submitted to Governor, Reserve Bank of India, https://www.rbi.org.in/Scripts/PublicationReportDetails.aspx?UrlPage=&ID=924 (2019). Last accessed 20 Apr 2020
2. Obi, J., Ibidunni, A.S., Tolulope, A., Olokundun, M.A., Amaihian, A.S., Borishade, T.T., Fred, P.: Contribution of small and medium enterprises to economic development: evidence from a transiting economy. Data Brief **18**, 835–839 (2018)
3. Singh, B.: Adoption of advance manufacturing technology in micro, small and medium enterprises of India. Global J. Enterprise Inf. Syst. **7**(2), 19–24 (2015)
4. Tahir, H.M., Razak, N.A., Rentab, F.: The contributions of small and medium enterprises (SME's) on Malaysian Economic growth: a sectorial analysis. In: Proceedings of the 7th international conference on Kensei Engineering and Emotion Research, pp. 704–711 (2018)
5. China Banking News, http://www.chinabankingnews.com/2018/06/22/smes-account-60-chinas-gdp-beijing-mulls-inclusion-loans-mlf-collateral/ (2018). Last accessed 03 Oct 2019
6. SMEs in UK, https://www.merchantsavvy.co.uk/uk-sme-data-stats-charts/ (2018). Last accessed 03 Oct 2019

7. Ndiaye, N., Razak, L.A., Nagayev, R., Ng, A.: Demystifying small and medium enterprises' (SMEs) performance in emerging and developing economies. Borsa Istanbul Rev. **18**(4), 269–281 (2018)
8. Wang, Y.: What are the biggest obstacles to growth of SMEs in developing countries? An empirical evidence from an enterprise survey. Borsa Istanbul Rev. **16**(3), 167–176 (2016)
9. Majama, N.S., Magang, T.I.T.: Strategic planning in small and medium enterprises (SMEs): a case study of botswana SMEs. J. Manag. Strategy **8**(1), 74–103 (2017)
10. Singh, B., Narain, R., Yadav, R.C.: Identifying critical barriers in the growth of Indian micro, small and medium enterprises (MSMEs). Int. J. Bus. Competition Growth **2**(1), 84–105 (2012)
11. Hill, T.J.: Teaching manufacturing strategy. Int. J. Oper. Prod. Manag. **6**(3), 10–20 (1987)
12. Prajago, D.I., Laosirihongthong, T., Sohal, A., Boon-itt, S.: Manufacturing strategies and innovation performance in newly industrialized countries. Ind. Manag. Data Syst. **107**(1), 52–68 (2007)
13. Khatri, P.: A study of the challenges of the Indian MSME sector. IOSR J. Bus. Manag. **21**(2), 5–13 (2019)
14. Sharma, P., Gupta, A., Malik, S.C., Jha, P.C.: Quality improvement in manufacturing process through six sigma: a case study of Indian MSME firm. Yugoslav J. Oper. Res. **29**(4), 519–537 (2019)
15. Hu, H.Y., Chiu, S.I., Yen, T.M.: Modified IPA for order-winner criteria improvement: a MICMAC approach. J. Appl. Sci. **9**, 3792–3803 (2009)
16. Laosirihongthong, T., Dangayach, G.S.: A comparative study of implementation of manufacturing strategies in Thai and Indian automotive manufacturing companies. J. Manuf. Syst. **24**(2), 131–143 (2005)
17. Singh, B.: Measures to accelerate the sustainable growth of small sectors: a literature review. IOP Conf. Ser. Mater. Sci. Eng. **691**, 012044 (2019)

Chapter 25
Application of Electrical Resistivity Method for Monitoring and Assessment of Cracks in Concrete Structures

P. Ashokkumar and D. Jegadeeswaran

Abstract The occurrences of cracks are unavoidable in concrete structures due to various reasons like settlement, overload, etc. so it is essential to identify and control the cracks to increase the adequate serviceability and durability of the structure. The analysis and identification of cracks are important for monitoring, diagnosis, and maintenance of concrete structures, to avoid malfunctioning or unsafe conditions in the structures. The most tedious process is to identify the exact time and location of a crack in the concrete structures. In this paper, the solution has been given to identifying the time and location of the crack by Electrical Resistivity Method which is one of the non-contact process methods without having any direct physical contact with the structure. Arduino UNO mega 2560 and resistors are used to find the occurrence of a crack in the structures. In the structure, the conductive paint coating has been given and it is separated into several blocks. Each block is connected with various values of resistors. The constant voltage has been given in the blocks and resistance values are noted. When the cracks occur, at that time conductive paint resistivity value will change. The change of resistivity value has been deducted by Arduino UNO mega 2560 microprocessor with coding. It gives the signal immediately after when the crack occurs in the structure. To manually interrupt the connection which is made by the conductive paint, switches have been fixed to validate the results. From this experiment, it is observed that the minute cracks are also deducted and exact results are given by Arduino.

25.1 Introduction

In the construction industry, many advanced technologies like the Internet of things, Argument reality, and virtual reality are entered and ruling the industry. By having the modern tools and technologies, the structural health monitoring techniques are become an easy task to measure the strength without any destruction and non-contact

P. Ashokkumar (✉) · D. Jegadeeswaran
Department of Civil Engineering, Sona College of Technology, Salem, Tamilnadu, India
e-mail: ashokdce@gmail.com

S. Yadav et al. (eds.), *Proceedings of Second International Conference in Mechanical and Energy Technology*, Smart Innovation, Systems and Technologies 290,
https://doi.org/10.1007/978-981-19-0108-9_25

methods. Even though various technologies are entered, but still some problems exist without a proper solution. In that, cracks play a major role in the building construction which desired the strength, durability, and life of the structures. [1–5]. In concrete structures, the occurrence of crack is one of the early indications which give the warning, that structures are starting to degrade. Sometimes due to the continuous extension of cracks leads to more damage to the structure as well as the occupant. It is necessary to take precautionary steps to arrest the cracks. Manually monitoring and assessing the type, nature, and width of the cracks is a tedious process, as well as this approach, will completely depend on knowledge and experience which leads to the improper assessment of the structures. To avoid these complications, the manual method was replaced by automatic image deduction. [3]. Even though by having the advanced method, the images taken for crack analysis by image processing methods will have some difficulties to make proper analysis because sometimes images may have low contrast, improper illumination and noise also generated. Crack detection on the standard Otsu method will not give expected outcomes. To improve the performance of Otsu method, a new model was developed to enhance the image algorithm called Min–Max Gray Level Discrimination (M2GLD). The newly constructed model is capable of identifying crack objects and analyzing their characteristics including the area, perimeter, width, length, and orientation. [6] To overcome the above difficulty, this paper explains about the monitor the surface of the beam to detect the crack and its location by passing a voltage through the conductive paint. This paper briefs the idea of monitoring the concrete by electrical resistivity method to find the crack. The resistors are used to vary the current in terms of an ohm to give the various outputs to find the crack and to develop corresponding health monitoring and damage identification strategies. 5V voltage has passed through the conductive paint that is applied on both sides of the concrete beam and the paint is connected to the breadboard that consists of resistors. To gives failure the indication, LED lights and buzzer were connected in the same breadboard. The whole setup is connected with the Arduino UNO mega 2560 board with help of a USB cable. When the crack has occurred in any one of the blocks, the output voltage will drop. For the dropped voltage the value ranges are already coded in the UNO mega 2560 which indicates the cracked block with the help of light, buzzer and display. By this, the exact location of a crack in the concrete beam will be identified within a fraction of a second and it will help to rectify the crack immediately [7–14].

25.2 Flextural Crack

Flexure is also known as "Bending". The crack due to bending of the beam always starts in the compression zone, i.e., the weakest part of the beam which is shown in Fig. 25.1. In a reinforced concrete beam the crack occurs at the straight bottom of the beam. Due to continuous loading, the width of the crack may get increased and become more uniform across the member.

Fig. 25.1 Flexural crack in RC beam

25.3 Materials Used

In this experiment, Concrete beam, Conductive paint, Wires and pins, Switch, Buzzer., Resistors with different values., Breadboard., LED lights (red and green), Arduino UNO mega 2560 board, and Laptop for display.

25.3.1 Concrete Beam

The experiment is about the detection of the crack in a concrete beam using the electrical resistivity method. The beam is cast and separated into several parts to find the exact location of the crack which is shown in Fig. 25.2. Then the beam is separated into eight equal parts. In that, four parts on the right edge and four parts on the left edge with the help of pins. Each separated location part are connected and soldered with the lugs to get the constant readings without any hindrance which is shown in Fig. 25.3.

Fig. 25.2 Beam separated by blocks

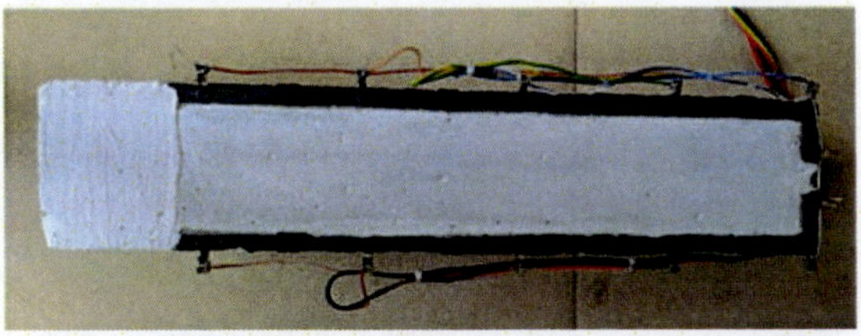

Fig. 25.3 Lugs attached in the beam

Fig. 25.4 Conductive paint

25.3.2 Conductive Coating

The conductive paint is coated on edges up to three coatings on both sides of the beam which is shown in Fig. 25.2. The conductive paint does not have any particular resistivity it will vary with the number of coating and thickness. The switches were placed on the conductive paint on each block which may be used to identify the beam whether the beam is healthy or cracked condition. When the switch is in ON position, it indicates healthy and if it is on OFF, then it indicates block is not healthy. It means that artificially induced the crack to know the status of a particular block. The conductive paint which is shown in Fig. 25.4 is used in this research and has good conductivity.

25.3.3 Breadboard

The breadboard has the setup which is shown in Fig. 25.5 consisting of 4 LEDs of green color and 4 LEDs of red color. The green color indicates a healthy condition and the red color indicates a failure condition. And it also consists of four resistors with values of 10k, 5k, 5k, 6k. The resistor is connected with the pins that place in the beam. The whole setup is replicated in the breadboard for the left and right sides of the beam.

Fig. 25.5 Bread board

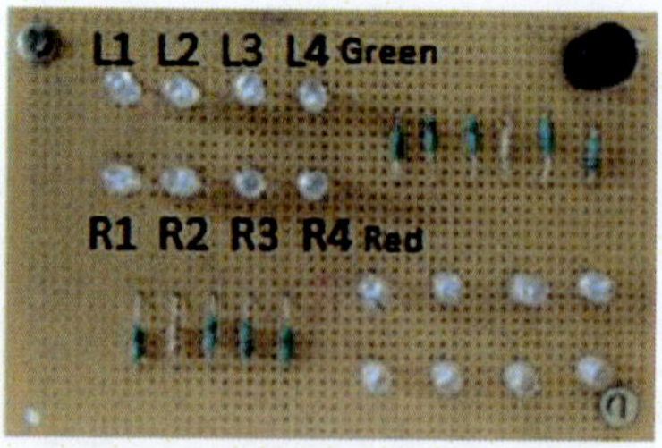

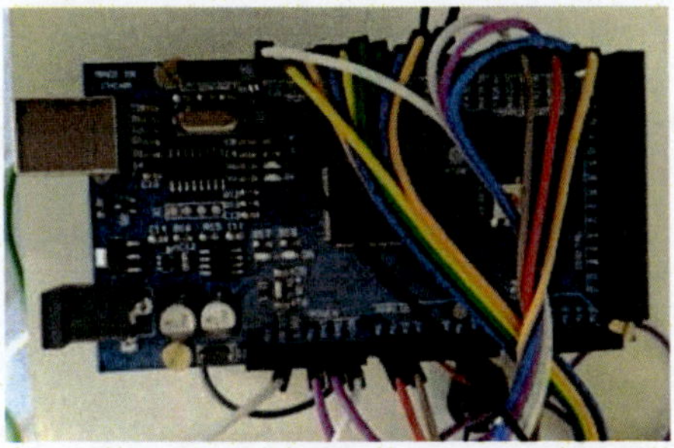

Fig. 25.6 Arduino board connection

25.3.4 Arduino Board

The lights and buzzer in the breadboard are connected to the UNO mega 2560 with the help of wires which is shown in Fig. 25.6. Arduino UNO mega 2560 board acts as a brain for crack deduction in the beam. The board is connected to the laptop using a data cable to monitor the data.

25.4 Overall Setup

The current of 5V voltage is passed through the conductive paint that is applied on both sides of the concrete beam. The beam is connected to the breadboard which consists of a resistor with the help of wire in eight separate blocks. For the indication LED lights and buzzer were connected in the same breadboard. When the crack has occurred in any one of the blocks the output voltage will drop. For the dropped voltage the value ranges are included in the coding and that coding has been uploaded in Arduino UNO mega 2560 which indicates the cracked block with the help of light, buzzer and display. By this, the exact location of a crack in concrete beam within a fraction of a second. The overall setup has shown in Fig. 25.7.

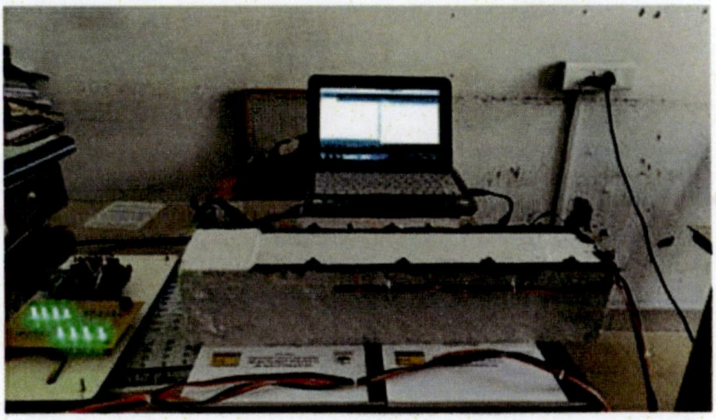

Fig. 25.7 Overall setup

25.5 Result and Discussion

To find the total resistance when the circuit is in parallel connection, the following formula is used to find the resistance value. In the beam, the parallel connection has been used as the circuit diagram which showed in Fig. 25.8. Equation (25.1) is used to find the difference in resistance value when the beam is in the crack. When the crack occurs, the resistance value will change and that particular change, Arduino will be deducted and output will be given. The value of total resistance in the parallel connection will not equal the sum of the total resistors in the circuit.

$$\text{TOTAL RESISTANCE}\text{“}1/R_t = 1/R_1 + 1/R_2 + 1/R_3 + \ldots\ldots\ldots\ldots + 1/R_n'' \quad (25.1)$$

The concrete beam is connected with wires and each end is properly coated with conductive paint and soldered with sleeves to avoid the damage and short of wires with each other which are shown in Fig. 25.9. The LED and buzzers are connected with Arduino to give the warning signal by light and sound when the cracks occur. For

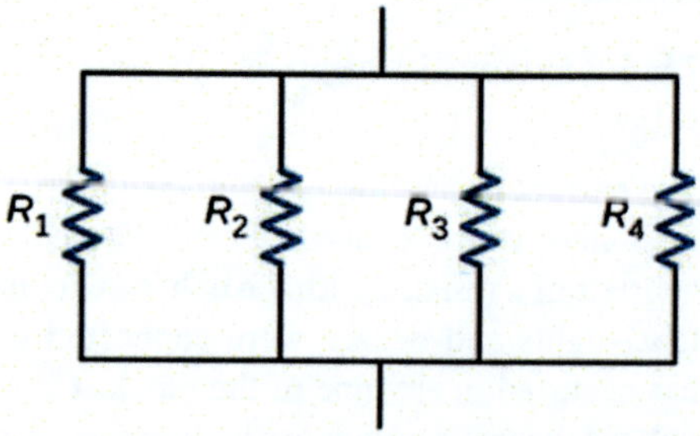

Fig. 25.8 Resistor connected in parallel

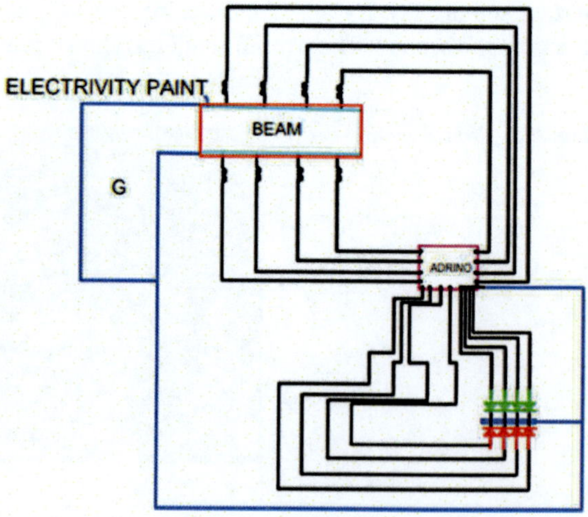

Fig. 25.9 Circuit connection diagram

each block, a separate LED has been arranged so that easily monitored and identified in which block crack occurs. The LED lights are arranged in different two colors green and red. The green color indicates that the block is healthy and red indicated damage.

The value of the voltage which gets as output is based on the resistors assigned for each block. When there is an absence of crack in block 1, then the output in the laptop shows as "Block 1 is Healthy" and when there is the presence of a crack in block 1, then it shows "Block 1 Is Failure Attention Required". This will repeat for all the four blocks on the left (LB1, LB2, LB3, LB4) and right (RB1, RB2, RB3, RB4) sides. The Green LED indicates the healthy or absence of crack condition with normal buzzer sound and Red LED indicates the failure or presence of crack condition with a continuous buzzer sound. All the conditions are work based on the resistivity value and coding incorporated in the Arduino board. Table 25.1 shows the resistance value, result condition, and buzzer condition for the various blocks.

25.6 Conclusion

Nowadays, structural health monitoring becomes one of the vital things in the Civil Engineering field to monitor and assess the structures to ensure safety and durability. It is easy to assess the health of the structure by using advanced technology. Normally to deduct the cracks in the concrete structures, need to use any non-destructive testing methods and any sensor technology which leads the high cost and uneconomical. To overcome this issue, the solution has been identified and an experiment has been carried out. From this experimental work, the continuous health monitoring of the structure is possible using the proposed Electrical Resistivity method. It is observed that the electrical resistivity method works well in beams for detecting the crack. The exact location of the crack in the beam can be identified within the fraction second when the crack occurs. Time will save for identifying the crack in the long-span beam. The electrical resistivity method is also an economical method and effective method which will enter into the Civil Engineering field to monitor the cracks.

Table 25.1 Resistance value, result condition, LED color and buzzer condition

Block	Resistance value greater than	Resistance value less than	Result condition	LED color	Buzzer condition
Left block 1	205	220	Left block 1 failure attention required	Red	Continuous sound
	Else		Healthy	Green	Normal sound
Left block 2	530	600	Left block 2 failure attention required	Red	Continuous sound
	Else		Healthy	Green	Normal sound
Left block 3	615	650	Left block 3 failure attention required	Red	Continuous sound
	Else		Healthy	Green	Normal sound
Left block 4	1000	1024	Left block 4 failure attention required	Red	Continuous sound
	Else		Healthy	Green	Normal sound
Right block 1	105	120	Right block 1 failure attention required	Red	Continuous sound
	Else		Healthy	Green	Normal sound
Right block 2	145	153	Right block 2 failure attention required	Red	Continuous sound
	Else		Healthy	Green	Normal sound
Right block 3	154	165	Right block 3 failure attention required	Red	Continuous sound
	Else		Healthy	Green	Normal sound
Right block 4	1	1000	Right block 4 failure attention required	Red	Continuous sound
	Else		Healthy	Green	Normal sound

References

1. Borosnyói, A., Balázs, G.L.: "Models for flexural cracking in concrete" **6**(2), 53–62 (2005), https://www.icevirtuallibrary.com/doi/abs/https://doi.org/10.1680/stco.2005.6.2.53
2. Beeby A.W.:"The influence of the parameter ϕ/ρeff on crack widths" **5**(2), 71–83 (2004), https://www.icevirtuallibrary.com/doi/https://doi.org/10.1680/stco.2004.5.2.71
3. Arun, M., Poobalb, S.: "Crack detection using image processing: a critical review and analysis. **57**(2), 787–798 (2018).https://doi.org/10.1016/j.aej.2017.01.020
4. Rimkus, A., Podviezko, A., Gribniak, V.: Processing digital images for crack localization in reinforced concrete members. Procedia Eng. **122**, 239–243 (2015). https://doi.org/10.1016/j.proeng.2015.10.031
5. Ashok Kumar, P., Jegatheesawaran, D., Karthick Raj, T., Iyappan, R.: "An experimental research on prognosis of compressive strength of concrete by using sensor", ISSN: 2278–3075, **9**(6) (2020). https://doi.org/10.35940/ijitee.F4076.049620
6. Hoang, N.D.: Detection of surface crack in building structures using image processing technique with an improved Otsu method for image thresholding. Adv. Civil Eng. (2018)
7. Tammo, K., Lundgren, K., Thelandersson, S.: Nonlinear analysis of crack widths in reinforced concrete. Source: Mag. Concrete Res. **61**(1), 23–34 (2009). https://doi.org/10.1680/macr.2009.61.1.23
8. Tammo, K., Thelandersson, S.: Crack opening near reinforcement bars in concrete structures. Source: Struct Concrete **7**(4), 137–143 (2006). https://doi.org/10.1680/stco.2006.7.4.137
9. Issa, M.A.: "Investigation of cracking in concrete bridge decks at early ages". J. Bridge Eng. **4**(2) (1999).https://doi.org/10.1061/(ASCE)1084-0702(1999)4:2(116)
10. Saleem, M.U.: "Cracking behaviour of RC beams strengthened with different amounts and layouts of CFRP". Accepted 7 Mar 2019, Published 12 Mar 2019. https://doi.org/10.3390/app9051017
11. Sheerin Sitara, N., Kavitha, S., Raghuraman, G.: "Review and analysis of crack detection and classification techniques based on crack types". ISSN 0973–4562 **13**(8), 6056–6062 (2018). http://www.ripublication.com
12. SP 25:"Handbook On Causes And Prevention Of Cracks In Buildings"
13. Yokoyama, S., Matsumoto, T.: Development of an automatic detector of cracks in concrete using machine learning. Open Access Procedia Eng **171**, 1250–1255 (2017). https://doi.org/10.1016/j.proeng.2017.01.418
14. Krishna Thulasi, T., Subathra, S., Meikandaan, T.P.: "An experimental study of crack patterns on reinforced concrete beam". IRJET 05(03) (2018). https://doi.org/10.1016/j.proeng.2015.10.031

Chapter 26
Using Machine Learning and Data Mining to Evaluate Modern Financial Management Techniques

Alim Al Ayub Ahmed, Sudha Rajesh, Sarika Lohana, Samrat Ray, Jnaneshwar Pai Maroor, and Mohd Naved

Abstract It is the question of determining patterns in big data sets that correlate to helpful data. It involves techniques that are at the confluence of machine learning, statistics, and legacy system, and it is also known as data mining. Machine learning is a branch of artificial intelligence that emerged from the areas of object recognition and artificial intelligence. It is concerned with the research and development of methods that can understand from assessment tools. The study shows financial institutions use of financial data performance and ensure precise management of consumer data in order to identify defaulters, to reduce the number of equipment failures associated, to process transactions quickly and efficiently, to reduce the number of incorrect judgments, to categorize potential customers, and to minimize the wastage of the financial organizations.

A. Al Ayub Ahmed
School of Accounting, Jiujiang University, 551 Qianjindonglu, Jiujiang, Jiangxi, China

S. Rajesh (✉)
Department of Computer Applications, BSA Crescent Institute of Science and Technology, Vandalur, Chennai, India
e-mail: drsudharajesh84@gmail.com

S. Lohana
State Bank Institute of Leadership, Kolkata, India

S. Ray
The Institute of Industrial Management, Economics and Trade, Peter the Great Saint Petersburg Polytechnic University, Saint Petersburg, Russia

J. P. Maroor
Department of Business Administration [JKSHIM], NMAM Institute of Technology, Nitte, India

M. Naved
Department of Business Analytics, Jagannath University, Delhi-NCR, India

S. Yadav et al. (eds.), *Proceedings of Second International Conference in Mechanical and Energy Technology*, Smart Innovation, Systems and Technologies 290,
https://doi.org/10.1007/978-981-19-0108-9_26

26.1 Introduction

A lot of academics wanted to uncover solutions in a huge amount of data to find a pattern. Statistics, machine learning, and computer intelligence techniques are utilized, gathered together under the framework known as data mining [1]. Data mining is a process used to detect interesting designs from a big number of data recorded in databases, data marts, warehouses, etc. Datamart and database server are solutions that assist in company information management. The information saved in the data warehouse is more significant for a particular organization in the period of the use of the data storage system by developing the data mart, however it is restricted to the use of any one of the departments but the Indian Academy of Science and Technology. The data storage is a data mart package that gives information on the many companies' businesses [2]. It can include information on the everyday activities of the corporation's different factions. Database systems can give transactional reporting and administration on certain divisions or businesses in the company as part of a data warehouse. In its database, each corporation can keep information about each department, such as the Finance Database, the Sales Database, the Manufacturing Database, and the Database Marketing. Database systems have a fundamental structure to construct data mining techniques. Data mining can be regarded in response to the natural development of multi-discipline information technology such as database or data warehouse technology, statistics, high-powered computing, machine learning, intelligent systems, information processing, data visualization, etc. [3].

26.2 Machine Learning in Finance

In the financial sector, machine learning and other principles of artificial intelligence (AI) are used in various ways. Some machine learning applications in finance also provide. Professional learning in finance is currently regarded to be fundamental element of many financial applications and services, such as investment management, risk assessment, credit value calculation, and even loan approval. Machine learning is a component of data science that provides the ability to learn and improve from experience without even being instructed. As an application of artificial intelligence, machine education is aimed at creating systems that acquire sources of data, and the system automatically changes its settings to improve meetings. Computer systems do background work and produce outcomes autonomously on the basis of how they are trained. Machine learning tends to be more accurate when large quantities of data are inputted properly to capture conclusions and forecasts. For example, the financial services sectors tend to discover huge amounts of data on the invoices, payments, sellers, and customers of day-to-day transactions that are perfect for machine learning. Many large Fin technology and financial services companies are already incorporating machine learning into their companies, making the process simpler, reducing risks, and improved approaches [4].

26.3 Administration of Portfolio (Robo-Advisors)

Robo consultants are online apps that are developed using machine learning and offer investors automatic financial advice. The apps utilize technology to build a financial portfolio based on the objectives and risk tolerance of an investor. Robo consultants need small minimum accounts and are typically cheaper than managers of human portfolios [5]. When utilizing robot consultants, customers must put their investment or savings objective into the system and the system determines automatically the best financial possibilities with the greatest returns.

26.4 Credit Undertaking

Corporations have access to millions of customer data in the banking and insurance sector, with which machine learning may be taught to address those concerns. Machine learning algorithms can quickly decide on borrowing and borrowing and save businesses time and money spent by people. Data analysts may train machines to evaluate millions of consumer information in order to match data files, to search for specific exceptions, and to choose if a customer is eligible to be lent or insured. For example, the algorithm is able to analyze consumer data, such as age, income, occupation, and credit behavior for the consumer default history, loans paid, foreclosure history, etc. so that any results can be detected which may determine whether the user is eligible to take out a loan or insurance plan [6].

26.5 Prevention and Detection of Fraud

For banks and financial services businesses, fraud is a significant issue; every year, it represents billions of dollars in damages. Finance firms often maintain a significant quantity of their data online and raise the danger of a violation of safety. Fraud in the financial sector is regarded as a significant danger to important data today, with growing technical progress. Fraud detection systems have been developed in the past on the basis of a number of criteria that contemporary criminals might simply circumvent. Most businesses are thus now using machinery for flagging and fighting illegal financial activities. Machine learning works by scanning huge data sets to identify unusual behaviors or abnormalities and lets security teams investigate them further. It compares a transaction to other datasets such as customer account history, IP address, location, etc. To evaluate whether the transaction is flagged parallel to the account-behavior holders. The system may then automatically reject withdrawal or acquisition according to the nature of a transaction until the person decides [7].

26.6 Trading in Techniques

The use of techniques to create better business choices refers to method trade. Traders often construct mathematical models to watch in real-time company news and trading activity to identify all variables that may lead to rising or falling securities prices. The program has a predefined list of guidelines for conducting trade without the active participation of the trader, on several parameters—such as time, price, amount, and other variables.

Machine learning, unlike human traders, can evaluate huge amounts of data at the same time and produce thousands of transactions per day. Machine learning produces quick commercial choices which offer traders an edge over the typical market [8].

Furthermore, automated trade does not make trade determinations based on emotions that are typical among traders that have emotional or individual goals as their judgment. The trading technique is primarily used to automate trading operations by Hedge Funds managers and financial organizations [9].

26.7 Machine Learning Models for Financial Management

26.7.1 Classification Techniques

Support Vector Machine (SVM) is a tool that may be used for categorization, depending on the functionality of the kernel. The most flexible kernels are the Gaussian kernel. The SVM classification results may be controlled using the width parameter Of the Gaussian kernel function. The Gaussian function may not only be utilized as an SVM kernel but also in some interesting neuro-fuzzy classifications.

Decision trees are classified as recursive components of the data space. Classification and Recovery Trees (CART) is a customization way of describing how variable Y distributes the metric after the prediction vector X has been assigned. The CART model utilizes a binary tree to split the predicted space into specific subsets on which Y is constantly assumed [10]. Artificial Neural Networks (ANNs) are a non-linear statistical model focusing on individual brain function. ANNs offer strong tools for data mining for modeling relationships with the data analyst.

ANNs can identify the complicated input-data pattern and can also accurately anticipate the result of the fresh independent input data. ANNs have the amazing ability to draw significance from complex or inaccurate data. It may be utilized for pattern extraction and trends utilizing specialized methods. ANNs are ideal for detecting patterns and are also ideal for predicting or predicting data. One of the most popular ANNs is the multi-layer perception, sometimes called the Back-Propagation Neural Network (BPN), which calculates the mistakes of each output layer after input data are processed [11].

It is an automated distinction method. BPN is distinguished by the backward reproduction of output mistakes: these errors are calculated on the output layer and the training is transferred back to previous layers in order to decrease the output errors46. Survival Analysis is a new credit score model methodology. A popular method for banks to distinguish client information when applying for a bank loan. Banks are can separating good information from inaccurate communication about the request for loans. The system can calculate profitable customers and also assess customers' profit margin [12]. Valuable Logistic regression can anticipate the length of the event in advance and estimate the likelihood of an event occurring.

This renowned data mining method was developed by the H2O team to analyze group data sets. The methods include Generalized Linear Models (GLM), Gradient Boost Method (GBM), and Random Forest Distributed (DRF). GLM is comparable to the model of linear regression. For regression analysis and data categorization, data mining methods are employed. GLM is popular because it is simple to understand and also a fast processing step for big data sets. GLM modeling is popular. The GBM model is a prediction tool that uses regression or classification. It is a set of tree models and offers significantly accurate results. The GBM model uses weak classification algorithms to gradually alter data and to establish a number of decision-making bodies. Finally, the DRF is a group of tree models, in which each tree is connected by other trees. DRF is by far the most powerful classification and regression method.

1. **What exactly is data mining?**

   Data mining is the process through which patterns, irregularities, and connections are uncovered in big data sets which may be utilized to forecast future trends. The primary aim of data mining is to extract useful data from the available data.

   Data mining is an interdisciplinary subject that combines informatics and statistics methods. Note that a misnomer is data mining [13]. It focuses on the finding of patterns and anomalies in datasets but is not linked to the extraction of the data themselves.
2. **Data mining and Knowledge Discovery from the Database (KDD):**

   Database data mining and knowledge discovery (KDD) are contemporary data management technology developments. KDD is a kind of data mining that extracts information from a huge number of data. The standardized data mining process based on Cross-Industry Standard Data Mining Process (CRISP-DM) has six stages.

- Business understanding phase: choice of goals, knowledge of business purposes, evaluation of the learning environment, and development of a project plan.
- Phase of data comprehension, consisting of data needs consideration and first data gathering, investigation, and quality evaluation.
- Data preparation phase, consisting of data selection, data integration, and formatting, transformation of data, and data purification.
- Modeling stage involving: selection of suitable modeling methods, creation and testing of various modeling techniques and characterization, and the determination of a model setting tuned according to an initial performance evaluation.

- Analytical study, equivalent to the evaluation of the outcomes of the model experiment.
- Deployment phase, reflecting the development of the system of a model report.

In order to manage large data, corporate efficiency, and business intelligence, data mining and machine learning are extremely important. Data mining has tremendous financial and banking significance. Banks have to discover the hidden patterns in the big data sets and may thus monitor data in their databases. Such data may indicate the financial position and financial conduct of the customer before and after the credit is issued. Most banks and financial institutions offer various helpdesk, such as data monitoring, for establishing a savings account for each customer's company [14]. The schedule provides clients with credit for operations such as the Mortgage Company, auto credit services, investment services, and insurance and investment services. Additional financial applications in data mining and engineering include forecasting future financial occurrences, such as stock market, currency exchange rates, bankruptcy, bank credit rating, predictive financial and investment analyses, future trading, bank comprehension, and financial risk management. As technology evolves, Artificial Intelligence (AI) technology is utilized in the administration of funds, asset management, and other financial organizations. Machine learning techniques are used to extract data from the big database and examine it. With this tool, certain patterns may be found and the result can be predicted. Furthermore, there have been various types of studies that may assist in decision-making using machine learning in the banking sector to predict future occurrences. Today, financial institutions and most banks engage in information technology to facilitate the datasets group by providing data mining and machine intelligence technologies in order to work with a competitive industry effectively. Figures 26.1 and 26.2 show the procedure of data mining.

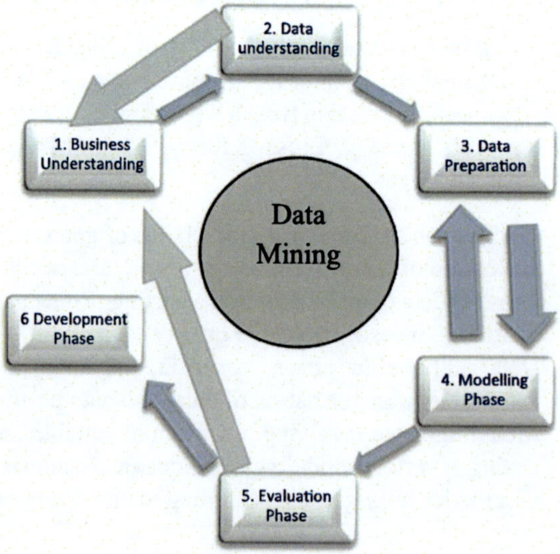

Fig. 26.1 Procedure of data mining

Fig. 26.2 Data mining techniques

Table 26.1 Data mining stages

Define the issue	Define the level and goals of the data exploration project's business issue
Examine the data	This phase involves exploring and gathering data to assist address the identified business issue
Preparation of data	To prepare it for future modeling processes, clean and arrange data gathered
Designing	Make a framework utilizing data mining methods to address the issue
Results interpretation and assessment	Obtain information and evaluate the correctness of the data model. Turn the findings into a company decision

The procedure may generally be split into the following stages as shown in Table 26.1.

The methods most frequently utilized in the field involve:

Clustering	Structures (clusters) identification in unstructured information
Anomaly detection	Identification in a dataset of anomalous data
Classification	Oversimplifying and applying the known structure to the data
Modeling of dependence	Find existing connections in a dataset. This often includes an examination of regression

26.8 Applications

Data mining provides numerous commercial applications. For instance, establishing appropriate data procedures (mining processes) may assist a business to reduce its expenses, boosting income, or gaining insights into its consumers' behavior and habits. It undoubtedly plays a significant part nowadays in the process of corporate decision-making [15].

Data mining is also used in finance constantly. For example, the applicable methods enable users to identify and evaluate the variables that affect financial securities price variations.

The field is developing fast. New data arrives at extremely high rates while technical progress makes it possible to address old issues more efficiently. In consequence, artificial intelligence and machine learning advances offer new pathways to accuracy and efficiency in the sector.

26.9 Bank Risk

The banks know the different risks. This can happen and can adversely impact the bank's enterprise. Banks examine the important risk factors. The quality of the risk analysis can influence the business' financial performance [16]. There are risks of many direct and indirect losses for all institutions and organizations. In the banks, three main risks relate to credit risk, business risk, and market risk. Financial organizations should, as appropriate, monitor credit risk management. Banks are required to maintain the credit risk individually compared to the credit management risk [17]. Productivity in credit risk management is important and vital to banks' long-term success. Credit scoring is a popular tool used to evaluate the credit risk of individuals. Credit scoring requires an active element assessment report. The exhibited increased method information about the status of credit risk data from credit offices and the reliability of credit provided in conjunction with the financial history and current financial situation of the borrowers [18]. Financial organizations have the option of removing undesirable characteristics in order to differentiate between excellent and poor credit risk management strategies of each business.

26.10 Conclusion

The study shows that machine learning technology-based data mining is a technique that may be used to evaluate current data, applications, and customer requirements to develop and sustain long-term consumer connections it may create trust for customers that make customer happiness and business the longer. In the credit payment forecast and customer credit policy development of the banking system, the use of machine learning methods for classification and clustering tasks is common. In this article, we presented data mining technology including two phases of handling. The grading step comprises multiple models: SVM, ANN, Decision Trees, and BPN. We discovered that the SVM model and the Decision Tree model are promising classification methods for financial applications. The above-mentioned methods may minimize human mistakes, speed up and save time, reduce judgments on the direct classification of clients, and can reduce the losses of financial institutions. In the classification step, K-means clustering is the most successful way of managing the credit score model for customers. The scoring techniques are used to assess the individual's creditworthiness. When credit loans and finances are in danger of failure, credit managers should create and use data mining techniques to handle and evaluate credit information to save time and minimize mistakes. Data mining (primarily based on machine learning methods) will be a difficulty for worldwide banking and financial research.

References

1. Adeoye, O.S., Ikemelu, C.R.K.: Industry wide applications of data mining. Int. J. Adv. Stud. Comput. Sci. Eng. **3**(2), 28 (2014)
2. Bishop, C.M.: Model-based machine learning. Philos. Trans. R. Soc. A Math. Phys. Eng. Sci. **371**(1984), 20120222 (2013)
3. Damrongsakmethee, T., Neagoe, V.E.: Data mining and machine learning for financial analysis. Indian J. Sci. Technol. **10**(39), 1–7 (2017)
4. Deshpande, S.P., Thakare, V.M.: Data mining system and applications: a review. Int. J. Distrib. Parallel Syst. (IJDPS) **1**(1), 32–44 (2010)
5. Han, J., Pei, J., Kamber, M.: Data mining: concepts and techniques. Elsevier (2011)
6. Hu, Z.G., Li, J.P., Hu, L., Yang, Y.: Research and application of data warehouse and data mining technology in medical field. In: 2015 12th International Computer Conference on Wavelet Active Media Technology and Information Processing (ICCWAMTIP), pp. 457–460. IEEE (2015)
7. Jain, A., Pandey, A.K.: Multiple quality optimizations in electrical discharge drilling of mild steel sheet. Mater Today Proc **4**(8), 7252–7261 (2017)
8. Jain, A., Pandey, A.K.: Modeling and optimizing of different quality characteristics in electrical discharge drilling of titanium alloy (grade-5) sheet. Mater Today Proc **18**, 182–191 (2019)
9. Jain, A., Yadav, A.K., Shrivastava, Y.: Modelling and optimization of different quality characteristics in electric discharge drilling of titanium alloy sheet. Mater Today Proc **21**, 1680–1684 (2020)
10. Jiang, N., Zhang, K., Ai, M., Du, X.: The momentum investment style in Chinese markets. Front Econ Manage **1**(7), 88–106 (2020)
11. Lin W.Y., Hu, Y.H., Tsai, C.F.: Machine learning in financial crisis prediction: a survey. IEEE Trans. Syst. Man Cybern. Part C (Appl. Rev.) **42**(4), 421–436 (2011)
12. Medelyan, O., Milne, D., Legg, C., Witten, I.H.: Mining meaning from Wikipedia. Int. J. Hum Comput Stud. **67**(9), 716–754 (2009)
13. Mohammed, M., Anad, M., Mzher, A., Hasson, A.: Meta-data and data mart solutions for better understanding for data and information in e-government monitoring. Int. J. Comput. Sci. Issues (IJCSI) **9**(6), 78 (2012)
14. Munawar, N.S., Ibrahim, R.: Quality oriented for physical design data warehouse (2006)
15. Nie, G., Zhang, L., Liu, Y., Zheng, X., Shi, Y.: Decision analysis of data mining project based on Bayesian risk. Expert Syst. Appl. **36**(3), 4589–4594 (2009)
16. Panwar, V., Sharma, D.K., Kumar, K.P., Jain, A., Thakar, C.: Experimental investigations and optimization of surface roughness in turning of en 36 alloy steel using response surface methodology and genetic algorithm. Mater Today Proc (2021)
17. Seah, B.K., Selan, N.E.: Design and implementation of data warehouse with data model using survey-based services data. In: Fourth Edition of the International Conference on the Innovative Computing Technology (INTECH 2014), pp. 58–64. IEEE (2014)
18. Singh, S.: Data warehouse and its methods. J Glob. Res. Comput. Sci. **2**(5), 113–115 (2011)

Chapter 27
A Survey on the Capability of Artificial Intelligence (AI) in Crime Prediction and Prevention

Akash Bag and Shailesh N. Hadli

Abstract The paper aims to compile reviews based on research which are conducted in the recent past about the usage of Artificial Intelligence (AI) in predicting crime and prevention for the same. The main question the survey tries to answer is *Are the current AI technologies capable enough to predict crime?* The literature review studies nine research papers based on this particular theme. The results of the current study suggested that the current technological AI which is in use by different crime prevention institutions is capable to predict and prevent crime. They are successful in finding patterns based on large data sets. Efficiency depends on the number of numerical data sets inserted. It increases with more data sets. The most common AI in this field is data mining. The other two popular kinds of AIs are Machine Learning along with Deep Learning. These are the most commonly used AI technologies in Crime detection and prevention. The present research showed the growth in the usage of AI technologies. But the field of AI needs further research. The subject of Ai in crime prediction needs an in-depth and thorough study to achieve higher levels of efficiency and accuracy, a huge amount of data sets is required. This will result in further testing and training of AI models which will eventually increase the percentage of accuracy in finding patterns. It is also the duty of criminologists to put more effort into the field so that they can implement the same in their respective studies.

27.1 Introduction

In the recent past, AI has started playing a vital role in all sectors of our society. We can see a substantial increase in the use of AI [1]. In everyday life, we can witness advanced technologies like Google Home, Apple Siri, or the smartphones that we

A. Bag (✉) · S. N. Hadli
Amity Law School, Amity University, Kharora, Chhattisgarh 493225, India
e-mail: akash.bag@s.amity.edu

S. N. Hadli
e-mail: Shailesh.hadli@cumail.in

S. Yadav et al. (eds.), *Proceedings of Second International Conference in Mechanical and Energy Technology*, Smart Innovation, Systems and Technologies 290,
https://doi.org/10.1007/978-981-19-0108-9_27

use, which are completely dependent on hi-tech AI technologies. We also witness use the of AI in the translation of languages and Biometric identification as well. This study is an interdisciplinary study that concerns both the fields of criminology and technology [2]. While conducting the review, the authors have taken care of the required ethical considerations [3, 4]. The study is strictly based on the review of papers that have varied to conclude exhaustively regarding AI and its capabilities of crime prediction and prevention [5]. The study is completely focused to show how in the field of AI and crime detection, the AI systems predict, detect, to prevent crimes with the help of resource allocation [3].

27.2 Definition of AI and Its Subfields

The term Artificial Intelligence is like a big umbrella that covers a wide variety of different applications [6]. We can categorize Ai into three different levels: Weak, medium, and strong [7–13]. The main goal of using AI in crime detection is to find a pattern based on the data provided and predict the time and place of future crimes and potential criminals by analyzing the large data sets [14–16]. The present review focuses on a study that shows how efficiently AI can detect crime and help to prevent it. No earlier study has focused on both of these aspects together. The chosen method has been chosen because there is extensive research which are been conducted on this particular area. The two main questions that the present research tries to answer are.

1. How much capable are the present AI technologies in predicting patterns of crimes?
2. How much capability do the present AI technologies have to prevent the crimes?

27.3 Method

Table 27.1 shows the papers which are included in the review and describes the intricacies of the research in various categories. The present study is completely based on previously conducted studies which the table depicts. Neither of the studies describes the number of primary studies which are taken into consideration.

27.4 Analysis and Result of the Reviews

The analysis was provided in the form of a narrative, which means it was expressed in words rather than numbers. This phase involved highlighting sections of the studies that assessed AI's ability to anticipate or prevent crime. These findings were collated and added to Table 27.2 in the columns of AI capacity, accuracy, AI potential, and

Table 27.1 Description of the included reviews

Name of Author(s), year	Purpose of study	Type of Crime	Research design	Number of primary studies	The primary studies' method	The primary studies' sample size
Prediction/Cybercrime/AIC						
Khairuddin, Alwee, Haron, 2019	The paper Compares various AI and Statistical models which can detect crime patterns	Hotspots in specific areas	Narrative style of review	16	Experiments and comparative studies on different statistical models and AI-models	Crime data ranges from 1960 to 2013
Saidi, Zeki, 2019	A review of how data mining and machine learning techniques can prevent and predict crimes	City crime, terrorist attack, cybercrime, and fraud	Systematic literature review	15	Experiments on data mining and machine learning	Not reported
Butt, Letchmunan, Hafinaz Hassan, Ali, Mubashir., Baqir, Husnain Raza Sherazi, 2020	A review of ML in crime hotspot and detection of the same	Sex crimes, pickpocketing, fraud, harassment, forgery, burglary, weapons and drugs	Systematic literature review	49	Experiments, comparative studies and mapping studies	Not reported
Hosseinkhani, Koochakzaei, Naniz, 2014	A review on mining useful information about suspicious information by means of data mining	Cybercrimes	Narrative literature review	Not reported	Experiments and reviews	Not reported
Shamsuddin, Ali, Alwee, 2017	A review of current implementations of crime prediction methods and them potential	Crimes against properties and crimes of aggression	Narrative literature review	Not reported	Experiments and comparative studies on different prediction methods	Not reported

(continued)

Table 27.1 (continued)

Name of Author(s), year	Purpose of study	Type of Crime	Research design	Number of primary studies	The primary studies' method	The primary studies' sample size
Kounadi, Ristea, Araujo, Leitner, 2020	An overview and evaluation of the current state in spatial crime forecasting	All crime types, such as burglary, theft, assaults	Systematic literature review	32	Reviews, conference papers, reports, and ISI articles	Ranges from 1400 incidents up to 6.6 million
Hayward, Maas, 2020	Review of AI's potential and its application by criminals	Cybercrime	Narrative literature review	Not reported	Surveys, reports, reviews, and workshop	Not reported
Caldwell, Andrews, Tanay, 2020	To identify possible applications of AI in the perpetration of crime and threats	AI-crimes, such as disrupting AI-controlled systems, large-scale blackmail, AI-authored fake news	Preparatory meta-review	Not reported	Workshop, reports, quantitative studies, and reviews	Not reported
Dilek, Çakir, Aydin, 2015	The study shows recent advancements made in the application of AI to detect and prevent crime	Cybercrime, cyber theft, and cyber-fraud	Systematic literature review	65	Experiments and comparative studies on different AI models	Not reported

Note The sample size for the primary studies is made up of crime data made up of criminal occurrences. When the authors do not provide the number of primary studies or the sample size of the primary studies, the term "not reported" is used

the need for more research. Data mining and machine learning could be used to anticipate where and when crimes will occur [17–21]. Five of six assessments on crime prediction indicated that AI systems can predict crime with a high degree of accuracy [18, 22–25]. Data mining is the greatest AI tool for crime prediction [17].

27.5 Conclusion

The First and foremost point is to identify the risk factors. By using more variables or we can say risk factors in this case on which the data sets depend which we insert in AI systems, this could give a positive outcome by providing a substantial increase in efficiency and accuracy of crime detection and prevention of the same. The

Table 27.2 Results of the reviews

Author(s), Year	Which AI- technology?	Result of review	Authors' conclusion	Is AI capable of predicting crimes?	With what accuracy?	The potential of AI	Needs further research?
Prediction/Cybercrime/AIC							
Khairuddin, Alwee, Haron, 2019	Machine learning	The capability of AI to capture non-Linear crime data patterns which the statistical models fail to capture	AI technologies are suitable for handling complex data distributions	Yes	Relatively high accuracy	Inconclusive	Yes
Saidi, Zeki, 2019	Data mining	Data mining is highly accurate in city crime prevention, terrorist attack prevention, cybercrime detection, and fraud detection	The use of historical data to enable prediction in crime prevention requires more research due to its potential	Yes	Relatively high accuracy	Great potential	Yes
Butt, Letchmunan, Hafinaz Hassan, et al. 2020	Data mining and machine learning	Crime hotspot detection should be scalable, deal with sparsity, underlying population, demographic factors, etc.	Crime hotspot detection and prediction need further investigation	Yes	Perform with an accuracy between 63 and 81%, relatively high accuracy	Great potential	Yes

(continued)

Table 27.2 (continued)

Author(s), Year	Which AI- technology?	Result of review	Authors' conclusion	Is AI capable of predicting crimes?	With what accuracy?	The potential of AI	Needs further research?
Hosseinkhani, Koochakzaei, Naniz, 2014	Data mining	The data mining techniques can identify patterns from unstructured and structured data	The investigator's expertise and experience are very important in the search for suspicious information while using data mining	Yes	Relatively high accuracy	Great potential	Yes
Shamsuddin, Ali, Alwee, 2017	Data mining	Using a wide variety of AI technologies can help to detect crime accurately and efficiently	Use of AI can solve major future crime detecting problems	Yes	Relatively high level	Inconclusive	Yes
Kounadi, Ristea, Araujo, Leitner, 2020	Data mining and machine learning	The current method of spatial crime forecasting lacks coherent terminology	Various recommended methods of AI help to fetch data of hotspots, rate of crimes, and kinds of crime	Yes	Relatively high accuracy	Great potential	Yes

(continued)

Table 27.2 (continued)

Author(s), Year	Which AI- technology?	Result of review	Authors' conclusion	Is AI capable of predicting crimes?	With what accuracy?	The potential of AI	Needs further research?
Hayward, Maas, 2020	Machine learning and deep learning	Criminologists must direct the conversation about AI-technology	Human expertise is important in computing since the use of AI is growing	Inconclusive	Inconclusive	Great potential	Yes
Caldwell, Andrews, Tanay, 2020	Machine learning	The most harmful crimes are audio/video impersonation, AI-authored fake news, learning-based cyber-attacks, large-scale blackmail, and disrupting AI-controlled systems	By identifying which crimes are the most harmful and hardest to defeat, AI-technologies can be useful	Yes	Relatively high accuracy	Inconclusive	Yes
Dilek, Çakir, Aydin, 2015	Machine learning and deep learning	The major issue of applying a solid model of AI in crime prediction is to differentiate between behaviors and attack	AI techniques have numerous applications in combating cybercrimes	Yes	Inconclusive	Positive potential	Yes

present study tried to underline a valuable point that in coming future various crime prevention institutions will increase the dependability on AI systems to detect and prevent crime. As discussed earlier the Medium Artificial Intelligence systems which are useful, they have the capacity to understand the world as any human intelligence will and also have the capability to perform multiple tasks at a single point in time like detection of crime, finding patterns in crime including behavioral patterns as well. But we know that when a technology comes with a boon it does carry curse with it. We cannot avoid multiple possibilities of Hazards that the institutions may face while implementing AI systems. It needs in-depth studies, testing and transparency should be analyzed before deciding the percentage of dependability on an AI system. Another major concern that the paper tried to highlight is the amount of trust we can show in the outcomes of these AI systems. Can we trust the results of predictions that are completely dependent on AI? What about the biasness rate? While in terms of ethical implications, there is a possibility that wrong individuals could get detected as criminals or incorrect criminal profiling, different locations may get detected which are out of the category of crime hotspots. The dilemma between AI and Human Intelligence will always raise issues such as how far can we trust AI and how could AI replace Human Intelligence? To what extent can we allow technological AI to replace human judgment?

References

1. Asaro, P.: What is an 'artificial intelligence arms race' anyway. J. Law Policy Inf. Soc. **15** (2019)
2. Benaich, N., Hogarth, I.: State of AI Report 2019—ONLINE.pdf (2019). Accessed: 31 Aug 2021. [Online]. Available: https://drive.google.com/file/d/1RE0I4VMLNoxswXNnleAyoWsWFGeXrz1F/view?usp=sharing&usp=embed_facebook
3. Liv, N.: Crime prediction technologies. International Institute for Counter-Terrorism (2019). Accessed: 31 Aug 2021. [Online]. Available: http://www.ict.org.il/Article/2408/Crime_Prediction_Technologies#gsc.tab=0
4. Lab, S.P.: Crime Prevention: Approaches, Practices, and Evaluations, 8th edn. Routledge, Waltham, MA, USA (2013)
5. Sarnecki, J.: Introduktion till kriminologi. 2, Straff och prevention - Upplaga 1 (9789144102658) | Studentapan, 1st edn., vol. 2. Studentlitteratur AB (2015). Accessed: 31 Aug 2021. [Online]. Available: https://www.studentapan.se/kurslitteratur/introduktion-till-kriminologi-2-straff-och-prevention-9789144102658
6. Piper, K.: The case for taking AI seriously as a threat to humanity. Vox, 21 Dec 2018.https://www.vox.com/future-perfect/2018/12/21/18126576/ai-artificial-intelligence-machine-learning-safety-alignment. Accessed 31 Aug 2021
7. Bello, I.: Beginners guide to artificial intelligence 'AI'. Medium, 19 July 2017. https://becominghuman.ai/beginners-guide-to-artificial-intelligence-ai-ec8a409b6424. Accessed 31 Aug 2021
8. AI Index, "AI Index," Steering Committee, Stanford HAI, Annual Report (2017). [Online]. Available: https://hai.stanford.edu/sites/default/files/2020-10/AI%20Index%202017%20Annual%20Report.pdf
9. Heath, N.: What is artificial general intelligence? | ZDNet (2018). https://www.zdnet.com/article/what-is-artificial-general-intelligence/

10. Yu, D.: PepsiCo sees future in artificial intelligence after launching snack-delivery robot. *Forbes* (2019). https://www.forbes.com/sites/douglasyu/2019/01/22/pepsico-sees-future-in-artificial-intelligence-after-launching-snack-delivery-robot/. Accessed 31 Aug 2021
11. IBM Cloud Education: IBM Cloud Education, 30 June 2021. https://www.ibm.com/in-en/cloud/learn/machine-learning. Accessed 31 Aug 2021
12. Broadhurst, R., Maxim, D., Brown, P., Trivedi, H., Wang, J.: Artificial intelligence and crime. In: Social Science Research Network, Rochester, NY, SSRN Scholarly Paper ID 3407779 (2019). https://doi.org/10.2139/ssrn.3407779
13. Yu, H., Liu, L., Yang, B., Lan, M.: Crime prediction with historical crime and movement data of potential offenders using a spatio-temporal cokriging method. ISPRS Int. J. Geo-Inf. **9**(12), Art. no. 12 (2020). https://doi.org/10.3390/ijgi9120732
14. Rademacher, T., Wischmeyer, T. (eds.): Artificial Intelligence and Law Enforcement. Springer International Publishing (2020). Accessed: 31 Aug 2021. [Online]. Available: https://www.springerprofessional.de/en/artificial-intelligence-and-law-enforcement/17451760
15. Danaher, J.: Philosophical disquisitions: the automation of policing: challenges and opportunities. Philosophical Disquisitions, 12 Oct 2018. https://philosophicaldisquisitions.blogspot.com/2018/10/the-automation-of-policing-challenges.html. Accessed 7 Sept 2021
16. Hajela, G., Chawla, M., Rasool, A.: A Clustering based hotspot identification approach for crime prediction. Procedia Comput Sci **167**, 1462–1470 (2020). https://doi.org/10.1016/j.procs.2020.03.357
17. Neill, D.B., Gorr, W.: Detecting and preventing emerging epidemics of crime, p. 1. Heinz School of Public Policy (2007)
18. Khairuddin, A., Alwee, R., Haron, H.: A review on applied statistical and artificial intelligence techniques in crime forecasting. IOP Conf. Ser. Mater. Sci. Eng. **551**, 012030 (2019). https://doi.org/10.1088/1757-899X/551/1/012030
19. Boland, A., Cherry, M.G., Dickson, R.: Doing a systematic review: a student's guide, 2nd edn. Sage, Los Angeles (2017)
20. Friberg, F.: Dags för uppsats: vägledning för litteraturbaserade examensarbeten - Att skriva - häftad (9789144115795) | Adlibris Bokhandel, 3rd edn. Studentlitteratur (2017). Accessed: 31 Aug 2021. [Online]. Available: https://www.adlibris.com/se/bok/dags-for-uppsats-vagledning-for-litteraturbaserade-examensarbeten-9789144115795
21. Eriksson Barajas, K., Forsberg, C., Wengström, Y.: Systematiska litteraturstudier i utbildningsvetenskap : vägledning vid examensarbeten och vetenskapliga artiklar. Natur & Kultur, Stockholm (2013)
22. Butt, U.M., et al.: Spatio-temporal crime predictions by leveraging artificial intelligence for citizens security in smart cities. IEEE Access **9**, 47516–47529 (2021). https://doi.org/10.1109/ACCESS.2021.3068306
23. Hosseinkhani, J., Koochakzaei, M., Keikhaee, S., Naniz, J.H.: Detecting suspicion information on the web using crime data mining techniques. Int J Adv Comput Sci Inf Technol **3**(1), 10 (2014)
24. Kounadi, O., Ristea, A., Araujo, A., Leitner, M.: A systematic review on spatial crime forecasting. Crime Sci. **9**(1), 7 (2020). https://doi.org/10.1186/s40163-020-00116-7
25. Shamsuddin, N.H.M., Ali, N.A., Alwee, R.: An overview on crime prediction methods. In: 2017 6th ICT International Student Project Conference (ICT-ISPC) (2017). https://doi.org/10.1109/ICT-ISPC.2017.8075335

Chapter 28
Role of Artificial Intelligence in Energy and Power Engineering

Hicham Mhamdi, Omar Kerrou, Chetan M. Thakar, Mohammed Bakraoui, and Mohammed Aggour

Abstract Over the last decade, the energy issue has been a major source of concern in many countries, and the usage of renewable energy has risen in importance internationally. Wind speed prediction is required to enhance the quantity of energy generated. The wind speed forecast balances the energy required and the energy generated. Voltage stability has recently received a lot of attention from academics due to the fact that it has become a key concern for modern power system operators. Several countries have experienced widespread blackouts as a result of voltage instability issues. Accurate forecasting of power demand and price is considered as one of the most significant research topics in electrical engineering in the present and future, with academics placing a strong focus on demand and price prediction in deregulated markets. The predictive nature of various machine learning algorithms makes them most suitable tool to deal with problems related to energy and power engineering. Machine learning techniques are capable of analyzing past data and on the basis of that analysis, these algorithms are capable of predicting future results. This article provides applications of machine learning in energy and power engineering.

28.1 Introduction

The energy crisis [1] has been a major worry in many nations over the last decade, and renewable energy use has grown in importance globally. Wind speed prediction must be performed in order to increase the amount of energy generated. The wind

H. Mhamdi · O. Kerrou · M. Bakraoui · M. Aggour
Laboratory of Electronic Systems, Information Processing, Mechanics and Energetics, Faculty of Sciences Kenitra, University Ibn Tofail Kenitra, Kenitra, Morocco

C. M. Thakar (✉)
Assistant Professor, Department of Mechanical Engineering, Government College of Engineering, Karad, Maharashtra, India
e-mail: cthakar12@gmail.com

S. Yadav et al. (eds.), *Proceedings of Second International Conference in Mechanical and Energy Technology*, Smart Innovation, Systems and Technologies 290,
https://doi.org/10.1007/978-981-19-0108-9_28

speed forecast provides a balance between the energy required and the energy generated. The extremely precise and dependable wind speed forecast models are used to optimize operating expenditures and improve the grid system's operational features.

Voltage stability [2] has lately gotten a lot of attention from academics since it has become a major problem for modern power system operators. As a result of voltage instability concerns, several nations have experienced extensive blackouts. Voltage instability problems develop when there is a continuous reduction in voltage as a result of irregular operation caused by a power supply disruption.

In the present and future, accurate forecasting of electricity demand and price [3] is recognized as one of the most important study areas in electrical engineering, with academics placing a high emphasis on demand and price prediction in deregulated markets.

Traditional techniques rely on fundamental algorithms based on linear analysis and, as a result, are incapable of accurately anticipating nonlinear demand series. The demand series has nonstationary means and variances due to stochastic characteristics. Academics are creating new artificial intelligence learning algorithms to solve these issues.

Intelligence typically involves the artificial manufacture of the human mind that can learn, plan, perceive or comprehend natural language. Computer systems theory and development is usually capable of doing human intelligence functions, for example, visual perception, voice recognition, decision making, and translation of languages. IT industry focuses mostly on equipment that is designed to function as a human being. John McCarthy (Daddy of AI) characterized AI in particular as knowledge of science and technology in the development of intelligent computer programmers.

Two of the most often used AI approaches are machine learning and deep learning. These models draw on data and are utilized by people, companies, and government agencies for forecasting. Machine learning models are currently being developed for the complexity and variability of data in the food sector.

28.2 Applications of Artificial Intelligence in Energy and Power Engineering

28.2.1 Machine Learning Approaches for Wind Speed Prediction in Renewable Energy Systems

Over the previous decade, in many countries, the energy problem has been a significant concern, and renewable energy usage has become more and more important worldwide. The power source is an important part of an expanding economy and the generation of energy from plentiful wind resources will lead to the creation of an energy model and help to optimize resource allocation. Wind is an extensively available natural resource and renewable energy. The pure and pollution-free source of electricity has been found to be wind power. The direction, speed, and time of occurrence are governed mostly by wind.

In order to obtain wind energy from natural wind, it is the force with which the wind passes or the speed of the wind. Wind strength or wind speed is typically nonlinear and naturally varied. Regardless of their original form, wind is able to supply the appropriate amount of energy for the regular demands of the country. In order to raise the quantity of energy generated, the wind speed prediction [4] has to be carried out. The forecast of wind speed [5] establishes a balance between the need for and the energy generated. The highly accurate and reliable wind speed forecast models serve as a tool to optimize operational expenses and improve the operational characteristics of the grid system.

For wind grid systems, the forecast of wind speed [6] is crucial. The wind speeds are all influenced by moisture and humidity, air pressure, temperature, precipitations, and other factors. Precise forecast of wind speed enables energy experts to plan energy requirements. Wind velocity forecasting has a variety of purposes, including grid electricity, satellite and rocket launch, agriculture, military control activities, etc. This prediction guarantees safe wind power generation and facilitates the integration of wind power in electrical grids.

For simulation task duration prediction, Chen and Wang [7] submitted the categorization of the Ensemble Artificial Neural Network (ANN). The proposed methodology considerably decreased the estimated time compared with the six other alternatives. The cloud-based, short-term solar irradiance forecast systems have been created by McCandless et al. [8] which offer a 15-min average clarity index estimate. A K-means algorithm predicts the cloud regime, based on surface weather and irradiance observations. The Artificial Neural Networks (ANNs) are trained to forecast the Clarity Index.

The ANN modeling technique for wind power curves utilizing six parameters was established by Pelletier et al. [9]. From over 50 parameters studied in variability between observed and anticipated power production, the selected six characteristics were chosen. Researchers [10] developed a novel hybrid model that combines the SOM-OPELM and time series techniques for the prediction of global solar radiation on the horizon. The proposed SOM-OPELM model performs in terms of efficiency and performance estimations other standard methods.

The UT Dallas planar MEA, built from low-cost material and methods, was described and characterized [11]. Detailed electrophysiological tests show that MEA types perform roughly the same for network characteristics, for instance, network spike rate and burst rate recording, as well as pharmacological reactions to a vast array of neuro-pharmacologic medications.

Authors [12] have proposed an artificial neural network that uses the estimated pulse rate to identify abnormal inter-beats. A deep neural network (DNN) joint synthesizer with tapped-delay input lines was proposed [13], which enables the model to gather context information in the pathway without any post-processing. The longitudinal dynamics of a flexible air respiratory hypersonic vehicle with control input constraints explored the design of a novel Adaptive neural controller [14].

Authors [15] submitted a wind energy analysis based on the Mycielski technique and the K-means classification. Wind power evaluation. The predicted results show that the clustering of K-means is more accurate than the Mycielski technique. Saleh et al. suggested a hybrid neuro-fuzzy wind power system [16]. Data clustering by means of modified Fuzzy C-Means determines the correct quantity (FCM). Four data subsets separately into four seasons of the year will be used for the predictive model.

28.2.2 Machine Learning Algorithms for Voltage Stability Assessment

Voltage stability has recently received a lot of attention from academics since it has become a serious concern to modern power system operators. Many countries have experienced widespread blackouts as a result of voltage instability issues. Voltage instability difficulties occur when there is a continual decrease in voltage as a result of irregular operation induced by a disturbance in the power supply. Based on power system disturbances, voltage stability is classed as minor or big disturbance voltage stability. When a tiny disturbance, such as a slight load increase, happens in the power system, small disturbance voltage stability defines the behavior of the power system. By linearizing the system equations, the influence of such a disturbance on node voltages may be studied. When a significant disturbance, such as a quick load shift or a severe line outage, happens in the power system, large disturbance voltage stability defines the behavior of the power system.

The SVM [17] is a supervised learning method that determines the optimal kernel function to execute the nonlinear transformation. Furthermore, it uses linear regression to determine the support vectors for optimum estimate. Using the support vectors, the SVM selects the best hyper-plane to accomplish the classification and regression tasks. In general, there are several power system variables that may be supplied into the computational intelligent model as input. If all of these factors are fed into the network, greater network complexity lengthens training time. As a result, it is necessary to use an appropriate approach to decrease the network input dimensions, such as feature selection or feature extraction. Using these approaches, any unnecessary and superfluous data may be removed.

In the power system, Chakrabarti [18] has developed an online voltage stability monitoring methodology employing ANN's reduced input properties deriving from the regression method for various scenarios. This paper offers a technique for building ANNs for an assessment of the power system's online voltage stability. A multilayer perceptron (MLP) neural network is established to calculate the voltage stability margin. As inputs to the ANN activated and reactive load power is used and the MW margin indicates the voltage stability level. A regression-based approach is used to choose the appropriate ANN features. The approach presented was shown by the 39-bus system in New England. For various circumstances, separate ANNs are used to control voltage stability. The ANN is taught with a conventional back-propagating approach for each network topology that takes a longer duration of training. This model generation approach requires more human effort and expenses since there are more adjustment elements.

28.2.3 Artificial Intelligence Techniques for Electricity Price Forecasting and Classification

Electrical is delivered to household and industrial users in a deregulated electricity market under the rules and regulations of deregulation standards. Indeed, government-controlled power sectors have been replaced with competitive power markets, and therefore monopolistic features of power sectors have been greatly decreased. The energy trading mechanism is based on contractual agreements between market players, based on which the spot electricity price, known as the market-clearing price (MCP), is set for consumers to ease auction and bidding procedures. As a non-storable special commodity, it is critical for the power sectors to keep electricity in a balanced state in terms of production and consumption. In the present and future, precise forecast of electricity demand and price is regarded as one of the key study topics in electrical engineering, with researchers placing a high value on demand and price prediction in deregulated markets [19].

Traditional approaches rely on basic algorithms based on linear analysis, and as a result, they are unable to properly anticipate nonlinear demand series. Because of stochastic features, the demand series exhibits nonstationary means and variances. To address these challenges, academics are developing new intelligent learning algorithms. Artificial neural networks (ANN) [20], support vector machines (SVM) [21], functional link networks [22], and rule-based systems [23] are examples of intelligent learning approaches.

28.3 Conclusion

Wind speed prediction is necessary to increase the amount of energy produced. The forecasted wind speed balances the energy required and the energy generated. Voltage stability has lately gotten a lot of attention from academics since it has become a major problem for modern power system operators. As a result of voltage instability concerns, some nations have experienced extensive blackouts. Accurate forecasting of power demand and price is regarded as one of the most important research problems in electrical engineering today and in the future, with academics focusing heavily on demand and price prediction in deregulated markets. The predictive nature of various machine learning algorithms makes them most suitable tool to deal with problems related to energy and power engineering. Machine learning techniques are capable of analyzing past data and on the basis of that analysis, these algorithms are capable of predicting future results. This article provides applications of machine learning in energy and power engineering.

References

1. Bou-Rabee, M., Lodi, K.A., Ali, M., Faizan Ansari, M., Tariq, M., Anwar Sulaiman, S.: One-month-ahead wind speed forecasting using hybrid ai model for coastal locations. IEEE Access **8**, 198482–198493 (2020). https://doi.org/10.1109/ACCESS.2020.3028259
2. Huang, X., Yan, Z.: Probabilistic static voltage stability assessment for distribution network considering correlated renewable energy. In: 2020 5th Asia Conference on Power and Electrical Engineering (ACPEE), pp. 289–293 (2020). https://doi.org/10.1109/ACPEE48638.2020.9136415
3. Jahangir, H., et al.: A novel electricity price forecasting approach based on dimension reduction strategy and rough artificial neural networks. IEEE Trans. Industr. Inf. **16**(4), 2369–2381 (2020). https://doi.org/10.1109/TII.2019.2933009
4. Santhosh, M., Sai, M.D., Mirza, S.: Ensemble deep learning model for wind speed prediction. In: 2020 21st National Power Systems Conference (NPSC), pp. 1–5 (2020). https://doi.org/10.1109/NPSC49263.2020.9331836
5. Long, Y., Zhang, R.: Short-term wind speed prediction with ensemble algorithm. In: 2020 Chinese Automation Congress (CAC), pp. 6192–6196 (2020).https://doi.org/10.1109/CAC51589.2020.9326917
6. Akash, R., Rangaraj, A.G., Meenal, R., Lydia, M.: Machine learning based univariate models for long term wind speed forecasting. In: 2020 International Conference on Inventive Computation Technologies (ICICT), pp. 779–784.https://doi.org/10.1109/ICICT48043.2020.9112534
7. Chen, T., Wang, Y.-C.: Estimating simulation workload in cloud manufacturing using a classifying artificial neural network ensemble approach. Robot. Comput.-Integr. Manuf. **38**, 42–51 (2016)
8. McCandless, T., Haupt, S., Young, G.: A regime-dependent artificial neural network technique for short-range solar irradiance forecasting. Renew. Energy **89**, 351–359 (2016)
9. Pelletier, F., Masson, C., Tahan, A.: Wind turbine power curve modelling using artificial neural network. Renew. Energy **89**, 207–214 (2016)
10. Wu, Y., Wang, J.: A novel hybrid model based on artificial neural networks for solar radiation prediction. Renew. Energy **89**, 268–284 (2016)
11. Charkhkar, H., Arreaga-Salas, D.E., Tran, T., Hammack, A., Voit, W.E., Pancrazio, J.J., Gnade, B.E.: Novel disposable microelectrode array for cultured neuronal network recording exhibiting equivalent performance to commercially available arrays. Sens. Actuators, B Chem. **226**, 232–238 (2016)
12. Yeh, B.-C., Lin, W.-P.: Using a calculated pulse rate with an artificial neural network to detect irregular interbeats. J. Med. Syst. **40**(3), 1–7 (2016)
13. Aryal, S., Gutierrez-Osuna, R.: Data driven articulatory synthesis with deep neural networks. Comput. Speech Lang. **36**, 260–273 (2016)
14. Bu, X., Wu, X., He, G., Huang, J.: Novel adaptive neural control design for a constrained flexible air-breathing hypersonic vehicle based on actuator compensation. Acta Astronaut. **120**, 75–86 (2016)
15. Goh, H., Lee, S., Chua, Q., Goh, K., Teo, K.: Wind energy assessment considering wind speed correlation in Malaysia. Renew. Sustain. Energy Rev. **54**, 1389–1400 (2016)
16. Saleh, A.E., Moustafa, M.S., Abo-Al-Ez, K.M., Abdullah, A.A.: A hybrid neuro-fuzzy power prediction system for wind energy generation. Int. J. Electr. Power Energy Syst. **74**, 384–395 (2016)
17. Adhikari, Naetiladdanon, S., Sagswang, A., Gurung, S.: Comparison of voltage stability assessment using different machine learning algorithms. In: 2020 IEEE 4th Conference on Energy

Internet and Energy System Integration (EI2), pp. 2023–2026 (2020). https://doi.org/10.1109/EI250167.2020.9346750
18. Chakrabarti, S.: Voltage stability monitoring by artificial neural network using a regression-based feature selection method. Expert Syst. Appl. **35**(4), 1802–1808 (2008)
19. Katkar, S.V., Dongale, T.D., Shinde, P.P., Kamat, R.K.: Calculation of electrical parameter of solar cell using java framework. AGU Int. J. Sci. Technol. (AGUIJST) **5** (2017). e-ISSN 2455-0493; p-ISSN 2455-3808
20. Hao, Q., Srinivasan, D., Khosravi, A.: Short-term load and wind power forecasting using neural network-based prediction intervals. IEEE Trans. Neural Netw. Learn. Syst. **25**(2), 303–315 (2014)
21. Ceperic, E., Vladimir, C., Baric, A.: A strategy for short-term load forecasting by support vector regression machines. IEEE Trans. Power Syst. **28**(4), 4356–4364 (2013)
22. Ren, Y., Suganthan, P.N., Srikanth, N., Amaratunga, G.: Random vector functional link network for short term electricity load demand forecasting. Inf. Sci. **367**, 1078–1093 (2016)
23. Arora, S., Taylor, J.W.: Short-term forecasting of anomalous load using rulebased triple seasonal methods. IEEE Trans. Power Syst. **28**(3), 3235–3242 (2013)

Chapter 29
Forest Fire Detection Using Satellite Images

Kakarapalli Lalitha and Geesala Veerapandu

Abstract Forest represents a complex ecosystem on earth that is a refuge to several living beings such as plants, animals, birds and also a huge resource of minerals, lakes and rivers. It almost covers 30% land of the earth and is highly necessary to balance the ecosystem along with the climate. Therefore, the loss of forests is a severe disaster. Along with deforestation, forest fire has a major impact. Although, the wildfires are not in control fully and most of the time in the various region of the earth, which not only harm the climate but also make a very bad impact on the ecosystem. As forest fires are a very common phenomenon, some of the preventive actions and units are already defined such as McArthur Forest Fire Danger Index (FFDI), the establishment of a separate section in the disaster team for monitoring and assessment of wildfire but intense research for preventing wildfire is highly necessary. Nowadays, satellites are used to scan the earth's surface. This technique can also be used to detect forest fires. With the help of the spatial high-resolution imagery system, the hot spot areas can be accurately located for determining the forest fire's locations.

29.1 Introduction

Wildfires are one of the most challenging global environmental issues that cause significant harm to natural and environmental resources. Most cases of fire in forests occurred during the summer seasons. Once the spark takes place anyhow in the forest, the large unmanaged area is filled with dry woods and leaves which act as fuel to this spark and due to this favorable and highly flammable environment, the spark will turn into the uncontrollable fire within a few moments. Estimating the risk of forest fires would be a great help to disaster management authorities in taking the appropriate actions in order to minimize the losses in terms of the evacuation of the area and saving lives of animals as well as local residents [1–3]. A huge number of commercial

K. Lalitha · G. Veerapandu (✉)
Aditya College of Engineering, Surampalem, AP, India
e-mail: veerapandu_ece@acoe.edu.in

S. Yadav et al. (eds.), *Proceedings of Second International Conference in Mechanical and Energy Technology*, Smart Innovation, Systems and Technologies 290,
https://doi.org/10.1007/978-981-19-0108-9_29

fire detection sensor systems are available in the market, but all of them have their own limitations. For use in wide-open areas such as forests, certain parameters of these alarming systems such as response time, maintenance requirements, cost, etc. must be critically taken into account. As forests are a very important factor to maintain our ecosystem, hence it is highly necessary to prevent these forests from any losses to maintain the ecosystem balance of nature. For taking one example forests are highly needed to maintain our global climate stability. However, due to natural (e.g., forest fire disasters) and human-caused disturbances, the worldwide forest area has changed considerably, resulting in a deterioration of forest covering and carbon stores in terrestrial ecosystems, where a variety of living organisms and resources coexist. The most serious uncontrollable threat in the woods is a forest fire [4, 5]. To identify the event of forest fires in a stimulated time, a variety of methods have been used. One of the conventional ways to measure the intensity of the wildlife fire is named as McArthur Forest Fire Danger Index (McArthur FFDI). It is developed by the scientist A. G. McArthur in 1960. This method necessitates a significant amount of data from numerous sources, as well as a lengthier computation time. McArthur FFDI is calculated using meteorological stations and ground data. However, collecting these diverse data sets for the assessment of fire hazards is quite challenging and also unavailable sometimes. These limitations motivate to pursue this research in pursuit of a universal approach to quickly assess fire threats and accounting for it as a huge disaster. Researchers worldwide are continuously working in the direction to minimize the losses due to the wildlife fire to save the climate. In this context, the goal of the study is to adapt the McArthur fire danger indicator by replacing its drought with a satellite-based indicator. Using satellite images to identify the wildfires are one of the potential techniques among all. In this technique, the images of the forest are captured by the satellite and with the help of image processing, these images are analyzed. As in the orbit of the earth, there are numerous satellites equipped with high-resolution cameras, this method is one of the most efficient techniques to detect the fire in forest in timely and help the disaster management team to make suitable plans, take appropriate preventive actions and analyse the risk factors in order to minimize the losses [6–8]. A block diagram of Fig. 29.1 stepwise approach of identification of the forest fire with the help of satellite imagery is given below.

In Fig. 29.1 all of these techniques consider weather factors while calculating a numerical index that measures fire danger. For fire and land managers, fire danger indices are critical indicators. The severity of a fire is usually defined as a number or an adjective. For analyzing the fire conditions in order to take preventive action, it is necessary to understand how fire affects the soil mantle, stems, and trees, as well as how to detect underground fires. All the influencing climatic variables such as air temperature, wind speed, relative humidity, the topography of the surface, etc. are also necessary to be taken care of. Using satellite images, help to make us understand the boundaries of fire regions and estimate their extent, these images also direct us toward the effects of fire in a specific circumstance, as well as the degree of difficulty to control it, which in turn enable us to take quick and cost-effective approaches to minimize the losses [9–13].

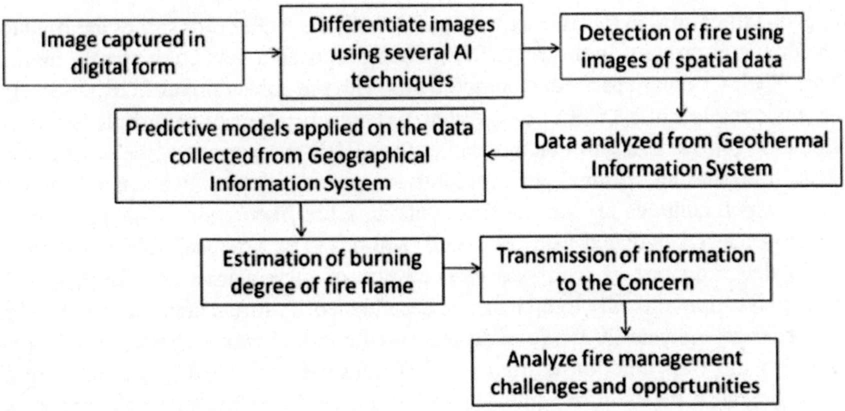

Fig. 29.1 Stepwise identification of detecting forest fire and its processing using satellite

29.2 Related Work

In the year 2003, Yong et al. concluded their study based on forest fire that occurred during May 4 to June 6, 1987, in northeast China and stated that the forests, as a complex ecosystem, provide a safe haven for all living creatures, including plants, animals, and birds, as well as numerous resources such as rivers, lakes, and minerals. Jungles encompass around 30% of the earth's surface. Satellites have been used to search for flames on the earth's surface. These satellites with identical orbits and having high-resolution spatial imaginary systems are used to capture the hot spot locations (using their zoom property) and correctly detect individual fires so that the type of cover impacted by the fire can be determined. Furthermore, satellite imaginary can be constructed at various scales, which is also helpful to estimate the factors accurately [1]. In the year 2007, Badrinath et al. performed their observation on the region of northeast India and presented their observations as tropical biomass ablaze and accompanying aerosol emissions into the environment contribute significantly to atmospheric disturbance and climate change. Using satellite data, this letter intends to investigate the influence of forest fires on aerosol concentration over this area. Also due to slash-and-burn agricultural techniques, the forests of this region are susceptible to catastrophic fire occurrences every year between January and May, the day-to-day active forest fire sites throughout the north-eastern region can be calculated by means of night-time satellite records from the Meteorological Satellite Program-Operational Line Scan system [2].

In reality, forest fire monitoring and assessment are critical components of good fire management. Estimating the risk of forest fires assists disaster management officials in taking the appropriate mitigation measures to reduce losses and evacuate local residents. Based on meteorological station parameters and ground datasets, fire hazard rating systems forecast the fire danger [3]. By using the image processing techniques for the identification of forest fires, at first, a model is required for fire pixel classification. In this rule-based color model, the RGB color spaces are used to develop an algorithm. In the year 2012, Vipin proposed a new color system model using Y Cb Cr color spaces and compare this with the conventional RGB model. In these two models, the YCbCr color system has the advantage of being able to differentiate brightness and chrominance better than RGB color space while comparing. These proposed algorithms' performances are evaluated using two sets of photos, one of which contains fire and another containing fire-like regions. The algorithm's performance is calculated using standard methods. The proposed method offers a higher detection rate as well as a reduced rate of false alarms. As the proposed system is computationally light, it can be used to identify forest fires in real-time [6].

Earlier, in the year 2005, Yu et al. presented a model using a genetic algorithm in which the cost function is minimized for unsupervised change detection in a forest fire with the help of multi-temporal satellite images. In this technique, at first, the multi-temporal satellite images are differentiated into two separate sections named "changed" and "unchanged", then the difference image is computed through the binary change detection mask realization with the help of a genetic algorithm. Now, determine the mean square error between the difference image values and its average for each of the regions and use the weighted sum of the MSE as a cost value for the corresponding change detection mask realization for both changed and unchanged regions. The final change detection is computed by applying the GA with the minimum cost. This minimum cost is evaluated using the initial realization of the binary change detection mask [7]. Apart from the summer season, the forest fire also occurs in the winter season. In the year 2017 Marchese et al. presented a new study in which the investigation has done on the winter fire regions. This study is basically a comparative study in which the RST-FIRES are compared with the well-established fire detection systems in terms of their appropriateness in identifying fires in the winter region. This investigation has done with the help of an Advanced Very High-Resolution Radiometer (AVHRR).

29.3 Forest Fire Detection Using Satellite

It is observed that despite the benefits of self-adaptive algorithms such as RST-FIRES at a local/regional level, the incidences of jungle ablaze in the winter season can easily be identified with the help of satellite images, which is a less complex process [8]. However, finding the smoke color, as a wildfire is identified in a smoke color, is a tedious job due to the variation in smoke color, environmental illumination and bad quality of images. It needed the appropriate combination of color space and pixel level. A variety of color space transformations are examined even from the non-smoke classes of pixels, between smoke and a mixture of color space and pixel-level for identifying an efficient smoke segmentation algorithm [9]. For detection and mapping of active forest fire using satellite data, two major algorithms are designed, which is based on brightness temperature band. From this study, it is observed that the short-wave infrared band is switched to the thermal brightness temperature band

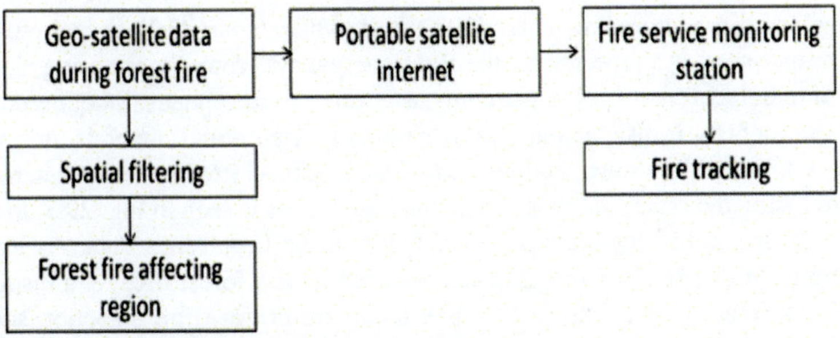

Fig. 29.2 Methods for detecting forest fire using satellite

if the data is collected during the night [10]. In Fig. 29.2 represents step by step approach of identification of the forest fire with the help of satellite imagery is given in Fig. 29.2.

As the forests are very wide-open areas and it is highly challenging to observe and control the fires in inaccessible and gigantic areas like a jungle. To address this issue Ganesan et al. proposed a new modified fuzzy c-means clustering approach in order to identify the probable region of wildfire using RGB and CIEL lab color space [11]. This early approximate detection of forest fire helps the forest management to make preventive action plans from spreading fire which causes severe ecological and financial damage. By using the negative relationship between the surface and land temperature, the threshold temperature can be calculated on which the probability of getting the forest fire is high. This computation is based on the differences in brightness temperature in remote-sensing data and an expected one derived from a regression model of brightness temperature and vegetation amount as indicated by the normalized difference vegetation index [12]. The experimental and technical research in the domain of fire monitoring is analyzed by Ruckchin et al. in the year 2015 and they proposed the algorithm for fire identification and its spreading rate measurement using satellite images. This technique is implemented in the smart telecommunication structures uses NM640X processor system which is equipped with the ability to learn the conditions of uncertainty [13]. Regardless, these advanced levels of findings and studies in this field are still needed improvement. This satellite-based wildfire detection is still immature and fails to a quick and efficient control forest fire. Due to the radiation beam and receiving signal direction and also because the intensity reduces with the inverse square of the distance, this technique suffers a great limitation as it could not identify forest fire at the initial stage.

29.4 Detection Result and Discussion

There were 140 firefighting agency reports in Japan airlines, 156 by USA and 52 flight reports in the Russian area that matched the satellite images. Its ground truth data and satellite pictures matched 4502 times, while Japan Airlines had only 450 such comparisons. As reported by the Russian agency, ground monitoring sites are not used to detect forest fires in incidents, according to the agency. Satellite photography can identify forest fires at a higher rate than traditional methods, despite its limitations. The oldest hotspot connected with each forest fire incident is determined by aggregating the comparative data, as illustrated. This is seen in Fig. 29.3. In other scenes, the forest fire region was not included in the time-series data. As long as there are hotspots, satellite imaging can help pinpoint a forest fire. As a result, we are able to detect forest fires earlier than local authorities. Fire detection success was characterized as a hotspot on a satellite image that corresponded to the fire. A comparable hotspot was identified within 10 km of the reported site for a period of 11 days surrounding the forest fire observation date. At least one hotspot on satellite imagery taken before the fire was seen qualifies as early discovery. The photographs we used were not geolocated due to technical concerns. Imagery from the Advanced Very High-Resolution Radiometer (AVHRR) provides an accuracy of up to 10 km without geolocation data. In addition, we defined the forest fire detection rate as the percentage of forest fires that were detected among the total number of forest fire reports made. Detection of forest fires was deemed a success by the number of fires that were noticed earlier than ground surveillance. According to our definition, successful early detection involves being able to detect reported fires before they are detected on the ground. We define early detection rate as the early detection rate of all reported fires. Forest fire detection rates are shown in Table 29.1.

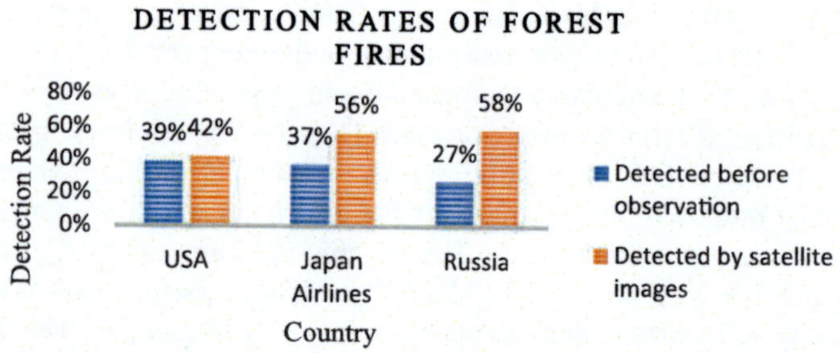

Fig. 29.3 Graphical representation of forest fire detection

Table 29.1 Detection rates of forest fires for each observation source

Fire observed by	No. of reports	Detected before observation	Early detection rate (%)	Detected by satellite images	Total detection rate (%)
USA	156	49	39	67	42
Japan Airlines	52	19	37	29	56
Russia	140	38	27	81	58
Total	348	106	31	177	52

Forest fire early detection rates were poor despite a relatively high fire detection rate, as seen in Fig. 29.3. Many of the pixels that match the reported flames are skewed and the detection analysis is limited. There was a considerable discrepancy between the observations of the firefighter and those of JAL's aircraft. Smaller flames can be seen by fire fighters' shorter observation distance of a few hundred meters in altitude. Location of nodes in a sensor network is optimized by minimizing the number of connections and computing costs. Estimation of the distance between regular nodes begins with head anchors and proceeds from there. A regular node's location is decided based on the anchor node's position. With a greater anchor ratio, location estimation becomes more accurate. There are a number of different ways to find grids, including anchor nodes with one and two hops and multiple hops. The average of the valid grids is used to identify the location of a normal sensor node. For the purpose of removing faulty grids and improving the accuracy of location estimates, head anchor information is used. A sensor network's accuracy in estimating position is increased by employing the ratio of anchor nodes to head anchors. The algorithm's accuracy grows as the number of anchors and anchors heads increases.

29.5 Conclusion

Wildfire is one of the major disasters for the ecosystem and the climate. If the ratio between the areas covered by forest and the human colony has misbalanced, the life on the earth will be in danger. This necessitates removing the forest fire at any cost. Among all of the conventional and modern techniques to identify and control forest fire, using satellite images is one of the efficient and effective methods. However, this technique has its own limitations but with the help of advanced image processing techniques and highly efficient complex algorithms, this technique has a major impact. Although, segmentation, illumination variation need to improve more in order to act this technique as a full proof and this study has still some deficiency but a mile to go with study and need to do a lot of research and improvements this method will act as a removal technique of wildfire.

References

1. Pang, Y., Guoqing, S., Li, Z., Che, X., Dong, Y., Zhongjun, Z.: Land cover change monitoring after forest fire in Northeast China. Int. Geosci. Remote Sens. Symp. 5(C).3383–3385 (2003). https://doi.org/10.1109/igarss.2003.1294790
2. Badarinath, K.V.S., Kharol, S.K., Chand, T.R.K.: Use of satellite data to study the impact of forest fires over the northeast region of India. IEEE Geosci. Remote Sens. Lett. **4**(3), 485–489 (2007). https://doi.org/10.1109/LGRS.2007.896738
3. Schwarz, D.A., et al.: Chronic, wireless recordings of large-scale brain activity in freely moving rhesus monkeys. Nat. Methods **11**(6), 670–676 (2014). https://doi.org/10.1038/nmeth.2936

4. Suresh Babu, K.V., Roy, A., Vanama, V.S.K., Prasad, P.R.: Assessment of forest fire danger using automatic weather stations and MODIS TERRA satellite datasets for the state Madhya Pradesh, India. In: International Conference on Advances in Computing, Communications and Informatics (ICACCI 2017), vol. 2017-Janua, pp. 1876–1881 (2017). https://doi.org/10.1109/ICACCI.2017.8126118
5. Celik, T.: Change detection in satellite images using a genetic algorithm approach. IEEE Geosci. Remote Sens. Lett. **7**(2), 386–390 (2010). https://doi.org/10.1109/LGRS.2009.2037024
6. Vipin, V.: Image processing based forest fire detection. Int. J. Emerg. Technol. Adv. Eng. **2**(2), 87–95 (2012)
7. Yu, L., Wang, N., Meng, X.: Real-time Forest fire detection with wireless sensor networks. In: International Conference on Wireless Communications, Networking and Mobile Computing (WCNM 2005), vol. 2, pp. 1214–1217 (2005). https://doi.org/10.1109/wcnm.2005.1544272
8. Marchese, F., et al.: Issues and possible improvements in winter fires detection by satellite radiances analysis: lesson learned in two regions of Northern Italy. IEEE J. Sel. Top. Appl. Earth Obs. Remote Sens. **10**(7), 3297–3313 (2017). https://doi.org/10.1109/JSTARS.2017.2670059
9. Krstinić, D., Stipaničev, D., Jakovčević, T.: Histogram-based smoke segmentation in forest fire detection system. Inf. Technol. Control **38**(3), 237–244 (2009)
10. Abuelgasim, A., Fraser, R.: Day and night-time active fire detection over North America using NOAA-16 AVHRR data. Int. Geosci. Remote Sens. Symp. **3**(613), 1489–1491 (2002). https://doi.org/10.1109/igarss.2002.1026158
11. Ganesan, P., Sathish, B.S., Sajiv, G.: A comparative approach of identification and segmentation of forest fire region in high resolution satellite images. In: IEEE WCTFTR 2016—Proceedings of the 2016 World Conference on Futuristic Trends in Research and Innovation for Social Welfare, pp. 1–6 (2016). https://doi.org/10.1109/STARTUP.2016.7583959
12. Huh, Y., Lee, J.K.: Enhanced contextual forest fire detection with prediction interval analysis of surface temperature using vegetation amount. Int. J. Remote Sens. **38**(11), 3375–3393 (2017). https://doi.org/10.1080/01431161.2017.1295481
13. Ruchkin, V., Kolesenkov, A., Kostrov, B., Ruchkina, E.: Algorithms of fire seat detection, modeling their dynamics and observation of forest fires via communication technologies. In: Proceedings of the 2015 4th Mediterranean Conference on Embedded Computing (MECO 2015)—Incl. ECyPS 2015, BioEMIS 2015, BioICT 2015, MECO-Student Chall, pp. 254–257 (2015). https://doi.org/10.1109/MECO.2015.7181916

Chapter 30
Evaluating the Role of Robotics, Machine Learning and Artificial Intelligence in the Field of Performance Management

Amgad S. D. Khaled, Dilip Kumar Sharma, T. Yashwanth, Vanga Mohan Krishna Reddy, Rumi Iqbal doewes, and Mohd Naved

Abstract Although intelligent automation academic production has evolved fast (e.g., artificial intelligence, robots) we currently lack a full knowledge of the effects on organizational (companies) and individual (workforce) levels of the use of these innovations in performance management (PM). The purpose of this study is to systematize scholarly input on smart automation to date and to elucidate their primary contributions to the performance management difficulties. We found 45 publications on artificial intelligence, robotics as well as other technologically advanced within PM settings in the international business (IB) and general management (GM) and information management (IM) journals. The results demonstrate that smart automation technologies provide a new strategy to managing personnel and improving company performance, providing various performance management opportunities and also major technological and ethical problems. Its influence is based on performance management techniques such as job substitution, work collaboration between human robots and AI, policy-making and apprenticeship possibilities, and performance management activities, namely recruitment, education, and job performance. This paper addresses these changes in-depth and the main contribution to future studies theories and practices.

A. S. D. Khaled
Al-Rowad University, Taiz, Yemen

D. K. Sharma
Department of Mathematics, Jaypee University of Engineering and Technology, Guna, MP, India

T. Yashwanth (✉)
Sathyabama Institute of Science and Technology, Chennai, India
e-mail: yashwanthtekumudi1@gmail.com

V. M. K. Reddy
Mechanical Engineering, Sathyabama Institute of Science and Technology, Chennai, India

R. I. doewes
Faculty of Sport, Universitas Sebelas Maret, Jl. Ir. Sutami, 36A, Kentingan Surakarta, Indonesia

M. Naved
Department of Business Analytics, Jagannath University, Delhi-NCR, India

S. Yadav et al. (eds.), *Proceedings of Second International Conference in Mechanical and Energy Technology*, Smart Innovation, Systems and Technologies 290,
https://doi.org/10.1007/978-981-19-0108-9_30

30.1 Introduction

Nowadays the worldwide landscape of performance management (PM) is being constantly reinvented using new technologies. Indeed, as artificial intelligence (AI) and other breakthrough innovations are expediting and widespread use, the interaction between companies, workers and consumers has fundamentally altered and the automating of the administration elements of PM operations and duties intensifies [1]. While technical developments in PM may be linked back to the industrial rebellion, advances in technology have changed physical and mental services simply. Nevertheless, nowadays, options to performance management have been progressively provided in tasks that typically need contact and communication between people, thus altering both the organizational structures and the type of job. For example, humanoid services robots and artificial intelligence bots draw more and more interest from the business. These intelligent entities have changed the conventional roles of human resources, offering increasing capabilities and possibilities for PM but also significant difficulties and expiration in employment. At the same time, in-depth learning algorithms, intelligent devices, and IoT are especially helpful for cross-border companies as they may promote more efficient coordination and collaboration [2]. Likewise, the advent of IT and other new technology in electronic performance management provides a number of ways to enhance and decrease the price of PM activities such as job evaluation candidates and employee performance assessments.

PM is progressively at the center of globally focused PM research, as a result of technical progress. Students notably highlight how IT changes PM-related practices via the introduction of e-recruitment, e-training, and e-competence management which makes a beneficial contribution to PM quality in the local and worldwide organizations. As new actors like social robots are introduced in these technologies, they are unlocking many opportunities and supporting different PM services.

30.2 Role of AI in Performance Management

Figure 30.1 shows the various steps in Artificial Intelligence.

30.3 Performance Assessments Automated

Corporate executives appreciate performance management driven by data. They and team leaders may concentrate on facts and conduct performance evaluations of employees.

Employee collaboration and decision-making are necessary. We can now assess the outcomes of these cooperation activities using AI more effectively [3].

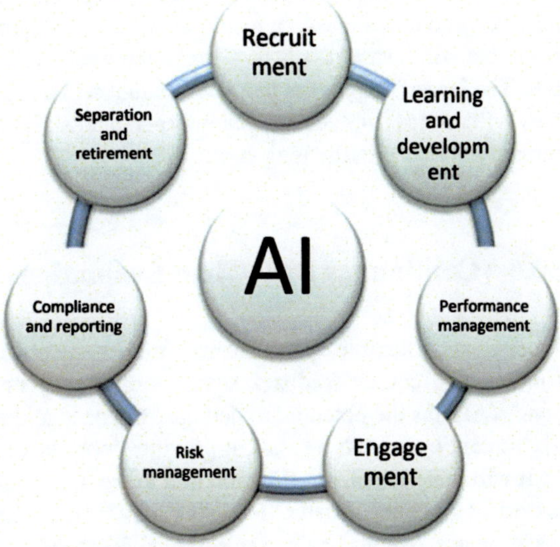

Fig. 30.1 Artificial intelligence steps

30.4 Three Significant Advantages

(i) Smooth data gathering. Instead, workers must ride on various levels of the workplace and seek information. It is OK if certain choices have enough time, but what about others that require quick action or agreement.

AI will thus guarantee that all data is kept in one location. This information or information from employees may assist leaders, stakeholders, and other members of the team to assess and communicate employee performance.

(ii) In actual research, a manager may quickly get insights. Because it is essential to maintain a frequent performance check with real-time analysis. No one needs obsolete information or data. Neither would anybody gain. The right data comes thus with actual research and helps to increase the performance of the employees.

(iii) It removes any psychological prejudice. Indeed, it is among the most significant reasons why AI manages, assesses, and assesses performance.

30.5 Learning and Development

Artificial intelligence can quickly identify the areas in which different workers need to progress. It may alert a management if an employee is falling short in any area and how they might improve in that area [4].

Artificial intelligence may assist workers in reinvesting their abilities long before they become outdated. As a result, artificial intelligence will not eliminate employment, but will instead assist workers in remaining relevant and improving their abilities. The fear of employees being redundant will be alleviated as a result of this since they will not [5]. It is certain that interpersonal skills and mathematical intellect will remain important in the long run.

30.6 Continuous Real-Time Evaluations

Trendy and flexible performance management techniques are beneficial in conducting frequent feedback sessions on a consistent schedule. Artificial intelligence (AI) has the potential to decrease the need and stress of rigorous performance assessment cycles. As a result, it provides more flexibility and more interface layer.

It will expedite the process and guarantee that feedback is received on time. This is critical since inconsistent feedback may have a negative effect on an employee's performance overall. As a result, quick feedback may instantly identify areas for improvement and help businesses increase their productivity [6].

30.7 Access to Information that Is Simple and Quick

According to a recent study conducted by Service Now, 30% of workers would prefer a Google-like interface to quickly get the assistance and information they need at work. Voice technology has taken the world by storm, especially in the average home. The growing popularity of voice-activated assistants such as Alexa, Google Home, and Siri provides sufficient evidence for the hypothesis with no need to do further research. As a result, voice help in the workplace is also an important thing to consider.

Aside from these concerns, a successful solution must be mobile, personalized, and offer the employee a variety of choices for how they want to interact with and be assisted by their company [7].

30.8 Improved Talent Management Performance

Finding the appropriate people for company is now one of the toughest problems for business executives and human resources professionals. There is enough talent on the market, but it is not obvious if the skill is beneficial for culture or not.

Business executives and recruiters increasingly concentrate on the use of AI-powered technologies to enhance recruiting quality. It helps them to find the appropriate talent pool and make the correct choice to hire. AI also reduces the need for repeated, time-consuming activities and automates all feasible jobs [8]. Finally, AI and machine learning will not only enhance performance management but also talent management and integration.

30.9 Role of Machine Learning (ML) in Performance Management

Machine learning in the management of the performance improves the experience of the end-users and IT experts, who monitor the performance of the network and the applications.

Performance management is a managerial process that includes data tracking and observing tools to detect problems that impede the performance of a network or application. The two most cited tools are application performance management (APM) and network performance management (NPM). The aim of these tools is to avoid the end-user from experiencing the unfavorable network/application. While incredibly helpful, these tools often overlook key variables which contribute to performance problems that IT workers can respond to the fact rather than be proactive. But these technologies will substantially reduce delays and network performance problems, along with machine learning [9].

The capacity of computers to continuously learn from data they collect without being scheduled to do so is machine learning. By detecting recurrent network issues, machine training aims to fix the dispute in real-time, either quickly before it improves productivity or alerts the IT admin to possible security risks.

The AppDynamics analytical platform purchased by Cisco, which identifies guidelines, is one of the few solutions presently using machine learning for the analysis of performance management data. AppDynamics couple's business measurements with machine learning noise cancellation capabilities to identify the underlying cause of business impact issues. Extra Hop's Addy technology detects cloud computing abnormalities and wire data metrics anomalies using machine learning [10]. The visibility architecture allows its product to understand changes and abnormalities in wire and, via this additional intelligence and automation, makes issue resolution and vulnerability detection much more intuitive, proactive, and fast.

Table 30.1 shows some of the advantages are helpful of machine learning in performance management.

30.10 Robotics

Automation, often known as robotics, is the design and construction of robots that can execute human movements and imitate human behavior (see Fig. 30.2). Briefly stated, robotics is a collection of disciplines that include artificial intelligence, machine

Table 30.1 Advantages of machine learning in performance management

S. No.	Key benefits
1	The use of predictor analysis to avoid problems that become a result of the obstruction in performance also improves the customer experience with the network or applications
2	Machine learning reduces the IT administrators' responsibilities of performance management. In his PowerPoint, he speaks in 100 simple stages on how effective, self-optimizing networks enable network managers to concentrate on more difficult problems inside their companies
3	Machine learning answers network and application performance problems in a timely manner
4	It can also detect and identify security risks which will notify the IT staff to further analyze and implement adequate protective measures

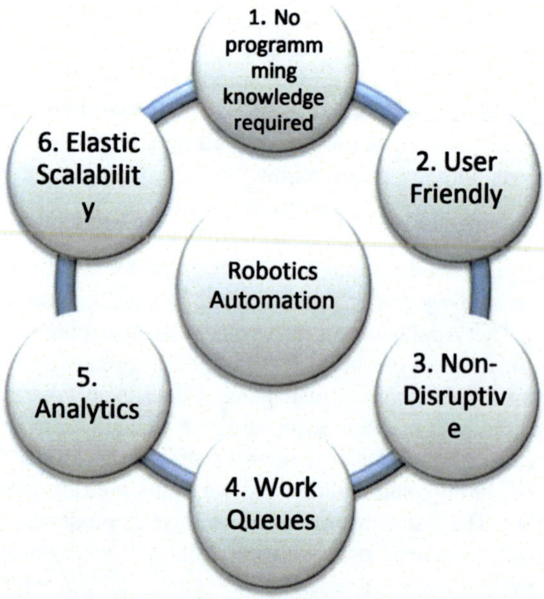

Fig. 30.2 Robotics automation steps

learning, electronics, nanotechnology, and a variety of other topics. In the discussion centered on the advancements in the field of robotic innovations, the consequences that robots will have on work and employment are highlighted, whereas there is considerable optimism about the learning and training opportunities that robots can provide for businesses and employees in organizations. As a result, research activities in robotic technology may be divided into three categories: job replacement, human–robot cooperation, and educational possibilities. Numerous professions, according to research on robotic innovations are expected to be eliminated or replaced by technology and robots in the near future [11]. Make a point of stating that the occupations that are most likely to be affected by job displacement are those that need welding and painting as well as those that require workers who are less educated, experienced, and competent. A further possibility for humanoid robots is that they may completely replace human front-line workers, such as robot waiters in restaurants and virtual assistants who offer advice to consumers via a company's website.

However, some research indicates that the effect of robots on PM, and particularly on unemployment, may be significant in the future; however, these studies do not specify when this could occur. In particular, given the way that artificial intelligence, digitization, and robotic technologies are being formed by socioeconomic and organizational forces, predictions of mass unemployment and substitution by robots are unlikely to be realized, according to the authors, who argue that situations defined by strong needs for compassion, in which creating original and creative alternatives is required, or that necessarily require high levels of social intelligence are unlikely to be realized. Moving away from the idea of job replacement, many academics have emphasized the importance of combining human skills with robotic technology in project management in order to provide more intelligent human resource solutions. In this respect, more competent and educated workers are required in the age of human–robot symbiosis and cooperation in order to reap the benefits of prospective possibilities and to counteract the dangers of potential threats. By providing possibilities for more technical roles that are either created or improved by robotic technology, robots may assist human workers in their jobs. The use of robots in surgery is an excellent example [12].

While, when used properly, robots may improve accuracy and decrease mistakes, human expertise continues to be a critical component of the process. A doctor's skills and expertise are needed, in addition to the technological aspects. This is especially true for children. A number of new learning possibilities have been created by robotic technology in the field of PM. Work in robotics explores the ways in which robotic technology may replace monotonous and regular tasks performed by human workers, allowing them to devote their time and energy to activities that allow them to utilize their talents more productively. However, this also provides workers with new learning opportunities that must be coupled with intensive training in order for them to fulfill their new duties and gain the skills necessary to work with a robot. Employees, on the other hand, may have varied attitudes about robots depending on their profession. It has been shown that highly competent workers have more favorable views about robots and their deployment since they provide them with chances to further develop their abilities and knowledge. Jobs created by artificial intelligence and robotic technology will, unavoidably, be fraught with uncertainty in the future. However, these innovations provide a chance for the development of problem-solving methods that will be very beneficial [13].

Table 30.2 Benefits of robotic technology

Durability	Robotics does not need sick days since they are available 24 h a day, seven days a week
Lifespan	Robotics projects typically last 9–12 months, with a return on the investment ranging from 30% to as high as 200% in the first year of operation
Scalability	When operating at scale, the cost per effort associated with robotics is insignificantly cheap, and it can be quickly scaled up or down in response to changes in demands in the network. Because of this, there is no waste of work or unavailability for the system overall
Small risk and easy to integrate	Unlike other technologies, robotics is low-risk and non-invasive, and it can be readily integrated into existing infrastructures. The development of advanced techniques and machine learning tools enables HR departments to develop a structure that will continue to grow as new technologies are introduced
Flexibility	Robotics is not limited to a single sector and may follow pre-programmed processes regardless of the area in which they are deployed
Consistency	In order to achieve flawless duplication and error-free performances, robotics is used, which removes output fluctuations over the course of an operating period
Correctness	As a result of its computing superiority, robotics can also provide accurate outputs and make judgments in a short period of time when it comes to data analytics, outperforming human counterparts by a wide margin
Higher productivity	Robotics can free up human workers to take on more high-value tasks such as employee engagement and retention tasks, and contribute directly to the overall strategic goals of the company thereby increasing overall productivity

30.11 Benefits and Impact

The opportunity to benefit company from robotic technology via performance management is enormous. And much of this ability has been shown. The main advantages of robots are stated as Table 30.2 in PM departments.

30.12 Conclusion

The purpose of this study was to evaluate studies on smart performance management technology. In 59 high-level PM, GM, IB, and IM publications, we looked for possibly relevant research to explain what is new in smart performance managing automation. During the selection process, 45 papers were identified, which provided a general panorama of the state of the art on this subject. While not comprehensive, this article highlights the effect on performance management of AI, robots, and other sophisticated technology. Particularly an international perspective to business was also given with recommendations for future study, which target possibilities for theoretical and empirical advancing the subject. We both look forward to making the contributions to our study more significantly widened and verified in practice to the upcoming generations of research.

References

1. Abraham, M., Niessen, C., Schnabel, C., Lorek, K., Grimm, V., Möslein, K., Wrede, M.: Electronic monitoring at work: The role of attitudes, functions, and perceived control for the acceptance of tracking technologies. Hum. Resour. Manag. J. **29**(4), 657–675 (2019)
2. Ancarani, A., Di Mauro, C., Mascali, F.: Backshoring strategy and the adoption of Industry 4.0: evidence from Europe. J. World Bus. **54**(4), 360–371 (2019)
3. Bondarouk, T., Harms, R., Lepak, D.: Does e-HRM lead to better HRM service? Int. J. Human Resour. Manage. **28**(9), 1332–1362 (2017)
4. Chao, G.T., Kozlowski, S.W.: Employee perceptions on the implementation of robotic manufacturing technology. J. Appl. Psychol. **71**(1), 70 (1986)
5. Jain, A., Pandey, A.K.: Multiple quality optimizations in electrical discharge drilling of mild steel sheet. Mater. Today: Proc. **4**(8), 7252–7261 (2017)
6. Jain, A., Pandey, A.K.: Modeling and optimizing of different quality characteristics in electrical discharge drilling of titanium alloy (grade-5) sheet. Mater. Today Proc. **18**, 182–191 (2019)
7. Jain, A., Yadav, A.K., Shrivastava, Y.: Modelling and optimization of different quality characteristics in electric discharge drilling of titanium alloy sheet. Mater. Today Proc. **21**, 1680–1684 (2020)
8. Loebbecke, C., Picot, A.: Reflections on societal and business model transformation arising from digitization and big data analytics: a research agenda. J. Strateg. Inf. Syst. **24**(3), 149–157 (2015)
9. Malik, A., Budhwar, P., Srikanth, N.R., Varma, A.: May the bots be with you! Opportunities and challenges of artificial intelligence for rethinking human resource management practices. In: Proceedings of the 33rd Annual International Conference of the British Academy of Management (BAM) 2019: Building and Sustaining High Performance Organisations During Uncertain Times (2019)
10. Marler, J.H., Parry, E.: Human resource management, strategic involvement and e-HRM technology. Int. J. Human Resour. Manag. **27**(19), 2233–2253 (2016)
11. Maurer, T., Hinck, G.: Cloud Security: A Primer for Policymakers. Carnegie Endowment for International Peace (2020)
12. Panwar, V., Sharma, D.K., Kumar, K.P., Jain, A., Thakar, C.: Experimental investigations and optimization of surface roughness in turning of en 36 alloy steel using response surface methodology and genetic algorithm. Mater Today Proc (2021)
13. Neha, E., Orlando, M.F., Suhaib, M., Shrivastava, Y.: An effort for resolving redundancy of a multi-finger robotic hand in straight-line motions. J. Braz. Soc. Mech. Sci. Eng. **42**(11), 1–27 (2020)

Chapter 31
Benefits of Digital Education Over the Impact of Pandemic on Indian Higher Education System

Viney Dhiman, Anupama Bharti, and Gaurav Gaur

Abstract Education is a valuable aspect of society, but COVID-19 made a drastic impact on the education system. It necessitated a flexible approach among institutions, Teachers, and students. For managing education effectively, institutions have initiated to apply electronic mode in the education system and comparable features are available on many platforms. It is argued that students are not aware of such platforms and they also pass from Fear, Anxiety, and control addiction. Further, Focus upon Training of Teachers with the articulation of particular Curriculum is necessitated and also course has not designed in a different language to increase their reach and more opportunities for the youth of rural India. Low Internet interpretation demands much investment for making sufficient digital infrastructure and there are certain societal barriers to online education. Still, faith in private players' investment in education is not supported due to their profit-oriented nature. Present chapters attract the adoption of online education to respond to the recent pandemics, particularly in Higher Education in India. For this purpose, some factors of digitalization that are in use in Indian Universities try to evaluate based on static analysis of Factors Awareness, Anxiety, Network Issues, Problems of students are facing not having required technology along with language barriers. On the other hand Training of Teachers, and the process of online education at Indian University. The present chapter attracts, how managing Education in pandemic based on an empirical study regarding paramount factors of higher education in the Indian Universities.

V. Dhiman (✉)
Post Graduate Institute of Medical Education & Research, Chandigarh, India
e-mail: viney.pu@gmail.com

A. Bharti
Department of Sociology & Social Work, Himachal Pradesh University, Shimla, India

G. Gaur
Centre for Social Work, Panjab University, Chandigarh, India

S. Yadav et al. (eds.), *Proceedings of Second International Conference in Mechanical and Energy Technology*, Smart Innovation, Systems and Technologies 290,
https://doi.org/10.1007/978-981-19-0108-9_31

31.1 Introduction

Education and Information Technology: Currently, technology is being used to enhance, assist, or expand teaching, learning, and creative inquiry in higher education globally. Students may assist one another and collaborate to better grasp the subject thanks to technological advancements. The ability to act as (supervised) instructors is something that they can do on occasion, and learning via teaching is very beneficial for understanding a subject and solving issues. While using Technology learners can have better understand the academic material and they also become more knowledgeable and easily solve the problems related to education.

31.2 Important Aspects of Technology in Education

- Drastic development in skills
- An enabler of others to better teaching and learning practices particularly those residing in remote areas
- Demands capacity building of teachers in the tune of technological advancement
- Build a platform for a large number of teaching material
- It enhanced the probability for Educators to become imaginative, flexible, and willing to renew their vision of teaching and learning
- It called into question to continuous training of teachers
- Focus upon invention and use of computers in various areas
- Improves the quality of education and leads to parallel and vertical changes
- A learner can Supervise teachers.

31.3 Status of Higher Education in India

Higher education: It is out of the ordinary India is in Higher Education made drastic development. Details can be visualized in Table 31.1.

After independence, the growth has been very impressive. India is the largest higher education system in the world in the potential of having universities ad colleges and Table 31.1 shows 50 times increase in academic institutions. The Kothari

Table 31.1 Status of higher education in India

Status of higher education	Number of universities	Number of colleges	Students enrolment at tertiary level
Beginning of independence	20	591	0.2 million
After independence	504	25,951	

Table 31.2 Higher education student population growth

Years	2017–18	2018–19
Gross enrolment ratio	25.8	26.3
Enrolment in absolute term (in crore)	3.66	3.74
No. of university	903	993
Educational institutional	49,964	51,649

commission set up in 1964 had recommended that the government should spend at least 6% of its gross domestic product (GDP) on education. However, at present India is able to achieve only half the target. Furthermore, the Knowledge Commission also recommends an increase of at least one and a half percent of GDP for higher education [1–6].

Challenges of Higher Education: Demand–Supply Gap and low quality of education are normal phenomena in Higher education. According to a recent government report, two-thirds of India's colleges and universities are below standard in the matter of quality of education. It is shocking that Low expenditure in Research and Development adversely affected the quality of education (MHRD, 2009–10). On the other hand, a huge Faculty Shortage is usually seen at every educational institution. Although, Poor infrastructure and excessive Political interference have distorted the system. This provides the opportunity to think that How regulation is made to attract Private Investment in the social Infrastructure [8]. Moreover, the present system of education has not found its ancient prominence. Thus, it has been felt to integrate the education system with global players to secure efficient Human Resource Management.

Present Status of Higher Education: It is a notable fact that Gross Enrolment Ratio in Higher Education increased between the years 2017 to 2019, however in absolute terms this increase account for only 8 billion. In this case, the number of University it has also increased in the same periods. While looking upon educational Institutional it also viewed increased. Details are also provided in the Table 31.2 and Fig. 31.1.

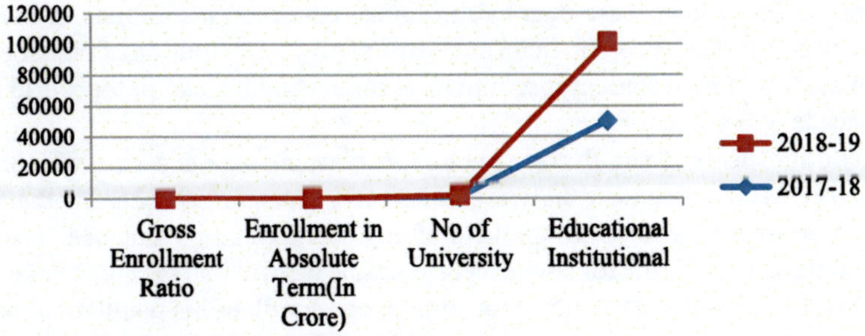

Fig. 31.1 Higher education student population growth

The second aspect attempt to explain the measure of how to convert pandemic into opportunity. In this sense, it is pertinent to mention here that a positive approach may have become fruitful in the exceptional situation while trying to control uncontrolled factors. At the same time, many administrative players report on how the situation may be understood and how teaching and learning can be continued. Politicians typically work with numerous professionals to deal effectively with the issue. Concerning social media, professionals and beginners may communicate their logical and irrational opinions [7–12]. Lockdowns, however, have numerous effects on pupils and underlie inequities as well as social and psychological stress. Parents and custodians of kids also had bad effects and tried to get engaged to improve the situation with social therapy and their children, especially those who are educated.

The third element shows how schools and instructors all around the globe transfer their labor from classrooms and classrooms to digital platforms. It also found flaws and weaknesses in this rapid shift. Here it is important to highlight that the scenario also offers business digital learning suppliers a new market opportunity. Thus, this reframing of policy along with the appropriate use and execution of private capital to initiate online education cannot be left out without stating [11, 13–17]. On social media, renowned experts have asked why some people, organizations, and businesses are so keen to provide advice and to consider whether they are motivated by commercial motives. Others caution that if the education system is placed in the hands of the capitalist entirely or without balance, undesirable results may be achieved [13, 14, 18–24]. However, in the form of monitoring the lives and dignity of students, online learning threatens privacy. Education becomes costly because of its business motivation when private actors are included in the education system. Pandemic also compels the educational organization, via the choice of the best alternative to the education technology, to consider ahead of time of the future problem. These decisions have to be adaptable in supporting the changing environment both for students and instructors. Educational organizations utilize previous knowledge and experience to overcome their shortcoming soon to minimize their potential negative effect [23–27].

Involvement of the Private Sector in Higher Education: Presumption about the economic success of the states is directly determined by their education systems now realized. The involvement of the private sector in higher education has grown drastically. Today over more than half of higher education institutions in India are promoted by the private sector. India has been the biggest institution of higher education in the world with student inscriptions at the second highest in the world in the past decade in particular [2].

Prominent Role of the Private Sector: At now, the private sector, which spans over and gives the appearance that it is continuously expanding quickly and that offers most of the training courses in the field of engineering and governance. There are many additional providers for laws except to enable them to join the area of education [6]. Despite increasing education expenditure, one-fourth of his population remains untrained; it is startling to discover that just 15% of Indian children finish high school and only 7% are graduates. The level of education is substantially low compared to the main developing countries of the globe, whether in elementary or higher education in India. By 2008, post-secondary India had a vacancy of one-quarter of all teaching jobs throughout the country, and it is also a disadvantage that fewer than 60% of university teachers participating in higher education lack either a master's degree at the NET or Ph.D. (Newsweek, 2011). As of 2011, 152 engineering colleges in India have been awarded an annual study intake of five lakhs, 82 thousand, including 1214 polytechnics, and a total yearly intake of 65 thousand lakhs (science and technology education, 2009). These schools, however, also face a significant lack of faculties [8]. By concentrating on the number of universities and colleges, the central government and the state governments strive to foster talent. In the past 40 decades, the number of universities has risen more than six times and more than 30 times between 1970 and 2011.

31.4 Online Education

Efficacy of online education: Technological added education system supported in designing innovative course materials and widening highly interactive student–teacher relationships.

Significance of Online Education: Online education emerged as downward learning where teachers have to learn innovative ideas to compete with new challenges due to technological advancement in the present education system. It is assisted that there should use of ICT in the education system and curriculum copes with desire innovation to make learning as interactive sessions). Now the world is giving priority to using ICT in education to avail benefits of efficiency and cost-effectiveness in educational business [11]. Thus, the education system is inclined to reshape itself and articulated with modernization where digitalization with innovative ideas stapled with knowledge). However, significance can be underlined as under:

Challenges of Implementation of Online Education: The basic challenges in the higher education system in India are resistance to change, motivation levels of students, technical skills of students, students' understanding of technology, and How to check student performance.

Basic Components of Online Education: The use of ICT is dependent on certain pre-requisites and we can only proceed to digital education after fulfilling these fundamental infrastructure conditions. Below are some of the key peripheral components in digital classrooms: (1) Smart Boards (2) Class Room PC (3) Projectors (4) Internet Connectivity.

Important online learning applications and tools: The digitalization of education pushes the academic institution to use these platforms for overcoming the problems of materials, tutors and getting better professional knowledge such as:

Factors Promoting Online Education in India: Some of the factors without that India cannot be able to the advancement of digital education culled and mentioned as: (1) Customized platforms for adaptive learning (2) E-learning two-way talks (3) Mobile-based Learning (4) Video-based Learning If mobile learning increases, it is predicted that it would increase the total Internet traffic by 80%.

Adverse Impact of Online Education on Students: Using ICT in education give away to some extraordinary adverse effects on social life and it has been cited as: (1) Addiction (2) Obesity (3) Social Disconnect (4) Reduced Face-to-Face Interaction (5) Unwarranted Information at Tender Age (6) Declining Writing Skills (7) Increasing Incidents of Cheating (8) Declining Mathematical Skills.

Major Initiative of Government of India to Digitalization of Education: After assessing the worth of online education the Government planned to invest in the Digitalization of education and details may be identified in the upcoming Table 31.3.

Table 31.3 depicts that from 2016 to 2018 concentrated to spend 64,676 crores for upgrading the online mode of education. India's makeover from present to future is only possible if cognitive therapy would grow in people of the country.

COVID-19 and status of Education in India: The effect of the COVID-19 epidemic is seen in many sectors across the globe. India's education sectors are severely impacted. It has caused the lockdown to have a very negative impact on the lives of the kids. About 32 billion students ceased moving schools, all educational activities were halted. Thus, it has been observed that pandemics brought not only some adverse experiences but also provided an opportunity for the articulation of the Education system to respond to tragedy [8] The Indian Government has been monitoring the impact of Covid-19 in many stages and the 5.0 shutdown has been declared effective from 30 April, 1 June to 30 June 2020.

Table 31.3 Overall major initiatives in Digital India

Scheme parameter	Current status	Target	Deadline	Budget
Broadband for rural gram panchayat	20 k villages connected under NOFN (April 2015)	2,50,000 Gram Panchayats	December 2016	32 k crores Rs
Universal Access to Mobile Connectivity	42,300 uncovered (June 2015)	Uncovered villages	March 2018	16 k crores Rs
National Information & Infrastructure	Under SWAN, NKN and NOFN, SWAN has made operational 34 villages	Nation-wide	March 2017	15,686 crores Rs
Wi-Fi enabled university	1038 institute connected on NKN 368 institutes connected on NMEICT	Over 1500 institutes	N/A	790 crores Rs
Skill Training in small/town villages for the IT sector	N/A	1 crores student	5 years	200 crores Rs

Table 31.4 Digital initiatives of UGC & MHRD for higher education for the exchange of information during COVID-19

Digital platforms for learners			
E-GyanKosh	Gyandarshan	Gandhara	SwayamPrabha provides Massive Open Online Courses (MOOCs)
E-PG Pathshala	E-Adhyayan	E-Pathya	National Digital Library of India (NDLI)
E-Yantra	Education Software that is Free/Libre and Open Source	Virtual Labs	E-ShodhSindhu
Shodhganga	VIDWAN	National Educational Alliance for Technology (NEAT)	SAKSHAT

The Effect on Higher Education and the Pandemic: Covid-19 has seriously destabilized all schooling. Different activities such as admission, testing, entry testing, competition exams by different institutions stay postponed and the authorities are intended to do the digital examination. All instructors and pupils are tempted to become more technological know-how. With the assistance of numerous conference technologies, such as Skype, YouTube Live, Google Meet Facebook Live, WebEx, etc. the responsible agency has launched several guidance programs and inductive meetings and counseling tools in support of the students. This project has established an excellent virtual learning environment and motivated students to participate in online activities. Teachers and students increased the usage of WhatsApp, Google Drive, Telegram, Twitter, and other electronic media to share information (Table 31.4).

31.5 Education and Impact of Covid-19

Education India after COVID-19 and Problems: It has been observed that in India at the Graduate and Post Graduate Level there has been adoption e-Learning mode to continue the educational process. However, Students stumble upon various Social psychological, and technological such as depression anxiety, poor internet connectivity, and unfavorable study environment at home. Despite this, Students from remote areas and deprived sections have faced other typical problems. It called upon the adoption of a tech-education system to provide proper education to students [12]. Again pandemic allowed policymakers to establish the education system which encourages social-psychological articulated model which based upon provided underneath (Fig. 31.2).

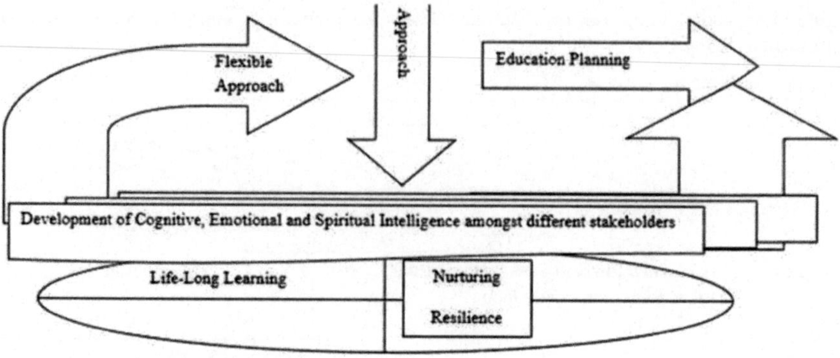

Fig. 31.2 Prerequisite of digital education

Internet penetration in India: Literature reveals that there is Low internet penetration in India as per requirement. There is also a gap visualized in Access Internet in Different sects, Gender and Regions, low frequency of electricity assisted with Network issues and low speed.

Reducing Investment in Digital Infrastructure: The funding for digital learning has been cut from Rs 604 crore in 2019–20 to Rs 469 crore in 2020–21. The main difficulty of remote learning is the difference in availability to PCs or cellphones from power and Internet connectivity. The Government of India has nonetheless made numerous initiatives to promote online education.

Recent UGC Guidelines: On 6th July 2020 University Grants Commission, UGC's Guidelines on University "Examinations and Academic Calendar" getting revised but the examination for final year students has not been canceled and decided to conduct by end of September 2020.

Judicial Intervention: Final Years Student Move to Supreme Court and rue cancelation of the examination. Supreme Court refused to give a stay on the examination but fixed the next hearing on 10 August.

31.6 University and Covid-19

Following the Lockdown, the University's teaching system is managed online. The authorities also encouraged instructors to use YouTube to broadcast lectures or presentations for pupils, and to respond to questions through video conferencing. It was also recommended that interactive meetings be held through Skype or Zoom. It has also been proposed that engineering courses utilize the National Program on Technology Enhanced Learning (NAPTEL) platform. They have completed an ongoing semester from the last week of March to 15th May 2020. At present only permanent faculty has been involved in teaching neither, contractual nor guest faculty have been involved still to teach an ongoing class. However, mid-semester students have been promoted to the next class. This state of affairs unemployment will be reflected valuably felt in University as some faculty members regularly wait for their joining.

31.7 The Factors Have Been Analyzed at Indian University

1. **Awareness of students**: It has been observed that students were not properly aware of two online support system in education.
2. **Problems Faced by Student**: Again it has been brought to notice that students were faced with Anxiety, encountered Network Issues, and an issue they had not Smartphone and 75% of them felt bottleneck of Language Barriers.

Teachers' Responses: Teachers' responses regarding upgrading the university infrastructure have been actualized and it seems that it requires Training of Teachers, demands Change in Curriculum, and Development of Infrastructure to cope with new demands of Digitalization of Education.

31.8 Discussion

Education is a valuable aspect of society, but COVID-19 made a drastic impact on the education system. It necessitated a flexible approach among institutions, Teachers, and students [6]. For strengthening online learning commercial digital learning platform providers are also engaged. But this initiative left the question mark if capitalist circumferences to support the education system then will it promote education as per holistic human growth and open the route for education in the desirable future [4]. The pandemic paved the door for the adoption of various digital learning systems in education.

Drop-in sessions, free webinars, blog postings, and emergency policy papers were made available as a result of the unusual circumstances. Assisting learning from experience [4]. Thus, effective policy implementation and using private capital are paramount factors to set in motion online education. Further, private players if interested to invest in the support of online education then, it is also necessary to know whether their motivation is profit-driven or value precedes them to take such action. These decisions may result in new power and control relationships, new kinds of student unfairness, and other unanticipated outcomes.

Again, previous experiences should be correlated with the current model, and re-shuffling should be done as needed [7]. Thus, online learning should not be seen as a 'thing' or method in and of itself, and empirical research may provide a positive push. Furthermore, including instructors and learners in the creation, application, and practice of educational expertise may influence how well technology may promote successful teaching and learning.

Involvement of the Private Sector in Higher Education: The involvement of the private sector in higher education has seen drastic changes. Today more than half the higher education institutions in India are promoted by the private sector [2]. Compared to other major emerging countries, India's elementary and secondary education is of low quality [1]. However, in India, there is a bigger opportunity to Secure Private Investment. Thus, the time has come to integrate the education system with global players. Since independence, India has progressed significantly in terms of the Gross Enrolment Ratio (GER) [17]. Online education should be attentive toward both urban and rural areas and it also requires articulation with demands of the present state of affairs. The basic challenges in the higher education system in India are Social and Infrastructural resistance [5].

Benefits and Flipside of Digital Education over Traditional Education System: Use of digital mode in Education makes students Smarter, Self-Motivated, and More Accountable because it implied with Greater Involvement of Educators Parents Learning tools and technologies. Moreover, it not only facilitates Better Information Sharing measures but also Increasing Students' Employability and cross Geographical Limitations while including more students in the educational grip. While digitalization accelerating education in prompt away but it also left some adverse impact such as Addiction, Obesity, Social Disconnect, Reduced Face-to-Face Interaction, Unwarranted Information at Tender Age, Lack of Concentration SMS and text messaging has become a favorite pastime of many students, Declining Writing Skills, Increasing Incidents of Cheating and upmost rally round Declining Mathematical Sklils [7]. In recent years India intent to make a huge expenditure to upgrade the online education system.

31.9 Solution

Based on the study followings solution may be culled out:

1. Awareness program for students and teachers regarding online Education facilitator
2. Counseling and social therapy to reduce fear, Anxiety and control addiction
3. Focus upon Training of Teacher with an articulation of particular Curriculum
4. Much investment for making sufficient digital infrastructure in the mode of Public–Private Partnership
5. Strategy for making an environment where socio and economical barriers can be minimized
6. Approach for convergence of offline education and online education in future
7. Adoption of Feedback versus Feed-forward technique
8. Acceptance of private investment in Online education
9. Adapting behavior and
10. Development of Community Participation.

Future Research: The study has been supported by primary data and statistical analysis was also made, but the sample size is very low. In future study problems of unemployment particularly for non-permanent teachers and students indebted due to loss of income of parents and too much cost required to go with online education can be conducted.

31.10 Conclusion

The following are the overall conclusions and highlights based on the research:

- The emphasis of this study was on the various implications of Covid-19 on higher education in India. The current epidemic has provided an opportunity to use technology-based learning at all levels of education. Post-Covid-19 education appears to be an alternative educational system that includes well recognized online/virtual education.
- In India, where students are unaware of the various platforms that aid online education, they are also anxious about how their performance will be rated, as well as poor Internet penetration and network concerns. Furthermore, due to their parents' indebtedness, students do not always have access to smartphones, and they occasionally face language hurdles.
- On the other hand, there is a need for teacher training and an imminent need for curriculum changes. Private player support can be used under strict conditions. As a result, policy for reducing fear, explaining bottlenecks with the cooperation of various stakeholders, and action should be determined with the joint efforts of the government, students, teachers, and parents, and ultimately the community's support would make it possible, and it cannot succeed without the use of information and communication technology.

References

1. Jena, P.K.: Challenges and opportunities created by Covid-19 for ODL: a case study of IGNOU. Int. J. Innov. Res. Multidiscip. Filed **6**(5), 217–222 (2020); Jena, P.K.: Impact of pandemic COVID-19 on education in India. Purakala 31(46):142–149 (2020)
2. UNESCO: COVID-19 Educational Disruption and Response. Retrieved on 3 June 2020
3. Apte, J.: The challenge of (online) education in India during the COVID-19 pandemic. Opinion South Asia, May 29, 2020. Retrieved From https://www.statecraft.co.in/article/the-challenge-of-online-education-in-India-during-the-covid-19-pandemic
4. Balachander, K.K.: Higher education in India: Quest for Equality and Equity. Mainstream.
5. Bates, T., Sangrà, A.: Managing Technology in Higher Education: Strategies for Transforming Teaching and Learning. Jossey-Bass, San Francisco (2011)
6. Bosamia, M.P.: Positive and Negative Impacts of ICT in our Everyday Life. Accessed from https://www.researchgate.net/publication/325570282

7. Bourdieu, P.: Distinction. A Social Critique of the Judgement of Taste (Trans. R. Nice). Harvard University Press, Cambridge, MA (1984)
8. Byker, E.J.: ICT in India's elementary schools: the vision and realities. Int. Educ. J. Compar. Perspect. **13**(2), 27–40 (2014)
9. Rashmi, R.: Challenges of quality in online learning, 6th May. Retrieved From https://timesofindia.indiatimes.com/blogs/edutrends-india/challenges-of-quality-in-online-learning/
10. Manthalkar, R., Gajre, S., Joshi, Y.: Education after COVID-19 Disruption (No. 3431). EasyChair (2020)
11. Nanda, A.: Information and Communication Technology, 1st edn. NGO Prints, Calicut (2005)
12. Dangwal, K.L., Singh, S.P.: Enhancing spiritualism in virtual world. Turkish Online J. Dist. Educ. **13**(2), 76–83 (2012)
13. Farrukh, S., Singh, S.P.: Teachers attitude towards use of ICT in technical and non-technical institutes. J. Educ. Soc. Res. **4**(7), 153 (2014)
14. Pardeep, T.: Role of ICT in professional development of teachers. Learn. Commun. **6**(1), 127–133 (2015)
15. Arulsumy, S., Sivakumar, P.: Application of ICT in education. Neelkamal Publications Pvt. Ltd. Hyderabad First Edition (2009)
16. Kumar, D.N.S.: Impact of COVID-19 on Higher Education. Retrieved on 25 May 2020
17. Jogan, S.N.: Higher education in India: a vision of 2030. Online Submission **6**(7), 365–370 (2019)
18. Venna, S.K.: Teacher education some qualitative consideration. Shikshak–ShikhaShodhPatrika **04**(1), 10 (2010)
19. Hiremath, S.S.: Current scenario of higher education in India: reflections on some critical issues. IRJSSH **1**, 73–78 (2017)
20. Editorial: e-Social work and digital society: re-conceptualizing approaches, practices and technologies. Eur. J. Soc. Work **21**(6), 801–803 (2018)
21. Fuchs, C.: Communication and Capitalism: A Critical Theory. University of Westminster Press, London (2020)
22. Gupta, D., Gupta, N.: Higher education in India: structure, statistics, and challenges. J. Educ. Pract. **3**(2) (2012)
23. St. Amour, M.: Privacy and the online pivot. Inside Higher Ed, 25 March (2020). https://www.insidehighered.com/news/2020/03/25/pivot-online-raises-concerns-ferpa-surveillance. Accessed 19 April 2020
24. Harwell, D.: Mass school closures in the wake of the coronavirus are driving a new wave of student surveillance. Washington Post, 1 April (2020). https://www.washingtonpost.com/technology/2020/04/01/online-proctoring-college-exams-coronavirus/
25. Herrington, J., Reeves, T.C., Oliver, R.: A guide to authentic e-learning. Routledge, London (2010)
26. Hodges, C., Moore, S., Lockee, B., Trust, T., Bond, A.: The difference between emergency remote teaching and online learning. EDUCAUSE Review (2020). https://er.educause.Edu/articles/2020/3/the-difference-between-emergency-remote-teaching-and-online-learning. Accessed 19 April 2020
27. Howland, J.L., Jonassen, D.H., Marra, R.M.: Meaningful Learning with Technology, 4th edn. Pearson Education, Boston (2011). https://en.unesco.org/covid19/educationresponse

Chapter 32
An Investigation on Impact of Machine Learning in Additive Manufacturing

Chetan M. Thakar, Randy Joy Magno Ventayen, Edwin Hernan Ramirez-Asis, Juan Emilio Vilchez-Carcamo, Misael Erikson Maguiña-Palma, and Abhishek Thommandru

Abstract One of the final phases in the product design process is prototyping or model creation. It is beneficial in the conception of a design. A model is produced and evaluated on a regular basis prior to the start of complete assembly. Historically, manual prototyping was employed to create a prototype. Additive manufacturing is a buzzword in the industrial and manufacturing industries. Initially, the CAD model of the components for the product was created in modeling software according to the specifications. Following the creation of a CAD model, the model is sliced by parallel planes equal to the layer thickness. As a result, the edges of these slices are quite sharp and squared, like a stair effect. These three-dimensional models will now be broken into small two-dimensional objects called slices. Simply said, a complicated three-dimensional problem has been reduced to a set of two-dimensional difficulties. These small two-dimensional files are known as STL files, and they are sent by tessellating the geometric three-dimensional model. Different surfaces of a CAD model are piecewise approximated by a sequence of triangles in tessellation, and the coordinates of triangle vertices and their surface normal are recorded. The predictive

C. M. Thakar (✉)
Assistant Professor, Department of Mechanical Engineering, Government College of Engineering, Karad, Maharashtra, India
e-mail: cthakar12@gmail.com

R. J. M. Ventayen
Acting Campus Executive Director, Pangasinan State University, Lingayen Campus, Philippines

E. H. Ramirez-Asis
Department of Economic and Administrative Sciences Faculty, Universidad San Pedro, Chimbote, Peru

J. E. Vilchez-Carcamo
Department of Management and Tourism Faculty, Universidad Nacional Santiago Antúnez de Mayolo, Huaraz, Peru

M. E. Maguiña-Palma
Department of Graduate School, Universidad Norbert, Wiener, Lima, Peru

A. Thommandru
Law Department, VIT AP UNIVERSITY, Andhra Pradesh, India

S. Yadav et al. (eds.), *Proceedings of Second International Conference in Mechanical and Energy Technology*, Smart Innovation, Systems and Technologies 290,
https://doi.org/10.1007/978-981-19-0108-9_32

nature of various machine learning algorithms makes them the best instrument for dealing with additive manufacturing challenges. Machine learning techniques are capable of evaluating previous data and predicting future outcomes based on that analysis. This article discusses machine learning applications in additive engineering.

32.1 Introduction

One of the final phases in the product design process is prototyping or model creation. It is beneficial in the conception of a design. A model is produced and evaluated on a regular basis prior to the start of complete assembly. Historically, manual prototyping was employed to create a prototype. As a result, manual prototyping was the earliest stage in the history of prototype development. In manual prototyping, a capable or experienced artisan creates a prototype that serves as the master prototype for the production of a number of comparable goods.

This prototype has generally been made from a range of materials, such as timber, thermoses, plastic, rubber, etc. The manual prototype contains several restrictions (for example, the building of a prototype takes time, a skilled person is needed and the delivery of the product to the end-user takes more time), which were the main obstacles for operation of high-speed mass production [1].

Today's industry is continuously seeking innovative methods to enhance efficiency while reducing both time and money manufacturing expenses. In the mid-1970s, a smooth and computerized model of the product was therefore displayed by curves, 3D surfaces, and solids, and so the second stage of prototyping began. Simulation, evaluation, and modifying these models utilizing comparable material and attributes of a component (electrical, mechanical, physical, etc.). The computerized design and engineering (CAD/CAE) sectors took up the mid-1970s. First, the machine was created and is widely used in industry. Numeric Controlled (NC) This led to a considerable rise in basic efficiency and product delivery to consumers in industries such as the computer controls (CNI) and high-speed processing technology [2].

Despite the fact that the aforementioned technologies offer numerous advantages, they face a few challenges. One notable challenge was producing the tiny size component. This would result in needless time for production of the component being wasted by planning the complete material required from raw material to completed product, including process planning and NC programming. So if you release your product one day later, the competition among competitive industries will not be as profitable as it may possibly have been due to that rivalry. In this way, the advanced technology is known as additive production (AM), the third and latest stage in prototyping [3], may overcome these barriers. It differs from subtractive machining in diametric terms, as its name suggests. Other names for AM are layer production (LM), rapid prototyping (RP), and solid freeform manufacturing (SFF).

Initially, the CAD model of the components for the product was created in modeling software according to the specifications. Following the creation of a CAD model, the model is sliced by parallel planes equal to the layer thickness. As a result, the edges of these slices are quite sharp and squared, like a stair effect. These three-dimensional models will now be broken into small two-dimensional objects called slices. Simply said, a complicated three-dimensional problem has been reduced to a set of two-dimensional difficulties. These small two-dimensional files are known as STL files, and they are sent by tessellating the geometric three-dimensional model. Different surfaces of a CAD model are piecewise approximated by a sequence of triangles in tessellation, and the coordinates of triangle vertices and their surface normal are recorded.

Machine learning is often used to indicate advancements in systems doing artificial intelligence tasks (AI). Examples of such tasks include recognition, diagnostics, scheduling, robot control, prediction, etc. The "changes" might either improve existing systems or synthesize new systems in the initial version.

The supervisor is the inspiration for the concept of supervised learning and is a teacher in the learning process. The machine learning job is known as monitored learning to derive a function from labeled training data. The training data are a set of training examples. Each example is a pair of an input item (typically a vector) and a desired output value in supervisory education (also called the supervisory signal). A supervised learning algorithm analyses training data and creates an inferred function for mapping new cases [4]. Ideally, for unknown cases, the algorithm can accurately anticipate the class labels. This requires the learning algorithm to generalize from the training information in "reasonable" ways to previously unknown conditions. Conceptual learning is a phrase used in psychology of humans and animals for simultaneous tasks. Monitored education produces a function that translates inputs to the desired outcomes (also called labels, because they are often provided by human experts labeling the training examples). The most often used learning paradigm is supervised learning. In fact, given the "experience" in the form of target function instances, that is, input–output pairs, it allows realistic solutions to be developed with a wide range of learning methods. The need for such a broad range of learning algorithms is partly determined by the enormous number of real-world learning challenges, which are covered by various tasks and problem-specific learning algorithms for solutions. These comprise classifying and regressing difficulties, including multi-label and multi-class, as well as multivariate regression issues, ranking problems, and common regression concerns (either label or instance rankings). The traditional technique is to translate these difficult concerns into a series of smaller, well-known factors and then add the resulting forecasts together. However, these approaches often have no consistent theory and/or much too many computer resources to be of value in applications in the actual world.

The difficulty of trying to find the latent organization in unlabeled data is called unchecked learning in machine learning. No error or bonus signal may be sent to evaluate a solution since the examples provided to the learner are not labeled. Uncontrolled education distinguishes itself from controlled learning [4]. The statistical problem of density estimation is intimately linked to unsupervised learning. In contrast, uncontrolled learning involves a large number of ways of summarizing and explaining important factual features. Many uncontrolled methods of learning are founded on data mining tools for preprocessing the data. This also applies in general as it requires no manual data classification by a human expert, rather than supervised learning. Labeled data is not just costly to get, but also contains some information and surely does not allow the parameters of sophisticated models to be properly assessed. Clustering is an unattended learning approach. Cluster analysis, also referred to as clustering, is the task of arranging collections of articles in a way that allows for objects in the same group (called cluster) (clusters).

32.2 Applications of Machine Learning in Additive Manufacturing

As shown in Fig. 32.1, machine learning algorithms are used in all the processes related to additive manufacturing. Table 32.1, summarizes applications of machine learning in additive manufacturing.

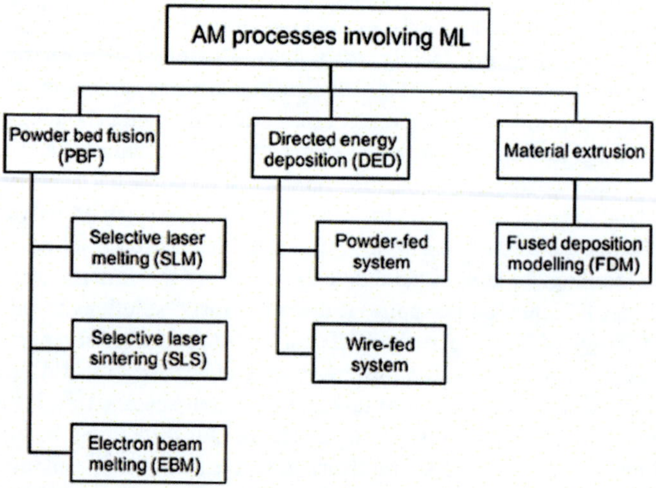

Fig. 32.1 Machine learning in additive manufacturing

Table 32.1 Applications of machine learning in additive manufacturing

1. Processing parameter optimization
2. Part property prediction
3. Geometric deviation control
4. Closed-loop control
5. Defect detection
6. Quality assessment
7. Quality prediction
8. Cost estimation

Nandhithaa et al. [5] investigated different feature extraction techniques for analyzing infrared thermal images obtained during online weld monitoring and concluded that the Euclidean distance-based image segmentation approach is best suited for the task. Lahiri et al. [6] employed IRT to detect and quantify weld flaws in friction stir welded aluminum and Tungsten Inert Gas (TIG) welded stainless steel connections. The depths of the defects were evaluated using the lock-in thermography approach. A digital X-ray radiography method was used to validate the findings. When compared to other non-contact NDT techniques, it can be stated that IRT is a useful approach for flaw identification in welded joints.

Lock-in thermography is preferable since it reveals the depth of the fault. Unnikrishnakurup et al. [7] created models to anticipate the heat cycles that occur during TIG welding. For his experiment, he used an infrared camera to detect temperature. Infrared thermography has been thoroughly reviewed. The history of IRT and IRT cameras, experiment methodology, and data processing methods were all thoroughly presented with a range of applications.

Mechanical system and tool condition monitoring have gained traction as a result of advancements in signal processing techniques and the simplicity of application of machine learning classifiers. Marwala [8] examined the applicability and significance of computational intelligence approaches for forecasting mechanical system conditions, such as Bayesian methods, neural networks, Hidden Markov Techniques, rough sets, Support Vector Machines, and Ensemble Methods. The major incentive for researchers to apply machine learning classifiers for predicting the behavior of welding processes is the quality of solutions achieved by applying machine learning classifiers for condition monitoring of mechanical systems. Roth et al. [9] discussed the significance of intelligent predictive techniques for monitoring tool conditions in drilling, milling, and grinding processes.

Martn et al. [10] utilized CART decision tree models and the random forest technique to categorize ultrasonic oscillograms of resistance spot-welded joints. The CART model was shown to have greater interpretability, while the random forest method lowered misclassification rates. To enhance prediction accuracy on noisy RSW data, Park and Kim [11] presented a methodology that combined bootstrap aggregation with support vector regression. Kim et al. [12] investigated inconsistent RSW weld data using k-nearest neighbor and generalized regression neural network with mean acceptable error measurements. Ahmed and Kim [13] investigated the efficacy of decision tree techniques to predict weldability in RSW. Chen et al. [14] used Support Vector Machines to investigate flaw classification in ultrasonic nondestructive evaluation (SVM). The wavelet coefficients from the ultrasonic signals were recovered, and the weld fault was classified. SVM was utilized by Wang et al. [15] to characterize the dynamic characteristics of a metal vapor plume generated during the laser 41 beam welding process. Six characteristics were retrieved and their classification accuracy rates and Pearson product-moment correlation coefficients were evaluated. Cocota et al. created a Hierarchical Support Vector Machine (HSVM) model for the SMAW process to predict the three weld bead classes using acoustic signature [16]. Mollayi and Eidi [17] used Support Vector Machines to study various kernel functions and predict weld bead shape in the GMAW process.

32.3 Conclusion

Traditionally, manual prototyping was used to build a prototype. In the industrial and manufacturing industries, additive manufacturing is a buzzword. Initially, the CAD model of the product's components was built-in modeling software in accordance with the specifications. After creating a CAD model, it is cut into parallel planes equal to the layer thickness. As a result, the edges of these slices are extremely sharp and squared, giving the impression of a stair effect. These three-dimensional models will now be subdivided into tiny two-dimensional objects known as slices. Simply said, a complex three-dimensional problem has been reduced to a series of two-dimensional challenges. STL files are tiny two-dimensional files that are delivered by tessellating the geometric three-dimensional model. A CAD model's various surfaces are piecewise approximated by a succession of triangles in tessellation, and the coordinates of triangle vertices and their surface normal are recorded. Because of their predictive character, machine learning algorithms are the perfect tool for dealing with additive manufacturing issues. Machine learning algorithms are capable of analyzing prior data and predicting future events. This article addresses the use of machine learning in additive engineering.

References

1. Li, H.: Machine-learning based modelling for AM processes. In: 2020 IEEE 10th International Conference Nanomaterials: Applications & Properties (NAP), pp. 02SAMA21-1–02SAMA21-4 (2020). https://doi.org/10.1109/NAP51477.2020.9309628
2. Li, Y., Yan, H., Zhang, Y.: A deep learning method for material performance recognition in laser additive manufacturing. In: 2019 IEEE 17th International Conference on Industrial Informatics (INDIN), pp. 1735–1740 (2019). https://doi.org/10.1109/INDIN41052.2019.8972334
3. Han, F., Liu, S., Liu, S., Zou, J., Ai, Y., Xu, C.: Defect detection: defect classification and localization for additive manufacturing using deep learning method. In: 2020 21st International Conference on Electronic Packaging Technology (ICEPT), pp. 1–4 (2020). https://doi.org/10.1109/ICEPT50128.2020.9202566
4. Mashhadi, F., Salinas Monroy, S.A.: Deep learning for optimal resource allocation in IoT-enabled additive manufacturing. In: 2020 IEEE 6th World Forum on Internet of Things (WF-IoT), pp. 1–6 (2020). https://doi.org/10.1109/WF-IoT48130.2020.9221038
5. Nandhitha, N.M., Manoharan, N., Rani, B.S., Venkatraman, B., Kalyanasundaram, P., Raj, B.: Comparative study on the suitability of feature extraction techniques for tungsten inclusion and hotspot detection from weld thermo-graphs for online weld monitoring. Indian J. Eng. Mater. Sci. **17**(1), 20–26
6. Lahiri, B.B., Bagavathiappan, S., Saravanan, T., Rajkumar, K.V., Kumar, A., Philip, J., Jayakumar, T.: Defect detection in weld joints by infrared thermography (2011)
7. Unnikrishnakurup, S., Krishnamurthy, C.V., Balasubramaniam, K.: Monitoring TIG welding using infrared thermography–simulations and experiments (2016). https://doi.org/10.15199/48.2016.04.02
8. Marwala, T.: Condition Monitoring Using Computational Intelligence Methods. Springer, London (2012)

9. Roth, J.T., Djurdjanovic, D., Yang, X., Mears, L., Kurfess, T.: Quality and inspection of machining operations: tool condition monitoring. J. Manuf. Sci. Eng. **132**(4), 041015 (2010). https://doi.org/10.1115/1.4002022
10. Martín, Ó., Pereda, M., Santos, J.I., Galán, J.M.: Assessment of resistance spot welding quality based on ultrasonic testing and tree-based techniques. J. Mater. Process. Technol. **214**(11), 2478–2487 (2014). https://doi.org/10.1016/j.jmatprotec.2014.05.021
11. Park, J., Kim, K.-Y.: Prediction modeling framework with bootstrap aggregating for noisy resistance spot welding data. J. Manuf. Sci. Eng. **139**(10), 101003 (2017). https://doi.org/10.1115/1.4036787
12. Kim, K.Y., Park, J., Sohmshetty, R.: Prediction measurement with mean acceptable error for proper inconsistency in noisy weldability prediction data. Robot. Comput.-Integr. Manuf. **43**, 18–29 (2017). https://doi.org/10.1016/j.rcim.2016.01.002
13. Ahmed, F., Kim, K.Y.: Data-driven weld nugget width prediction with decision tree algorithm. Procedia Manuf. **10**, 1009–1019 (2017). https://doi.org/10.1016/j.promfg.2017.07.092
14. Chen, Y., Ma, H.W., Zhang, G.M.: A support vector machine approach for classification of welding defects from ultrasonic signals. Nondestructive Test. Eval. **29**(3), 243–254 (2014)
15. Wang, T., Chen, J., Gao, X., Li, W.: Quality monitoring for laser welding based on high-speed photography and support vector machine. Appl. Sci. **7**(12), 299 (2017). https://doi.org/10.3390/app7030299
16. Cocota, J., Garcia, G., da Costa, A., de Lima, M., Rocha, F., Freitas, G.: Discontinuity detection in the shield metal arc welding process. Sensors **17**(12), 1082 (2017). https://doi.org/10.3390/s17051082
17. Mollayi, N., Eidi, M.J.: Application of multiple kernel support vector regression for weld bead geometry prediction in robotic GMAW process. Int. J. Electr. Comput. Eng. **8**(4), 2310–2318. https://doi.org/10.11591/ijece.v8i4.pp2310-2318

Chapter 33
A Short Study on IoT-Based Cellular Network

M. V. Sai and G. Rama Naidu

Abstract Nowadays, the requirements of the world getting change rapidly in the field of communication. With the evaluation of the internet, almost everything is supported by the network. Huge amounts of information are transferred every time, which enhance the necessity of an efficient and secure network system. As the networks are also required to address the real-time issues, it is needed that the network must be highly efficient in terms of response time and adaptability. Critical information is sensed and analyzed and transmitted through the network which has complex structure and scalability. Apart from that in today's world, data theft is one of the major issues. As the conventional network fails to fulfil these requirements, a new but potential technique is evolving. This new technology is a combination of the cellular network data transmission technique and internet of things (IoT) technology. The combination of these techniques makes this technology adaptable, scalable, highly responsive and extremely secure. This presented paper highlights the basic structure of a cellular network and the IoT-based cellular network along with its major breakthrough and potential.

33.1 Introduction

The presented paper mainly deals with the concept of the internet of things with the multilayer method of transporting data from end-to-end communication through a cellular network securely. The complete implementation of this type of network provides an extremely secured network pathway to communicate. Nowadays, as the uses of the Internet of things intensify, it brought with it a heightened awareness of the security issues that are associated with linked devices. There are numerous ways but although, all of these systems/devices are utilizing cellular connectivity to interconnect. Rather than the cellular connectivity-based solutions, some more options can be utilized such as Ethernet or wireless Wi-Fi-based using which the

M. V. Sai · G. Rama Naidu (✉)
Aditya Engineering College, Surampalem, AP, India
e-mail: ramanaidu.gangu@aec.edu.in

S. Yadav et al. (eds.), *Proceedings of Second International Conference in Mechanical and Energy Technology*, Smart Innovation, Systems and Technologies 290,
https://doi.org/10.1007/978-981-19-0108-9_33

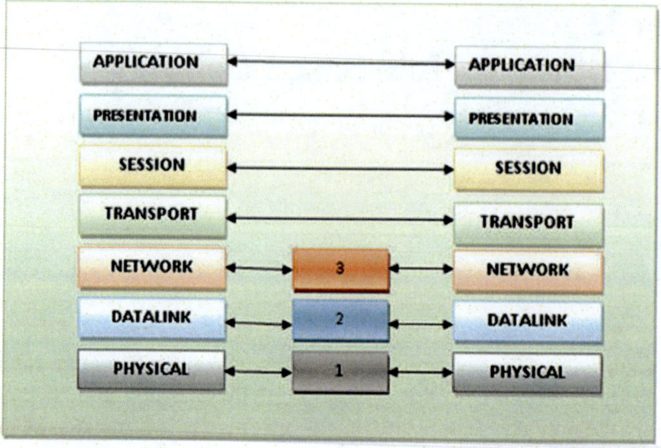

Fig. 33.1 Layers in ISO layers

data is transported from the edge devices to the host. All of these alternate solutions also offer remote monitoring and tracking of devices. However, the security threat risk is very high. Various security breach incidents have happened in which Wi-Fi over the public internet was used as a means of gaining access to the data stream and altering the operation of internet of things devices. Therefore, it is necessary to secure communication between two or more linked devices. This study proposed that the multilayer method of transportation in the Internet of things (IoT) communication is one of the best potential candidates that protect and ensure security to the data which is exchanged between connected devices.

Cellular networking has a high capacity in voice and data communication with increased multimedia and coherent roaming facilities for supporting cellular devices. It is well known and presented in Fig. 33.1 that from source to destination it has seven layers process occurs application, presentation, session, transport, network, data link, and physical layers on both the sides of source and destination. This process is done for transporting the data from source to destination securely. The complete process of communication required some basic characteristics. These essential necessities are highlighted below.

33.2 Methodology Discussion

33.2.1 Infrastructure

The whole communication network areas are bifurcated into several small cells. To connect all base stations available in the region, these small cells necessitate a complicated infrastructure. Switches for call forwarding, location registers, and other infrastructure are also highly required. These all the requirements make the single-cell structure too complex.

33.2.2 Handover Process

As these communicating base stations are dynamic and there is a high probability that one base station can move from one cell to another while communicating, hence it is necessary for the mobile station that the process of the handover must be efficient and uninterrupted when the base station is switching from one cell to another.

33.2.3 Proliferation of Antenna

Nowadays, the number of antennas is greatly increased (transmitting as well as receiving antennas) on a cell tower and cell phone in modern cellular networks, which has boosted the performance of the network significantly. However, this arrangement required large space for the establishment, which has its own limitations. Therefore, a way must be required to optimally fulfil this challenge. In this scenario, the use of MIMO (multiple-input, multiple-output) seems to be a potential solution. The customers who can receive simultaneous signals from many transmitters have seen a huge improvement due to the MIMO technique. However, there are two side effects of MIMO that reduce the performance for some users. First, MIMO has reverse distance effects, in other and simple words MIMO mostly favors individuals who are nearest to the cell tower, which means that others who are further distant from the cell tower have fewer quality connections accessible. Second, MIMO works best with non-adjacent cellular channels which means at the time of high call rate MIM will enhance the signal strength for one channel at the cost of the strength of the other channel, which affects the customers who are on the weaker channel; they may be facing lowering quality issues.

33.2.4 Shape of the Cellular Footprint

The strength of the signal around the cell tower is not identical. If it is measured at any cell site and plot signal strength, it can be easily observed that the footprint of a cell tower resembles that of an amoeba, with the signal traveling a very short distance in certain directions and traveling much farther in others. Also, this footprint is not static in nature, it is highly influenced by the temperature, humidity and user density present in the region of the cell footprint. Due to this, it is critical to distribute the broadcasting of the wireless services as the delivery footprint is continuously moving, often radically. That is why, it is hard to predict the absolute nature of the point-to-point connection between the mobility data center and the destination host, which provide the most efficient and reliable services. This connection might be an IPsec VPN tunnel, MPLS, frame relay, or any number of other landlines secure point-to-point connectivity methods provided by the carrier's outbound side. Generally, the IPsec VPN route seems to be the most common, with Cisco firewall VPN equipment, which is used at both the carrier and host data sites. The other possibility to establish a close secure connection between radio modules of the IoT device through the tower and the carrier's mobility data center transmitting through the GGSN/PDN-GW is to combine two functional elements delivered by the custom APN and IP sec VPN tunnel. It is also designed in such a way that the data connection is routed out of the mobile data center via a special off-the-internet secure pipe to the enterprise customer's host, which is terminated at the IP sec VPN router. Even the different IoT devices cannot talk of communicating to each other directly as the common misconception has spread out. This misconception is might be due to the other recognition of IoT which is "Machine to Machine". This also follows the same technique that the communication through the defined protocols using all the layers presents in the design or system. Direct machine-to-machine communication through any carrier or router is prohibited. In reality, the communication between the IoT devices is only routed through the application layer within the backend host that manages IoT solutions. If the situation demands the transformation of data between two IoT devices, this can be accomplished via the application layer on the backend host server, rather than a direct data packet transfer between the two devices. Although it is technically possible to establish an APN to route device-to-device traffic via the carrier's data center, this method violates the security measures outlined above. If it is permissible to exchange the data directly from one IoT device to other via the carrier's data center, then, no record or footprint would be left in the customer's router or backend host system. In other words, the devices may be communicated back and forth without the enterprise customer being aware of or realizing it, then one cannot able to investigate the communication for malicious behavior, which increase the security risk to a great extent. However, through the appropriate protocols followed, the device can communicate with each other efficiently with no risk. Therefore, it can be said that the devices are now the user of the data they collected. As communication between the devices is an inherent part of IoT, the device can be able to exchange information automatically without any presence of the monitoring body or system.

Figure 33.2 represents a general multi-hop interconnected intelligent IoT system. This designed system can transfer the data with its peer devices, acquired information and transmit them via the multi-hop method, utilizing neighboring networks. The proposed structure of the network will ensure that information exchange within and between clusters is trustworthy [1]. For this purpose, various networks have been designed to transform data security and reliability. Some of the major breakthroughs are highlighted here for reference and a better understanding of the real scenario.

In the case of IoT equipped uplink cellular network, where a cellular user can able to gather the information of mobile data for a cluster of IoT devices, the user can send the sensory data directly or via cellular communication to the base station. This process is accomplished by machine-to-machine communication. This idea was

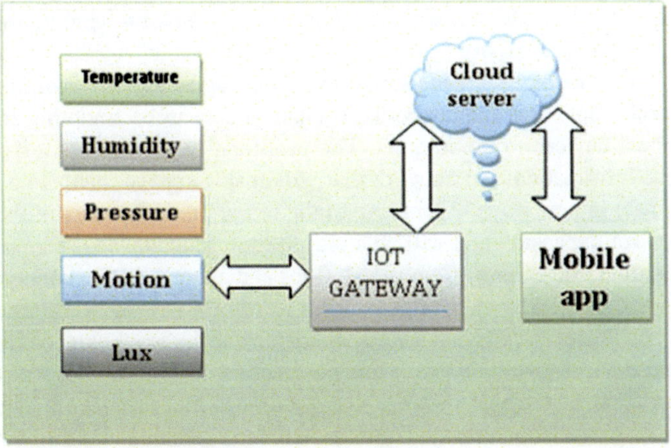

Fig. 33.2 A multi-hop interconnected intelligent IoT system

proposed by Zhang et al. in the year 2019, according to their study, this proposed IoT scheme can accommodate a greater number of IoT devices than the allowed spectrum alone. The presented scheme for communication in a cluster-based network equipped with an IoT facility makes the use of an unlicensed spectrum for machine-to-machine communication in clusters. As data traffic develops, more internet of things devices is connected to the base station, resulting in reduced interference [2]. In the same year, Khan et al. presented another model IoT network keeping goal to make it cost-efficient and spectrum effective utilization. This designed model is also low power IoT which is unrestrained and unhindered interference in nature. This model addressed the effective spectrum utilization problem with the help of cognitive radio networks, which is also observed as a low-cost solution. The observation of this study demonstrated that the IoT network is better and efficient as compared to the conventional networking strategies. The designed model also supports the complex data with the improved character of throughput and delay while comparing with the traditional models for varieties of traffic [3]. As the IoT technology is getting improved, it is observed that this technology has tremendous features that can be used to solve real-life issues such as environmental contamination but this necessitate a real-time monitoring system to be made. The entire system consists of the hardware module equipped with efficient sensors that sense the environmental parameters such as temperature, humidity, etc. These acquired data is transmitted through the wireless router under the instruction set of wireless fidelity protocol. A remote server is also deployed and this receives data via the internet from the sensor node and saves these data in the database. To make the system more reliable and efficient, finite state machines and asynchronous communication protocols are implemented in the system. After analyzing the data it is concluded that at a certain height the concentration is more near the ground and after that, it reduces with the increase in attitude. This entire model was developed to address the real-time pollution issue by Xu et al. in the year 2016 [4].

The IoT encompasses a wide range of technologies and applications, as well as commercial potential and concerns. It enables data to be smoothly transported from physical devices to the internet. The proliferation of intelligent devices will establish an information-rich network that will enable supply chains to accumulate and interrelate in new ways. According to the research, the internet of things has a lot of room for expansion and will be a big revenue generator [5]. This technology has the capability of finding effective solutions and adoptability to connect with the new and various types of networks in order to form a new network structure and going beyond the technological barriers which are due to the inherent confines of the constrained devices usually used in this perspective. This technology also enables the development of scalable network architectures which is capable of collecting and processing data from restricted devices in order to provide valuable services and applications to end-users [6].

33.3 Conclusion

This study mainly focused on the end-to-end data communication using IoT. It is observed that this network has the capability to provide an extremely secure network. This network is also capable of effective interconnects using cellular connectivity. As the new world requirements are the out of the way network structures in order to make the communication efficient and secure along with the quick response to address the real-time issues, the IoT-based network is a much better performance as compared to the conventional. Its adaptability makes this technique unique as complex algorithms and structures can be easily implemented built using this, in other words, this technology has a high order of flexibility and scalability. These all characteristics make IoT-based networks a future potential network to address our unsolved and upcoming issues in terms of data collection, communication, etc. Along with these, due to its ability to build an information-rich network, it has high commercialization value too. However, there is a mile to go and need intense study and research in this field in order to cope with the limitations and challenges but it can be easily predicted that IoT-based communication network is our future.

References

1. Bello, O., Zeadally, S.: Intelligent device-to-device communication in the internet of things. IEEE Syst. J. **10**(3), 1172–1182 (2014)
2. Zhang, H., Di, B., Bian, K., Song, L.: IoT-U: cellular Internet-of-Things networks over unlicensed spectrum. IEEE Trans. Wireless Commun. **18**(5), 2477–2492 (2019)
3. Khan, F., Rehman, A.U., Jan, M.A., Rahman, I.U.: Efficient resource allocation for real time traffic in cognitive radio internet of things. In: 2019 International Conference on Internet of

Things (iThings) and IEEE Green Computing and Communications (GreenCom) and IEEE Cyber, Physical and Social Computing (CPSCom) and IEEE Smart Data (SmartData), pp. 1143–1147 (2019)
4. Xu, B., Zheng, J., Wang, Q.: Analysis and design of real-time micro-environment parameter monitoring system based on Internet of Things. In: 2016 IEEE International Conference on Internet of Things (iThings) and IEEE Green Computing and Communications (GreenCom) and IEEE Cyber, Physical and Social Computing (CPSCom) and IEEE Smart Data (SmartData), pp. 368–371 (2016)
5. Singh, S., Singh, N.: Internet of Things (IoT): security challenges, business opportunities & reference architecture for E-commerce. In: 2015 International Conference on Green Computing and Internet of Things (ICGCIoT), pp. 1577–1581 (2015)
6. Davoli, L., Belli, L., Cilfone, A., Ferrari, G.: From micro to macro IoT: challenges and solutions in the integration of IEEE 802.15. 4/802.11 and sub-GHz technologies. IEEE Internet Things J. 5(2), 784–793 (2017)

Chapter 34
Decomposition Makes Better Rain Removal: An Enhanced Attention-Guided Image De-raining Using Deconvolutions Network

Saurabh Arora, Sushant Bindra, Musheer Ahmad, and Tanvir Ahmad

Abstract Downpour streaks noticeable all around show assorted qualities with various shapes, headings, densities, even the complex covered marvel, causing extraordinary difficulties for the de-pouring assignment. As of late, profound learning-based picture de-pouring techniques have been broadly examined because of their amazing exhibition. In any case, the majority of the current calculations actually have constraints in eliminating precipitation streaks while saving rich textural subtleties under muddled downpour conditions. To this end, we propose to disintegrate downpour streaks into numerous downpour layers and separately gauge every one of them along the organization stages to adapt to the expanding abstracts. To all the more likely portray downpour layers, a further developed non-neighborhood block is intended to abuse the self-similitude of downpour data by learning the all-encompassing spatial element relationships while lessening the estimation intricacy. Also, a blended consideration instrument is applied to direct the combination of downpour layers by zeroing in on the neighborhood and worldwide covers among these downpour layers. Broad tests on both manufactured blustery/downpour cloudiness/raindrop datasets, certifiable examples, the murkiness, and low-light situations show generous enhancements both on quantitative pointers and special visualizations over the present status-of-the-craftsmanship advances.

34.1 Introduction

Extreme climate conditions, such as downpour, haze, and snow, unfavorably influence the visual nature of images delivering them futile for additional use and sharing. In addition, such debased images radically influence execution of visual frameworks.

S. Arora (✉) · M. Ahmad · T. Ahmad
Department of Computer Engineering, Jamia Milia Islamia, New Delhi 110025, India
e-mail: sa2550arora@gmail.com

S. Bindra
Department of Computer Engineering, Manipal University, Jaipur, Jaipur 303007, India

S. Yadav et al. (eds.), *Proceedings of Second International Conference in Mechanical and Energy Technology*, Smart Innovation, Systems and Technologies 290,
https://doi.org/10.1007/978-981-19-0108-9_34

Hence, it is critical to take care of the issue of single image de-coming down where data mining plays an integral role to organize the data and selecting the most effective one [1]. In this paper, we explore computational methodology to the single picture de-coming down issue, unlike the existing methodologies. It has been broadly recognized that eccentric impedances like light, clamor, and serious climate conditions (for example downpour, snow, and haze) unfavorably influence the visualization of numerous image calculations like recognition, arrangement. This is due to the reason that the preliminary calculations are prepared to utilize images that are caught under controlled environmental or climatic conditions [2]. The presence of weighty downpour significantly impedes visual nature of the image, consequently delivering face identification and check calculations insufficient for such debasements. A potential technique to resolve the issue is to incorporate images under unconstrained conditions in the image preprocessing and the interaction of the image computation in the data mining association rule, which is used as a base of the research in [3]. Though, it may not be useful to gather such pictures for all classes in the preparation set due to enormous scope of image settings. Moreover, in this time of pervasive cell phone use, images captured by cell phone cameras under troublesome climate conditions go through debasements that definitely influence the visual nature of pictures making the pictures futile for sharing and utilization. To improve the general nature of such noise corrupted images for better visual allure and to guarantee upgraded execution of image processing, it is fundamental to eliminate the unfortunate rarities of the image due to the troublesome climate conditions as explained earlier. We propose a solitary picture-based de-pouring/de-snowing calculation utilizing a convolutional-deconvolution structure for outwardly improving images that have gone through debasements because of downpour [4]. A stormy image can be visualized as the superposition of two images—one relating to rain streaks and the other compared to the unmistakable foundation picture. Thus, the blustery image can be represented as

$$x = Y + W \tag{34.1}$$

where $\boldsymbol{Y}$ addresses the unmistakable foundation image and $\boldsymbol{W}$ addresses the downpour streaks [5]. Similar to the image de-noizing and image partition, image de-coming down can be seen as the issue of isolating two segments from a blustery image which have been connected on different requisition domains that incorporate telecommunication networks, market analysis, hazard official suite, account control, and many others.

34.2 Association Rule Mining

The proposed design for preparing the model is as per the following constraints:

- Each preparing image is a 256 × 256 pixels' picture (resized).

- We accept the contribution as 256 × 256 × 3 where 3 indicates the channels of RGB picture.
- The input is taken care of to a convolutional layer of 2 × 2 channel followed by a ReLU layer.
- The yield of the past layer is taken care of into the Batch Normalization layer.
- The yield of the past layer is again taken care of into a convolutional layer of 2 × 2 channel followed by a ReLU layer.
- The next layer is max pooling layer with channel size of 2 × 2.
- The yield of the past layer is again taken care of into a convolutional layer of 3 × 3 channel followed by a ReLU layer.
- The next layer is max pooling layer with channel size of 2 × 2.
- The yield of the past layer is again taken care of into a convolutional layer of 2 × 2 channel followed by a ReLU layer.
- The yield of the above network is taken care of into a dropout layer with a likelihood of every neuron of 0.5.
- The next layer is max pooling layer with channel size of 2 × 2.
- The yield of the past layer is taken care of into the Batch Normalization layer.
- The next layer is an upsampling of 2 × 2 layer.

The yield of the layer is taken care of into a de-convolution layer [6]. The issue of expulsion of downpour also called draining of the image as well as recordings is an important issue in the present day and age where we are utilizing PC vision and example acknowledgment in an assortment of fields, face acknowledgment being perhaps the main application. The presence of downpour goes about as a commotion for these applications [7].

In other related work, the author proposed the content to discourse acknowledgment based on AI and ML [8], here the author addresses the exactness of text to discourse strategy plan that accomplishes the joining of cutoff rightness affirmation and data botch limitation.

34.3 Dynamic Exploration Has Been Going on in This Field

In [9], the authors investigated the elements of downpour just as the photometer developed a relationship model that catches the elements of downpour and a material science-based movement obscure model that clarifies the photometry of downpour. In light of these models, we foster productive calculations for identifying and eliminating precipitation from recordings [10]. A novel calculation for deraining of recordings by utilizing Temporal Correlation and Low position lattice finishing in which the authors chipped away at the perception that downpour streaks are excessively little and move too quickly to even consider influencing the optical stream assessment between continuous edges [11]. Here in [11], an underlying precipitation map by deducting transiently distorted edges from a current edge. They eliminated the downpour distinguished after the deduction and eliminated downpour streak pixels utilizing low position lattice consummation [12]. The high recurrence picture has additionally deteriorated into downpour and non-downpour parts dependent on the understanding that learned word reference particles can scantily address clear foundation picture and downpour streak picture independently [13]. In [14], a novel single picture de-pouring technique was proposed, which is referred to as Image De-Pouring Conditional General Adversarial Network (ID-CGAN), and shows better outcomes than [15].

34.4 Method Development

A Convolutional Neural Network (ConvNet/CNN) is a Deep Learning computation that can take in a data image, distribute importance (learnable burdens and tendencies) to alternate points of view/objects in the image, and have the choice to isolate one from the other. The pre-getting ready required in a ConvNet is a ton lower when stood out from other request computations. While in rough strategies channels are hand-planned, with enough getting ready, ConvNets can acquire capability with these channels/characteristics [16, 17].

34.5 Generative Adversarial Network:

Generative Adversarial Networks (GAN) is one of the most encouraging ongoing improvements in Deep Learning. GAN, presented by Ian Good fellow in 2014, tackles the issue of solo learning via preparing two profound organizations, called Generator and Discriminator as shown in Fig. 34.1, that contend and help out one another [18].

34.6 Implementation

The proposed methodology is implemented on keras 1.1, trained on NVIDIA GTX960M. The model was trained using a training set of 700 images and a testing set of 200 with 10 epochs each with four batch sizes [19]. The simulation of the regression in dot net is shown in Fig. 34.4. Figure 34.4a shows the implementation of regression for optimization in dot net environment.

To analyze the suitability of the proposed methodology and to compare the results with other methodologies, few metric measures are considered in this paper. As shown in Fig. 34.2, the PSNR of images is calculated for the proposed methods. The PSNR is calculated using

$$\text{PSNR} = 10\ _{10}\left(\frac{I^{2}}{\text{MSE}}\right) = 20\ _{10}\left(\frac{I}{\text{MSE}}\right) \tag{34.2}$$

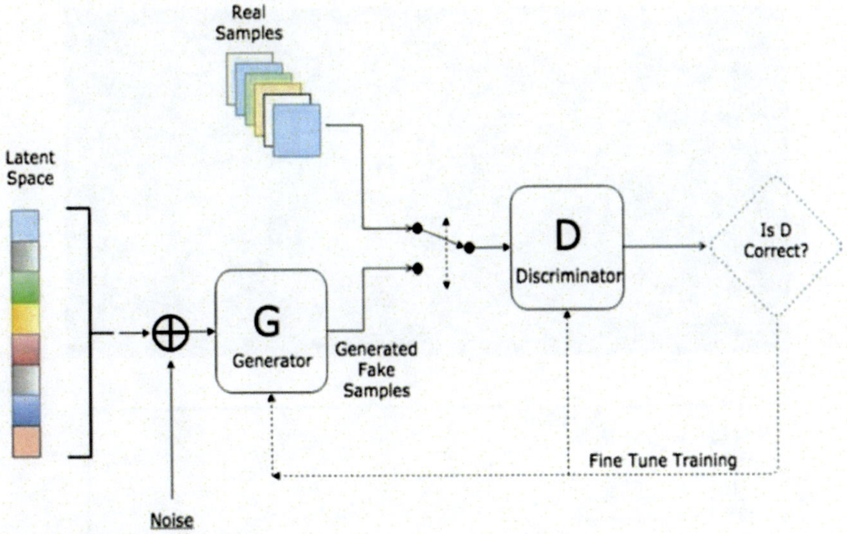

Fig. 34.1 GAN model

where MSE is defined as,

$$\text{MSE} = \frac{}{\text{mn}} \sum^{m-1} \sum^{n-1} |(\quad)|^2 \tag{34.3}$$

Similar to PSNR measure, as shown in Fig. 34.3, the metric measure SSIM is defined as follows with standard nomenclatures of image processing,

$$\text{SSIM}(\quad) = (\quad)(\quad)/(\mu 2.X + \mu 2.Y + C1)(\alpha 2.X + \alpha 2.Y + C2) \tag{34.4}$$

$$\text{Info}(\,) = \sum^{n} - \quad _i \quad _i \tag{34.5}$$

$$(\quad) = 1/P \sum^{n} \sum \text{SSIM}_j \tag{34.6}$$

The UIQI and SSIM are more accurate and consistence than MSE and PSNR despite they cost more, here

$$(\,) = 1/M \sum^{n} Qj \tag{34.7}$$

And the Variation Inflation Factor (VIF) is defined as,

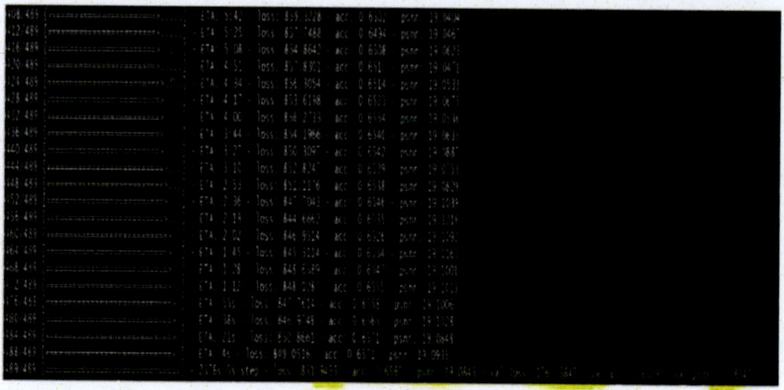

(a)

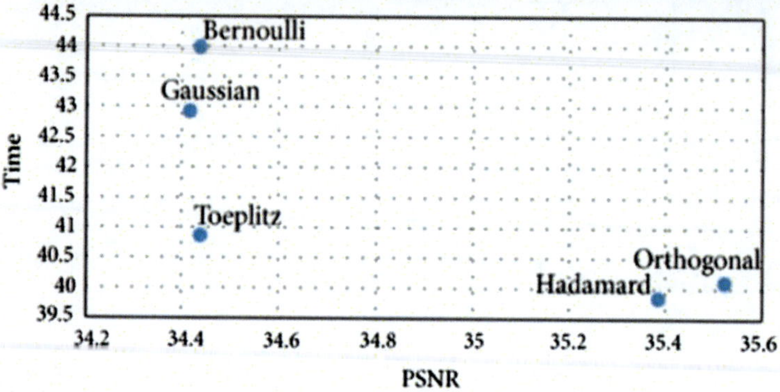

(b)

Fig. 34.2 Calculation of PSNR: a calculation environment of the proposed method and **b** shows the comparative study of PSNR

No.	X2	X3	X4	X5	X6	X7	X8	X9	X10	X11	X12
X1	0	0	0	0	0	0	0	0	0	0	0
X2	1	0	0	0	0	0	0	0	0	0	1
X3	1	1	1	1	1	0	0	0	1	0	1
X4	1	0	1	1	1	0	0	0	1	0	1
X5	1	0	1	1	1	0	0	0	1	0	1
X6	1	0	1	0	1	0	0	0	1	0	1
X7	1	1	1	1	1	1	1	1	1	1	1
X8	1	1	1	1	1	0	1	1	1	0	1
X9	1	1	1	1	1	1	0	1	1	0	1
X10	1	0	1	0	0	0	0	0	1	0	1
X11	1	1	1	1	0	1	1	1	1	1	1
X12	0	0	0	0	0	0	0	0	0	0	1

Fig. 34.3 SSIM: The SSIM index is calculated on various windows of an image

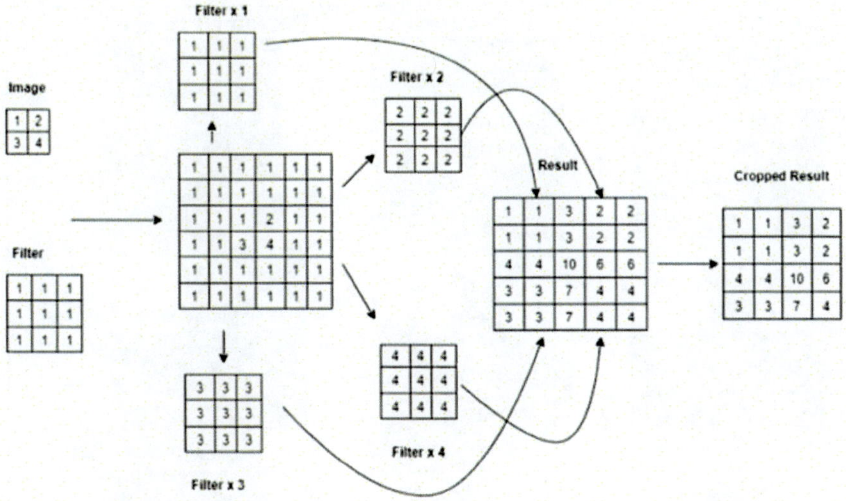

Fig. 34.4 Convolution 2D Transpose

$$( ) = \frac{}{1} - R2i \tag{34.8}$$

34.7 Result Analysis with Simulation

The proposed model is trained and tested over a database of more than 700 images, the MNIST Multilayer Neural Network and Dropout Iteration over the entire Dataset is shown in Fig. 34.4. It is interesting to see that the training cost reduces exponentially with the number of iterations over the entire data set. As shown in Fig. 34.5, the proposed model was able to output a de-rained image with little blurring effect for most of the images. However, there are certain inputs for which the blurring effect was considerable. Using the proposed De-Conv algorithm, the system is able to fetch the particular required image even in the rainy condition. The detrained image smoothens out the effect of rain as shown in Fig. 34.5 and hence the proposed methodology shows encouraging results for the visual confirmation of the proposed algorithm.

The results of the proposed methodology, (De-Conv) Algorithm, is compared to other various De-Redraining algorithms using different metric like PSNR (34.2), VIF (34.8), and SSIM (34.4), and the proposed algorithms show encouraging results compared to others, as shown in Table 34.1.

Fig. 34.5 The results of the proposed methodology as detrained images

Table 34.1 Comparison of different algorithms using PSNR, VIF, SSIM

Metric/Model	SPM	DSC	De-Conv
PSNR(dB) (34.2)	18.88	18.56	19.08
VIF (34.8)	1.21	1.10	1.06
SSIM (34.4)	0.8	0.75	0.82

34.8 Conclusion/Future Scopes

In this paper, we proposed a Convolution-Deconvolution-based calculation for the evacuation of downpour streaks structure a solitary image. In contrast with the present methodologies which take care of the de-coming down issue in image disintegration structure by utilizing previous data. Our proposed network is capable of effectively removing heavy rain in single images where the strong rain streaks and fog are compounded. Conv-Deconv's PSNR is more prominent than some best in class strategies versus Sparse word reference-based strategy (PSNR: 18.88) and Discriminative inadequate coding-based method (PSNR: 18.56). Though the Conditional General Adversarial Network (GAN) shows better PSNR (PSNR: 22.73), GAN experiences a couple of disadvantages it neglects to eliminate the white-round downpour particles. Since the proposed Convolution-Deconvolution methodology shows encouraging results without use of GAN, we expect to expand our research in the Deconvolution with GAN system for better results and to examine the chance of utilizing comparable constructions for tackling related issues. The artificial neural network and convolutional neural network BASED IOT application were also determined for future application. The image retrieval is also determined using MATLAB simulate tool also used for FOG computing, respectively.

References

1. Chen, M.S., Han, J., Yu, P.S.: Data mining: an overview from a database perspective. IEEE Trans. Knowl. Data Eng. **8**(6), 866–883 (1996)
2. Tudor, I., Universitatea Petrol-Gaze din ploeiesti: Association rule mining as a data mining technique. Bd.Bucuresti 39, ploeiesti, Catedra de Informatica **LX**(1) (2008)
3. Goswami, D.N., Anshu, C., Raghuvanshi, C.S.: An algorithm for frequent pattern mining based on apriori. Int. J. Comput. Sci. Eng. **02**(04), 942–947 (2010)
4. Wu, X., Kumar, V., Ross Quinlan, J., Ghosh, J., Yang, Q., Motoda, H., McLachlan, G.J., Ng, A., Liu, B., Yu, P.S., Zhou, Z.-H., Steinbach, M., Hand, D.J., Steinberg, D.: Top 10 algorithms in data mining. Knowl. Inf. Syst. **14**(1), 1–37 (2007)
5. VLDB J., 507–521 (2007)
6. Li, N., Zeng, L., He, Q., Shi, Z.: Parallel implementation of apriori algorithm based on MapReduce. In Proceedings of the 13th ACIS International Conference on Software Engineering, Artificial Intelligence, Networking and Parallel & Distributed Computing (SNPD '12), Kyoto, IEEE, pp. 236–241 (2012)
7. Lin, M., Lee, P., Hsueh, S.: Apriori-based frequent itemset mining algorithms on MapReduce. In: Proceedings of the 16th International Conference on Ubiquitous Information Management and Communication (ICUIMC '12), Article No. 76. ACM, New York, NY, USA (2012)
8. Klosgen, W., Zytkow, J.M. (eds.): Handbook of data mining and knowledge discovery. OUP, Oxford (2002)
9. Garg, K., Nayar, S.K.: Detection and removal of rain from videos. Proc. CVPR **1**, 528–535 (2004)
10. Provost, F., Fawcett, T.: Robust classification for imprecise environments. Mach. Learn. **42**(3), 203–231 (2001)
11. Kim, J.-H., Sim, J.-Y., Kim, C.-S.: Video deraining and desnowing using temporal correlation and low-rank matrix completion. TIP **24**(9), 2658–2670 (2015)
12. Scheffer, T.: Finding association rules that trade support optimally against Confidence. Unpublished manuscript
13. Larose, D.T.: Discovering knowledge in data: an introduction to data mining. Wiley, New York (2005)
14. Zhang, H., Sindagi, V.A., Patel, V.: Image de-raining using a conditional generative adversarial network. IEEE Trans. Circ. Syst. Video Technol. (2020)
15. Kantardzic, M.: Data mining: concepts, models, methods, and algorithms. Wiley, New Jersey (2003)
16. Aher, S.B., Lobo, L.M.R.J.: Combination of clustering, classification and association rule based approach for course recommender system in e-learning. Int. J. Comput. Appl. **39**(7) (2012). 0975-8887
17. Tunedit Machine Learning Repository. Available http://tunedit.org/repo/UCI/spect_test.arff
18. Shridhar, M., Parmar, M.: Survey paper on association rule mining and its approaches. Int. J. Comput. Sci. Eng. (IJCSE) **5**(3), 129–135 (2019). E-ISSN 2347-2693
19. Yadav, K., Parmar, M.: Review paper on data mining and its techniques and Mahatma Gandhi National Rural Employment Guarantee Act. Int. J. Comput. Sci. Eng. (IJCSE) **5**(4), 68–73. E-ISDN 2347-2693

Chapter 35
Review Based on Named Entity Recognition for Hindi Language Using Machine Learning Approach

Rita Shelke and Sandeep Vanjale

Abstract Named Entity Recognition (NER) is a critical job in machine learning that automatically recognizes and explains Named Things in writing, such as an Individual, a Position, or an Association. NER has played a critical role in numerous applications, including information removal and recovery, machine transformation, question answering (Q–A), and writing summarization. While much research has been conducted on NER in Hindi, no instrument with high accuracy has been created yet, as per the Literature Review. Developing a NER system for Hindi is very difficult due to the language's ambiguity, morphological richness, and resource scarcity. We provide a state-of-the-art review of different natural language processing methods (NER) for the primary language of Hindi in this article.

35.1 Introduction

The process of identifying Named Entities (NEs) from a textual document and classifying them into different conceptual categories (Name, Place, Party, Designation) is an important step in the task of Natural Language Processing (NLP). This process is called Named Entity Recognition (NER). In this age of World Wide Web, information is available in abundance but in Indian Languages, there is very less work done on NLP and Analytics per se. NER is especially a crucial proviso in applications of Information Retrieval, Machine Translation, Question Answering, Text Summarization, Efficient Search Engines, etc. [1–7]. NER development for Indian languages remains a challenge owning to syntactic and semantic complexities of Indian Languages and lack of processing tools. Since Hindi is the official language of India, it has been selected for this project to develop NER tool with Hybrid NLP Ontology and rule-based and Machine Learning-based approaches to gain better accuracy than available tools. Since Hindi is part of Indo-Aryan family, the current approach can be used

R. Shelke (✉) · S. Vanjale
Department of Computer Engineering, Bharati Vidyapeeth (Deemed to be University) College of Engineering, Pune, India
e-mail: rita.shelke@cumminscollege.in

S. Yadav et al. (eds.), *Proceedings of Second International Conference in Mechanical and Energy Technology*, Smart Innovation, Systems and Technologies 290,
https://doi.org/10.1007/978-981-19-0108-9_35

for other Indo-Aryan languages such as Gujarati, Bengali, Marathi, etc. as optimal NER tools are not present in the aforesaid languages as well. There are three main ways to perform NER. Using Linguistic Rule-sets along with Ontology, Machine Learning, or a Hybrid approach comprising both. Machine Learning has been the most successful in predicting unknown entities and Rule-based systems give highest accuracy [8–15].

Thus, Hybrid NER systems are most efficient for Indian languages. This system was developed to improve upon some of the issues in the current systems. Some of the issues realized are as follows:

- The accuracy of NER systems varies as per texts.
- Rule-based systems have higher accuracy but are not customizable
- Limited tag set is considered in existing systems.
- Context of word is not taken into account especially when it comes to ambiguously named entities. (Location and Person Name being same)
- Proper nouns are used as common nouns in certain cases.

In the current system, these problems are addressed by combining with indigenously built NLP algorithm using Linguistic rules and ontology depicted through RDF in tandem with the Machine Learning (ML) model of Conditional Random Fields (CRF) [16–24]. RDF has been used for creating a list of standard named entities compiled from various sources as cue to NLP algorithm and ML. 10 tags for NER have been considered whereas most of the existing systems only consider the three main tags- person, location, organization. The named entities are also phrase marked to better elicit information from the text. Bangare et al. presented cloud computing research [25–30]. Deep learning approaches were described by LMI Leo Joseph et al. [26], Athawale et al. [27] given security concerns.

35.2 Challenges

The following are the difficulties and problems that must be resolved while applying the NER tool in Hindi.

35.2.1 No Capitalization

The capitalization characteristic is critical for recognizing NEs in the English language and other foreign languages. For instance, in the phrase "Rajesh is traveling to Agra," the English NER tool would classify "Rajesh" and "Agra" as proper nouns owing to the capitalization feature, but the Hindi language lacks this notion.

35.2.2 Morphologically Rich

The morphological analysis aims to separate a word's stems and affixes. Due to the complex morphology of Indian languages, root identification is very difficult.

35.2.3 Ambiguity

An appropriate equivocal noun is a given appellation that is also a legitimate vocabulary term that takes on its meaning when employed in the text.

35.2.4 Lack of Standardization and Spell Variations

One of the primary difficulties with Indian dialects is that various individuals in India spell similar things differently.

35.3 Previous Work and Gap Analysis

Machine Learning (ML) techniques and hybrids of Rule-Based and ML algorithms are the most often used approaches to NER. The Stanford NER tagger is an open-source NER tool that is already accessible. Stanford's classification algorithm is based on the CRF framework.

5400 Hindi words were used in the tool's analysis and testing. Accuracy and memory of 0.45 and 0.5, correspondingly, were achieved. An F-score of 0.47 was obtained. In contrast, English NER has an accuracy of 0.9223.

Chopra et al. [7] implemented NER in Hindi using the Hidden Markov Model (HMM). As previously stated, HMM aids in the development of language-independent NER systems. Scaling and analysis of these systems are equally straightforward. On 2343 training tokens and 105 testing tokens, an F-Measure of 97.14% was achieved. HMM, on the other hand, suffers from label bias. Sinha [1] makes extensive use of a small training and testing corpus.

Sinha [3] examined a total of 29 characteristics, including context words, word prefixes, word suffixes, point-of-sale information, and gazetteer lists—lists of people, places, and organizations. They get an overall accuracy of 72.78%, recall of 65.82%, and F-score of 70.45%. Sinha et al. [3] focus on applying CRF to de-ambiguate Ambiguous Proper Names (APN) in Hindi. A common noun is a name that is also an APN.

Sinha [3] also discuss the need of deriving a relevant corpus in their work. Following that, a relevant corpus was generated and classified into three categories:

names (NEP), dictionary terms not used as designations (NNE), and other confrontations (OTH). When CRF and Rule-Based outputs are merged, the resulting yield is OR-ed, a combined F-Score of 71.16% is obtained. While conducting a survey for NER in Marathi, discovered that.

Patil et al. [4] solve the issue of correctly classifying each term using a probabilistic Hidden Markov Model (HMM) trained on a manually annotated language corpus. When no pre-processing is used, the planned structure in [4] information an overall F1-Score of 62.70%, while when pre-processing is used, the proposed system reports an overall F1-Score of 77.79%. The method described in [4] correctly identified individuals, places, numbers, and units of measurement, but not other named things. The reference [4] improves efficiency via pre-processing methods such as lemmatization, which may be very expensive. The purpose of suggestion rule removal is to discover guidelines based on a collection of contacts. It assists in establishing relationships among variables in a big database.

Patawar and Potey [9] implemented NER on Marathi linguistic tweets using a hybrid method. As a widely utilized social media network, Twitter is naturally available in a variety of languages. Patawar et al. used CRF and K-Nearest Neighbor to create their method. First, we give a confidence value "cf" to the normalized tweets using a K—the value of 4. Following that, a label is assigned using the CRF labeler. The probability is calculated using CRF; however, the base label is assigned directly if "cf" is greater than. The given token is stored in the clusters and is used for further training by the system. If not, it is given a CRF label. They identify just Locations and Names, with an accuracy of 39.80 and a recall of 85.11 for location tags and 59.72 and a memory of 25.28 for designation tags. CRF simplifies the process of adding prefixes and suffixes required for NER in Indian idioms. SP has created a deep neural network-based term article recognition and classification system for the English philological. IoT and user interface work was presented by Kamal Gulati et al. [16, 25]. Bangare et al. [20] highlighted scaling challenges in machine learning.

35.4 Hindi NER Systems

The following classifications apply to the NER systems created for Hindi in the literature.

35.4.1 NER Rule-Based System

The rule-based method necessitates the creation of a set of rules by language experts. For each named entity class, a massive gazetteer list is generated using this technique. Rule-based NER arrangements are exact since they require extensive knowledge of a given linguistic and domain in order to construct syntactic-lexical pattern-based directions. Combining a rule-based and a list-based strategy, a NER arrangement for the Hindi dialectal was produced [4]. Their algorithm discovered three new things: the money worth, the course value, and the animal/bird entity. Their organization improved previous rules by implementing a new rule called the "no-name entity rule." The list lookup method included creating several tables in the record and extracting named things from these tables. Their method has a 95.77% accuracy rating.

35.4.2 ML Based Supervised Hindi NER System

The following are the Hindi NER systems that are supervised learning-based.

35.4.2.1 Hidden Markov Model (HMM)

The authors of [5] used HMM to carry out NER in Hindi, Bengali, and Telugu. The writers used two corpora for the Hindi text: (1) At Banasthali Vidyapith, a tourist domain was established, and (2) Indian corpora of the NLTK. This method is composed of three distinct stages. The first step is annotation, in which text data is labeled. The second step is to train the HMM, which is responsible for computing the HMM's three most critical parameters: start possibility, emission possibility, and transition possibility. The third step was the HMM test stage, during which specific test phrases provided by the user were tested. Using HMM parameters, the Viterbi method calculated the optimum state sequence for the given test phrase. In [6], the authors implemented the NER system for seven distinct Indian languages using a statistical HMM-based model: Bengali, English, Hindi, Marathi, Punjabi, Tamil, and Telugu. For the Hindi language, this method achieved F-values of 0.7520. NER in Hindi was also modeled using HMM [7]. On 2343 and 105 tokens, respectively, the system was trained and tested. Their method obtained 97.14% F-measures.

35.4.2.2 Maximum Entropy Model (MaxEnt)

A NER model based on MaxEnt was created for the Hindi dialectal [8], in which the writer used a variety of characteristics, including orthographic geographies (decimal, numerals), affixes, left and right background arguments, and part-of-speech features. The authors utilized eight gazetteer lists for performance purposes, including month and weekday names, society end term lists, individual preface confrontations lists, place names lists, first names lists, middle names lists, and surnames lists. The system's presentation was compared in contradiction of a blind test set comprised of four modules: person, society, place, and date. This scheme obtained an F-Measure of 81.52%.

35.4.2.3 Support Vector Machine (SVM)

The support vector machine built a NER structure for Bengali and Hindi for two Indian dialects [9]. The system utilized many language-independent characteristics to identify named things, including background words, word affixes, expression frequency, and part-of-speech information. For Hindi, tenfold validation tests revealed recall of 80.48%, an accuracy of 74.54%, and an F-Score of 77.39%. The use of vocabulary background patterns resulted in a substantial increase of 5.13% in the F-Score. The authors of [10] created a NER structure for Hindi and the biomedical area by implementing a new kernel purpose for support vector machines. The kernel computed the weighted detachment between features based on strings.

35.4.3 ML Based Unsupervised Hindi NER System

A NER system was created to compare the co-occurrence of terms in English and Indian [18]. The authors used Wikipedia's intrinsic structure, with unstructured sheets, infoboxes, summaries, and titles, to recognize named things in Indian languages. The writers used English Wikipedia information to bootstrap the documentation of named things in Hindi and Marathi, resulting in a list of NEs. Future, this NE list was utilized to enhance the populating of multilingual entities. The NER job was completed in four steps: comparable grouping information, documentation of NEs from the infobox, documentation of NEs from the intellectual, and documentation of NEs from the text's captions. Their method was tested on a dataset of 3853 English and 2935 Hindi Wikipedia apprenticeships.

35.4.4 Deep Learning-Based Hindi NER Systems

The authors' model was created in two phases. In order to learn word embeddings using the skip-gram and glove techniques, the authors used an unlabeled dataset in the first stage. In the second stage, they employed bidirectional LSTMs. The system was qualified end-to-end utilizing labeled information after initializing the system's embedding layers with learned word vectors for each word. Their program identified Hindi properly 77.48% of the time.

35.5 Conclusion

Recognizing named entities is significant since it aids in the pre-processing stage of many natural language processing requests. This comprehensive review of Hindi NER is an excellent place for novice scholars to begin. The research demonstrates that when a language expert is available, rule-based methods work optimally. In contrast, methods based on machine learning are more robust and straightforward to build for a language.

References

1. Mittal, N., Agarwal, B., Chouhan, G., Pareek, P., Bania, N.: Discourse based sentiment analysis for Hindi reviews. In: PReMI'13, pp. 720–725 (2013)
2. Greenwood, M., Gaizauskas, R.: Using a named entity tagger to generalise surface matching text patterns for question answering. In: Workshop on Natural Language Processing for Question Answering (2007)
3. Sinha, R.M.K.: Learning recognition of ambiguous proper names in Hindi. In: 2011 10th international conference on machine learning and applications and workshops, Honolulu, HI, pp. 178–182 (2011)
4. Patil, A.S., Patil, N.V., Pawar, B.V.: HMM based named entity recognition for inflectional language. In: Comptelix, 2017 (International Conference on Computer, Communications and Electronics 2017), pp. 565–572
5. Morwal, S., Jahan, N., Chopra, D.: Named entity recognition using Hidden Markov Model (HMM). Int. J. Nat. Lang. Comput. **1**(4), 15–23 (2013)
6. Gayen, V., Sarkar, K.: A HMM based named entity recognition system for indian Languages: The JU System at ICON 2013. In arXiv:1405.7397 (2014)
7. Chopra, D., Joshi, N., Mathur, I.: Named entity recognition in Hindi using Hidden Markov Model. In: 2nd International Conference on Computational Intelligence Communication Technology, pp. 581–586 (2016)
8. Saha, S., Sarkar, S., Mitra, P.: A hybrid feature set based maximum entropy Hindi named entity recognition. IJCNLP, 343–349 (2008)
9. Patawar, M.L., Potey, M.A.: Extending hybrid conditional random fields approachof named entity recognition for Marathi tweets. In: 2016 ICCUBEA
10. Saha, S., Narayan, S., Sarkar, S., Mitra, P.: A composite kernel for named entity recognition. Pattern Recogn. Lett. **31**(12), 1591–1597 (2010)
11. Dandapat, S., Way, A.: Improved named entity recognition using machine translation-based cross-lingual information. Int. J. Comput. Sci. Appl. **20**(3) (2016)
12. Abhinaya, N., Neethu, J., Barathi, B., Kumar, A. Soman, K.P.: AMRITA_CEN@FIRE-2014: named entity recognition for Indian languages using rich features. In: The forum for the Information Retrieval Evaluation, pp. 103–111 (2014)
13. Saharia, N.: Phone-based identification of language in code-mixed social network data. J. Stat. Manag. Syst. **20**(4), 565–574 (2017)
14. Devi, G.R., Veena, P.V., Anand Kumar, M., Soman, K.P.: Entity Extraction of Hindi-English and Tamil-English code-mixed social media text. In: Forum for Information Retrieval Evaluation. Text Processing, pp. 206–218 (2018)
15. Ekbal, A., Bandyopadhyay, S.: A conditional random field approach for named entity recognition in Bengali and Hindi. Linguistic Issues in Language Technology-LiLT **2**(1), 1–44 (2009)
16. Gulati, K., Boddu, R.S.K., Kapila, D., et al.: A review paper on wireless sensor network techniques in Internet of Things (IoT). Mater. Today Proc0 https://doi.org/10.1016/j.matpr.2021.05.067
17. Bangare, S.L., Patil, S.T., et al.: Implementing tumor detection and area calculation in MRI image of human brain using image processing techniques. Int. J. Eng. Res. Appl. **5**(4), (Part-6), 60–65 (2015). ISSN 2248-9622
18. Bangare, S.L., Patil, S.T., et al.: Reviewing Otsu's method for image thresholding. Int. J. Appl. Eng. Res. **10**(9), 21777–21783 (2015)

19. Bangare, S.L., et al.: Regenerative pixel mode and tumor locus algorithm development for brain tumor analysis: a new computational technique for precise medical imaging. Int. J. Biomed. Eng. Technol. **27**(1/2) (2018)
20. Bangare, S.L., et al.: Neuroendoscopy adapter module development for better brain tumor image visualization. Int. J. Electr. Comput. Eng. (IJECE) **7**(6), 3643–3654 (2017)
21. Kadam, R.R., Bangare, M.L.: A survey on security issues and solutions in live virtual machine migration. Int. J. Adv. Found. Res. Comput. (IJAFRC) (2012). ISSN 2348-4853
22. Chavan, S.K., Bangare, M.L.: Secure CRM cloud service using RC5 algorithm. Int. J. Comput. Trends Technol. **4**(3), 325–330 ()
23. Bangare, M.L., Joshi, S.A.: Kernel interpolation-based technique for privacy protection of pluggable data in cloud computing. Int. J. Cloud Comput. **9**(2–3), 355–374 (2020)
24. Bangare, S.L., Gupta, S., Dalal, M., Inamdar, A.: Using node. Js to build high speed and scalable backend database server. In Proceedings of NCPCI Conference, vol. 2016, p. 19 (2016)
25. Gulati, K., Sriram, V.P., Sharma, M., Parul, Eliyas, S., Bangare, S.L.: Use for graphical user tools in data analytics and machine learning application. Turk. J. Physiotherapy Rehab. **32**(3). ISSN 2651-4451, e-ISSN 2651-446X
26. Leo Joseph, L.M.I., Bangare, S.L., et al.: Methods to identify facial detection in deep learning through the use of real-time training datasets management. EFFLATOUNIA—Multidiscip. J. **5**(2), 1298–1311. ISSN 1110-8703
27. Athawale, S., Giri, V., Bangare, S.L.: Collateral extension in provocation of security in IoT. Int. J. Future Gener. Commun. Netw. **14**(1), 3703–3716. ISSN 2233-7857, Web of Science
28. Bangare, S.L.: Brain tumor detection using machine learning approach. Design Engineering (7), pp. 7557–7566. ISSN 0011-9342, Scopus Index- Q4
29. Bangare, M.L., Bangare, P.M., Apare, R.S., Bangare, S.L.: fog computing based security of IoT application. Design Engineering (7), pp. 7542–7549. ISSN 0011-9342, Scopus Index-Q4
30. Bangare, S.L., Bangare, P.S., Patil, K.P.: Internet of things with green computing. Turk. J. Physiotherapy Rehab. **32**(3), 12494–12497. ISSN 2651-4451, e-ISSN 2651-446X

Chapter 36
A Critical Analysis of Machine Learning's Function in Changing the Social and Business Ecosystem

V. P. Sriram, A. V. L. N. Sujith, Anupama Bharti, Sanjeeb Kumar Jena, Dilip Kumar Sharma, and Mohd Naved

Abstract Machine learning is an automization based technique that learns automatically about something without specific programming of the task. It is used in a variety of fields. The capabilities of Data-driven modeling (DDM) have recently been expanded by advances in machine learning, allowing artificial intelligence to infer system behavior by correlating computing and exploiting between variables that were observed within them. The use of auto-generated high volume business data can be enabled by machine learning algorithms and aided by applying models of ecosystem services across scales, allowing the flow of these services to be analyzed and predicted to disaggregated beneficiaries. Machine learning is a very advanced field with numerous applications in a wide range of business environments. Currently, in the field of information science, data processing techniques such as machine learning have been developed and applied in a variety of areas for practical applications.

V. P. Sriram (✉)
Department of Business Administration, Acharya Bangalore B School (ABBS), Bengaluru, Karnataka, India
e-mail: dr.vpsriram@gmail.com

A. V. L. N. Sujith
Department of Computer Science Engineering, Gates Institute of Technology, Gooty, Anantapur, Andhra Pradesh, India

A. Bharti
Department of Sociology and Social Work, Himachal Pradesh University, Shimla, India

S. K. Jena
Department of Commerce, Rajiv Gandhi University, Doimukh, Arunachal Pradesh, India

D. K. Sharma
Department of Mathematics, Jaypee University of Engineering and Technology, Guna, Madhya Pradesh, India

M. Naved
Department of Business Analytics, Jagannath University, Delhi-NCR, India

S. Yadav et al. (eds.), *Proceedings of Second International Conference in Mechanical and Energy Technology*, Smart Innovation, Systems and Technologies 290,
https://doi.org/10.1007/978-981-19-0108-9_36

36.1 Introduction

No one can deny that ML is the most influential and powerful technology in today's world. It is critical that we recognize and comprehend the potential of machine learning. Without a doubt, machine learning will continue to make headlines in the coming months and years. The information is converted into knowledge using a tool known as machine learning. Data has exploded in the last fifty years. If these data are not analyzed and we are unable to discover the hidden patterns in the data, this mass of data will become completely useless. Machine learning techniques were used to automatically detect the valuable and hidden patterns that exist within the complex data. Without the use of machine learning, it would have been extremely difficult to discover this data. The future events can be predicted by using hidden patterns, and the decision-making process can be facilitated by having knowledge of the problem.

Machine learning has become a mainstay of the IT sector over the last two decades, and it has also become a central and hidden part of our lives. Because the amount of data available to every sector is increasing by the day, it is becoming a very good reason to say and believe that smart data analysis will become more prevalent. Because of the abundance of data available, it is becoming an essential component for the advancement of technologies to intelligently analyse it. The tasks that are performed because of system changes and are associated with "artificial intelligence" (AI) are typically referred to as machine learning. The tasks that are referred to as machine learning involve recognition, diagnosis, planning anything, controlling the robot, and predicting anything. Changes in the system because of machine learning can be improvements to the performance of existing systems or the synthesis of new systems. It can be more precisely explained as displaying the architecture of a traditional "artificial intelligence agent." The agent perceives and models the environment, and the actions are appropriately computed; perhaps the effects are also anticipated by the agent. Changes made to any component of the system can be categorized as machine learning. Different learning mechanisms can be applied to the system, depending on the subsystem in which the changes have been made. Work in machine learning is converging from a wide range of sources. Different traditions bring different methods and vocabulary, which are then assimilated in a much more disciplined discipline. Figure 36.1 shows the process and mechanism of machine learning.

Machine learning is classified into four types: "supervised learning", "unsupervised learning", "semi-supervised learning", and "reinforcement learning". Dey [1] discovered that, whether we realize it or not, we are all in the habit of using machine learning today, from getting a recommendation for a specific product while shopping for it online to updating our status on social networking sites. There are different approaches for each type of machine learning, all based on the same underlying process and theory. Supervised learning: Supervised machine learning algorithms always require external assistance. The input data set is divided into train and test data sets. The variable output of the train set of data must be predicted and classified. The algorithms learn all kinds of patterns from the training set of data and then apply

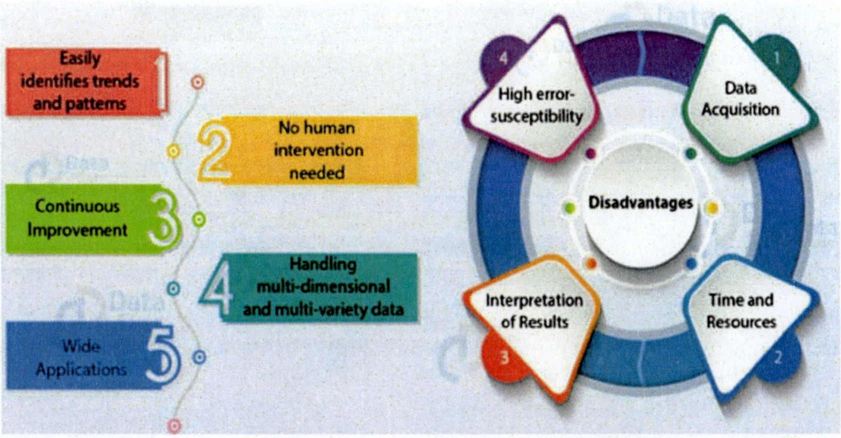

Fig. 36.1 Mechanism of machine learning

them to the test dataset for prediction and classification. Unsupervised Learning: In unsupervised learning, the algorithm learns only a few features from the data, and when new data is introduced, it uses the previously learned feature to recognize the new class of data. The main application of unsupervised learning is feature clustering and reduction.

Semi-Supervised Learning: The semi-supervised learning algorithm combines the powers of supervised and unsupervised learning. There are some areas of machine learning and data mining where there is already a large amount of unlabeled data and the process of obtaining new labeled data is time-consuming. In these cases, semi-supervised learning can be very useful. Figure 36.2 shows how data is managed and processed through machine learning.

Reinforcement Learning: In this, actions are taken based on decisions made to achieve a more positive outcome. Until and unless a learner is presented with a situation, he lacks the knowledge to act. The learner's actions, as well as their future actions, can have an impact on the situation. There are only two criteria on which reinforcement learning is solely dependent; these are trial and error and delayed outcome. Multitask Learning: The simple goal of multitask learning is to assist other learners in performing better. When the multitask learning algorithm is applied to a task, it remembers how it solved this particular problem previously and reached the conclusion. Ensemble Learning: Ensemble learning is the combination of various types of individual learners to form a single learner. In supervised learning, there is a target or an outcome variable or a dependent variable for which prediction must be made using the predictors of a given set or independent variable. These variables are used to create a function that maps the inputs to the desired output. Unsupervised learning is used in the clustering of different groups in order to segment customers into different groups for intervention in a specific manner, e.g., Apriority algorithm, K-means. The reinforcement learning algorithm is used to train the machine to make the

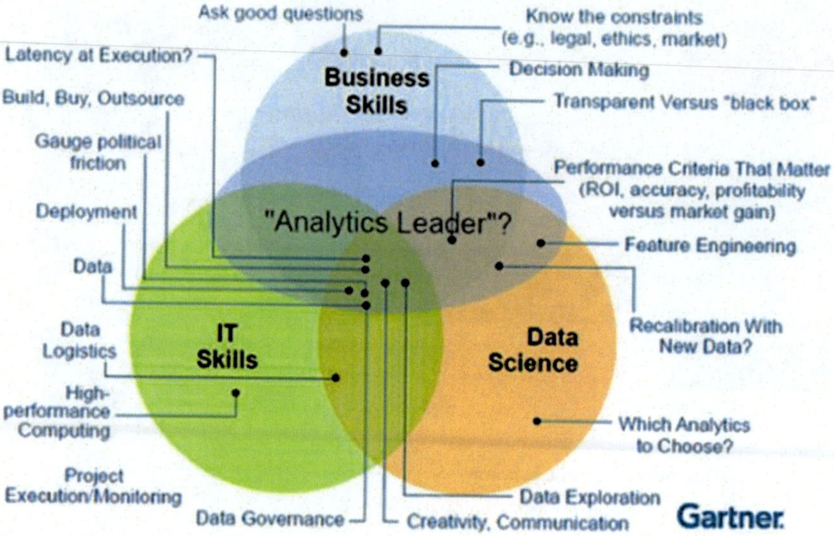

Fig. 36.2 Skills and responsibilities of data science leader. *Source* Data Science Learning

appropriate decisions. When exposed to any environment, the machine continuously trains itself using trial and error in the process of reinforcement machine learning. This type of machine has always attempted to capture the knowledge that is best for it in order to make accurate business decisions by using its previous or previous experiences, for example, the Markov Decision Process.

36.2 Applications of Machine Learning

Traditionally, weather forecasting was done using physical models of the atmosphere; however, these physical models were very unstable to perturbations, so forecasting was inaccurate for a long time. For a long time, weather forecasting has been done potentially by machine learning techniques for more accurate forecasts because machine learning is more robust to perturbations. The physical model is governed by a system of ordinary differential equations, which is unstable under perturbations, and the initial measurements of atmospheric conditions had uncertainties, and the accuracy of weather forecasting is inaccurate and limited to a period of ten days due to incomplete understandings of the complex atmospheric system. Machine learning, on the other hand, is relatively resistant to perturbations, and there is no need to fully comprehend the physical processes that govern the atmosphere. As a result, machine learning represents a viable alternative to these physical models in weather forecasting. Jakaria et al. [2] discovered that machine learning models accurately predict weather features and that they are accurate enough to compete with existing traditional models. The weather in any given area can be easily predicted by using historical data from the surrounding areas. This method of forecasting is far more effective than focusing solely on the area for which forecasting is required.

Many scientific disciplines are taking integrative approaches to the problems of the planets, and this number is growing by the day. The problems of the planets include global climate change, food security, and human migration. There are numerous methods for miming data and analyzing it, and many of these methods make use of machine learning algorithms. The process of fitting a model to a set of data using learning and training is known as machine learning algorithm. Willcock et al. [3] concluded that data-driven modeling (DDM) plays a clear role when the services of modeling ecosystem are helping to produce interdisciplinary models and providing holistic solutions to the complex issues of socioecology. Recent advances in machine learning have expanded the capabilities of data-driven modeling (DDM). Recknagel [4] discovered that machine learning techniques are inductive and deal with data-driven modeling approaches. Machine learning techniques, which can be represented by machine learning techniques for ecological issues, require a reasonable complex database. Artificial Neural Network (ANN) techniques have been shown to be very effective for nonlinear ordination and visualization, as well as for multiple regressions and time series models. It can also be seen that ANN is effective for image recognition and classification. The trained ANN model, in conjunction with sensitive analysis, reveals the nature of the relationship between the driving and output variables in an ecosystem. GA techniques can be used to hybridize deductive ecological models. This GA technique allows the causal rules to evolve and process the equations, as well as optimize the process parameters. The adaptive agents provide a novel framework for discovering and forecasting the emergent structures of ecosystems and their behaviors in response to environmental changes. Rana and Miller [5] discovered that no policy has an average impact on local contextual conditions. There is significant heterogeneity, which has a conditional impact on the local context. The region under study is influenced by a variety of climatic and biophysical factors, which are shaped by different policies for joint forest management. Cooperative forest management, on the other hand, performs better in areas where existing grazing-based livelihoods are preserved. Instead of having so much potential, the approaches have so many restrictions and limitations, such as no estimation of valid precision for heterogeneity estimates and issues with stability estimates.

Traditional and common methods for protecting against fraud were used in the banking industry; these methods were used where the rules were defined by humans. It is a fact that nearly 90% of the banking industry and financial institutions rely on and follow these riles and methods. Because traditional methods may not be scalable and sustainable in the future, and the number of frauds in the banking and financial industry is increasing daily, there are many people and banks adopting new technologies to protect themselves from fraud. Every day, customers of the banking sector file a large number of complaints about false-positive fraud. False-positive frauds are non-fraudulent transactions that appear to be fraudulent. As a result, there is a loss of millions of dollars in transactions, and the rule-based method is a major contributor to this type of problem. Furthermore, the pattern of the fraud is not constant, and it changes its behavior over time, making a rule-based system cumbersome and quickly obsolete. Shirgave et al. [6] reviewed a number of machine learning algorithms that can detect fraud in credit card transactions. All of these techniques' performance is analyzed and tested using accuracy, precision, and specificity metrics. Random forest supervised learning techniques are used to classify alerts as either fraudulent or authorized. These classifiers will be trained using feedback and delayed supervised samples. In order to detect the alerts, each probability will be aggregated. It is proposed that the approach be ranked by learning and that the priority be used to rank the alerts. This proposed method will resolve the class imbalance as well as the concept drift problem. Semi-supervised learning methods will be used in the future to classify alerts in FDS. Lima [7] discovered that the implementation of rule-based or supervised learning-based algorithms is impractical because the number of scenes with the possibility of fraud is growing by the day. The modern fraud detection system will be able to react in real-time, and this modern fraud detection system will provide more accurate results for new types of fraud scenarios. The unsupervised algorithm, like deep learning algorithms, can detect this new type of fraud because it does not require any human classification and can ingest a large volume of data. This study focuses on DL algorithms such as autoencoders, which are the most promising, and restricted Boltzmann machines, which focus on anomaly detection rather than misuse detection. Saragih discovered that credit and debit card fraud is increasing every day over the last few years. It has been observed that there are numerous incidents of fraud occurring as a result of a lack of knowledge about credit cards and their use. Credit card users frequently share their personal information, credit card information, and even their one-time password with bogus and unknown calls. The money from the appalling loss of the fake trades is used by the individual in a blackmail activity, whether honestly or indirectly, and the owner even considers this to be a trade. These types of frauds and cheats are committed on a regular basis using credit or debit cards. Because the security structure is very weak, these types of frauds are taking place.

Machine learning is used to discover patterns and data on its own and to derive conclusions from them. Machine learning has the potential to enable personalized care, also known as precision medicine. The field of healthcare is advancing daily thanks to the use of machine learning methods. Machine learning is expected to bring about significant change in the healthcare sector within the next few years. Machine learning and Artificial Intelligence have the potential to transform the entire healthcare system; however, the quality of machine learning and the decision support system of Artificial Intelligence must address the issues that patients and health experts face in the effective diagnosis procedure. According to Jabbar et al. [8], it is not possible to have a spontaneous increase in efficient healthcare providers.

With the use of machine learning and artificial intelligence technologies, healthcare costs can be reduced, more patients can be served in less time, and healthcare outcomes can be improved. According to Bhardwaj et al. [9], the healthcare sector is the fastest growing in today's economy because more and more people require healthcare, which is becoming increasingly expensive. Now, the government is spending a lot of money on the healthcare sector, and there is a need to connect patients and

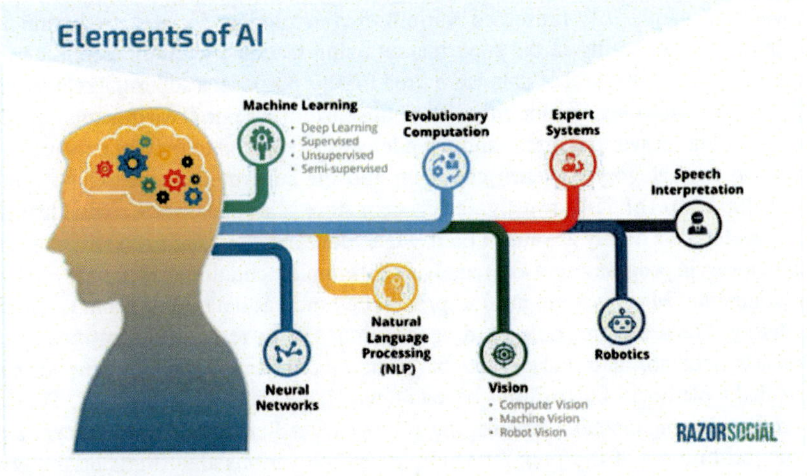

Fig. 36.3 Machine learning as an elements of AI. *Source* RazorSocial

doctors. Big data and machine learning technologies had the potential to provide all the assistance required for patients and healthcare providers to provide and receive better care at lower costs. There are numerous companies and organizations that have entered this industry and begun assisting patients by assisting them with transitions and demonstrating patient-centered care to them. There is so much data already available there that one must figure it out and interpret it in a specific channel. Companies are stepping forward to take that single step to gain a better understanding of it (Fig. 36.3).

Tziridis et al. [10] discovered that by analyzing historical fare data, it is possible to predict flight fares. The results of the experiments show that machine learning models are a good tool for predicting airline prices. It is critical to collect data and select features for airfare prediction to draw some conclusions that will be useful for this process. Wang et al. [11] discovered the effectiveness of machine learning algorithms and techniques; they also compared the performances of different machine learning classifiers and discovered the best machine learning algorithm for the prediction task of airfare prices by leveraging information from the DBIB and T-100 data sets. There are numerous factors that influence the price of airline tickets, such as the distance of the flight, the time of purchase, the price of fuel, and so on. Each carrier has its own set of proprietary rules and algorithms in place to determine its airfare. Because of advancements in machine learning and artificial intelligence, it is now possible to make price variations by interfering with their rules and model. According to Lu [12], to solve some specific problems, Ada Boost Decision Tree Classification is said to be the perfect model, and it also performs well. Ada Boost Decision Tree Classification outperforms others in the random purchase strategy by nearly 61.35 percent, and it has a small performance variance relative to the other eight routes. Companies also prefer the Uniform Blending Classification and Q-learning methods due to their high performance.

Kanavos et al. [13] presented various methodologies for the prediction and modeling of purchasing in the supermarket using machine learning techniques. To be more specific, two sets of data were used for this purpose: a supermarket database and an Amazon database containing all of the customer's purchase information. The Amazon data set was analyzed, and a model was developed to predict new products based on the category for each customer and the supermarket customer's preferences. Ramesh et al. [14] investigated grocery store sales patterns. Machine learning is thought to be a highly advanced field. Machine learning is said to be able to enable applications in programming after analyzing its implementations in so many business environments. Machines are used in procedures such as data mining and predictive modeling. The use of machine learning techniques in the real world identifies trends, forecasts behavior, and makes facts based on recommendations. Machine learning provided a plethora of algorithms based on which data sets were analyzed. The data is predicted using machine learning algorithms to predict the sales of a grocery store in the third month of the year. After implementing various algorithms on the given sets of data, it is discovered that linear regression produces the best results, indicating that data from the previous month can be used to predict grocery store sales for the coming month.

According to Thessen [15], machine learning methods provide a wide range of techniques that can be accessed by researchers individually. These arrays of techniques are ideal for the complex sets of data generated by earth and ecology science. More accurate models are provided, and scientific progress is accelerated because these machine learning models have the potential to improve the quality of research in a scientific manner. These machine learning methods widen the necks of the bottles, fill gaps in the data, and improve understanding of the system's operation. If society wants to reap the benefits, the applications of these methods in ecology and earth science must be expanded. Encourage the use and adoption of these methods, communication should be improved, interdisciplinary collaboration should be encouraged, formal and informal education should be expanded, and machine learning researchers should be supported. The best source of knowledge transfer is collaboration between companies and organizations that demonstrate an interest in environmental issues. Random forest is an excellent and introductory machine learning method. The random forest is simple to implement and produces excellent results. Willcock [3] concluded that data-driven modeling can be used to scale up ES models so that they can provide more policies and make more relevant decisions. Data-driven modeling allows big data to interact with one another, as well as the creation of interdisciplinary models and the provision of holistic solutions to complex socio-ecological issues.

Machine learning is a very advanced field with numerous applications in a wide range of business environments. At the moment, we all use machine learning in our daily lives, whether knowingly or unknowingly, whether while shopping online or surfing social networking sites. Today, machine learning has applications in almost every field. It is significantly reducing the burden of workload in the processing of bid data. Currently, in the field of information science, data processing techniques such as machine learning have been developed and applied in a variety of areas for practical applications.

Machine learning is useful in the healthcare sector for improving healthcare and lowering costs, in the prediction of sales in supermarkets and grocery stores, in protecting the banking sector from financial frauds, and weather forecasting is done potentially by machine learning techniques for more accurate forecasts because machine learning is more robust t The capabilities of Data-driven modeling (DDM) have recently been expanded by developments in machine learning; machine learning is said to be a satisfactory tool for predicting Airfare prices. Machine learning is producing fruitful and satisfactory results in all sectors.

36.3 Conclusion

Machine learning is a very advanced field and it has many implementations with so many aspects of business environments. At present each and everyone is using the machine learning in our day to day life knowing or unknowingly, whether while shopping online or during our surfing on the social networking sites. The machine learning has its applications in each and every field today. It is playing a great role in lowering the burden of workload in the processing of bid data. At present, in the field of information science, the techniques of data processing such as machine learning were developed and applied in different areas for practical uses.

The machine learning is useful in the healthcare sector for improving the healthcare and lowering down the expenses, in the prediction of sales in the supermarkets and grocery stores, in protecting the banking sector from the frauds of financial issues, weather forecasting is done potentially by the techniques of machine learning for more accurate forecasts since machine learning is more robust to the perturbations. The capabilities of Data-driven modeling (DDM) were expanded by the developments in machine learning in recent days, machine learning is said to be a satisfactory tool for the prediction of Airfare prices. The machine learning is giving the fruit-full and satisfactory results in all the sectors.

References

1. Dey, A.: Machine learning algorithms: a review. Int. J. Comput. Sci. Inf. Technol. **7**(3), 1174–1179 (2016)
2. Jakaria, A.H.M., Hossain, M., Rahman, A.: Smart Weather Forecasting Using Machine Learning: A Case Study in Tennessee Conference Paper. Tennessee Tech University Cookeville, Tennessee (2019)
3. Willcock, S., Martínez-López, J., Hooftman, D.A.P., Bagstad, K.J., Balbi, S., Marzo, A., Athanasiadis, I.N.: Machine learning for ecosystem services. Ecosyst. Serv. (2018)
4. Recknagel, F.: Applications of machine learning to ecological modelling. Ecol. Model. **146**, 303–310 (2001)

5. Rana, P., Miller, D.C.: Machine learning to analyze the social-ecological impacts of natural resource policy: insights from community forest management in the Indian Himalaya. Environ. Res. Lett. **14**(2), 024008 (2019)
6. Shirgave, S.K., Awati, C.J., More, R., Patil, S.S.: A review on credit card fraud detection using machine learning. Int. Conf. Sci. Technol. Res. **8**(10), 1217–1220 (2019)
7. Lima, S: Deep Learning for Fraud Detection in the Banking Industry. Human IST Institute, University of Fribourg, Switzerland (2018)
8. Jabbar, M.A., Samreen, S., Aluvalu, R.: The Future of Healthcare: Machine Learning. Int. J. Eng. Technol. 7 (4.6), 23–25 (2018)
9. Bhardwaj, R., Nambiar, A.R., Dutta, D.: A study of machine learning in healthcare. In: 2017 IEEE 41st Annual Computer Software and Applications Conference (COMPSAC) (2017)
10. Tziridis, K., Kalampokas, Th., Papakostas, G.A., Diamantaras, K.I.: Airfare prices prediction using machine learning techniques. In: European Signal Processing Conference (2017)
11. Wang, T., Pouyanfar, S., Tian, H., Tao, Y.: A framework for airfare price prediction: a machine learning approach. In: International Conference on Reuse and Integration for Data Science, pp. 200–207 (2019)
12. Lu, J.: Machine learning modelling for time series problem: Predicting flight ticket prices. Computer Science, EPFL (2018)
13. Kanavos, A., Iakovou, S., Sioutas, S., Tampakas, V.: Large scale product recommendation of supermarket ware based on customer behaviour analysis. Big Data Cogn. Comput. **2**(2), 11 (2018)
14. Ramesh, K., Mathew, J., Hemalatha, N.: Machine learning approach for the predictive analysis of sales in grocery store. Int. J. Latest Trends Eng. Technol. 095–098 (2017)
15. Thessen, A.E.: Adoption of Machine Learning Techniques in Ecology and Earth Science One Ecosystem, pp. 2–38 (2016)

Chapter 37
Image Forgery Detection Techniques: A Brief Review

Pydipalli Sai Achyuth and Vella Satyanarayana

Abstract As the digital camera came into the picture, it is very common practice to capture the image and share it with the help of the internet. Even nowadays documents are also shared with the same methods, and there comes the need for security. As the research says that most people are not able to discriminate the real and fake images or edited documents, therefore the chances of forgery improved. In this digital era, the system will arrange tools or weapon in the form of software and make this forgery easy and cheap. However, there are also various techniques and software developed by researchers from all over the world to counter this forgery. In this present paper, the process of image forgery will be highlighted and their various types are discussed. Numerous detection techniques are also conferred in this presented paper.

37.1 Introduction

A well-known idiom i.e. "A photograph is really well worth one thousand words" [1]. Nowadays, all of us are aware of the essence of this phrase. But as there are numerous tools available for the manipulation of the picture, the image can be tempered very easily with the little knowledge of a computer. Therefore, it is very hard to predict the originality/authenticity of any image. This image manipulation is known as image forgery or it can be defined as "falsely and fraudulently altering a digital image". This image forgery is not a new issue, as the beginning of the photography technique or even before that in the era of painting, this tempering or making duplicacy exists. As the advancement comes, the technique of image forgery has also evolved.

The first case of image forgery, which attracts the attention of the world, is created by a French photographer Hippolyte Bayard named "Self Portrait as a Drowned Man", in the year 1840 (Fig. 37.1). In this image, he declared that he attempted suicide [2].

P. Sai Achyuth · V. Satyanarayana (✉)
Department of Electronics and Communication Engineering, Aditya Engineering College, Surampalem, Andhra Pradesh, India
e-mail: vasece_vella@aec.edu.in

S. Yadav et al. (eds.), *Proceedings of Second International Conference in Mechanical and Energy Technology*, Smart Innovation, Systems and Technologies 290,
https://doi.org/10.1007/978-981-19-0108-9_37

Fig. 37.1 First forged image [2]

After that there are several cases of image forgery came into the light, some of the popular incidences are highlighted here. A picture was launched in which Iran has shown to test its missiles. In this image, it is observed that four missiles are heading to the sky at once. Later it is found that this image is fake [3]. Even an image from the American civil war in which it was showed that General S. Grant was sitting on the horse and leading his battalion to Virginia, was later found bogus [4] and recently in the year 2017, July, a picture was very popular in which the Russian president Vladimir Putin shared the stage with the president of United State of America, Donald Trump at the G20 summit 2017. This image got too much recognition from all over the world but later found to be a fake image [5]. From these of some cases, one can easily understand that this forgery is nowadays very common and no one could be spear from this type of cheating, whether how much one is popular or powerful. As stated above in the digital era, it is easy to make the image forgery with the help of software. This forgery is too intense as a new branch in the field of image processing came into the existence named Digital Image forensics. The study is dedicated to finding out the quantitative proof of the originality and the source of any image [6]. The Picture has an exceptional part in different regions like criminal and criminological examination, knowledge and reconnaissance frameworks, sports, legitimate administrations, clinical imaging, protection guarantee, and news coverage etc. This issue influencing the general public on different perspectives because of the picture imitation, counterfeit individuals are going into the world instead of the right individuals. False documentation is raising easily and due to this, the genuine individual enduring a great deal. Therefore, the researchers from all over the world have been working continuously to make or modify the technique which detects the image forgery efficiently.

37.2 Methodology for Forgery Identification

The image forgery detection techniques are mainly classified into two categories: Active and Passive. The pictorial representation of these techniques with their classification is given below (Fig. 37.2).

Active Image Forgery: Inactive forgery detection method's previous information about the picture is necessary for verification. In this technique, some data is intentionally hidden at the time of the creation of the image. Checking this code confirms the inventiveness of the picture. Active verification techniques are further characterized into two sorts advanced Digital watermarking and Digital signature. These codes are implanted into the pictures at the hour of picture securing or in the handling stage and advanced marks insert some optional data, ordinarily separated from the picture, at the securing end into the picture. The primary disadvantage of these approaches says that they are to be embedded into the pictures at the hour of making it, therefore utilizing some extraordinary tools can able to extract the earlier data about the picture [7]. Digital Watermarking: Watermarking is also utilized for the detection of the willing full distortion made in the pictures. In this method with the use of the checksum pattern, some additional information is added into the last most important bit of the pixels. In order to determine the watermark, one needs to apply a maximal length linear shift register sequence to the pixel data and compute the spatial cross-correlation function of the sequence as well as the watermarked image. These watermarks are intended to be imperceptible or to mix in with regular camera or scanner noise. There is also an existence of visual watermarks. Moreover, an additional imperceptible watermarking pattern is also present which can detect the adjustment in single pixels and it can discover anyplace if the changes happen [8].

Digital Signature: In this method, the originality of the image is secured from altering by utilizing a mathematical format called digital signature. In this digital

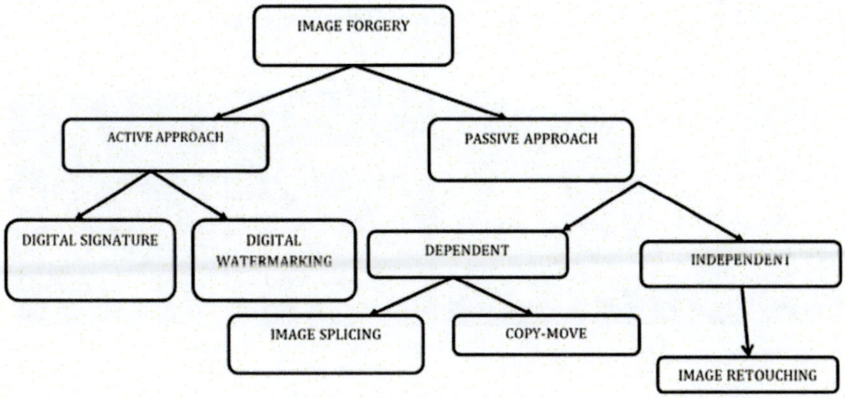

Fig. 37.2 Classification of image forgery techniques

signature technique, at first, robust bits are removed from the original picture. A picture is then apportioned into squares of 16 × 16 pixels. A secret key is utilized to get a fixed number (say N) of arbitrary matrices with sections consistently circulated in the span of [0, 1]. A low pass channel is applied on each arbitrary matrix oftentimes to get N random smooth pattern [9]. In this way, the system generated the digital signature by put on these marking measures on the picture.

Active methods have a few drawbacks since they required some human inclusion or uniquely prepared.

Passive Image Forgery: Passive way of image verification is also known as Image Forensic. It is a blind method of detecting the image forgery. In this way, there is no necessity for previous information about the image. Passive strategies depend on the understanding that despite the fact that altering may not leave any visual follow however they are probably going to modify the basic measurements due to noise inconsistency. It is these irregularities that are utilized to recognize the altering [7].

Passive techniques are again divided into dependent and independent forgery methods. The dependent techniques are intended to distinguish for some definite kind of falsifications such as copy-move and splicing which are subject to the kind of fabrication did on the picture whereas the independent methods are used to recognize the type of falsification on the image artifact traces left during interaction of resampling and because of lighting irregularities. Mainly the passive methods used to differentiate a given picture as unique or altered [11].

Dependent Passive Image: In this category mainly two techniques are available namely copy-move and image splicing.

A.1. Copy-move: It is the most general and is quite possibly the most boundless picture altering method, additionally it is exceptionally hard to recognize this sort of alteration as the replicated picture is taken from a similar picture. In other words, in this tempering the additional section which is going to embed in the targeted picture is taken from the same target picture, which means an image is taken, some particular region of that image is copied and then pasted again on that same picture. It is generally used for hiding some information or providing some false information through the image. This forgery can be detected by majorly three methods (Fig. 37.3).

(a) (b)

Fig. 37.3 Copy-move forgery **a** providing false information; **b** hiding information [7]

Brute Force Method: In this method, the detection process involves comprehensive search and autocorrelation. In a thorough examination of the image, at first, try to find out the similar sections in the image using circularly moved adaptations way. As in this process, a huge number of examination and comparison are made, therefore the computational unpredictability in this method is very high. The autocorrelation is used to confirm the alteration in area. Block-Based Method: This method used various algorithms for detecting the forgery in the image. Some of the popular algorithms that are used for it are Discrete Wavelet Transform (DWT), Discrete Cosine Transform (DCT), Principle Component Analysis (PCA), and Singular Value Decomposition (SVD) [8–10]. Key Point-Based Method: This technique utilizes scale and rotation invariant element finder and descriptor calculations which are Speeded-up Robust highlights (SURF) and Scale Invariant component Transform (SIFT).

Image Splicing: Apart from the copy-move technique, in this method, the alterations are done using a different image. As clear that this method involving at least two different images for the alteration in the picture. Due to the utilization of different images in this forgery, it is highly difficult and tedious work to make the borders and boundaries incoherent, therefore some specified tools are used such as Photoshop to make this authentic.

Independent Passive Image: Image retouching is the technique used in this category.

Image Retouching: In this type of forgery the alterations are done on the features of the targeted image. As per the requirement, the properties of the image either improve or dull (Fig. 37.4). These changes are done using rotation, stretching and/or scaling of the various features and then combined to make a duplicate image. It is highly difficult to identify the original one due to the absence of any fundamental alteration in the different section of the picture [9].

Common steps for image forgery identification: Fig. 37.5, here represents the general stepwise process to detect the image forgery.

Fig. 37.4 Image retouching forgery [9]

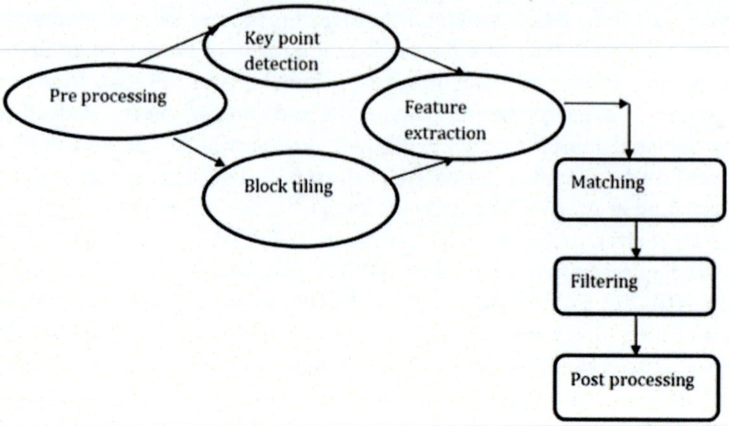

Fig. 37.5 General steps for forgery detection

Image Pre-processing: The first step to detect the image forgery is image pre-processing. This is performed using the process such as RGB to grey scale transformation, image enrichment, image filtering etc. [12].

Feature Extraction: The picture set is separated from other classes by the features specified for each class, but the picture set remains consistent for the class chosen. The appealing aspect of the selected collection of attributes is the minute measurement, which reduces the computational complexity while providing a wide distinction from other classes [13].

Selection of Classifier: The appropriate classifier is either picked or composed based on the feature set acquired during feature extraction. Due to the huge number of training sets, the classifier performance will be enhanced [14].

Classification: The goal of the classification process is to only determine whether the picture is real or not. To find the originality of the image LDA [15], Neural systems [16], and SVM [17] classifiers are utilized.

Post-processing: A few image falsifications may necessitate post-processing that includes alterations such as confinement copy locale localization [18].

37.3 Conclusion

Nowadays image forgery is a huge risk. As there are several types of image alterations available it is very difficult to make a common tool that can detect all the forgery. In this digital era, due to the system software making these types of counterfeits is very easy and cheap. If intentionally one will do the alteration in the picture with expertise level, it is almost impossible to identify the falsification and various times this causes pain to the genuine. Many of the forgery cases happened with the help of this image forgery. However, there are several techniques available and with the help of these developed methods numerous image forgery cases are exposed but the techniques available are not completely trustworthy. This technique necessitates more development and evolvement in terms of efficiency.

References

1. Meena, K.B., Tyagi, V.: Image forgery detection: survey and future directions. In: Data, Engineering and Applications, pp. 163–194. Springer, Singapore (2019)
2. Revolvy.com: Hippolyte Bayard (French, 1801–1887). https://www.revolvy.com/topic/HippolyteBayard&item_type=topic
3. Photo Tampering Throught History. http://pth.izitru.com/
4. Loc.gov: Civil War Glass Negatives and Related Prints. https://www.loc.gov/pictures/collection/cwp/mystery.html (2008)
5. Tait, A.: How a badly faked photo of Vladimir Putin took over Twitter. http://www.newstatesman.com/science-tech/social-media/2017/07/how-badly-faked-photo-vladimirputin-took-over-twitter (2017)
6. Sharma, V., Jha, S., Bharti, R.K.: Image forgery and it's detection technique: a review. Int. Res. J. Eng. Technol. (IRJET) **3**(3), 756–762 (2016)
7. Singh, M.K., Singh, A.K., Singh, N.: Disguised voice with fast and slow speech and its acoustic analysis. Int. J. Pure Appl. Math **11**(14), 241–246 (2018)
8. Fridrich, J: Robust bit extraction from images. In: Proceedings IEEE International Conference on Multimedia Computing and Systems, vol. 2, pp. 536–540 (1999)
9. Singh, M.K., Singh, A.K., Singh, N.: Multimedia utilization of non-computerized disguised voice and acoustic similarity measurement. Multimedia Tools Appl. 1–16 (2019)
10. Easow, S., Manikandan, L.C.: A study on image forgery detection techniques. Int. J. Comput. (IJC) **33**(1), 84–91 (2019)
11. Nath, V.V., Gaharwar, G.K.S., Gaharwar, R.D.: Comprehensive study of different types image forgeries (2015)
12. Singh, M., Nandan, D., Kumar, S.: Statistical analysis of lower and raised pitch voice signal and its efficiency calculation. Traitement Signal **36**(5), 455–461 (2019)
13. Ibrahim, R.W., Moghaddasi, Z., Jalab, H.A., Noor, R.M.: Fractional differential texture descriptors based on the Machado entropy for image splicing detection. Entropy **17**(7), 4775–4785 (2015)
14. Prasanna, G.S., Pavani, K., Singh, M.K.: Spliced images detection by using Viola-Jones algorithms method. Mater. Today Proc. (2021).
15. Kanchana, V., Nath, S., Singh, M.K.: A study of internet of things oriented smart medical systems. Mater. Today Proc. (2021)
16. Balaji, V.N., Srinivas, P.B., Singh, M.K.: Neuromorphic advancements architecture design and its implementations technique. Mater. Today Proc. (2021)
17. Punyavathi, G., Neeladri, M., Singh, M.K.: Vehicle tracking and detection techniques using IoT. Mater. Today Proc. (2021)
18. Veerendra, G., Swaroop, R., Dattu, D.S., Jyothi, C.A., Singh, M.K.: Detecting plant diseases, quantifying and classifying digital image processing techniques. Mater. Today Proc. (2021)

Chapter 38
The Role of Blockchain Technology on Human Rights Management and Business Ethics—Utopia or Dystopia

Akash Bag, S. M. Aamir Ali, Anuttama Ghose, Prateek Mishra, Bhanu Pratap Singh, and Samrat Datta

Abstract Blockchains are a type of technology that uses cryptography's unique capabilities to securely store data. They do not rely on a central source to preserve records, instead of disseminating data over a network, unlike traditional types of record-keeping. Blockchains are the technological underpinnings of digital currencies like Bitcoin. Beyond transactions, blockchains, both private and public, have a variety of present and proposed applications. They can be used to verify people and data, as well as serve as a foundation for file systems and content distribution. The Paper explains why public permissionless blockchains exist in a world apart from the real world. Several architectural elements of such blockchains make it partially law-proof, or, to a degree, a technology that creates a realm in which the law cannot be implemented as it is today. For starters, these blockchains use pseudonymity to protect their users' privacy. When using blockchain platforms and services, users show their "public key," which is an encrypted identification, rather than their real-life identity. Second, because of their distributed and decentralized character, these blockchains are immutable, posing a substantial obstacle to enforcement. A promising technology has gotten a lot of attention over the last decade: blockchain. It has a lot of promise, according to predictions and testimony. Despite this proliferation of material, there is still a shown need to comprehend and overcome the technology's limits, as well as

A. Bag (✉)
Amity School of Law, Amity University, Raipur, Chhattisgarh 493225, India
e-mail: akash.bag@s.amity.edu

S. M. Aamir Ali
Symbiosis Law School, Symbiosis International (Deemed University), Pune, Maharashtra 411014, India

A. Ghose
School of Law, MIT-WPU, Pune, Maharashtra 411038, India

P. Mishra · B. P. Singh
KIIT School of Law, KIIT University, Bhubaneswar, Odisha 751024, India

S. Datta
Seedling School of Law and Governance, Jaipur National University, Jaipur, Rajasthan 303012, India

S. Yadav et al. (eds.), *Proceedings of Second International Conference in Mechanical and Energy Technology*, Smart Innovation, Systems and Technologies 290,
https://doi.org/10.1007/978-981-19-0108-9_38

inadequate empirical research of its stated values. This Paper particularly examines two contemporary blockchain applications for human rights business practice and argues that established legal and regulatory functions and systems of accountability must not be replaced with technology for blockchain's actual value to be realized. We conducted a qualitative analysis using a literature study and primary research to highlight avenues to examine for future uses of the technology in supply chain initiatives, as well as existing evolutions of blockchain's current implementations.

38.1 Introduction

Technology has been pivotal in furthering human rights such as access to information, services, and education [1, 2]. It has been perceived as either a tool to promote or suppress human rights, depending on how it is implemented and managed [2, 3]. Blockchain, also known as distributed ledger technology, was originally conceptualized as a use case for a secure database for the transfer, issuance, and verified ownership of cryptocurrency, has now been recognized as a "peer-to-peer network that sits on top of the internet [4]." The implementation of blockchain technology has shown the potential to revolutionize transparency and accountability in business practice. As a result, preliminary questions arise that we explore in this thesis:

1. What does technology provide, or what value does it add to human rights?
2. Why would corporate actors want to leverage technology for human rights interests? And why would they have human rights interests at all?
3. What unique role, then, does blockchain technology play for corporate supply chain business practice?

There are gaps in how the technology may seek to confront ethical behavior and human rights violations in the diamond mining industry. The maturity of blockchain and its potential as a tool for the protection of human rights is dependent on its implementation even being described as "the fourth industrial revolution [5, 6]."

38.1.1 *The Main Argument*

The two case studies analyzed in this paper illustrate that blockchain technology is a tool for transparent management of transactions and interactions to fortify and secure communication in a complex information environment [7].

A decentralized system means that it operates without the need for a central authority or method to process or validate transactions. A distributed system means there are still proprietary authorities, but no single central one (Fig. 38.1).

Open-sourced blockchains such as IBMs Hyperledger allow for transparency and accountability in business-to-business and business-to-customer transactions through

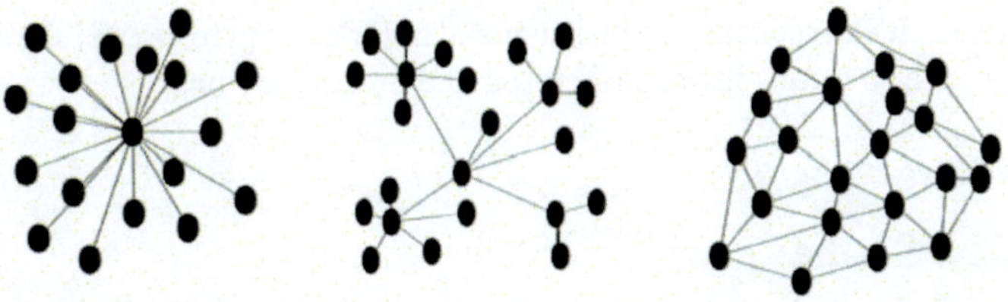

Fig. 38.1 Centralized, decentralized, and distributed ledgers as described by Hyperledger

"smart contracts, digital assets, record repositories, a decentralized consensus-based network, and cryptographic security [8–12]."

There are different kinds of blockchains operating today and can range from private to public (open source) and permissioned or permissionless.

- Public and permissionless: anyone can read the data, but only those with a key can write the data.
- Public and permissioned: anyone can read and write the data.
- Private and permissionless: only those with keys can read the data, and only those with an additional key can write the data.
- Private and permissioned: only those with a key can read the data, but anyone with that key can also write the data.

38.2 Previous Work

Globalization and technology have generated new understandings of human rights abuses and new approaches to them [13]. Technology paves a road for a unique opportunity. At the same time, however, it presents unique pitfalls that must be addressed with appropriate instruments [3]. Globalization has pushed the progression of international institutions and organizations, directly magnifying the impact on and responsibility for human rights [14–17].

Regulatory pressures for businesses to respect human rights go further. Since 2003, the United Nations has indicated an awareness of corporate activities impacting human rights through their operations [18–20], the International Organization for Standardization (ISO) [21], the OECD Guidelines for Multinational Enterprises [22], the United Nations Global Compact [23], the United Nations Norms on the Responsibilities of Transnational Corporations [24], and most recently, the United Nations adopted Guiding Principles on Business and Human Rights [1, 25–27].

By providing the right technologies, current business policies and expectations can be enhanced to be able to meet socially responsible global standards in supply chain management [28, 29].

Among lack of political will and expertise, limits for the regulatory impact such as treaties include lack of continuity within the recording system [30]. For example, when data on an operation is being inputted in a system by different individuals at different times, it can often be inconsistent and unreliable. In response, she advocates for open-source solutions instead of closed and proprietary ones.

38.3 Case Studies

With its wide range of potential to revolutionize industries such as banking and finance, business, government, and other industries such as the internet of things, blockchain's existing real-world application has been overinflated by its hype [4, 21, 23]. In early 2016, members of the Blockchain Council of the Dubai Prime Minister's office held a series of talks preceding a pilot project to discuss how blockchain can be used in the Kimberley Process.

38.3.1 Everledger: Raw Diamonds

In 2009, Human Rights Watch interviewed more than one hundred witnesses and victims of human rights abuses within military-controlled diamond fields, operating under President Robert Mugabe's rule. Their documentation on the Zimbabwean military killing, torturing, and forcing children to labor provided a stark depiction of a system of abuse and suffering for the diamond-revenue-generating profit of power-hungry actors.

Although not explicitly stated in public information by either Everledger or The Kimberley Process documents, a pilot project using blockchain to restrict conflict diamonds from markets has been announced [30]. As such, it is beneficial to analyze the business model of Everledger and how it may be seen within the KPCS. On June 22, 2015, Everledger began its process of laser inscription registry, being able to track the characteristics of each diamond through various unique identifiers consisting of 40 metadata points and multi-step verification factors.

38.3.2 Provenance: Tuna in Thailand

In recent years, the topic of human trafficking and forced labor has garnered a major focus in human rights literature. Labor trafficking, the human rights violation focused in this case study analysis, the International Labor Organization (ILO).

In a study conducted by the ILO in 2012, it was found that out of the estimated 20.9 million victims of trafficking, 14.2 million were people trafficked for forced labor primarily in the fields of agriculture, construction, domestic work, manufacturing, mining, and utilities.

Phase 1: The First Mile

We put up fishermen with keys on the blockchain, linked to their mobile numbers, key verified data, and catch estimations in collaboration with a local NGO. Local fisherman used simple SMS messages to register their daily catches, creating new blockchain assets and notifying suppliers.

Phase 2: All The Way Down The Chain

To ensure that what goes into a plant is the same as what comes out, we connected with ERP systems and other supply chain tools that were already digitizing factory data. The conversion variables from raw fish to completed product are handled by a smart contract.

Phase-3: End-User Environments

At the end of the supply chain, shoppers have access to the information obtained at the point of origin and along the supply chain. Brands and merchants can use this technology to replace the clutter of traditional printed communication with mobile-accessible information about the product's makers, suppliers, and procedures.

38.4 Discussion

Despite numerous whitepapers, media articles, books, and research papers on the topic of blockchain technology as a tool for addressing human rights problems, there remains a lack of empirical analysis. Blockchain technology has been distinguished for its lucrative operation as a plat-form for digital currency and democratization of asset exchange through Bitcoin.

- Understanding of the technology: Limited number of developers, and poor management of expectations resulting in "branding issues" of the technology.
- Acceptance and cooperation as the ultimate source of authority: The need for blockchain to be the ultimate source of authority to prevent lack of scalability, oligopolies, and scalability.
- Political overhead and lost-standing industry systems: the need for functional human systems to be in place before a technology system is built into them.
- Laws and regulations: Issues of immutability and privacy.

Insufficient understanding of the technology itself, and where it adds value can cause businesses and organizations to do too much too soon with powerful technology. This could limit its potential and growth. With a distributed ledger system, data is no longer held by a third party. Rather, it is distributed, encrypted, and anonymous with the ability to encrypt contracts to govern its management.

38.5 Conclusion

Questions of the scalability of blockchain technology partially come from the fact that it was originally only thought of as a platform for digital cryptocurrencies. Due to latency (time of transaction) protocols of smart contracts, peer-to-peer consensus can be cumbersome if the blockchain platform is not preprogrammed and built for handling large amounts of data and multiple interactions. Digital identity is the most important factor of the validation process and determining the performance of the network. The lack of digital identification could present difficulties for victims of human trafficking and forced labor through the identification of victims, as well as fears of criminalization, deportation, or harm from traffickers. There is high importance for a balance of governance and principles both within the blockchain ledger and the system it is trying to manage. While many companies are claiming that blockchain can change the very way economies are organized, proper systems must be implemented first. The need for functional human systems must always precede that of technical ones.

References

1. Metzl, J.F.: Information Technology and Human Rights. Human Rights Q. **18**(4), 705–746 (1996) [Online]. Available: https://www.jstor.org/stable/762590. Accessed 15 Sept 2021
2. Weeramantry, C.G., United Nations University, NetLibrary, Inc, (eds.): The Impact of Technology on Human Rights: Global Case-Studies: Studies on the Affirmative Use of Science and Technology for the Furtherance of Human Rights, Commissioned as a Special Project by the United Nations University, Following a Reference to the University by the United Nations Human Rights Commission. United Nations University Press, Tokyo (1993) [Online]. Available: http://ezaccess.libraries.psu.edu/login?url=http://www.netLibrary.com/urlapi.asp?action=summary&v=&bookid=20549. Accessed 15 Sept 2021
3. Brysk, A. (ed.): Globalization and Human Rights. University of California Press, Berkley (2002)
4. Iansiti, M., Lakhani, K.R.: The truth about blockchain. Harvard Bus. Rev. (2017) [Online]. Available: https://hbr.org/2017/01/the-truth-about-blockchain. Accessed 15 Sept 2021
5. Wright, C.: The Kimberley Process Certification Scheme: A Model Negotiation? p. 8 (2012) [Online]. Available: https://elr.info/sites/default/files/181-187_wright.pdf
6. Schwab, K.: The Fourth Industrial Revolution. Crown Publishing Group, USA (2017)
7. Drescher, D.: Blockchain Basics, Original Text: A Non-Technical Introduction in 25 Steps. Apress, New York (2017) [Online]. Available: https://www.springer.com/de/book/9781484226032. Accessed 15 Sept 2021
8. Lee, T.B.: WorldWideWeb, the first Web client. https://www.w3.org/People/Berners-Lee/WorldWideWeb.html. Accessed 15 Sept 2021
9. Buchanan, B.G.: A (Very) brief history of artificial intelligence. AI Mag 26(4) (2005). Article no. 4. https://doi.org/10.1609/aimag.v26i4.1848
10. Berger, I.W.: Content Curation Tool with Influencer Marketing | Social Media Management. Gotcha. https://gotchamobi.com/stream-articles/artificial-intelligence/ (2017). Accessed 15 Sept 2021
11. Monetary Authority of Singapore: Project Ubin: Central Bank Digital Money using Distributed Ledger Technology. https://www.mas.gov.sg/schemes-and-initiatives/project-ubin (2017). Accessed 15 Sept 2021

12. Hyperledger: Hyperledger Blockchain Performance Metrics White Paper. Hyperledger. https://www.hyperledger.org/learn/publications/blockchain-performance-metrics. Accessed 15 Sept 2021
13. Graham, A., Behlendorf, B.: Applying Blockchain Technology for Social Impact—RightsViews. https://blogs.cuit.columbia.edu/rightsviews/2017/04/17/applying-blockchain-technology-for-social-impact/ (2017). Accessed 15 Sept 2021
14. Ruggie, J.G.: Constructing the World Polity: Essays on International Institutionalisation, 1st edn. Routledge, London; New York (1998)
15. Business & Human Rights Resource Centre: Business and human rights: The role of business in weak governance zones'. Paper submitted by IOE, ICC & BIAC to John Ruggie. https://www.business-humanrights.org/en/latest-news/business-and-human-rights-the-role-of-business-in-weak-governance-zones-paper-submitted-by-ioe-icc-biac-to-john-ruggie/ (2006). Accessed 15 Sept 2021
16. e-Estonia: Building blocks of e-Estonia. https://e-estonia.com/solutions/. Accessed 15 Sept 2021
17. Martin, J., Bravo, K.E. (eds.): The Business and Human Rights Landscape: Moving Forward, Looking Back. Cambridge University Press, Cambridge (2015). https://doi.org/10.1017/CBO9781316155219
18. UNDP: Take Action for the Sustainable Development Goals. United Nations Sustainable Development. https://www.un.org/sustainabledevelopment/sustainable-development-goals/ (2015). Accessed 15 Sept 2021
19. Burchell, J., Cook, J.: Confronting the 'corporate citizen': shaping the discourse of corporate social responsibility. Int. J. Sociol. Soc. Policy **26**, 121–137 (2006). https://doi.org/10.1108/01443330610657188
20. Posner, M.: Shared Responsibility: A New Paradigm for Supply Chains [Online]. Available: https://www.stern.nyu.edu/experience-stern/global/new-paradigm-supply-chains (2015). Accessed 15 Sept 2021
21. ISO: About us. ISO. https://www.iso.org/about-us.html. Accessed 15 Sept 2021
22. OECD: Guidelines for multinational enterprises. https://www.oecd.org/corporate/mne/. Accessed 15 Sept 2021
23. What we do. Everledger. https://everledger.io/. Accessed 15 Sept 2021
24. University of Minnesota Human Rights Library: Norms on the Responsibilities of Transnational Corporations and Other Business Enterprises with Regard to Human Rights. http://hrlibrary.umn.edu/links/norms-Aug2003.html. Accessed 15 Sept 2021
25. OHCHR: OHCHR | OHCHR and business and human rights. https://www.ohchr.org/en/issues/business/pages/businessindex.aspx (2011). Accessed 15 Sept 2021
26. Deva, S., Bilchitz, D. (eds.): Human Rights Obligations of Business: Beyond the Corporate Responsibility to Respect? Cambridge University Press, Cambridge (2013). https://doi.org/10.1017/CBO9781139568333
27. Webster, F.E.: Determining the characteristics of the socially conscious consumer. J. Consumer Res. **2**(3), 188–196 (1975) [Online]. Available: https://www.jstor.org/stable/2489054. Accessed 15 Sep 2021
28. The World Bank: Human Rights Impact Assessments: A Review of the Literature, Differences with other forms of Assessments and Relevance for Development. Nordic Trust Fund [Online]. Available: https://documents1.worldbank.org/curated/en/834611524474505865/pdf/125557-WP-PUBLIC-HRIA-Web.pdf (2013)
29. OHCHR: Human Rights Indicators. https://www.ohchr.org/en/issues/indicators/pages/hrindicatorsindex.aspx. Accessed 15 Sept 2021
30. Heyns, C., Viljoen, F.: The impact of the United Nations Human Rights treaties on the domestic level. Human Rights Q. **23**(3), 483–535 (2001) [Online]. Available: https://www.jstor.org/stable/4489346. Accessed 15 Sep 2021

Chapter 39
Fake News Identification and Detection: A Brief Review

K. Sushma and M. Neeladri

Abstract Nowadays, bogus news affects a lot of humans in social and moral ways. A common person can't identify the difference between real and fake news and make his/her opinion based on that news. Due to several reasons, there is a high probability of the news that comes to him/her or spread drastically is bogus. Due to which the opinion based on bogus news ultimately harms the personal and social character. Therefore, it is necessary to take some step to detect this falsified news. The linguistic approach and the network-based approach are two methods for detecting fake news. Here it is presented that the operational principles for a feasible false news detecting system, even though developing a false news detector is not an easy task. The technique of detecting whether the news is fake or not is referred to as falsified news detection.

39.1 Introduction

Falsified news could be a representation of data that comprises intentional data or deceptions unfurl by means of antiquated print and broadcast print media or online social media. Fake news is composed and printed with the aim to misdirect in order to wreck work environment, element, or individual, as well as gain monetarily or strategically, normally with publicizer, overstated, or clearly bogus features that grab attention. By choice, dishonest and deceptive fake news is totally different from obvious caustic remark or parody that is meant to humor instead of misleading its audience. Fake news typically employs attention-getting headlines or entirely invented news stories to extend audience, online sharing, and net click revenue. Within the latter case, it's kind of like sensational online headlines and depends on advertising revenue generated from this activity, in spite of the truthfulness of the printed stories. Fake news additionally undermines serious media coverage and

K. Sushma · M. Neeladri (✉)
Electronics and Communication Engineering Department, Aditya Engineering College, Surampalem, Andhra Pradesh, India
e-mail: neeladri.m@aec.edu.in

S. Yadav et al. (eds.), *Proceedings of Second International Conference in Mechanical and Energy Technology*, Smart Innovation, Systems and Technologies 290,
https://doi.org/10.1007/978-981-19-0108-9_39

makes it tougher for journalists to hide vital news stories. There are primarily two ways to identifying fake news: the first is the linguistic approach, and the second is the network approach.

39.1.1 Linguistic Approach

Under this approaches the substance of deceiving communications is isolated and studied in order to link language design tendencies to fraud. Most of the falsifiers utilize their language deliberately to try not to be caught. Despite all the arrangement carried out by the management, generally, the language "spillage" occurs with definite verbal angles that square measure exhausting to watch like frequencies and examples of the closed-class word, combination, and negative inclination word utilization [1]. The objective of the phonetic approach is to appear in situations of escape or purported "predictive deception cues" discovered within the text of a communication.

39.1.2 Demonstration of Data Statistics

Maybe the best technique to represent the text is "bag of words" style that treats every word as a solitary, similarly imperative unit inside the "bag of words" method, singular words or "n-grams" (multiword) frequencies region unit mass and dissected to uncover signals of misdirection more labeling of words into individual linguistically signs, e.g., segments of dialogue or "shallow syntax" [2], emotive measurements [3], or area-based words [4] region unit all ways that of giving frequency sets to uncover semantic signals of duplicity as shown in Figs. 39.1 and 39.2. The straightforwardness of this representation furthermore brings about its greatest deformity. Furthermore, to depending totally on language, the strategy relies upon separated n-grams, regularly separated from supportive setting information during this strategy,

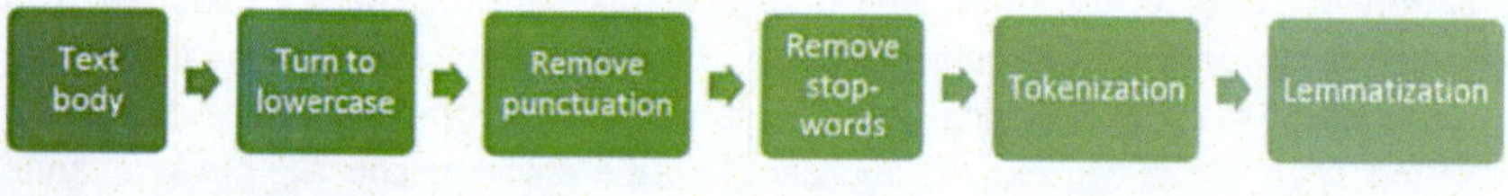

Fig. 39.1 Pre-processing of text [3]

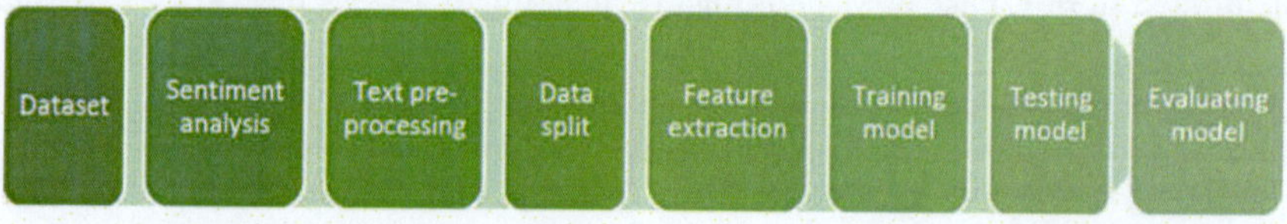

Fig. 39.2 Analysis setup for programing [3]

any goal of equivocal implied stays non-existent [5]. Various fraud finding analysts have observed this technique supportive combined with entirely unexpected, integral examination [6, 7].

39.1.3 Deep Syntax

In most cases, analyzing word usage alone is insufficient to anticipate deceit. To forecast instances of dishonesty, deeper language patterns (syntax) are examined. Probability Context-Free Grammars (PCFG) allow for in-depth syntactic study. Sentences are rebuilt to a gathering of change governs (a separate tree) to clarify grammar arrangement, for instance, thing and action word express, that is progressively modified by their grammatical constituent components [8]. A definitive arrangement of modifications yields a separate tree with a specific chance. This system is utilized to recognize rule classes (lexicalized, unlexicalized parent hubs, and so on) for trickiness unearthing with 85–91% exactness (contingent upon the standard class utilized) [8]. Apart from this for automation, some other third-party apparatuses are utilized like the Stanford computer program [9], Auto Slog-TS syntax instrument [10]. Also, it may be possible that alone syntax analysis has not the ability to differentiate fraud appropriately. Therefore, it is observed that the utilization of this approach occurred while combining it with other linguistic or system interpretation techniques [1, 8].

39.2 Semantic Analysis

As another to duplicity prompts, signs of trustworthiness have also been investigated and accomplished by portraying the level of similarity between private expertise (e.g., A structure audit) when contrasted with a substance "profile" got from a bunch of comparable to information. This methodology expands the n-gram and sentence structure model by fusing profile similarity choices, showing the expansion impressively improves order execution [1]. The instinct is that a misleading creator with no mastery with an incident or article (for example, ne'er visited the building being referred to) might epitomize logical inconsistencies or oversight of realities present in profiles on comparative subjects. For item audits, an author of an honest survey is extra apparently to frame comparative remarks concerning parts of the product as various honest commentators. Separated substance from similar remarks comprises of trait: descriptor attempt. By situating profiles and furthermore the portrayal of the author's very own aptitude, veracity appraisal might be a work of the similarity scores:

1. Similarity with the presence of some particular side (e.g., A craftsmanship safe near the lodging);

2. Compatibility with the portrayal of some broad side, similar to area or administration.

Using this way, the expectation of lie can be demonstrated up to 91% right with the system.

39.2.1 Classifiers

Word-set along with class frequencies region unit are accommodating for analyzing the machine-driven mathematical examination. One basic utilization of instructing of "classifiers" as in "Support Vector Machines (SVM)" [6] and "Naïve Bayesian models" [10]. Basically, when a numerical archetypal is adequately prepared from previously available coded models in one among two classes, it is used to forecast the cases of forthcoming misdirection on the reason of numeric bundle and distances. The work with dissimilar methods and distance capacities among the data focuses structure of the precision of SVM [11], which solicitations new experimentation on the web effect of those factors. Credulous mathematician calculations make the characterizations that supported the amassed proof of the relationship among the provided variable (for example, language structure) and consequently the elective factors are treated as a helping factor inside the model [12].

39.2.2 Network Approach

In these methods the web-based information, similar to message data or organized information network inquiries, are regularly controlled to supply mixed trickiness measures. In every structure, most of the part consolidates with AI procedures for instructing classifiers to suit the examination. It's mandatory for scientists to know these totally various regions, yet there is no known compartmentalization of techniques that exists inside the present writing. The objective is to deliver a review of the current investigation while offering a crossover technique, for at most execution of a false news location device.

Imaginative and shifted, exploitation network properties and conduct region unit manners by which to improve content-based methodologies that acknowledge tricky language and release prompts to foresee trickery. Timeframe content on recent developments is continuously enhanced through miniature contributing to blog applications like Twitter which misdirection examination apparatuses region unit every one of the tons of vital.

39.2.3 Linked Data

The utilization of data organizations could address a major advantage toward the versatile interaction certainty checking ways.

For certain data, bogus "authentic proclamations" will address a type of duplicity since they'll be separated and analyzed close by findable articulations in regards to the remarkable world. This methodology uses Associate in Nursing's current group of aggregate human information to evaluate the truth of the most recent assertions. The strategy relies upon questioning existing information organizations, or openly out their organized data, as "DBpedia metaphysics", or the "Google Relation Extraction Corpus (GREC)".

39.2.4 Social Network Behaviors

Confirmation of character via web-based media is abrogating to the thought of belief. The propagation of stories inside the kind of recent developments via general innovations such as "micro-blogs" welcomes the ways that of learning the differentiation among the bogus and honest matter. Apart from the investigation of data, the work of information and obvious conduct of problematic bases is necessary [13]. The new utilization of Twitter in impacting political discernments [14] is one situation any place certain information, especially the consideration of hyperlinks or related information, is aggregated to determine honesty appraisals. Focusing reverberation investigation (CRA), a method of organization-based content examination, addresses the matter of enormous arrangements of writings by recognizing the first fundamental words that connect various words inside the organization. This was utilized by Papachariss I and Oliviera to spot content examples in posts about Egypt's election (2012). Consolidating notion and conduct contemplates having incontestable the competition that conclusion centered surveys from singleton benefactors essentially influence internet positioning [15], which is regularly a partner pointer of "pushing" or helpful false audits to unnaturally damage the ranking.

39.3 Fake News on Different Platforms

39.3.1 Bogus News on Conventional News Media

Counterfeit information is in itself another issue. After some time, the media environment of falsified information has shifted from printed media to radio/TV and, currently, internet news and web-based media. To address this, one of the precaution steps is to designate the fabricated news as "customary fake news" before web-based systems administration impacts influenced its creation likewise, spread. Then, some of the mental and social underpinnings that represent the influence of false news on both the personal and social levels of information climate are delineated.

39.3.2 Psychological Foundations of Falsified News

Generally, people are not able to accurately and logically differentiate between certified and counterfeit news. There are a couple of mental and scholarly speculations that can explain this miracle and the compelling energy of falsified news. Two essential issues make buyers normally weak against counterfeit news: (i) Native Realism: clients prefer to assume that their perception of the facts is the basic precise perspectives, while those who are not having a similar opinion are perceived as uneducated, foolish, or uneven; and (ii) Confirmation Bias: customers want information that supports their current viewpoints. Fake news is frequently seen as genuine by clients due to these mental characteristics that are inherent in human nature. It is often difficult to correct a misconception after it has been created. Brain science research shows that replacing bogus information (for example falsified news) with the pre-impression of verifiable, authentic material isn't only ineffective in reducing misperceptions, but can sometimes significantly increase them, particularly in philosophical social affairs.

39.3.3 Social Foundations of the Falsified News Ecosystem

Considering the entire news usage climate, some of the similar social movement can be depicted that adds to the extension of falsified news. The Prospect speculation depicts essential authority as a system by which people make choices considering the relatively their personal impact and setbacks when diverged from their present status. This needs to increase the value of a decision also relates to social gains, such as continuing with confirmation from others in a customer's brief social connection. As depicted by the social character theory and regularizing sway speculation, this tendency for social affirmation and declaration is central to a man's character additionally, certainty, making customers inclined to pick "socially sheltered" decisions while believing and scattering news data, keeping the principles developed in the gathering even in case the news being shared is falsified information.

This prudent speculation of fake news correspondence can be demonstrated from a monetary redirection theoretical point of view by computing the news age and using the cycle as a two-player strategy redirect. In order to explain counterfeit news, two types of important actors in the information organic local region are expected: distributers and purchasers. The methodology of information conveying is shown as a planning from novel banner "s" to resultant news report "a" with an effect of twisting predisposition "b", i.e., $s^{b \to a}$, where $b = [-1, 0, 1]$ shows [left, no, right] tendencies take impacts on news dispersing measure. Normally, this helps in getting that how much a bogus story may be uneven or bent to make falsified information [16]. There are two perspectives on which it can be useful to the distributor: (i) here and now utility: the desire to increase advantage, which is directly proportional to the customers served; (ii) Long-term utility: their track record for information trustworthiness. Client usefulness is divided into two categories:

(i) Data utility: obtaining verifiable and appropriate data (commonly additional endeavor cost required);
(ii) Brain research utility: Accepting news that meets their past judgments and social needs, such as affirmation inclination and prospect theory. In this news system, both the distributor and the consumer want to increase their total utility. It is clear that fake news occurs when the current time and place utility guidelines, i.e., a distributer's overall utility, and brain research utility principles, i.e., the buyer's overall utility, are maintained in balance. This describes the social stream that creates an information natural structure conducive to the spread of falsified information.

39.3.4 Bogus Information on Web-Based Media

This section highlights the specific pattern of bogus information on social media. Precisely the focus is on how any falsified news is spread by social media and how it affects or converts the attitude of common people. Note that the previously mentioned attributes of customary phony news are likewise appropriate to online media.

While numerous clients via web-based media are genuine, online media clients may likewise be pernicious, and now and again are not even genuine people. The minimal expanse of making web-based media account likewise energizes malevolent client accounts, for example, "social bots", "cyborg clients", and savages. A "social bot" is a web-based media account. This type of accounts is controlled and run with the help of computer algorithm in order to distribute content and make connection with people (or other bot clients) through online media. "Social bots" can evolve into spiteful substances with the intentional intention of causing harm, like to manipulate and disseminate through falsified information via web-based media. According to reports, "Social bots" are responsible for manipulating the 2016 U.S. official political decision online dialogues on a massive level.

Trolls, physical human being clients who intend to upset online communities and incite customers into an enthusiastic reaction, are additionally assuming a significant part in getting out counterfeit information via web-based media. For example, some evidence suggests that 1000 paid Russian trolls disseminated false information about Hillary Clinton. Trolling practices are exceptionally influenced by individuals' mindsets and the setting of online conversations, which empowers the simple scattering of bogus news among in any case "normal" online communities. The impact of trolling is on human's internal negative feelings, as hatred and terror, bringing about uncertainty, doubt, and silly conduct. Finally, cyborg customers can obtain fake information that combines automated operations with human input. In general, individuals create cyborg profiles to use as a disguise and set up automated projects to do tasks in social web-based media. The ability to easily swap between human and bot functions gives cyborg clients intriguing freedoms in terms of obtaining false information. Essentially, these extremely dynamic and sectarian harmful recordings via internet media become incredible sources and disseminators of false information.

39.3.5 Echo Chamber Effect

Online media gives another worldview of data creation and utilization for clients. The data chasing and utilization measure are transforming from an interceded structure (for example through the columnists) to a high level of disruptive intervention method. Buyers are specifically presented to particular sorts of information in light of the manner in which news source shows up on their landing page in web-based media, enhancing the mental difficulties to dispersing counterfeit news recognized previously. For instance, clients on Facebook consistently follow similar individuals and, in this way, get news that advances their supported existing accounts. Consequently, clients via online media will in general frame bunch containing similar individuals where they then, at that point enrapture their feelings, bringing about a reverberation chamber impact.

39.4 Conclusion

Linguistic and network-based techniques have demonstrated highly accurate outcomes in order errands inside restricted spaces. This conversation drafts fundamental compartmentalization of strategies out there for greater refinement and investigation and gives a premise to the arranging of exhaustive artificial news recognition apparatus. Strategies emerging from different methodologies are additionally utilized alongside in an extremely crossover framework, whose alternatives region unit summed up:

- Linguistic processes should be established on several layers for maximum performance, beginning with "word/lexical analysis" and progressing to the highest "discourse-level analysis".
- By identifying reputable sources as a feasible substitute to solely content-based methods, network behavior needs to be connected with the "trust" component.
- To keep the design of the tools, in such a way that it supplements, not replace, human verdict, the links between the machine's output and the operations should be clear.
- The gold standard dataset, which is easily available for the general contribution needs to be organized consistently in order to verify the current reality.

References

1. Feng, V.W., Hirst, G.: Detecting deceptive opinions with profile compatibility. In: Proceedings of the Sixth International Joint Conference on Natural Language Processing, pp. 338–346 (2013)
2. Markowitz, D.M., Hancock, J.T.: Linguistic traces of a scientific fraud: the case of Diederik Stapel. PloS ONE **9**(8), e105937 (2014)
3. Kostoska, T.: Effects of emotive language on performance of fake news detection model (2020)
4. Hancock, J.T., Woodworth, M.T., Porter, S.: Hungry like the wolf: a word-pattern analysis of the language of psychopaths. Leg. Criminol. Psychol. **18**(1), 102–114 (2013)
5. Singh, M.K., Singh, A.K., Singh, N.: Acoustic comparison of electronics disguised voice using different semitones. Int. J. Eng. Technol. (UAE), **7**(2.16), 98 (2018)
6. Zeng, J., Zhang, Y., Ma, X.: Fake news detection for epidemic emergencies via deep correlations between text and images. Sustain. Cities Soc. **66**, 102652 (2021)
7. Singh, M.K., Singh, N., Singh, A.K.: Speaker's voice characteristics and similarity measurement using Euclidean distances. In: 2019 International Conference on Signal Processing and Communication (ICSC) (pp. 317–322). IEEE (2019)
8. Feng, S., Banerjee, R., Choi, Y.: Syntactic stylometry for deception detection. In: Proceedings of the 50th Annual Meeting of the Association for Computational Linguistics (Volume 2: Short Papers), pp. 171–175 (2012)
9. Singh, M., Nandan, D., Kumar, S.: Statistical analysis of lower and raised pitch voice signal and its efficiency calculation. Traitement du Signal **36**(5), 455–461 (2019)
10. Oraby, S., Reed, L., Compton, R., Riloff, E., Walker, M., Whittaker, S.: And that's a fact: distinguishing factual and emotional argumentation in online dialogue. arXiv preprint arXiv: 1709.05295 (2017)
11. Stahl, K.: Fake news detection in social media. California State University Stanislaus **6**, 4–15 (2018)
12. Balaji, V.N., Srinivas, P.B., Singh, M.K.: Neuromorphic advancements architecture design and its implementations technique. Mater. Today Proc. **51**, 850–853 (2022)
13. Singh, M.K., Singh, A.K., Singh, N.: Multimedia utilization of non-computerized disguised voice and acoustic similarity measurement. Multimedia Tools Appl. **79**(47), 35537–35552 (2020)
14. Cook, D.M., Waugh, B., Abdipanah, M., Hashemi, O., Rahman, S.A.: Twitter deception and influence: Issues of identity, slacktivism, and puppetry. J. Inf. Warfare **13**(1), 58–71 (2014)
15. Prasanna, G.S., Pavani, K., Singh, M.K.: Spliced images detection by using Viola-Jones algorithms method. Mater. Today Proc. **51**, 924–927 (2022)
16. Zhou, X., Zafarani, R.: A survey of fake news: fundamental theories, detection methods, and opportunities. ACM Comput. Surv. (CSUR) **53**(5), 1–40 (2020)

Chapter 40
Consumer Behaviour on Digital Marketing Platforms—Specifically in Terms of Consumer Loyalty Using Machine Learning

V. P. Sriram, Asmat Ara Shaikh, B. K. Sumana, Anuj Kumar, Viney Dhiman, and Mohd Naved

Abstract This article aims to review consumer behaviour on digital platforms and how machine learning influences customer loyalty. The researcher adopted a qualitative method to understand consumer behaviour and how Machine Learning (ML) helps customer loyalty. A correlational research design is used to conduct this research. In addition, Secondary data is used in the study to obtain reliable conclusions. The researcher analyses all journals in the general marketing literature. Furthermore, this research develops a less complex process to predict consumer purchasing behaviour on digital platforms. It has been found that customer loyalty can also be tracked using machine learning and the ability to collect customer preference data on digital platforms. In addition, finding depicts machine learning is better at identifying customer loyalty.

V. P. Sriram (✉)
Department of MBA, Acharya Bangalore B School (ABBS), Bengaluru, Karnataka, India
e-mail: dr.vpsriram@gmail.com

A. A. Shaikh
Bharati Vidyapeeth's Institute of Management Studies and Research, Navi Mumbai, India

B. K. Sumana
PG Department of Business Administration, Maharanis Womens Commerce and Management College, Mysore, Karnataka, India

A. Kumar
Apeejay School of Management, Dwarka, Delhi, India

V. Dhiman
Postgraduate Institute of Medical Education and Research, Chandigarh, India

M. Naved
Department of Business Analytics, Jagannath University, Delhi-NCR, India

S. Yadav et al. (eds.), *Proceedings of Second International Conference in Mechanical and Energy Technology*, Smart Innovation, Systems and Technologies 290,
https://doi.org/10.1007/978-981-19-0108-9_40

40.1 Introduction

Nowadays, almost every brand is marketing its products on digital platforms, making it more convenient for consumers to shop and compare online [3]. With the help of digital platforms such as Instagram, Facebook, Snapchat, and more, marketers offer products and services according to the taste and preference for enhancing customer satisfaction, which also sustains customer loyalty for the long term. The study of consumer behaviour is significant because it helps to understand what influences consumers' buying decisions. For example, how the target audience sees the Facebook post, how does it develop a potential lead. Digital platforms can either build or destroy the brand image of the company. Hence the strategy should be specially designed as per the customer preference [16].

From SMEs to large enterprises, it is found that all need to have a digital presence in the market to grab consumers' attraction towards them. It is true that marketing influences consumer's purchase decisions. In the present scenario, consumers are focusing more on digital marketing platforms for buying a product [8]. With the help of such platforms, companies also increase consumer engagement and, at the same time retaining customer loyalty. It is evaluated that every customer in the market is different and has a different perception when purchasing the product [1, 2].

It is recognized as easy to gain customer loyalty as long as the company is offering products conveniently and quickly. It has been identified that organizations designed several strategies and adopt technology in marketing to maintain or insights on customer loyalty. In this context, machine learning is a very effective tool that helps marketers or companies make data or report on customer loyalty. Machines can quickly and efficiently analyse large amounts of data without human error. Hence, companies consider machine learning to interpreted customer loyalty fast. It has been seen that different techniques are adopted for predicting customer loyalty by the marketers that include support vector machines (SVM), which help analyse loyalty by customers. With the help of ML, marketers can quickly determine the retention rate of consumers in different digital platforms. In has also found that consumers also shift frequently from one platform to another, so this becomes difficult to consider exact prediction regarding a particular digital platform. ML makes it easier for companies to identify which channel is more effective for promoting their products [3].

Machine learning is a practical approach that helps to get accurate and reliable results in the end [10]. Machine learning algorithms use real-time data to refine statistics and information stored of customer preference [11]. It allows marketers to prepare loyalty card systems. Through which they will maintain loyalty of the customers. Most importantly, the business will get to know which digital platform is impacting the customers more [4]. Figure 40.1 shows the machine learning journey.

Furthermore, Researches shows that Marketing automation reduces marketing overhead. It is entirely possible with machine learning, as it takes a lot more personal to generate reliable data.

It includes churn prediction models, regression analysis, and more.

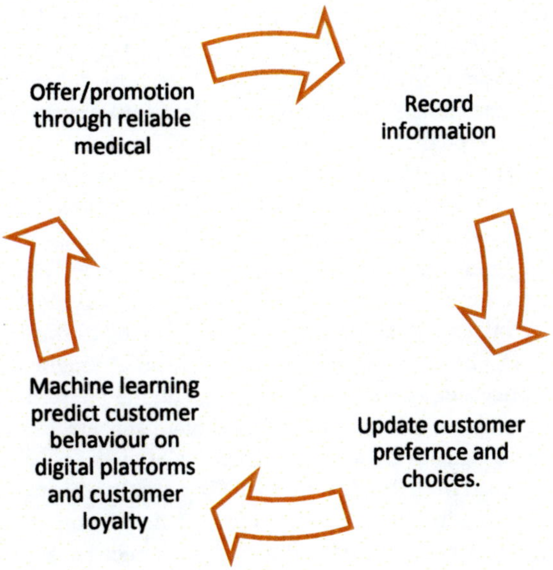

Fig. 40.1 Machine learning journey

Furthermore, to understand these concepts more deeply, this research is conducted. This independent research report describes consumer behaviour on digital platforms. The purpose of this research is to identify how machine learning affects customer loyalty to digital platforms. This research deepens the theoretical understanding of machine learning across the consumer loyalty chain. Understanding consumer buying behaviour is key to attracting and retaining customers and converting them into potential leads [5].

40.2 Literature Review

In the views of Ballestar et al. [4], Machine learning helps in making a smart decision by using historical data to predict future trends, ultimately sustaining customer loyalty. There are many instances of how machine learning can be applied to marketing efforts: audience segmentation, influencing customer behaviours, and more. It is found that consumers require a consistent and personalized experience for remaining loyal to the brand for an extended period. Understanding the requirements of the target audience is at the core of making a profitable marketing strategy. With ML, marketers get insight into their behaviour and predict what they might see Dzyabura et al. [6].

Most organizations are adopting machine learning techniques for evaluating loyalty amongst consumers, such as Regression analysis. It is an effective technique under machine learning through which customer loyalty can be assessed efficiently. The model depicts in evaluating independent variables that further contribute to forming binary dependent variables. In this case, the independent variable is considered customer loyalty that can be gained by enhancing customer satisfaction. So, these are some tactics taken into account by companies for evaluating the retention of consumers with the particular brand. Moreover, machine learning prevents less reliable predictions made manually [7].

The Impact of Digital Marketing on Consumer Behaviour—Chain Reaction [14] states that the role of digital marketing platforms is very effective for influencing the behaviour of consumers in the present market situation. For example, the trends on Instagram attract more consumers to buy. There are many small businesses which growing slowly but steadily. In this context, digital platforms (Instagram) are influencing more compared to others platforms [8].

On the other side, Consumers are also leveraging technology to accurately assess products or services. So, this shows digital platforms influence changing behaviour of customers. In addition, Digital platforms are accessible to all. Hence, people are engaging actively with brands on social media pages [9].

Apart from this, Galvão et al. [7] argue that companies also do customer segmentation to understand consumer loyalty. It helps companies develop strategies to promote products and services to influence customer behaviour. It also aids them in making modifications that will empower consumers to make better buying decisions.

Jain and Pamula [9] states, the role of machine learning and artificial intelligence in influencing consumer behaviour should not be undervalued. It is a prevailing medium that empowers consumers to make better choices. Digital marketing is driven efficiently and feasibly by technology like artificial intelligence and machine learning. Machine learning is undoubtedly going to significantly impact consumer behaviour, as brands that such technology is presenting their products and services more professionally compared to those that don't. In addition, machine learning also helps brands improve customer support to maintain friendly relationships with customers [10].

40.3 Research Questions

- How is machine learning helpful in assessing customer loyalty in digital platforms?
- How digital marketing influence consumer behaviour?

40.4 Methodology

In order to achieve the research objectives, a research methodology was devised. It is an essential part of which was the setting of goals that would seek to bridge the gap in the range of knowledge and achieve the research objective. Secondary data is used to study the importance of the subject. This research aims to examine the literature on this topic to elaborate research in relation to digital marketing and consumer behaviour. Collectively, this research shed light from many diverse aspects on how consumers are influenced by digital platforms, machine learning on customer loyalty. As the world moves towards the digital age, digital channels are increasing the sales of any brand's products. Hence the present study attempted to reveal the impact of digital marketing on the customer purchase decision.

The rise of digital marketing has totally changed the market scenario. It has provided an equal opportunity to the companies to promote their brand and make the customers shop easily. However, the impact of digital marketing is not limited to business only, but also helps maintain consumer loyalty. In modern times, consumers expect a more consistent and personalized experience [5]. In this context, companies actively and digitally market their unique consumer offerings to suit modern tastes and attract modern customer. To remain competitive in the current market scenario, marketers need to understand every factor of digital marketing. An in-depth understanding allows marketers to set up a customized and hardcore marketing campaign, enabling companies to accomplish their goals. Modern marketing campaigns should be quick and more about building long-term relationships with the target audience. In addition, Digital marketing has recently been recognized as one of the best tools to bridge this gap and interact directly with the consumer. Therefore, with the trend of digital marketing, more attention is being paid to the use of digital platforms as a means of effective advertising to consumers.

This study draws on advanced research on machine learning to evaluate consumer loyalty. Machine learning has been presented as a method of analysing the satisfaction and loyalty data of consumers. In tough competition and rising consumer expectations, companies are trying to retain existing customers rather than enticing new ones [2]. Satisfying existing consumers becomes an important goal. It has been found that satisfaction is directly linked with customer loyalty. This means that companies remain loyal to them if they meet the needs and requirements of the customers. To monitor customer behaviour and loyalty, companies implement various techniques such as machine learning. ML allows taking into account the heterogeneity of the original data-set. Therefore, it helps to deliver better insight into the nature of consumer loyalty. By using customer purchase history and other data, ML techniques can make precise and surprisingly accurate product recommendations and loyalty [11].

Machine learning identifies customer needs and recommended products according to their preferences. Machine learning can enhance customer experience, improve customer interactions, and reinforce buying habits that will strengthen customer loyalty. Machine learning can help businesses design a personalized market strategy that delights customers into potential leads [1]. Understanding the requirements of the target audience is at the core of every successful advertising strategy. With ML, the company can gain insight into customer behaviour and predict what they might see.

Furthermore, Predictive marketing analytics is a technique that uses all the data to predict future buying of customers and inform the purchase behaviour [6]. It integrates various methods from data mining, statistics, modelling to analyse and different data sets to develop a forecast marketing strategy. Using this strategy,

Company can classify potential threats that could affect customer buyer journey. It gives information in advance, so that firm can act on time. It brings together customer information with determined information based on ML analysis, including time spent browsing, clicks, scrolling, and determining choices, and interest levels. Companies like Starbucks use predictive analytics to collect and analyse customer data [12–14].

Therefore, it is necessary to predict, it will prevent customer from losing. It has been determined that the marketer also uses the churn prediction model as with the help of this model probability of using the product by the consumer can be evaluated effectively. This also provides an insight into how long the consumer is using products and then shift to another platform in the market. This strategy constitutes maintaining customers for an extended period and providing them valuable information along with satisfaction. Reliable and predictable is extremely valuable for taking the proper steps to prevent customers at risk.

A decision tree is another model which is often used in companies. It makes a tree-like structure that demonstrates the sequential decision-making procedure. Features in a dataset are signified as internal nodes in the tree. Decision tree algorithms are highly flexible. It precisely forecasts which non-loyal customers are most possibly become loyal in the future and for current loyal customers.

With the help of Machine Learning, marketers have a real advantage of establishing practical forecasts. Through ML, systems can collect and examine data that will aid in answering such questions like:

- Which target segment can respond to certain proposals?
- What sort of products would the customer want to buy?

With real-time data, ML can evaluate the user's behaviour to gain insight into their preferences and loyalty and then modify the website as per the collected material. By generating a personalized experience, Businesses can take consumer experience to the upper level. This personalized experience eventually builds robust relationships between customers and brands. This also makes the shopping experience enjoyable and retains loyalty. A more fruitful consumer experience, of course, means a higher retention rate. Brands leverage ML to effectively use this data to enable loyalty marketing and positively impact customer behaviour.

Further, it has been found that Machine learning application gives systems the ability to analyse mechanically and progress from experience [16]. Machine learning is a powerful and innovative technique that drives patron relationships and enhances loyalty programs. In addition, ML helps to answer two critical questions. First, what are the actual needs of consumers? And second, how does the marketing approach meet their requirements?

Machine learning can accomplish and automate repetitive tasks and accelerate current business procedures [15]. It helps the firms to get new opportunities that improve their marketing competency. It also strengthens the strategy to deliver the facilities at the right time to the target customers through the proper digital channels. Moreover, such offerings help companies to build stronger relationships with their

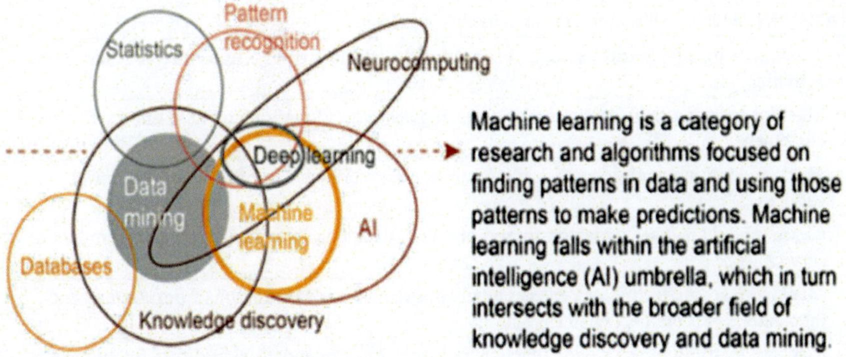

Fig. 40.2 Machine learning in customer loyalty

customers and ultimately their loyalty [15]. Figure 40.2 shows the machine learning customer loyalty.

40.5 Result

Researchers considered several literature reviews to gain knowledge about digital marketing and ML. The result shows that the study of consumer behaviour is essentially the key to understanding purchasing patterns and decision processes. Information about consumer behaviour is significant to businesses. This is because they are constantly planning to enhance their buying experience and, ultimately, their sales. It is found that consumers are looking for convenience, and digital platforms make it all that easy. Findings indicated that marketers use digital marketing to increase brand credibility. In addition, the study found that marketers adopt technology like machine learning to monitor customer loyalty. Table 40.1 is given below to show how machine learning helps assess customer loyalty in digital platforms:

40.6 Conclusion

It has been concluded from the above research that Consumer behaviour is continually evolving in the digital age. Knowing consumer behaviour's taste and trends helps companies stay visible and relevant in the marketplace. The study of consumer behaviour allows companies to understand the preferences and choices of consumers. The advancement of digital marketing has fundamentally influenced customer buying behaviour and organized a customer-centric digital network. Further, most people prefer to shop through digital channels, and this study has shown that consumer

Table 40.1 Machine learning aids in customer loyalty

• Machine learning algorithms look at consumers' past buying patterns to forecast their future activities
• Machine learning reduces the cost of acquiring customers by giving reliable information about customers
• Machine learning analyses which platforms are more effective in offering products and services
• Churn model analyses which customers and with what possibility will stop using product or service
• Machine learning technology mechanically draws analysis assumptions from the data and then determines the best business actions
• Machine learning tools provide practical real-time recommendations and data for customers' online or offline visits

behaviour has been affected by digital marketing. The digital transformation of businesses has a significant impact on all sectors, especially marketing, where people are most volatile.

Apart from this, it has been found that many businesses are leveraging machine learning to examine large customer databases and identify customer loyalty. Machine learning has laid the basis for concluding customer behaviour patterns to drive value from data, forecast future purchase decisions, and generate a personalized customer experience. Machine learning tools will help businesses gain more reliable information and progress a successful customer engagement journey. Understanding customer loyalty includes customer behaviours, which are very helpful to the business to make strategies.

As technology progress and new solutions emerge, the developments in ML are forming the customer experience in the digital era. Furthermore, brands are focusing their efforts on enhancing the customer experience to encourage loyalty in a competitive environment. ML is one of the most acceptable ways to attain this through the analysis of historical and new customer data. With ML, marketers can mechanize the assortment of data and accomplish more detailed segmentation.

Marketers are facing tough challenges in attracting new customers as well as retaining existing ones. It is difficult for marketers who use traditional marketing methods to develop a brand amongst a market pool. Hence, ML helps marketers retain consumers for a more extended period of time. Here, digital transformation also plays an important role.

40.7 Future Scope

The study helps marketers decide how to present their products to put a maximum impact on consumers. Several directions for future research are looking forward to encouraging researchers to consider a wide range of phenomena to conduct this study. Due to time constraints, many different optimizations and experiments have been left for the future. For example, experiments with accurate data are often very time-consuming, requiring many days to complete the study. Future work deals with in-depth analysis of specific mechanisms, new proposals to try diverse approaches, or simply interest. This research presents the relationship between digital platforms influencing consumer sentiments. However, more work is needed on each of these selections: norms and the interrelationship between them. There are many other factors in the context of machine learning that could be used to advance this research. In addition, the relationship between consumer behaviour, satisfaction, and loyalty is uncertain, and additional research is still desired.

References

1. Aluri, A., Price, B.S., McIntyre, N.H.: Using machine learning to cocreate value through dynamic customer engagement in a brand loyalty program. J. Hosp. Tourism Res. **43**(1), 78–100 (2019)
2. Antons, D., Breidbach, C.F.: Big data, big insights? Advancing service innovation and design with machine learning. J. Serv. Res. **21**(1), 17–39 (2018)
3. Bala, M., Verma, D.: A critical review of digital marketing. Int. J. Manage. IT Eng. **8**(10), 321–339 (2018)
4. Ballestar, M.T., Grau-Carles, P., Sainz, J.: Predicting customer quality in e-commerce social networks: a machine learning approach. RMS **13**(3), 589–603 (2019)
5. Batat, W.: Experiential Marketing: Consumer Behavior, Customer Experience and the 7Es. Routledge (2019)
6. Dzyabura, D., Yoganarasimhan, H.: Machine learning and marketing. In: Handbook of Marketing Analytics. Edward Elgar Publishing (2018)
7. Galvão, M.B., de Carvalho, R.C., de Oliveira, L.A.B., de Medeiros, D.D.: Customer loyalty approach based on CRM for SMEs. J. Bus. Ind. Market. (2018)
8. How Does Digital Marketing Affect Consumer Behaviour?—Talentedge Learning Series. Retrieved 13 June 2021, from https://talentedge.com/blog/how-digitalmarketing-affect-consumer-behaviour/ (2021)
9. Jain, P., Pamula, R.: A systematic literature review on machine learning applications for consumer sentiment analysis using online reviews. Retrieved 13 June 2021, from https://arxiv.org/abs/2008.10282 (2021)
10. Kumar, A., Kabra, G., Mussada, E.K., Dash, M.K., Rana, P.S.: Combined artificial bee colony algorithm and machine learning techniques for prediction of online consumer repurchase intention. Neural Comput. Appl. **31**(2), 877–890 (2019)
11. Machine Learning Techniques for Predicting Customer Loyalty. Retrieved 13 June 2021, from https://heartbeat.fritz.ai/machine-learning-techniques-for-predicting-customer-loyalty-9fb090015bdb (2021)
12. Papazisi, M., Papazisi, M.: mi—QIVOS Blog. Retrieved 13 June 2021, from https://blog.qivos.com/the-vital-role-of-machine-learning-in-customer-loyalty/mi (2021)
13. Shinde, P.P., Oza, K.S., Kamat, R.K.: Big data predictive analysis: using R analytical tool. In: 2017 International Conference on I-SMAC (IoT in Social, Mobile, Analytics and Cloud) (I-SMAC), pp. 839–842 (2017). https://doi.org/10.1109/I-SMAC.2017.8058297
14. The Impact of Digital Marketing on Consumer Behaviour—Chain Reaction. Retrieved 13 June 2021, from https://www.chainreaction.ae/blog/the-impact-of-digital-marketing-on-consumer-behaviour/ (2021)

15. Thobani, S.: Improving e-Commerce Sales Using Machine Learning. Retrieved 13 June 2021 from https://dspace.mit.edu/handle/1721.1/118511 (2021)
16. Ways to Revitalize Customer Loyalty With Artificial Intelligence. Retrieved 13 June 2021, from https://dangingiss.com/customer-loyalty-artificial-intelligence/ (2021)

Chapter 41
Machine Learning for the Manufacturing of Digital Marketing Techniques and Its Impact on Health Care System

Prabhakara Rao Kapula, Amarendra Singh, Shaba Parveen Khan, Sonu Kumar, V. V. Devi Prasad Kotni, and Harikumar Pallathadka

Abstract The study explores among the most significant developments in the digital marketing sector has been the emergence of machine learning and artificial intelligence technologies to assist in the streamlining of marketing operations and the increased effectiveness of companies. The manner in which digital marketers collaborate with machine learning-based technologies will determine the direction of marketing in the next. As machine learning and artificial intelligence grow increasingly prevalent in the digital marketing environment, it is critical that the finest digital companies understand how to use machine learning into their digital marketing initiatives in order to remain competitive. As a healthcare organization, you are undoubtedly aware of how difficult it can be to keep up with the digital revolution in the healthcare industry. Selecting which new technologies are worthy the investment in and convincing your staff to embrace changes is frequently the most difficult aspect of the process. Look, adjusting to the digital world necessitates a change in perspective toward one that is more adaptable and threat. It entails letting go of antiquated corporate procedures and placing faith in the fact that disruptions will produce significant benefits.

P. R. Kapula
Department of ECE, B V Raju Institute of Technology, Narsapur, Telangana, India

A. Singh
Department of Computer Science & Engineering, Kamla Nehru Institute of Technology, Sultanpur, U.P., India

S. P. Khan (✉)
Computer Science and Engineering, Medicaps University, Indore, India
e-mail: ershaba@gmail.com

S. Kumar
Ignite, Bhumi, Chennai, India

V. V. D. P. Kotni
Department of Marketing, Gitam Institute of Management, Gitam Deemed-to-be University, Visakhapatnam, India

H. Pallathadka
Manipur International University, Imphal, Manipur, India

S. Yadav et al. (eds.), *Proceedings of Second International Conference in Mechanical and Energy Technology*, Smart Innovation, Systems and Technologies 290,
https://doi.org/10.1007/978-981-19-0108-9_41

41.1 Introduction

Machine Learning in digital marketing has given a fresh start to the healthcare sector by increasing the exposure of your hospitality sector in the online arena, as per the World Health Organization. Using digital marketing, you may connect out to sufferers and the healthcare seekers all over the globe, establish faith among your target community, and accelerate the expansion of your healthcare facilities. The use of machine learning in digital marketing services for your healthcare industry is becoming more important. When it comes to promoting your healthcare industry, now will be the time to adopt digital marketing methods if you haven't done so before. It is the procedure of advertising your services to potential patients via the use of digital marketing tools in the healthcare industry. Digital marketing may be a profitable investment choice for your digital medical providers if done correctly. In the U.S., about 88% of individuals search online for health information, and approximately 70% of people say that the health information they obtain affects their choice regarding treatment. Digital marketing is an excellent tool for keeping you in touch with existing and prospective patients anytime they are in need of your services. It is the beneficial effect of technologies on healthcare that has been shown by machine learning in digital marketing transformations.

Just few tangible instances of digital modernization in healthcare include machine learning (ML)-enabled diagnostic equipment and block chain-based electronic health records, which are totally trying to shape how we communicate with medical experts, how our information is communicated among suppliers, and how choices are taken regarding our treatment regimens and wellness results. Excellent internet and digital interactions are the focus of this project, with the ultimate aim of simplifying doctors' job while also optimizing processes and increasing outcomes for patients while also decreasing personal mistake and cutting costs. However, when it refers to adopting digital marketing tactics, the healthcare and pharmaceutical sectors have fallen beyond the rest of the world.

41.2 What is the Meaning of Machine Learning (ML)?

Both machine learning and artificial intelligence (AI) are distinct concepts that occur to be complementary to one another [1]. In contrast to artificial intelligence (AI), which seeks to control some elements of the "thinking" brain, machine learning (ML) is aimed at assisting people in solving issues more efficiently. As a component of artificial intelligence, machine learning (ML) utilizes information to teach itself how to finish a task with the assistance of AI characteristics. As a sophisticated technology that utilizes information to offer efficient answers to a wide range of difficult digital marketing issues, machine learning (ML) may benefit companies by assisting them in discovering secret information in accessible customer information and streamlining marketing operations [2].

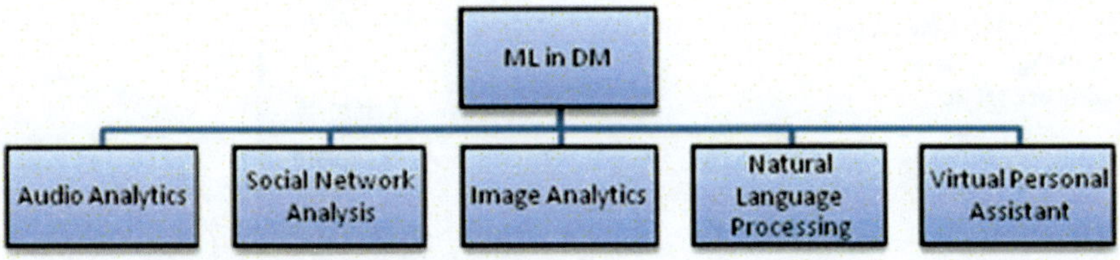

Fig. 41.1 Machine learning in digital marketing

41.3 Machine Learning in Digital Marketing

In what areas of digital marketing might machine learning be applied? There are many possibilities, including the ability to control and track social media, generate truly customized information, handle leads, ensure information sharing, and manage campaigns, to mention a few examples. Incorporating machine learning into digital marketing would open the door to new options such as making instantaneous content choices for particular consumers and real-time forecasts, similar to Uber's mentioned pricing strategy. Machine learning may also be used for real-time data intake by digital marketers, who could include performance information as it is streamed into the system. This performance information may then be utilized for real-time model development [3], which is currently not possible. As machine learning advances over period, continual algorithms assessment and optimization enables us to immediately react to pieces of information, brand identity, content, and any distinct trends that emerge among these (Fig. 41.1).

41.4 The Role of Machine Learning in Digital Marketing in Your Healthcare Industry

All around world, machine learning is being used in digital marketing organizations. Among its ramifications are the use of data, information, and online platforms to improve efficiency while also assisting digital entrepreneurs in better understanding their targeted audiences [4, 5]. These are a few instances of how machine learning (ML) may be integrated into your digital marketing methods to help you extend the development of your medical providers in the internet arena, as collected by the professionals at Smart Analytics (Fig. 41.2).

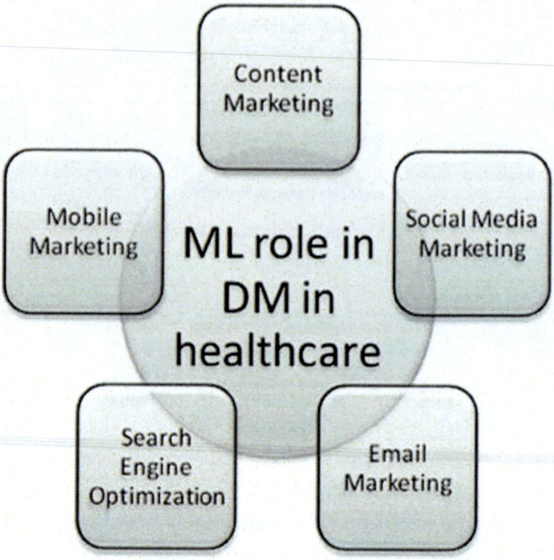

Fig. 41.2 Machine learning role in digital marketing in healthcare sector

41.4.1 Content Marketing

Content marketing, such as blogging, visualizations, videos, e-books, and so on, reveals your knowledge to the appropriate audience. It is simple to share your knowledge and connect with individuals when you have high-quality material. For example, if you are an accomplished dietitian, you may create blogs on subjects such as obesity in children, nutritious diet, exercise advice, and so on. Publishing papers on subjects relevant to your area allows you to offer the finest answer to your users while also showcasing your knowledge to your community [5].

Ensure your material focuses on delivering quality to your patients rather than being a self-promotional component. Additionally, by eliminating medical terminology in your material, you may make it more comprehensible to your viewers. We've compiled a few facts to demonstrate why content marketing is critical for your healthcare service [6].

- Healthcare content marketing generates three to four times as many leads as outbound advertising and expenses 62% below most marketing channels.
- Healthcare multimedia advertisers get 66% more lead generation each year.
- 83% of health companies use content marketing.
- Blog posts are used in the digital marketing of 58% of medical advertisers.

41.4.2 Social Media Marketing

Social networking is growing in the healthcare sector, and it's a wonderful way to remain engaged with your patients even when things aren't going well. Social media marketing allows you to target patients based on their age, gender, medical problems, health concerns, and so on [7]. For example, the Indian government created a Messenger App chat bot that successfully answers queries and leads people to the appropriate resources by combating disinformation. On either hand, Province or territory medical health officials have successfully utilized social media and medical communications. During the COVID-19 epidemic, they routinely provide formal data. As a healthcare professional with an engaging activity on social media, you may raise public awareness about keeping healthy throughout pandemics, offer advice for maintaining overall mental and physical health, and clarify patients' concerns via your knowledge, among other things. The figures shown below demonstrate the significance of social media in the healthcare sector [8].

- In the United States, 94.41% of healthcare facilities have a Facebook page, 50.82% use Twitter, 99.14% use Yelp, and 99.41% have a Foursquare profile.
- Sixty percent of social media consumers believe content made by health experts and physicians on social media.
- 83% of social media users have looked for wellness info online, such as psychological health, illness management, vaccinations, and so on.
- 91% of participants stated online social media groups help them make healthcare choices.

41.4.3 Email Marketing

Email marketing is yet another low-cost method to keep in contact with your audiences, offer a customized service, build a connection, and keep on your patients' thoughts. When you begin email marketing, you must categorize your email subscribers depending on characteristics such as age, interests, health problems, and so on. Customize your emails by including an attractive subject line, a call-to-action (CTA), and simple login choices.

Use email marketing management techniques to sending powerful emails on a regular basis with little effort. You can successfully recruit new healthcare seekers for your digital medical facility and maintain existing patients via email marketing. Here are some data that demonstrate the benefits of email marketing for your digital medical facility [9].

- Email is the best successful delivery method for fresh material for 74% of digital healthcare advertisers.
- Email is 40 times more successful in attracting new patients than Facebook and Twitter combined.

- Email is the main method of lead production for 89% of digital healthcare advertisers.
- The estimated return on investment for email healthcare marketing is $44.25 for every $1 invested.

It is obvious that digital marketing is critical to the expansion of your digital healthcare services. You may convert your customers into engaged participants in your healthcare facility by correctly utilizing these digital marketing channels.

41.4.4 Search Engine Optimization

Search engine optimization (SEO) and customer experiences (UX) are inextricably linked. If you become an effective care SEO expert, you must provide your patients the greatest customer experience possible. The user experience is determined by factors such as flexible layout, voice-search optimization, good content, stunning architecture, quick navigation, website load speed, and so on. SEO is a clever digital marketing strategy that improves your digital visibility. You can enhance your patients' customer experience, boost your presence in search engine results, and generate more inbound links to your healthcare website with the assistance of SEO professionals. We've compiled a few facts that demonstrate the significance of SEO for sustaining your digital medical facility [10].

- 77% of digital health searchers use Google, Bing, as well as other google search to get health-related data.
- Consumers looking for physical therapists, care facilities, and eye doctors are most likely to contact healthcare providers after doing a search.
- More than 70% of health-related searches click on first results page.
- 47% of web users look for information on physicians or other healthcare provider.

41.4.5 Mobile Marketing

Mobile marketing is a low-cost method for reaching healthcare aspirants on their cellphones, tablets, as well as other smart applications using clever and responsible marketing techniques. Several mobile marketing concepts that will completely suit your needs are software advertising, in-game mobile advertising, location-based advertising, mobile search advertising, mobile picture ads, and SMS messaging. You may focus local audiences with mobile marketing and urge consumers to utilize your services as soon as possible. Here are some facts that demonstrate the effect of mobile advertising on the development of your healthcare solutions [11].

- 44% of patients who do hospital research on a smart phone make an appointment.
- A mobile device is used to open 47.6% of healthcare sector emails.
- Around 19% of those polled had fitness and health applications on their cellphones.

41.5 The Impacts of Machine Learning on the Healthcare Sector

Several manufacturing industries have elegantly accepted machine learning digital marketing in recent years, and as a result, they are ascending the ladder of success at an alarming rate. Because it is a highly controlled sector, digital marketing in healthcare is a little slower than in other industries because it cannot take any new advertising trends at face value. Numbers, on the other hand, can never be disregarded. As per the results of the study, 72% of internet users have looked for wellness information available on the web in the last year. It is the third most common internet activity [12] behind shopping and socializing. Now that consumers are searching for answers online, even healthcare experts and different medical organizations have taken the initiative and begun developing strategies for digital marketing in hospitals and other healthcare facilities (Table 41.1).

41.6 The Future of Machine Learning in Healthcare Digital Marketing

In the late twentieth century, you could rely on your advertisements to be seen in newspapers, magazines, television or radio, and direct mail. The situation now is totally different, as we must adapt to a variety of new technology.

In Table 41.2. The following are some new contributions to the mixture of goods and services that may assist any doctor, physician, hospital, healthcare system, or pharmaceutical business in better targeting their customer [14].

41.7 Conclusion

An individual must be motivated, enthusiastic, and ready to respond to new professional environments in order to start a career in digital marketing. With an apparently limitless online material stream and information sources, the work of a digital marketer has evolved from that of a company storyteller to that of a technical manager. To simplify procedures and boost productivity, digital marketers—both current and future—must start using ML technologies to automate operations and make the best use of information. Digital marketing is a thriving business full of possibilities and

Table 41.1 The following are the general effects of digital marketing

Personalizations	With so much information accessible on the web, healthcare marketing firms are ensuring they not only connect out to individuals but also engage with them. The goal is to provide the appropriate messages to the appropriate individual at the appropriate moment about their health problems. They utilize customer information to not only determine the patient 's choices, but also to have the maximum effect on them. The availability of personal, social, behavioral, and economic information on the web has aided healthcare marketing firms in running a customized strategy. It not only directs inbound links and the appropriate patients to them, but it also aids in patient retention in the long term
Performances that have been analyzed	One benefit of the whole marketing campaign, as per healthcare services digital marketing organizations, is that you can evaluate your success right away. You no longer need to look for different benchmark to determine the performance and defeat of a marketing effort. With digital marketing, you can not only evaluate project success but also learn what works and what doesn't. With the assistance of digital marketing, the ROI of a marketing effort can be readily shown. Healthcare experts and organizations may now make educated decisions on how to improve their healthcare offerings. It is much simpler to manage your advertising spending when you have precise statistics. You may invest more funds in items that are successful and provide excellent outcomes while investing less budget in less efficient methods
Reaching	The earliest and most significant effect of medical marketing on the healthcare industry has been an expansion in the accessibility of healthcare practitioners and organizations. Because most individuals search for healthcare solutions on the web, physicians and hospitals are seeking out to them there. As per to one study, about 77% of individuals search digitally before scheduling a meeting with a healthcare practitioner [13]. Digital marketing has not only reduced the price of marketing, but it has also assisted physicians in reaching their target patients

difficulties, and it shows no signs of slowing down in the near future. In this topic we show the meaning of machine learning, after that we discuss about the machine learning in digital marketing. In this study we examine the role of machine learning in digital marketing in your healthcare industry. After in this topic we are discussing the impacts of machine learning on the healthcare sector. And last in this study we examine the future of machine learning in healthcare digital marketing.

Table 41.2 The following are some new contributions

1	Assistant for voices	We are all familiar with Siri in iPhone, which was the first popular voice assistant to be released. Google Assistant, which was introduced in May 201 6, is created to assist you to conduct internet searching more quickly [15]
2	Push notifications	These opt-in digital emails and messages for mobile applications are increasing rapidly in number across the globe, and they may assist marketers in increasing engagement, loyalty, and conversions. Opt-in rates from Engage in 2014 revealed that 46% of iPhone people choose to get push notifications 46% of the time, whereas 100% of Android people chose to receive push notifications when they downloaded an application. The most important step in increasing the number of new customers who opting in is to clearly describe the advantages of the data they will get
3	Automated marketing	This kind of services for healthcare advertising allows healthcare marketers to specifically target part of your community in order to communicate a critical message to that part of your crowd. For instance, you might send emails or a letter to males over the age of 50 who are currently patients at your clinic. All of these individuals are a colonoscopy possibility [16]
4	Local SEO	Regionalized organic marketing in healthcare has mostly overtaken Yellow Pages advertising. This search-based method of locating local companies is particularly useful for doctors and physicians. By initially adding your identity, location, and phone number data on your website, you may be discovered online using local healthcare SEO. You then connected it to any local web-based accounts, such as Google +, Bing Local, or Yahoo Local. Local SEO is much less expensive and much more popular among internet potential patients
5	Video streaming in real time	2 different healthcare is available and becoming more popular. As per the 2017 HIMSS Insights Telemedicine Survey, 71% of healthcare companies currently provide 2 different videos and webcams

(continued)

Table 41.2 (continued)

6	Virtual reality	Customers may utilize this innovation, similar to 360-degree films, to present prospective patients your office lobby, newest treatments, healthcare labs, and personnel. This is particularly useful for any private practice dealing with a hospital to demonstrate that they have the most up-to-date technology [17]
7	High-definition video tours and 360-degree images	Consider becoming capable of giving any potential patient a 360-degree tour of your office, complete with pictures and videos. This function was launched by Facebook in June 2016 for iPhone and Android Devices mobile phone users. Offices may use applications like Street View or Google Cam to create 360-degree pictures or movies, which can then be shared on Facebook
8	Ad blocking	In the 12 months leading up to June 2015, ad blocking increased by 48% on average, exceeding 45 million monthly customers. As a consequence, your healthcare marketing spent for online sponsored search advertising would be affected. Because more prospective patients are obstructing your online efforts, you must develop a media plan that combines offline and online advertising

References

1. Cauvin, B., Benning, P.: Machine learning. Int. J. 3-D Inf. Model. **6**(3), 1–16 (2017). https://doi.org/10.4018/ij3dim.2017070101
2. Jain, A.Y, Shrivastava, Y.: Modelling and optimization of different quality characteristics in electric discharge drilling of titanium alloy sheet. Mater. Today: Proc. **21**, 1680–1684 (2020). https://doi.org/10.1016/j.matpr.2019.12.010
3. Miklosik, A., Kuchta, M., Evans, N., Zak, S.: Towards the adoption of machine learning-based analytical tools in digital marketing. IEEE Access **7**, 85705–85718 (2019). https://doi.org/10.1109/access.2019.2924425
4. Panwar, V., Kumar Sharma, D., Pradeep Kumar, K., Jain, A., Thakar, C.: Experimental investigations and optimization of surface roughness in turning of en 36 alloy steel using response surface methodology and genetic algorithm. Mater. Today: Proc. (2021). https://doi.org/10.1016/j.matpr.2021.03.642
5. Pikuła-Małachowska, J.: Content marketing as a modern tool of market communication. Marketing i Zarządzanie **47**, 291–299 (2017). https://doi.org/10.18276/miz.2017.47-27
6. Rowley, J.: Understanding digital content marketing. J. Mark. Manage. **24**(5–6), 517–540 (2008). https://doi.org/10.1362/026725708x325977
7. Social media: the future of marketing. Int. J. Contemp. Res. Rev. (2017). https://doi.org/10.15520/ijcrr/2017/8/07/268
8. Alves, H., Fernandes, C., Raposo, M.: Social media marketing: a literature review and implications. Psychol. Market. 33(12):1029–1038 (2016). https://doi.org/10.1002/mar.20936

9. Hartemo, M.: Email marketing in the era of the empowered consumer. J. Res. Interact. Mark. **10**(3), 212–230 (2016). https://doi.org/10.1108/jrim-06-2015-0040
10. Berman, R., Katona, Z.: The role of search engine optimization in search marketing. SSRN Electron. J. (2011). https://doi.org/10.2139/ssrn.1745644
11. B.S.: Marketing, mobile and future of mobile advertisement: life changing through mobile. Int. J. Psychosoc. Rehabil. **23**(4), 1623–1630 (2019). https://doi.org/10.37200/ijpr/v23i4/pr190487
12. Miklosik, A., Evans, N.: Impact of big data and machine learning on digital transformation in marketing: a literature review. IEEE Access **8**, 101284–101292 (2020). https://doi.org/10.1109/access.2020.2998754
13. Jain, A., Pandey, A.: Multiple quality optimizations in electrical discharge drilling of mild steel sheet. Mater. Today: Proc. **4**(8), 7252–7261 (2017). https://doi.org/10.1016/j.matpr.2017.07.054
14. Jain, A., Kumar Pandey, A.: Modeling and optimizing of different quality characteristics in electrical discharge drilling of titanium alloy (grade-5) sheet. Mater. Today: Proc. **18**, 182–191 (2019). https://doi.org/10.1016/j.matpr.2019.06.292.
15. Patil, A., Samant, S., Ramtekkar, M., Ragaji, S., Khanapuri, J.: Intelligent voice assistant. SSRN Electron. J. (2020). https://doi.org/10.2139/ssrn.3568721
16. Goeldner, C.: Automation in marketing. J. Mark. **2**(1), 53 (1962). https://doi.org/10.2307/1249632
17. Aylett, R., Louchart, S.: Towards a narrative theory of virtual reality. Virtual Real. **7**(1), 2–9 (2003). https://doi.org/10.1007/s10055-003-0114-9

Chapter 42
Advancements in Controlled Atmosphere Storage Technology—A Review

Md Ozair Arshad, Yash Chauhan, Prateek Singh, Prakhar Srivastav, Manoj Gupta, and Narayan Patwa

Abstract Controlled-atmosphere storage (CAS) is one of the most important breakthroughs in post-harvest technology, as the composition of the gas in the storage has an impact on the shelf life of the stored product. The Control atmosphere storage is used to prevent harvested fruits and vegetables from perishing during storage, transport, and distribution. This article presents a brief overview of the current CAS technology and also reviews at how new post-harvest methods are being utilised and indicates areas where more research is needed. The article identifies Dynamic Controlled Atmosphere (DCA) storage and Synthetic Plant Growth Regulator (SPGR) integration as two most promising field in the domain of CAS. The article moves on with the overview, advantages, and limitations of said technologies. DCA is shown to prevent various physiological disorder as well as maintain physio-chemical properties such as firmness of the product. The high chances of low O_2 and CO_2 injuries continue to be concerning and warrant further research. As for the integration of SPGR, the biggest hurdle is the market perception of chemical-based preservation method which is proven beneficial, but in today's organic centric market its use is not widely accepted. Improvement in the currently used SPGR, 1-methylcyclopropene as well as development of alternatives such as silver thiosulfate can improve market perception.

42.1 Introduction

One of the pressing issues today is maintaining food security for the world's expanding population whilst guaranteeing sustainable development for the long term. According to the Food and Agriculture Organization (FAO), food production will need to increase by ~70% by 2050 to feed 9+ billion people. A worldwide endeavour

M. O. Arshad (✉) · Y. Chauhan · P. Singh · P. Srivastav · M. Gupta · N. Patwa
Department of Mechanical Engineering, Galgotias College of Engineering & Technology, Uttar Pradesh, Greater Noida 201310, India
e-mail: ozairarshad4jmi@gmail.com

S. Yadav et al. (eds.), *Proceedings of Second International Conference in Mechanical and Energy Technology*, Smart Innovation, Systems and Technologies 290,
https://doi.org/10.1007/978-981-19-0108-9_42

to ensure sustainable food production and consumption needs an integrated and creative strategy [1, 2].

The most successful technology brought to the food industry in the twentieth century is controlled atmosphere (CA) storage. When refrigeration is insufficient to increase the shelf life of seasonal perishable produce, CA storage is employed. To preserve fruit for a prolonged time, the fruit's natural ripening must be delayed without compromising the eating quality. First, the temperature of the fruit is lowered to the lowest level possible without causing damage due to freezing or low-temperature breakdown. To further delay ripening, the atmosphere in the storage chamber is altered by lowering oxygen levels and allowing carbon dioxide levels to rise. The widespread usage of CA storage these days is due to its unmatched advantages over refrigeration. It can help prevent certain storage disorders and diseases in fruits and vegetables [3]. Temperature, oxygen, and carbon dioxide levels required to maximise storage life and minimise storage diseases vary. Produce, cultivars, growing conditions, maturity, and post-harvest treatments all influence them [4]. Controlled atmosphere storage technology is undergoing a rapid change in order to, include the vary many horticultural produces, increase shelf life of stored commodity and retain quality of the produce. Dynamic Controlled atmosphere storages continuously monitor the storage environment according to the commodity physiology through sensors and then process the data in a control system to adjust atmosphere in the storage room [5]. It enables to store Apples in the highest quality without the use of various scald controlling chemicals [6]. Synthetic plant growth regulators like 1-methylcyclopropene and silver thiosulphate help in controlling ethylene production. Ethylene is a plant hormone which causes ripening of the fruit. Delaying ethylene production leads to a prolonged shelf life [7].

42.2 Controlled atmosphere—An Overview

The technology for storing diverse food items after harvest has advanced significantly over time. There has been a concerted effort in the last two decades to minimise the use of chemical treatments such as diphenylamine (DPA) and 1-methylcyclopropene (1-MCP) in the post-harvest treatment of fruits and vegetables in order to control physiological problems and increase the shelf life of the product. These compounds inhibit the synthesis of ethylene as well as the creation of other volatiles such as α-farne-sene and conjugated trienols (CTols). These volatiles degrade the stored commodity's quality. However, new research has suggested that DPA is carcinogenic to humans and hazardous to aquatic life. [8, 9].

Controlled atmosphere is a storage method for extending shelf life of produce whilst retaining fruit qualities like, firmness, colour, and texture. The senescence of fruits and vegetables can be pushed by slowing down the rate of respiration and by maintaining a high humidity environment around the horticultural produce to prevent huge moisture loss. Fruits and vegetables naturally undergo the chemical oxidation of sugars to liberate carbon dioxide, water, and heat. The rate at which this chemical change occurs inside the produce is known as respiration rate. The respiration rate is influenced by factors like temperature and the concentration of gases in the surrounding environment. Similarly, high moisture loss, i.e., high transpiration rate, is also affected by temperature and relative humidity around the horticultural produce. To provide such a storage environment, CA storage is generally employed. CA storages are capable of checking the rate of respiration and transpiration by controlling the environmental variables such as temperature, concentration of CO_2 and O_2 and relative humidity (RH). The various control parameters are:

Carbon Dioxide (CO_2) The effect of a high CO_2 atmosphere on the respiration rate of horticultural produce varies with the crop, cultivar, and the stage of development of horticultural produce [10]. During respiration the CO_2 is produced by the product stored. Ripening of fruit is generally delayed by concentrations greater than 1% in the storage [11]. Increased CO_2 level also checks the growth of microorganisms deemed harmful to our product [12].

Oxygen (O_2) Oxygen supports respiration and other metabolic reactions. Any reduction in oxygen concentration leads to a reduction in the rate of respiration. Ripening is observed to slow down when the O_2 concentration decreases below 8% [13]. Reduced production of ethylene is identified as the cause of slowed ripening when the O_2 level is lowered. Ethylene production gets reduced to half at oxygen levels of around 2.5% [14, 15].

Ethylene (C_2H_4) All plant tissues and microorganisms produce ethylene, which is a natural plant hormone. Ethylene is the hormone that begins the process of ripening in the fruit [16]. In climacteric fruits, a low concentration of 0.1–1.0 μl is enough to start the ripening process. The production of ethylene accelerates the process of ripening in climacteric fruits [17]. The ethylene synthesis can be repressed by keeping low oxygen and high CO_2 concentrations [18].

Temperature Temperature affects the rate of metabolic reactions. As the temperature increases, the metabolic reaction also increases. Since the increased metabolic rate leads to undesirable faster ripening, keeping the temperature low is suggested. To delay the ripening rate, the temperature must be lowered but not below critical storage temperature, i.e., temperature below which Chilling Injury starts occurring in the fruit [19, 20]. Increasing the temperature leads to increased transpiration rate and a simultaneous decrease in moisture content, i.e., relative humidity [21]. Conversely, a decrease in temperature causes a sedentary enzymatic activity, leading to curtailment in ethylene production [22, 23].

Relative Humidity (RH) The generally recommended relative humidity level of 85% to 95% stores fresh produce. It is a trade-off that prevents excessive weight loss whilst providing some degree of microbial spoilage control. For fresh produce preservation, the commonly suggested RH levels of 85–95% [24] constitute a compromise between preventing severe weight loss and controlling microbial deterioration. After exposure to a certain RH, any material's moisture content approaches an equilibrium level by gaining or losing water. However, depending on the system, equilibrium can take days or even weeks to achieve [25].

42.3 Development in Atmosphere Control Techniques

42.3.1 Carbon Dioxide (CO_2) Scrubbing

Scrubbing is the process of separating a particular gas from the room atmosphere. A wide variety of scrubber systems are commercially available for scrubbing CO_2 out of the CA storage. Different scrubbers use different reagents to extract CO_2. An aqueous solution of Sodium Hydroxide (NaOH) was excessively used in earlier designs. The solution was made to pass through open tubes sequestering CO_2 from the atmosphere [26]. Later, hydrated lime gained much acceptance due to the risk involved in the handling and disposal of sodium hydroxide and its corrosive nature [27]. In the hydrated lime method, several lime sacs are directly placed inside the storage room.

Activated charcoal can also be used to adsorb CO_2 in scrubber systems. Another type of scrubber is a molecular sieve scrubber (MSS) which uses zeolites (porous material) to extract CO_2. Some materials that have been used in MSS are sodium and Aluminium silicate zeolites. However, the application of MSS is limited due to the high energy requirement for regeneration.

Besides using scrubbers, CO_2 can also be removed by flushing N_2 inside the storage chamber [28]. The advantage of this method is that it can be used to control both CO_2 and O_2 concentration in the chamber independently. Another method of removing CO_2 involves using semi-permeable membranes of silicon rubber placed inside a gas diffusion panel [29].

42.3.2 Oxygen (O_2) Scrubbing

The most straightforward approach to reduce the O_2 concentration is to let the natural respiration of the post-harvest produce take place in the initial stages of storage, which would use up the oxygen from the chamber [30, 31]. Due to the significant amount of time required for natural respiration to reduce the O_2 concentration to the desired level, propane burners were used to consume the O_2 in the chamber. Thus, resulting in reduced O_2 levels by burning propane. But since the combustion of propane gives out Ethylene (C_2H_4) as a byproduct, whose higher concentration is undesirable in the CA storage chamber, such 'open flame' systems are no longer used in CA storage [32].

Another way to remove O_2 from the chamber, which does not include combustion and hence doesn't risk accumulating unwanted byproducts, is to flush Nitrogen

(N_2) gas in the storage chamber. Generally, a generator produces N_2 gas on-site in commercial storage systems by segregating N_2 and O_2 from the air [31, 33].

These generators are available in two different variants (1) hollow fibre membrane (HFM) generator (2) pressure swing adsorber (PSA). In an HFM system, a selectively permeable membrane allows O_2 and CO_2 to diffuse faster than N_2. Therefore, a higher concentration of N_2 is achieved continuously. In a PSA system, molecular sieves extract O_2 and CO_2 from air passing through it, which creates a system that can filter out N_2 [31].

42.3.3 Ethylene (C_2H_4) Scrubbing

Ethylene is the gas generated by the produce (post-harvest cultivar), responsible for ripening of fruits and vegetables. A higher than a desired ethylene concentration in the storage chamber may cause unwanted ripening or several physiological disorders. Hence, in controlled atmosphere storage, a kind of ethylene scrubbing system is essential. There are two types of C_2H_4 scrubbing systems that are used in commercial CA storages: (1) Oxidation type and (2) Absorbing bead type [34]. In the oxidation type C_2H_4 scrubber, several techniques can be used to oxidise the C_2H_4 from the storage chamber, thus reducing its concentration to a lower level. Two of the widely used techniques are ultraviolet radiation and ozone gas techniques, respectively. Whilst in a C_2H_4 absorbing type scrubber, tiny beads made out of porous material like zeolite and silica gel infused with potassium permanganate extract C_2H_4 from the atmosphere inside the storage chamber [35].

42.4 Dynamic Controlled Atmosphere (DCA) Storage

The conventional (or static) CA storage reduces the ethylene biosynthesis and respiration rate to delay the ripening [36]. However, disorders such as scald and greasiness have been observed under low O_2 conditions [37]. Low oxygen damage and a high incidence of off-flavours in fruits were also reported [36, 38]. Attempts were made to overcome these shortcomings by introducing dynamically controlled atmosphere (DCA) technology [36, 39].

A dynamically controlled atmosphere (DCA) works precisely like a conventional CAS. However, the only added feature is that it uses sensors that monitor the real-time gas concentration inside the storage chamber. Hence, it can actively alter the concentration levels during storage depending on the sensor readings. This additional ability in the system allows the storage to react to dynamically changing conditions which prevent physiological disorders from occurring in the product stored.

42.4.1 Types of DCA Systems

The DCA systems are categorised based on the sensor they use for monitoring oxygen concentration. Currently, three types of sensors are used in DCA technology; (1) chlorophyll Fluorescence (CF) sensor, (2) Respiration Quotient (RQ) sensor, and (3) Ethanol (ET) sensor. Hence, there are three corresponding types of DCA systems available: (1) DCA-CF, (2) DCA-RQ, (3) DCA-ET. However, very little research has been done on DCA-RQ and DCA-ET. As a result, DCA-CF is the most commonly used system in the fruit industry.

Chlorophyll Fluorescence (CF) Sensor When storing a cultivar in any storage system, the most crucial factor is the health of the cultivar itself. Since the environment in CA storage is drastically different from the outside, health monitoring plays a vital role. The general method to monitor the product's health inside the storage chamber is to measure chlorophyll fluorescence (CF). CF can be measured by well-tested methods without any harmful effects on the produce [40].

Respiration Quotient (RQ) Sensor Respiration Quotient sensors are deployed to monitor the stored produce's respiration process continuously. It is used to ensure that the product won't ever reach the point of anaerobic respiration throughout the storage. The ratio of CO_2 produced to O_2 consumed is computed. This ratio is called the Respiration Coefficient ($RQ = CO_2$ produced/O_2 consumed). Whenever the O_2 concentration drops, this increases the value of RQ, through which the gas regime can be monitored and controlled [41, 42].

When a product is partially exposed to sun, the exposed parts exhibit higher metabolic activity depending on the canopy orientation. Since CF depends on metabolic activity. CF can differ within the particular produce depending on the area from where we measure CF. This leads to an error in estimating the low oxygen limit, which diminishes product quality and shelf life. DCA-RQ was developed to overcome this shortcoming of the DCA-CF [43, 44].

Ethanol (ET) Sensor Ethanol sensors were developed as an alternative method to measure the lowest oxygen level for cultivars with low natural Chlorophyll Fluorescence. Low oxygen levels lead to anaerobic respiration, which leads to ethanol production in the produce. DCA-ET continuously measures ethanol production, the level of oxygen is decreased to the point where the ethanol growth reaches a threshold value set for maintaining food quality [36].

42.4.2 Advantages of DCA Over Conventional CA

Less Physiological Disorders Studies have demonstrated that, unlike other storage technologies that must be complemented with chemical treatments, DCA can be used as a 100% non-chemical post-harvest treatment. And due to active control of the gas concentration DCA have also been proven to prevent various physiological disorders such as superficial scald [45, 46], Internal browning [39, 47], and Coreflush [48].

Enhanced Physio-Chemical Quality Fruit firmness is shown to be maintained by DCA technology throughout long-term cold storage. DeLong et al. [46] found that DCA stored 'Delicious' apples had consistently higher fruit firmness than CA stored fruit and better overall fruit quality. Similarly, research on 'Golden Delicious' [45] and 'Granny Smith' apples [41, 49–51] found that DCA storage retained firmness better than CA or regular air storage. Fruit stored in DCA had superior firmness than fruit held in static CA, according to [52–54].

42.4.3 *Limitation of DCA Storages*

Whilst DCA is relatively successful at managing physiological diseases, research has shown that DCA stored apples have a lot of low O_2 and serious CO_2 damage. For example, DeLong et al. [46] found that 'Delicious' apples kept in DCA had a greater incidence of low O_2 injury characterised by purpling of the skin than those stored in conventional CA. Another common disorder in DCA kept apples is CO_2 injury, which manifests as internal browning disorder and is accompanied by cavity formation [45, 47]. External harm caused by the condition can be evident as partly sunken rough brown lesions in the skin of the infected fruit in extreme situations.

42.5 Integration of Synthetic Plant Growth Regulators (SPGRs)

Synthetic Plant Growth Regulators (SPGR) are compounds that can be used to regulate/control the production of various plant hormones, and since ethylene is one of the hormones synthesised in the produce, a certain SPGR that can inhibit the production of ethylene can be used to control its concentration in the chamber. One such SPGR is 1-methyl cyclopropane (1-MCP). Physiological processes like ripening and senescence directly influenced by ethylene production can be delayed using 1-MCP [7, 55–59].

42.5.1 *Types of SPGR Used in CA Storages*

Two SPGR are mainly used in CA Storages (1) 1-MCP and (2) Silver Thiosulfate ($Ag_2S_2O_3$). Out of which, only 1-MCP is commercially used (Marketed as SmartFreshTM). In comparison, Silver Thiosulfate is still being researched extensively.

42.5.2 *Advantages of Using SPGR*

Less Physiological Disorders In a study on d'Anjou pears (a variety of pear) stored in CA after treatment with 1-MCP, it was seen that the treatment not only resulted in reduced cases of physiological disorders like superficial scald and decay, but the pears were seen to have more firmness, Chlorophyll content, and better colour profile compared with produce stored without 1-MCP treatment [60].

Less Electric Energy Consumption Additionally, in a study conducted on 'galaxy' apples, it was observed that after treating the produce with 1-MCP, the product could be stored at higher temperatures whilst having the desired characteristic, which leads to a reduction in the electric energy consumption and a more efficient CA storage system [61].

Improved Shelf Life and Physio-Chemical Properties A recent study on Hayward and Qihong 'KiwiFruit' showed that the room temperature shelf life was increased by an additional eight days for Hayward and ten days for Qihong. Along with this, the 1-MCP significantly delayed loss of firmness and did not affect the solid soluble content of the fruit [62]. Similar results were obtained for Braeburn apples treated with 1-MCP in another recent study [63].

42.5.3 *Limitation of SPGR*

A high dose of 1-MCP caused atypical ripen with a softer and greener peel than the control fruit, according to a study on banana fruit [64] which suggested that the use of SPGR whilst prevents common physiological disorders can also induce peculiar ripening characteristics in the product. Another study on 'galaxy' apples indicated that using 1-MCP reduced volatile chemical buildup in various storage settings, particularly ester formation [65], which affected the fruity smell of the product that can damage the marketability of the fruit. And overall, the current combinational use of 1-MCP and CA is expensive and leads to rejection from highly lucrative organic fresh produce markets.

42.6 Conclusion and Future Prospects

A decade of research proves that the CA Storage technology is not only an academic interest but is an essential tool in today's food processing economy. Hence, further development in the field is just a question of time instead of motive. Whilst the primary themes in the field are highly developed, such as the effect of varying gas (O_2, CO_2 or N_2) concentration, the introduction of new techniques and technology like DCA and SPGR act as an incentive to attract new researchers into this domain. The DCA technology has seen significant advancement over time, yet there are still challenges associated with this technology. Analysis of this review shows that highly underdeveloped technologies like DCA-RQ and DCA-ET might get investigated extensively in the coming time. Also, the development of O_2 and CO_2 injuries in DCA stored fruit remains a big concern. Thus, future research aimed at refining DCA technology to ensure that such injuries are avoided is warranted. Although the influence of DCA on fragrance and flavour volatiles has been investigated, further research into instrumental analysis and consumer sensory perception of DCA stored fruit is required. As for the SPGR integration in conventional CA systems, the use of 1-MCP is relatively new in the market and hasn't yet been widely accepted due to sheer ambiguity. The trend of using organic products has also provided hurdles in the acceptance of SPGR. At the same time, alternatives for 1-MCP like silver thiosulfate are still under investigation. Further research is instrumental to the growth of this sub-domain in the field of CA technology.

References

1. Nellemann, C., MacDevette, M., Manders, T., Eickhout, B., Svihus, B., Prins, A.G., Kaltenborn, B.P. (eds).: The environmental food crisis—The environment's role in averting future food crisis. UNEP, Nairobi (2009)
2. FAO, Global food losses and waste: extent, causes and prevention (2011)
3. Prange, R.K., DeLong, J.M., Daniels-Lake, B.J., Harrison, P.A.: Innovation in controlled atmosphere technology. Stewart Postharvest Rev. **3**(9), 1–11 (2005)
4. Thompson, A.K.: Postharvest Technology of Fruit and Vegetables. Blackwell Science, Oxford (1996)
5. Gasser, F., Eppler, T., Naunheim, W., Gabioud, S., Höhn, E.: Control of critical oxygen level during dynamic CA storage of apples. Agrarforschung **15**, 98–103 (2008)
6. DeLong, J.M., Prange, R.K., Leyte, J.C., Harrison, P.A.: A new technology that determines low oxygen thresholds in controlled-atmosphere-stored apples. Hort. Technol. **14**, 262–266 (2004)
7. Watkins, C.B.: The use of 1-methylcyclopropene (1-MCP) on fruits and vegetables. Biotechnol. Adv. **24**, 389–409 (2006)
8. Chervin, C., Raynal, J., Andre, N., Bonneau, A., Westercamp, P.: Combining controlled atmosphere storage and ethanol vapors to control superficial scald of apple. HortScience **36**(5), 951–952 (2001)
9. Watkins, C.B., Nock, J.F., Whitaker, B.D.: Responses of early, mid and late season apple cultivars to post-harvest application of 1-methylcyclopropene (1-MCP) under air and controlled atmosphere storage conditions. Postharvest Biol. Technol. **19**(1), 17–32 (2000)
10. Rama, M., & Narasimham, P.: Controlled-atmosphere storage. In: Effects on Fruit and Vegetables (2003)
11. Watada, A.E., Ko, N.P., Minott, D.A.: Factors affecting quality of fresh-cut horticultural products. Postharvest Biol. Technol. **9**, 115–125 (1996)
12. Bennik, M.H., Vorstman, W, Smid, E.J., Gorris, L.G.: The influence of oxygen and carbon dioxide on the growth of prevalent Enterobacteriaceae and Pseudomonasspecies isolated from fresh and controlled-atmosphere-stored vegetables. Food Microbiol. ISSN 0740-0020 (1998)
13. Sandhya, K.V.K.: Modified atmosphere packaging of fresh produce: current status and future needs. LWT Food Sc. Technol. **43**, 381–392 (2010)

14. Satyan, S.H., Scott, K.J., Graham, D.: Storage of banana bunches in sealed polyethylene tubes. J. Horticult. Sci. **67**, 283–287 (1992)
15. Ščetar, M., Kurek, M., Galić, K.: Trends in fruit and vegetable packaging—A review. Croat. J. Food Technol. Biotechnol. Nutrit. **5**(3–4), 69–86 (2010)
16. Watkins, C.B.: Principles and practices of post-harvest handling and stress In: Ferree, D.C., Warrington, I.J. (eds.) Apples: botany, production and uses, pp. 585–614. CAB Publishing. UK (2003)
17. Kader, A.A.: Influence of preharvest and post-harvest environment on nutritional composition of fruits and vegetables. In: Quebedeaux, B., F.A. Bliss (eds.) Horticulture and human health, contributions of fruits and vegetables, pp. 18–32. Prentice-Hall, Englewood Cliffs, NJ, USA (1988)
18. Kader, A.A.: Biochemical and physiological basis for effects of controlled and modified atmospheres on fruits and vegetables. Food Technol. **40**(5), 99–104 (1986)
19. Kader, A.A.: Post-harvest biology and technology: an overview. In: Kader, A.A. (eds.) Postharvest Technology of Horticultural Crop, publication number 3311, pp. 39–48. Regents of the University of California, Oakland, CA, USA, pp. 39–48 (2002)
20. Thompson, J.F., Mitchell, F.G., Kasmire R.F.: Cooling horticultural commodities. In: Kader A.A. (eds.) Postharvest Technology of Horticultural Crops, 3rd edition, publication number 3311, pp. 97–112. Regents of the University of California, Division of Agricultural and Natural Resources, Oakland, CA, USA (2002)
21. Kang, J.S., Lee, S.: A kinetic model for transpiration of fresh produce in a controlled atmosphere. J. Food Eng. **35**, 65–73 (1998)
22. Larrigaudiere, C., Graell, J., Salas, J., Vendrell, M.: Cultivar differences in the influence of a short period of cold storage on ethylene biosynthesis in apples. Postharvest Biol. Technol. **10**, 21–27 (1997)
23. Zhou, H.W., Dong, L., Ben-Arie, R., Lurie, S.: The role of ethylene in the prevention of chilling injury in nectarines. J. Plant Physiol. **158**, 55–61 (2001)
24. Hardenburg, R.E., Watada, AE, Wang, C.Y.: The commercial storage of fruits, vegetables, and florist and nursery stocks. United States Department of Agriculture Handbook, vol. 66, Washington, DC (1986)
25. Labuza, T.P.: Application of chemical kinetics to deterioration of foods. J. Chem. Educ. **61**(4), 348 (1984)
26. Raghavan, G.S.V., Vigneault, C., Gariépy, Y., Markarian, N.R., Alvo, P.: Refrigerated and controlled/modified atmosphere storage. In: Barett, D.M., Somogy, L., Ramaswamy, H. (eds.) Processing Fruits, pp. 23–52. CRC Press, Boca Raton, FL (2005)
27. Osterloh, A.: Lagervervahren and Planung, Bau und Ausrüstung von Obstlägern. In: Osterloh, A., Ebert, G., Held, W.-H., Schulz, H., Urban, E. (eds.) Lagerung von Obst und Südfrüchten, pp. 113–141. Verlag Eugen Ulmer, Stuttgart, Germany (1996)
28. Bishop, D.: Controlled atmosphere storage: a practical guide. David Bishop Design Consultants, Heathfield, East Sussex, England (1996)
29. Marcellin, P., Leteinturier, J.: Premières applications industrielles des membranes en caoutchouc de silicone à l'entreposage des pommes en atmosphere contrôlée. Inst Int. Froid. Congr. Intern. Froid. Madrid, Espagne, pp. 1–9 (1967)
30. Allen, D.: Controlled atmosphere storage buildings: construction and operational techniques. Nova Scotia Fruit Growers' Association, Agricultural Center, Kentville, Canada (1998).
31. Malcolm, G.I.: Advancements in the implementation of CA technology for storage of perishable commodities. Acta Hort. **682**, 1593–1597 (2005)
32. Coquinot, J.P., Chapon, J.F.: Entreposage frigorifique des pommes et des poires. Tome 1 EquipementCtifl, Paris, France (1992)
33. Dilley, D.R.: Development of controlled atmosphere storage technologies. Stewart Postharvest Rev. **6**(5), 1–8 (2006)
34. Wojciechowski, J.: Ethylene removal from gases by means of catalytic combustion. Acta Hort. **258**, 131–141 (1989)

35. Wills, R.B.H., Warton, M.A., Ku, V.V.V.: Ethylene levels associated with fruit and vegetables during marketing. Austr. J. Exp. Agric. **40**, 465–470 (2000)
36. Wright, A., Delong, J., Arul, J., Prange, R.: The trend toward lower oxygen levels during apple (Malus×domestica Borkh) storage. J. Hortic. Sci. Biotechnol. **90**(1), 1–13 (2015)
37. DeLong, J.M., Prange, R.K., Harrison, P.A.: The influence of 1-methylcyclopropene on 'Cortland' and 'McIntosh' apple quality following long-term storage. HortScience **39**(5), 1062–1065 (2004)
38. Imahori, Y., Uemura, K., Kishioka, I., Fujiwara, H., Tulio, A., Ueda, Y., Chachin, K.: Relationship between low-oxygen injury and ethanol metabolism in various fruits and vegetables. Acta Hort. **682**, 1103–1108 (2005)
39. Weber, A., Brackmann, A., Both, V., Pavanello, E.P., Anese, R.d.O., Thewes, F.R.: Respiratory quotient: Innovative method for monitoring Royal Gala apple storage in a dynamic controlled atmosphere. Scientia Agricola **72**(1), 28–33 (2015)
40. Prange, R.K., DeLong, J.M., Harrison, P.A., Leyte, J.C., McLean, S.D.: Oxygen concentration affects chlorophyll fluorescence in chlorophyll-containing fruit and vegetables. J. Am. Soc. Hortic. Sci. **128**(4), 603–607 (2003)
41. Bessemans, N., Verboven, P., Verlinden, B., Nicolaï, B.: A novel type of dynamic controlled atmosphere storage based on the respiratory quotient (RQ-DCA). Postharvest Biol. Technol. **115**, 91–102 (2016)
42. Gabioud, S., Bozzi Nising, A., Gasser, F., Eppler, T., Naunheim, W.: Dynamic CA storage of apples: monitoring of the critical oxygen concentration and adjustment of optimum conditions during oxygen reduction. X International Controlled and Modified Atmosphere Research Conference, vol 876 (2009)
43. Feng, F., Li, M., Ma, F., Cheng, L.: Effects of location within the tree canopy on carbohydrates, organic acids, amino acids and phenolic compounds in the fruit peel and flesh from three apple (malus × domestica) cultivars. Horticulture Res. (2014)
44. Sun, S., Xin, L., Gao, H., Wang, J., Li, P.: Response of phenolic compounds in 'Golden delicious' and 'Red delicious' apples peel to fruit bagging and subsequent sunlight re-exposure. Sci. Hortic. **168**, 161–167 (2014)
45. Zanella, A., Cazzanelli, P., Panarese, A., Coser, M., Chistè, C., Zeni, F.: Fruit fluorescence response to low oxygen stress: Modern storage technologies compared to 1-MCP treatment of apple. Acta Hort. **682**, 1535–1542 (2005)
46. DeLong, J.M., Prange, R., Harrison, P.: Chlorophyll fluorescence-based low-O_2 CA storage of organic 'Cortland' and 'Delicious' apples. Acta Hort **737**, 31–37 (2007)
47. Lafer, G.: Storability and fruit quality of 'Braeburn' apples as affected by harvest date, 1-MCP treatment and different storage conditions. In: International Conference on Ripening Regulation and Post-Harvest Fruit Quality, vol 796, 1, pp. 79–184 (2007)
48. Köpcke, D.: 1-methylcyclopropene (1-MCP) and dynamic controlled atmosphere (DCA) applications under elevated storage temperatures: Effects on fruit quality of 'Elstar', 'Jonagold' and 'Gloster' apple (Malus domestica borkh.). Eur. J. Horticult. Sci. **80**, 25–32 (2015)
49. Mditshwa, A., Fawole, O.A., Vries, F., van der Merwe, K., Crouch, E., Opara, U.L.: Minimum exposure period for dynamic controlled atmospheres to control superficial scald in Granny Smith apples for long distance supply chains. Postharvest Biol. Technol. **127**, 27–34 (2017)
50. Mditshwa, A., Fawole, O.A., Vries, F., van der Merwe, K., Crouch, E., Opara, U.L.: Repeated application of dynamic controlled atmospheres reduced superficial scald incidence in Granny Smith' apples. Sci. Hortic. **220**, 168–175 (2017)
51. Brizzolara, S., Santucci, C., Tenori, L., Hertog, M., Nicolai, B., Stürz, S., Tonutti, P.: A metabolomics approach to elucidate apple fruit responses to static and dynamic controlled atmosphere storage. Postharvest Biol. Technol. **127**, 76–87 (2017)
52. Tran, D.T., Verlinden, B.E., Hertog, M., Nicolaï, B.M.: Monitoring of extremely low oxygen control atmosphere storage of 'Greenstar' apples using chlorophyll fluorescence. Sci. Hortic. **184**, 18–22 (2015)
53. Kittemann, D., McCormick, R., Neuwald, D.A.: Effect of high temperature and 1-MCP application or dynamic controlled atmosphere on energy savings during apple storage. Eur. J. Hortic. Sci. **80**(1), 33–38 (2015)

54. Thewes, F.R., Brackmann, A., Both, V., Weber, A., de Oliveira Anese, R., dos Santos Ferrão, T., Wagner, R.: The different impacts of dynamic controlled atmosphere and controlled atmosphere storage in the quality attributes of 'Fuji Suprema' apples. Postharvest Biol. Technol. **130**, 7–20 (2017)
55. Li, F., Zhang, X., Song, B., Li, J., Shang, Z., & Guan, J.: Combined effects of 1-MCP and MAP on the fruit quality of pear (Pyrus bretschneideri Reld cv. Laiyang) during cold storage. Scientia Horticulturae, **164**, 544–551 (2013)
56. Wang, Y., Sugar, D.: Internal browning disorder and fruit quality in modified atmosphere packaged 'Bartlett' pears during storage and transit. Postharvest Biol. Technol. (2013)
57. Dong, Y., Wang, Y., Einhorn, T.C.: Post-harvest physiology, storage quality and physiological disorders of 'Gem' pear (Pyrus communis L.) treated with 1-methylcyclopropene. Scientia Horticulturae **240**, 631–637 (2018)
58. Spotts, R.A., Sholberg, P.L., Randall, P., Serdani, M., Chen, P.M.: Effects of 1-MCP and hexanal on decay of d'Anjou pear fruit in long-term cold storage. Postharvest Biol. Technol. **44**(2), 101–106 (2007)
59. Xie, X., Song, J., Wang, Y., Sugar, D.: Ethylene synthesis, ripening capacity, and superficial scald inhibition in 1-MCP treated 'd'Anjou' pears are affected by storage temperature. Postharvest Biol. Technol. **97**, 1–10 (2014)
60. Guo, J., Wei, X., Lü, E., Wang, Y., Deng, Z.: Ripening behavior and quality of 1-MCP treated d'Anjou pears during controlled atmosphere storage. Food Control **117**, 107364. 10.1016 (2020)
61. Both, V., Brackmann, A., Thewes, F.R., Weber, A., Schultz, E.E., Ludwig, V.: The influence of temperature and 1-MCP on quality attributes of 'Galaxy' apples stored in controlled atmosphere and dynamic controlled atmosphere. Food Pack. Shelf Life **16**, 168–177 (2018)
62. Chai, J., Wang, Y., Liu, Y., Yong, K., Liu, Z.: 1-MCP extends the shelf life of ready-to-eat 'Hayward' and 'Qihong' kiwifruit stored at room temperature. Scientia Horticulturae **289** (2021)
63. Schmidt, S.F.P., Schultz, E.E., Ludwig, V., Berghetti, M.R.P., Thewes, F.R., Anese, R.D., Both, V., Brackmann, A.: Volatile compounds and overall quality of 'Braeburn' apples after long-term storage: Interaction of innovative storage technologies and 1-MCP treatment. Scientia Horticulturae **262** (2020)
64. Song, Z., Qin, J., Yao, Y., Lai, X., Zheng, W., Chen, W., Zhu, X., Li, X.: A transcriptomic analysis unravels key factors in the regulation of stay-green disorder in peel of banana fruit (Fenjiao) caused by treatment with 1-MCP. Postharvest Biol. Technol. 168 (2020)
65. Thewes, F.R., Rogerio Oliveira Anese, R.O., Thewes, F.R., Ludwig, V., Klein, B., Wagner, R., Nora, F.R., Rombaldi, C.V., Brackmann, A.: Dynamic controlled atmosphere (DCA) and 1-MCP: Impact on volatile esters synthesis and overall quality of 'Galaxy' apples. Food Pack. Shelf Life **26** (2020)

Chapter 43
A PLS-SEM Approach for Analysing Job Satisfaction and Human Resource Practices in Indian Banking Sector

Neeru Gupta and Pawan Kumar Verma

Abstract In the economic growth of every country like India, the Indian financial sector plays a crucial role. However, employees in this industry are mainly not left to their job pleasure. This article seeks to identify the current work satisfaction criteria for bank employees in order to ensure the stable economic growth of India, and to advocate that regulatory authorities advance their degree of satisfaction. A total of 220 workers were randomly selected as examples for this study from the bank of India. The investigation was completed via a structured questionnaire. SPSS software has been utilised for the analysis of demographic information. Multivariate methods such as Structural Equation Modeling (SEM) were also utilised to analyse data to discover characteristics which are critical for the work satisfaction of bank employees using SmartPLS software. This survey shows that performance assessment, salaries, and bonuses significantly affect Bangladeshi banking workers' job satisfaction. This article argues that both the management and regulatory bodies should guarantee performance evaluation and that the regulatory bodies for this industry should profit from salaries.

43.1 Introduction

The performance of companies nowadays by taking into account their vital role in the growth of the domestic economy. The organisational performance is derived from the notion of how efficiently and flexibly companies manage physical (capital) and non-tangible resources of their policies (employees). It is important for the company to build its non-tangible resource and tangible resources. The fundamental idea is that human resources are the single variable impossible to duplicate by other organisations. It was created to obtain competitive advantage as one of the important

N. Gupta (✉)
Assistant Professor, Maharaja Agrasen University, Baddi, Himachal, India
e-mail: neerugupta01@gmail.com

P. K. Verma
Controller of Examination, Shri Khushkal Das University, Hanumangarh, Rajasthan, India

S. Yadav et al. (eds.), *Proceedings of Second International Conference in Mechanical and Energy Technology*, Smart Innovation, Systems and Technologies 290,
https://doi.org/10.1007/978-981-19-0108-9_43

resources available for organisations. Organisations have more and more recognised that the human resources of the firm are highly important and may be a single source of success. But management of human resources is an organisation's most essential responsibility as human resources are an organisation's wealth that may help achieve its objectives. As a result, scholars and practitioners have focused on the study of every way they may gain the most from HR. Number of studies have shown that HRM practices have been positively related to organisational performance financial factors like pay, bonuses, incentives, and other financial benefits such as health care, sick leaves, and so on. However, other non-financial factors such as work, employment guidance, non-biased work performance evaluation, education, and employment satisfaction have improved. In order to balance wages and perks, employers should be focused on families and friends such as picnics, transit facilities for employees, childcare facilities, etc.

43.2 Literature Review

Anupama Gupta is the subject of this article (2010) When considering present economic conditions, human resource management concerns have been highlighted and the challenges that must be overcome. This essay focuses on the challenges of attracting and retaining top talent in the short term. This shows how HR management handles the challenges that come up regularly. Essential advice was also given for resolving problems. Saini R.R. describes the United Commercial Bank Management, a Chandigarh-based case study that explains HRD policies and regulations (2010). This paper primarily outlines the challenges associated in formulating regulations and putting them into effect [1–3]. HRD policies and standards were analysed by a sample of 100 persons in this study. To put it another way, a candidate's qualifications are critical in all levels of management hiring. A business needs to find the best candidates for management positions so that it may maximise productivity whilst also retaining key staff. Kundu. A paper by Subhash and Malhan (2009), entitled "Human Resources Management Practices in Insurance Corporations: The Study was undertaken in Indian and MNC," emphasises that human resources only benefit the business. However, one of the advantages of Human Resource Management is the provision of insurance for human resources [4–6]. According to the report, both national and international insurance businesses must strengthen their HR practices, including performance review, human resource planning, and recruitment. Human resources in the firm are being utilised to their utmost potential, according to Tripathy (2008). Commitment, culture, and competence are the three Cs that they adhere to. The organisation benefits greatly from adopting these three C cultures in the workplace. A surge in financial growth is expected as a result of this strategy. He is known as Singh (2008) Maintaining employee involvement is critical to great productivity and exceptional Organisational growth, as this study demonstrates. Having a strong work ethic helps the employee succeed and increases the quality of his work. The company should help the individual improve his interpersonal skills whilst he is being trained. Individual advancement should be a joint effort between the employer and the employee [7, 8]. All of Kallinath S. Patil's names are Patil (2007). The development of the facilities management industry is essential to the country's growth. One of the largest service providers, Life Insurance Corporation, constantly provides high-quality service. The Transmission takes a long time to arrive. Workers have made tremendous efforts to serve the field of service as a result of this. Changes in the Human Resources function were examined by Rao (2007) in his research. There has been a major shift in the Human Resources department's responsibilities. To ensure that all employees are treated equitably, regardless of their caste, religion, or gender, they should be working within the organisation. It is the calibre of a company's workforce that determines the productivity of that company. According to Sharma and Jyoti (2006), the best and most effective motivation for an employee to remain in a job is Job Satisfaction [9–12].

Emotional involvement helps employees provide the best possible results for their employer, they suggest. In the workplace, employee discontent is one of the most common causes of turnover. As a result, a person's ability to advance in their profession is directly related to their level of job satisfaction. The recruiting and selection process of the right person for the right function helps employees work well, according to Subramanian (2005). A job-specific interview process will help to arouse the interest of potential candidates. Using data from a study of 120 State Bank of India branches in Andhra Pradesh, researchers from Chalam and Srinivas (2005) sought to understand the fundamental differences between men and women in the workplace. He learns that female employees are significantly more concerned than male employees about the situation [13–16]. The company's Human Resources department has a higher level of respect amongst female employees. He shows how human resource development has grown through a number of procedures such as recruiting and selection as well as payroll processing and adhering to office rules and norms in Maitin's (2003) research. In order to get the most out of your staff, it's important to treat your personnel with respect. An author by the name Salokhe (2002) Human resources conducted a study to determine the best ways to help people advance personally inside the company. To enhance earnings, it helps the company execute a better and more efficient work. A company's most valuable asset is its workforce, which generates profits. An empirical study by Mishra and Bhardwaj (2002) was undertaken on the nature of work in private enterprise's Human Resources Department. There were a total of 107 managers selected at random from various levels of management. For this purpose, they were asked to fill out a questionnaire to learn more about the Human Resources department. HRD is currently in an acceptable state. In his book, a human resource audit is described by Rao (1999) as defining the procedures and principles that must be adopted. His method helps to estimate a variety of elements, including core competency, culture, and organisational worth. Examining how well an employee gets along with others is one way they might make up for some of the system's flaws [17–19]. In addition, the company's higher management includes individuals who are amenable to hearing the concerns of the company's lower and middle management levels. Each person in the company thus contributes to the company's success by boosting productivity. Employees' and the company's personal growth can be facilitated by the Human Resources Department, according to a study by Pareek and Rao (1999). Throughout his article, he covers topics such as the importance of human resource management (HRM), organisational behaviour (OB), and strategic planning (SP). A wide range of educational institutions and a significant number of professions are also included in this resource. Venkateswaran (1997) shows how the Human Resource Department helps the company's financial success in his study. In order to do this, he conducted an empirical study with the help of 132 executives from a privately held corporation. This study sheds light on the inner workings of HRD in a company and the advantages it provides to employees. As a result, HRD is able to operate more successfully since it is more comfortable with the organisation's rules and laws. This also helps to keep the workplace pleasant [20].

43.3 Objectives of the Study

The objectives of this study are:

- To identify the factors of job satisfaction of bank employees in India.
- To give some suggestions for the improvement of the satisfaction level of the bank employees in India.

43.4 Methodology of the Study

This study aims to determine the elements which influence the employment of employees of private trade banks in India with a job satisfaction model for service quality. The data were obtained from primary sources to carry out the investigation. Six distinct private trade banks in Bangladesh obtained the primary data.

43.4.1 Determination of Sample Size

For this survey, the respondents chose their subjects heterogeneous. From the past survey, we showed that more than 10 workers of Lac Bank work in other banks of Bangladesh. However, this sample has been obtained from six individual commercial private banks in India. Prior research suggests that 100–200 samples are generally a suitable starting point for route modelling [12]. Therefore, the participants sampled may be identified by utilising the following format that Yamane has found out (1967). The following formulation is illustrated in the following:

$$n = (1 + 2)$$

where

n = Sample size N = Population e = Level of precision.

The following assumptions have been used in determining the sample size, n = 204. For this reason, data from six different banks in India have been gathered, which amounted to over 204 utilising the aforementioned sample approach.

Population size is > 1,000,000 workers.

Level of precision is 7%

43.4.2 *Questionnaire Design and Test of Reliability*

For questions linked to the effect of human resource determinants on work satisfaction of private business banks in Bangladesh a structured questionnaire has been designed at the five-point Likert scope. The questionnaire included a five-point scale of 1 to 5 with 1 showing severe disagreement, and 5 showing strong agreement. 5. Table 43.2 displays the questionnaire's reliability coefficient. It demonstrates that the composite reliability of cronbach alpha and the average variance retrieved from the questionnaire is displayed in Table 43.3, which is satisfactory as per Nunnally and Berstein (1994), Hair et al. (1998), Fornell and Larcker (1981), Henseler et al. [10] respectively.

43.5 Hypothesis of the Study

The next hypothesis has been developed with regard to the aims of these studies, which include the five key variables for employee satisfaction of banks in India, including safety of work, performance assessment, relationship, wage & benefits and working conditions.

H_1 Job Security has no significant impact on Job Satisfaction of bank employees.
H_2 Performance Appraisal has no significant impact on Job Satisfaction of bank employees.
H_3 Relationship has no significant impact on Job Satisfaction of bank employees.
H_4 Salary and Benefits have no significant impact on Job Satisfaction of bank employees.
H_5 Working Environment has no significant impact on Job Satisfaction of bank employees

43.6 Conceptual Framework

This study aims to examine the job satisfaction of the bank's staff in India on the basis of a variety of functions, including job security, performance assessment, relationship, wages, and employment conditions. In the literature, the associated

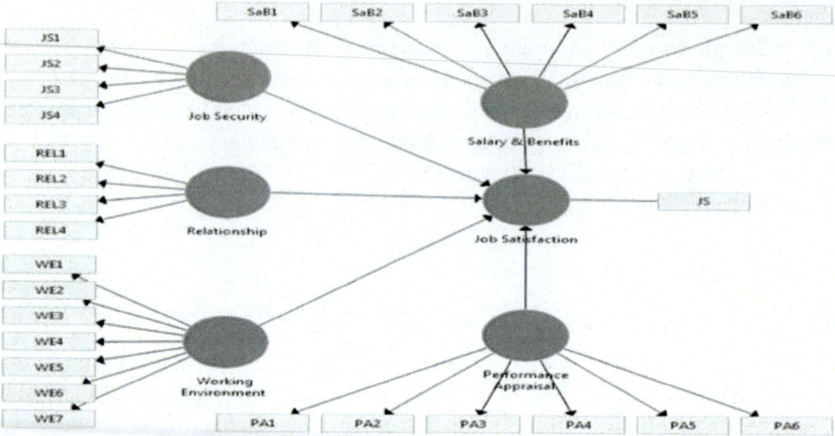

Fig. 43.1 The conceptual framework of human resources practices factors and job satisfaction of the private commercial bank employees in India

research shows that the different kinds of path-model applications in six various private commercial institutions include occupational safety, performance evaluation, connection, wage and benefits, and workplace conditions. Figure 43.1 presents the theoretical model. For each of the assumptions in the following we will examine the theoretical model.

43.7 Data Collection and Data Analysis

Six private commercial banks have been surveyed in India. Before the interview was resumed, interviewees were adequately instructed on questions that represented the questionnaire. In addition to descriptive statistics, the data analysis was conducted using SPSS (Statistical Package for Social Science) and SmartPLS to analyse inferential statistical techniques such as factor analysis and structural equation modelling (statistical software). In order to determine the important elements, structural equation modelling has been performed; the quality of service of the banks of the study influenced the factors (Table 43.1).

43.8 Data Analysis, Findings and Discussion of the Bank Employees in India

Table 43.2 reveals that about 68% of respondents were men and confirms that bank staff are male dominants in this industry. The age distribution amongst respondents

Table 43.1 Skewness and kurtosis

	Dimensions	N	Skewness	Kurtosis
Job Security	After certain periods of time our jobs become permanent	220	−0.52	−0.35
	My job is highly secured	220	0.22	−0.04
	I never feel vulnerable at my job	220	0.70	0.11
	The company has a faster career growth opportunity	220	0.50	0.18
Performance Appraisal	My employer values the contributions I make to my department	220	1.09	0.48
	My organisation has a fair system of rewarding employee performance	220	1.00	0.13
	My company offers promotion based on performance	220	0.93	−0.17
	My company gives promotions fairly	220	0.99	0.10
	Performance feedback is communicated properly in my company	220	1.00	0.23
	My company is concerned with an employee's well-being	220	1.04	0.23
Relationship	I enjoy working with my coworkers	220	0.66	−0.31
	My co-workers respect each other's opinions	220	0.98	0.16
	My supervisor is fair and reasonable	220	1.10	0.72
	I have a good working relationship with my supervisor	220	1.01	0.08
Salary and benefits	I am provided with adequate salary by the company	220	0.74	−0.12
	My salary has a match with my experience	220	0.98	0.18
	My salary has a match with my training, skill I received	220	1.05	0.21
Working environment	There is no gender discrimination in my company	220	1.01	0.16
	The fire protection facilities are modern in our company	220	0.93	−0.07
	My company does not force me to stay after the office time	220	1.06	0.22
	The rules and regulations are convenient with me	220	1.01	0.15

Table 43.2 Personal Information of the respondents of bank employees in India

Private commercial Bank			
Gender	Male	150	68
	Female	70	32
	Total	220	100
Age of the respondents	18–30 years	90	41
	31–45 years	65	30
	45 years and above"	65	29
	Total	220	100
Name of the bank	Basic Bank Limited	37	17
	HDFC Bank Limited	36	16
	ICICI Bank Limited	37	17
	Federal Bank Limited	37	17
	IndusInd Bank Limited	36	16
	IDBI	37	17
	Total	220	100

as described in Table 43.2 shows that 90 (41%) of employees were aged between 18 and 30 years, whilst the majority of employees were between 18 and 30 years; instead, 30% and 29% of employees in the bank whose ages are between 31 and 45 and 45 years and older respectively. We have picked six Private Business Banks in India to collect data on the same basis.

43.9 Exploratory Factor Analysis

EFA is a commonly used and frequently employed social science statistical technology. This section analysed a total of 220 valid survey responses. The technique of factor analysis was employed to analyse the connection between several elements that satisfy the employment situation of bank employees in Bangladesh.

Five factors from the rotated matrix have been identified. In the next paragraph these variables have been explored.

Factor-1 (Job Security) It covers four variables like: "I am not susceptible at my job, my employment is extremely secure" and "The company provides a faster career advancement possibility," which are the main elements. This contains four variables: It therefore provides a framework for the conceptualisation of a dimension identifiable as a component of occupational safety.

Factor-2 (Performance Appraisal) This contains four elements including a fair system in which my organisation offers rewarding employee performance, my firm fairly promotes and my company communicates performance feedback correctly, which are the main aspects. It therefore offers a framework on which a dimension may be conceptualised and defined as a performance assessment element.

Factor-3 (Relationship) This covers four parameters, including "I have good working connections with my supervisor," which are essential considerations. My colleagues respect one another's ideas. It therefore gives a framework for conceiving a dimension that may be identified as a factor in relation.

Factor-4 (Salary and Benefits) This comprises four criteria such as "My pay is consistent with my experience, my pay corresponds to my training, my skills, my wages and payments are paid on time," which are the main considerations. It therefore gives a framework on which to conceptualise a dimension which may be recognised as a pay and benefit component.

Factor-5 (Working Environment) This comprises four variables like "there is no discrimination between the sexes in my workplace" and "My firm does not compel me to remain after office hours," and "Rules and regulations are suitable for me," which are the major elements. It gives a framework to conceptualise a dimension that may be identified as a component in the working environment.

Table 43.3 also shows that all of the *T*-Statistic are larger than 2.33 at the 1% level of significance, we can say that the outer model loadings are highly significant. So, our SEM model is accepted for above evidence in this study.

Generally, A global fit measure (GOF) was conducted for path modelling; it is defined as the geometric mean of average commonality and average Q^2 (especially endogenous variables) [4] (see the formula). In this study, GOF value was 0.78 ($Q^2 = 0.509$, average AVE = 0.614 for job satisfaction of the bank employee). So, the value of GOF exceeded the largest cutoff value (0.36), and it was indicated that the proposed model of this study had better explaining power than that based on the recommended value of $GOF_{small} = 0.1$, $GOF_{medium} = 0.25$, and $GOF_{large} = 0.36$ [1].

43.10 Results of Multivariate Analysis—Partial Least Square (PLS)

In order to determine the important factors of work satisfaction from the components found by factors analysis, the multivariate analytic approach "Partial Least Square (PLS)" was employed. By utilising SEM, the road map for employee happiness in India shows that salaries and benefits have the greatest influence on the work satisfaction of employees. The hypothesised trajectory connection between independent variables such as salaries and benefits, performance assessment and job satisfaction is a positive 1% association. This indicates that if performance assessment is enhanced by one unit, bank employees' work satisfaction is improved by 0.492 and if compensation and benefits are increased by 1 unit, bank workers' job satisfaction is increased by 0. 167. However, the safety of jobs, relations, and work conditions

Table 43.3 Factor Analysis of the job satisfaction of the private commercial bank employees in India

	Variables	Factor loading	Sample mean	SD	t-value	Alpha	CR	AVE	Q^2 value
Job security	Permanent after certain periods of time	0.85	0.85	0.04	20.80	0.79	0.86	0.61	0.35
	Secured job	0.76	0.74	0.10	7.91				
	Not feel vulnerable	0.72	0.69	0.13	5.73				
	Faster career growth opportunity	0.78	0.78	0.07	10.93				
Performance appraisal	Fairly rewarding system	0.71	0.71	0.07	9.58	0.69	0.83	0.62	0.25
	Fairly promotions	0.85	0.85	0.04	21.30				
	Performance feedback	0.79	0.78	0.05	16.08				
Relationship	Respect from the co-workers	0.69	0.68	0.09	8.12	0.62	0.80	0.57	0.20
	Relationship with supervisor	0.66	0.66	0.08	7.96				
	I have a good working relationship with my supervisor	0.89	0.88	0.04	21.66				
Salary and benefits	My salary has a match with my experience	0.79	0.77	0.08	10.10	0.79	0.87	0.70	0.38
	My salary has a match with my training skill I received my salary has a match with my training skill I received	0.89	0.89	0.04	21.61				
	My salary and payments are made timely	0.83	0.82	0.07	11.53				
Working environment	Gender discrimination	0.74	0.74	0.07	11.24	0.63	0.80	0.57	0.17
	Stay oblige	0.72	0.70	0.11	6.35				

(continued)

Table 43.3 (continued)

	Variables	Factor loading	Sample mean	SD	*t*-value	Alpha	CR	AVE	Q^2 value
	Convenient rules and regulations	0.79	0.78	0.07	11.27				

AVE = Average Variance Extracted, C.R = Composite Reliability, Cronbach's alpha, VIF = Multicollinearity Statistic.
Note AVE > 0.50 [5, 10], Composite Reliability > 0.70 (Hair et al. 1998), Cronbach's alpha ≥ 0.60, (Nunnally and Berstein 1994), Indicator Reliability ≥ 0.4 (Hulland 1999), Q^2 Value > 0 [7, 19]

have no important effect on employment satisfaction, since the p-value for those items is over 5%. All VIF values have demonstrated that these variables have no multicollinearity effect. In this study, the politicians and authorities involved should also concentrate more on aspects such as salaries and perks and the performance assessment in the banking industry of India.

43.11 Hypothesis Testing

The hypothesis test was conducted by evaluating the path coefficients (beta) between latent structures and their meaning. The bootstrapping approach was employed with a 500 re-sampling in order to justify the importance of the route coefficients. The R2 values of the latent endogenous construct show that the model is predictive. The results and testing of hypotheses are shown in Table 43.4. The data demonstrate that the H2 and H4 hypotheses have been rejected, given that the value at the 1% importance is greater than 2,33. That is, performance assessment, wages/benefits had a substantial influence on employment satisfaction, whilst H1, H3, and H5 were statistically less important at 5%, as their value did not exceed 1.96%. This indicates stability in employment; relationships and work environments do not have any substantial effect on employee happiness in India.

Table 43.4 Summary results of the model constructs

	Original sample	Sample mean	SD	*T* statistic	*P*-value	VIF
Job security --> Job satisfaction	0.046	0.059	0.087	0.531	0.59	1.386
Performance appraisal --> Job satisfaction	0.492	0.479	0.108	4.535	0.00	1.95
Relationship – > Job satisfaction	0.01	0.015	0.106	0.097	0.92	2.143
Salary and benefits --> Job satisfaction	0.167	0.186	0.079	2.113	0.04	1.31
Working environment --> Job satisfaction	0.172	0.175	0.108	1.59	0.11	1.788
R^2	0.509					
R^2 Adjusted	0.483					

Collinearity Statistic (VIF) The rules of thumb for the VIF is as follows: VIF < 3; no problem, VIF > 3; potential problem, VIF > 5; very likely problem, VIF > 10; definitely problem

	Null hypothesis	Accepted/Rejected
H_1	Job security has no significant impact on job satisfaction of bank employees	Accepted
H_2	Performance appraisal has no significant impact on job satisfaction of bank employees	Rejected
H_3	Relationship has no significant impact on job satisfaction of bank employees	Accepted
H_4	Salary and benefits has no significant impact on job satisfaction of bank employees	Rejected
H_5	Working environment has no significant impact on job satisfaction of bank employees	Accepted

43.12 Conclusion

The main aim of this study is to identify influential elements such as employment security, performance evaluation, relationship, salary & benefits, working environment, and relationships on employee satisfaction at the banks. In addition, this study focuses on factors affecting job satisfaction for private employees at business banks and in India. Sarker et al. [16] suggested that non-financial factors such as job description, employment guidance, unbiased job performance assessment, training, job security, effective support, a sound workplace related to job satisfaction amongst bank employees, and other factors, such as wages and benefits, should concentrate on friends and family benefits. According to the foregoing findings, only three criteria, such as a performance assessment and wages, really connect to the job happiness of bank workers who have been supported by prior employment in India.

References

1. Akter, S., D'Ambra, J., Ray, P.: An evaluation of PLS based complex models: the roles of power analysis, predictive relevance and GoF index
2. Ambrose, S., Huston, T., Norman, M.: A qualitative method for assessing faculty satisfaction. Res. Higher Educ. **46**(7), 803–830
3. Anderson, J.C., Gerbing, D.W.: Structural equation modeling in practice: a review and recommended two-step approach. Psychol. Bull. **103**(3), 411–423 (1988)
4. Chin W.W.: How to write up and report PLS analyses. In: Handbook of Partial Least Squares, pp. 655–690. Springer, Berlin, Heidelberg (2010)
5. Fornell, C., Larcker, D.F. (1981). Structural equation models with unobservable variables and measurement error: algebra and statistics. J. Mark. Res. 382–388
6. Gefen, D., Straub, D., Boudreau, M.C.: Structural equation modeling and regression: guidelines for research practice. Commun. Assoc. Inf. Syst. **4**, 1–70 (2000)
7. Geisser, S.: A predictive approach to the random effect model. Biometrika **61**(1), 101–107 (1974)

8. Hair, J.F., Ringle, C.M., Sarstedt, M.: PLS- SEM: Indeed a silver bullet. J. Mark. Theory Pract. **19**(2), 139–152 (2011)
9. Hair, J.F., Sarstedt, M., Ringle, C.M., Mena, J.A.: An assessment of the use of partial least squares structural equation modeling in marketing research. J. Acad. Mark. Sci. **40**(3), 414–433 (2012)
10. Henseler, J., Ringle, C.M., Sinkovics, R.R.: The use of partial least squares path modeling in international marketing. In New challenges to international marketing, pp. 277–319. Emerald Group Publishing Limited (2009)
11. Henseler, J., Ringle, C.M., Sarstedt, M., Okazaki, S.: (2012). Using partial least squares path modeling in advertising research: basic concepts and recent issues. In: Handbook of Research on International Advertising, pp. 252–276
12. Hoyle, R.H.: Structural Equation Modeling: Concepts, Issues, and Applications. Sage (1995)
13. Hulland, J.: Use of partial least squares (PLS) in strategic management research: A review of four influence on job satisfaction and turnover intentions. J. Bus. Ethics **78**(4), 547–557 (1999)
14. Pushpakumari, M.D.: The impact of job satisfaction on job performance: An empirical analysis. City Forum **9**(1):89–105
15. Ringle, C., Wende, S., Will, A.: Smart-PLS Version 2.0 M3. University of Hamburg (2005)
16. Sarker, M.A.R., Afroze, R.: Can HRM practices improve job satisfaction of Ready Made Garment (RMG) workers in Bangladesh? An alternative solution to recent unrest. Int. J. Bus. Manage. **9**(10), 185 (2014)
17. Shabnam, S., Sarker, A.R.: Impact of CSR and internal marketing on employee job satisfaction and organisational commitment: a case study from export-oriented SMEs in Bangladesh. World J. Soc. Sci. **2**(7) (2012)
18. Spector, P.E.: Job Satisfaction: Application, Assessment, Causes, and Consequences, vol. Sage Publications (1997).
19. Stone, M.: Cross-validatory choice and assessment of statistical predictions. J. R. Stat. Soc. B **36**, 111–147 (1974)
20. Yamane, T.: Elementary Sampling Theory Prentice Inc. Englewood Cliffs. NS, USA (1967)

Chapter 44
Role of Data Mining in Detecting Theft and Making Effective Impact on Performance Management

Vipin Jain, Alim Al Ayub Ahmed, Vijesh Chaudhary, Divya Saxena, Muthukumar Subramanian, and Md. Khaja Mohiddin

Abstract This paper analyses the effectiveness of DM (Data Mining) classification techniques for company detection. Different approaches of data mining are accessible for the process of data mining. In many applications, data mining techniques are used to identify and prevent various kinds of theft. While data mining research and possible measures for the detection and identification of different forms of theft are already in progress, there is limited study that reconstructs many elements of theft that employ data mining methods. The use of data mining techniques to detect theft is a responsibility. We are also classifying theft into three groups with respect the use of data mining as an aid for the identification and prevention. This report examines the efficacy in determining false financial statements of policymakers, artificial neural networks, and Bayesian beliefs networks. Management theft, Theft against Customers, and computer-based theft are all three kinds of theft.

V. Jain (✉)
Teerthanker Mahaveer University, Uttar Pradesh, Moradabad, India
e-mail: vipin555@rediffmail.com

A. Al Ayub Ahmed
School of Accounting, Jiujiang University, 551 Qianjindonglu, Jiujiang, Jiangxi, China

V. Chaudhary
Central University of Himachal Pradesh, Shahpur, India

D. Saxena
Master of Science in Computer Science, University of Texas at Arlington, Alington, TX 76019, USA

M. Subramanian
Nagarjuna College of Engineering & Technology, Bangalore, India

Md. K. Mohiddin
Bhilai Institute of Technology, Raipur, Chhattisgarh, India

S. Yadav et al. (eds.), *Proceedings of Second International Conference in Mechanical and Energy Technology*, Smart Innovation, Systems and Technologies 290,
https://doi.org/10.1007/978-981-19-0108-9_44

44.1 Introduction

Retrieving or mining data from vast amounts of data refers to data mining. There are a variety of data mining techniques such as segmentation, pattern recognition, and extrapolation. Here, just few other data mining approaches that are regarded to be significant for detection of theft are discussed. (1) The Bayesian network, the risk group classification system and (2) The decision tree, which is responsible for constructing each risky group's descriptive model. The economic contribution of these approaches in law enforcement is constrained since substantial collections of data on theft training are almost nonexistent. Data Mining is an efficient technique for the identification of theft, among its many diverse purposes. Organizations, because of theftulent corporate practices that may be described as a disappointment created for personal or group profit through many ways, such as unscrupulous or unlawful presentation, lose significant income in organizations [1]. Some of the primary forms of theft are insurance theft, theft on credit cards, theft in calls, and check to forge. Theft in car insurance, travel, is widespread. The Bayesian network is the probabilistic classifier utilized. Segmentation identifies which of the classes a certain information relates to given a group of preset categorization classifications. Descriptive models are created using decision trees. The technique of data mining pulls from vast datasets new trends and patterns that may be utilized to make significant corporate choices. Algorithms such as decision trees, artificial neural networks, and Bayesian belief networks can bé used to accomplish the data mining algorithms [2].

Objective: In this topic, we deeply talk about effective control for detecting theft, Artificial Neural networks, Decision Trees, Bayesian Belief Networks, Theft based on Computers, Theft against customers, Theft against Management. Table 44.1 shows the most effective tools fraud.

Effectively control for detecting theft: The existing theft assessment strategy was included by an artificial neural network with optimal input parameters [3].

The result indicated the probabilities of theft and the causes for a false insurance payout. This method had a somewhat improved outcome. In order to detect those who do not follow the norm, the EFD system merged expert knowledge with statistical information. Expanded the areas near approach to create and study the rules using

Table 44.1 Most effective tools fraud

S. No	Tools	Percentage (%)
1	CVN (Card Verification Number)	54
2	Biometric indicators	53
3	Customer order history	52
4	Two factor phone authenticity	50
5	Payer authentication (3-D secure)	49
6	Merchant negative list	39

genetic algorithms. The credit card theft model recommended a theft/legal classification approach and a cluster, accompanied by a theft/legal-free categorization technology [4].

The strategies of categorization were highly efficient in detecting theft thus may be used by individuals with pre-existing crimes. A realistic cost model for evaluating C4.5, CART, and Bayesian basic classification models is used in the distributed data mining model. Card payments were subject to the procedure. The method of neural data mining is based on rules to mine symbolic and Radial Basic Function neural data. The method examines the relevance of using non-numeric information for the identification of theft. SAS Enterprise Miner software is based on feature selection for the case of theftulent claims, grouping recognition, and classification methods [5].

The STAGE method for utilized in the studies of the Bayesian Belief Network (BBN) and the Artificial Neural Networks (ANN). STAGE combines two search phases repeatedly: the original search technique runs on optimization problems and hill-climbing runs to improve the cost function. The outcome reveals that BBNs were trained significantly quicker but were applied more slowly in new cases. Software Theft Focus evaluates all allegations instantly. The results are ranked in a fall in order of the possibility to make false claims and provide detailed rules. For the theftulent usage of credit cards. In order to educate user profiles and user profiles history, the ASPECT group focuses on neural networks. The profile and background of a contact are matched to likely theft. Use an occurrence technique to assign theft scores to identify theft in order to build on the adaptive theft detection framework. It can also detect theftulent kinds utilizing regulations. This architecture has been utilized for theft sensing systems both wirelessly and wired. Dynamic BBN's were used to identify theftulent claims using the Suspicion Building Tool, a rule builder [6].

44.2 Artificial Neural Networks

ANN is an advanced technique that has defined theory and acknowledged applications in various fields. ANN consists of many linked neurons as shown in Fig. 44.1. The processing units are interlinked. A numerical number, termed weight, is connected with each link. Every neuron accepts data from linked neurons and calculates the total incoming signal. The overall input of neuron j is $u_j = Rw_{ij} * x_i$, where x_i is the input of neuron I and w_{ji} are the weight of the neuron j-neuron link. The resulting value transforms via the differential equation of the neuron and eventually the neuron fires if the total input signal intensity reaches a specified value. The neurons are layered organized. At least one (first) and one (final) output layer is a multilayered network. There are one or more hidden layers between the input and output layers. Various types of ANNs have various layers. Only the inlet and outlet layer are in the self-organizing maps (SOM), while the activation function ANN also contains one or even more hidden units. The system must be informed after the

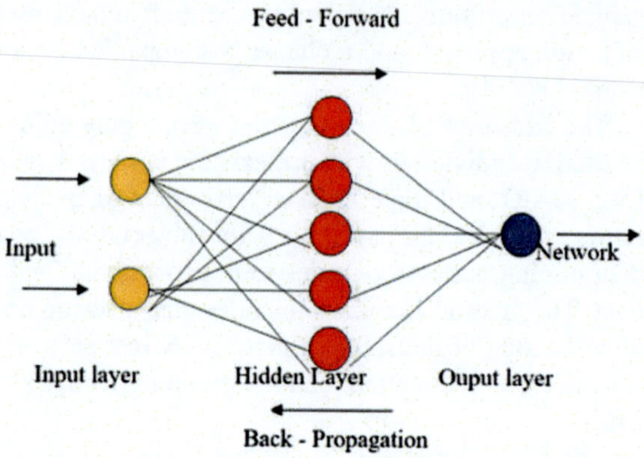

Fig. 44.1 Artificial neural networking

architectural style is designed. The pattern is applied to the input layer in backpropagation networks and an end output is computed on the output layer. The outputs are determined by the maximum result and mistakes in the ANN are spread by adjusting the interconnections' strengths. This procedure will continue up to a reasonable error rate. Artificial neural networks do not assume independence for characteristics, are able to handle noisy or inconsistent information, and offer an appropriate alternative to issues when an algorithm solution cannot be used [7].

44.3 Decision Trees

Feature selection is artificial intelligence approach that describes a tree-formed architecture that offers a collection of judgments with independent characteristics and reliant characteristics. Classification algorithms are IF–THEN statements where all the tests are ANDed rationally, if each rule is to be produced, and all the checks are to be successful. The methods linked to drug trafficking, government financial trading, and customs reporting theft are analyzed for more severe crimes such as killings related to drugs, serial sex crimes, and home safety (see Fig. 44.2). Methods of information mining have fixed things of safety and crime analysis. Subjects examined for system administrators, criminal justice investigations, counter-intelligence agents, theft experts, and security professionals: smart agents, design dictates, text analytics, decision-making treasures, personality maps, and machine learning and artificial neural network. C4.5 is used for the separation of data in sections and for the creation of descriptive rules for categorization of a training node. C4.5 can assist anticipate and developing patterns of crime. It creates rules from trees and processes numerical characteristics, null values, tape, and error detection. As regards predictive accuracy, C4.5 is somewhat better than CART and ID3. The categorization and learning phases are usually quick. However, when C4.5 is applied to big datasets, efficiency decreases might occur. C5.0 provides some enhancements in the induction of decision tree [8–10].

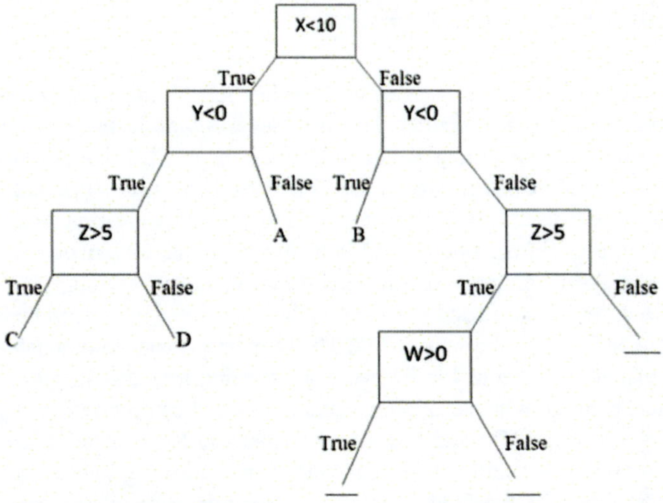

Fig. 44.2 Examples of decision tree

44.4 Bayesian Belief Networks

The Bayesian Identity Networks offer a graphical model of causative interactions, on which the odds of participation are projected to be legitimate or theftulent. Bayesian Classifier categorization presumes that, given the target attribute, an instance's attributes are self—reliant as shown in Fig. 44.3. The goal is to give the highest posterior likelihood to a new instance. Relative to C4.5, decisions trees, and background dispersion the method is extremely efficient and can provide greater classification accuracies. Unfortunately, the prediction accuracy is lowered when the qualities are duplicated [11].

Fig. 44.3 Bayesian belief networks

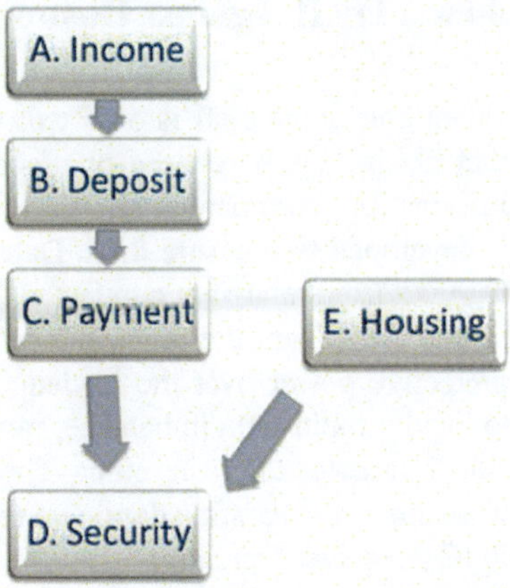

44.5 Theft Based on Computers

Lapses in computer safety may give a tool for the theftster to perpetrate theft. Scam perpetrators use computer security defects. Uses machine learning and predictive mining to find dangerous binaries and categorize them. That would remain unnoticed without or before or eliminating any obstruction on a user's external hard disk. N-grams have been used to encrypt legitimate and malignant binaries. After choosing the associated n-grammar, a number of inductive approaches were discovered, such as naive Bayes, decision trees, vector support machines, and boosting. These results indicated that the increased decision-making bodies exceeded other models and had a region under the ROC curve of 0. 996.The three important accomplishments: (1) detection of harmful executable by means of existing text classification methods; (2) empirical study of wild harmful executable code and categorization using inductive approaches. The SVM and Multivariate Adaptive Regression Spline artificial neural networking ensembles were also superior to existing approaches to categorization accuracy in the field of intrusion detection. The information from the Lincoln Lab Institute of Technology in Massachusetts and the information gathered by the (DARPA) identified five separate design categories. The results have shown that, provided proper intelligent paradigms are adopted, 100% prediction accuracy may be attained. For example, realistic techniques for organizational information system security selection and implementation have been presented (ISS). The three business information systems models have been presented, including ISS offense model, ISS defensive model, and safeguarding model. The paradigm of attack and defensive shows objects that aid infringes security of the system and avoid violations. The protection model examines the link between access control factors, such as the demand for corporate security, security approaches, and checks available for security improvements. Corporations must assess the loss of their information systems from a possible security violation [12, 13].

44.6 Theft Against Customers

When consumer theft is detected, the customer either does not pay for the items that he/she has purchased or misled organizations. A large quantity of payment information, commercial questionnaires, health records, and information on health is developed to measure theft. Data mining utilizes methods of categorization and learning from related or connected data. If there are considerably different instances in one class from the other, the detecting theftulent technique tends to yield poor predictive power over the inadequately. It addressed the use of a novel approach to handle multi-data imbalance, in which the one category in the targeted relationship is greater than the other. The findings showed that the proposed technique was superior to existing dominant machine learning techniques, especially when the ROC (Receiver Operating Characteristics) Curve and AUC were highly imbalanced

Table 44.2 Theft records

Rank	Categories	Number of Reports
1	Credit card	55,526
2	Online services	63,984
3	Associated investment	15,900
4	Health care	47,450
5	Software and computer components	20,550
6	Electronic media and broadcasting	40,335
7	Proposals of foreign money and falsified check scammers	28,445

(Area Under the Curve). In the same way, credit card theft detection (CCF) systems were developed utilizing the artificial neural network technology, with the rule-based element, and in place of the two-stage model often employed with theft detection techniques, employing four clusters (low, high, risk, and high risk). The model was developed utilizing the self-organizing map method to recognize the conduct of a cardholder and evaluate the transaction characteristic to detect theftulent transactions. Implementing the artificial neural network developed with the uncontrolled teaching strategies has produced a number of predictions. This generating was done to provide an accurate outcome and to reduce the mistaken categorization of a legitimate transaction. Innovative approaches and frameworks for the identification of theft were also previously suggested. Detecting theft framework by using simulation data mining algorithms to provide user profiles for the identification of client behaviors in the online detection of theftulent transactions via a number of association rules. By correlating inbound user transactions with the person's profile depending on its recent transactions, abnormalities were found. The trials have demonstrated the proper interpretation by the suggested algorithm of discrepancies here between abnormality and the monitored user conduct [14, 15]. Table 44.2 shows the number of reports based on various categories.

44.7 Management Theft

A theftster might be categorized depending on an employee, a customer, a supplier, a shareholder, or management party. For, e.g., the corporation's senior management is guilty of management theft and may take action, such as fabrication or manipulation of costs, invoices, revenues, taxable items, etc. The management might provide the shareholders with false information. In particular, management theft is a purposeful and illegal conduct by public businesses that uses materially misleading financial statements, which might harm customers, creditors, and the economic market, commonly termed financial statements theft. Accountants can discover any errors in financial data from top management through management theft. The connection of theft with the features of the supervisory board. In fact, a thorough examination of this association indicated that a financial expert was present on the supervisory board and the chance of financial theft was reduced. Search and make a comparison between both the achievements of several elements related to the financial statement theft using a sample of 76 Greek manufacturers. They employed three methods of data mining that were decision trees, artificial neural networks, and networking of the Bayesian belief.

The Bayesian belief networks were the best among these three models with the 90.3% accurate pass categorization followed by the 80% success neural networks and finally the 73.6% rate of success decision trees. Even though both theft prevention and theft investigation are an effective anti-Theft strategies of an organization, avoidance of theft is considerably cheaper than identification, and therefore should prevail as regards the distribution of resources. Was using a generalized theft prevention data mining architecture and a theft risk reduction for theft in the financial statement. The mining of information was separated into two groups: prediction and informative tasks. Prospective computer vision and machine intelligence have helped to improve the detection of theft.

44.8 Conclusion

Throughout this paper, we investigate the successful use of several approaches for data mining to identify and prevent theft and classified theft. In particular, we talked about data extraction strategies for three kinds of theft: management theft, client theft, and computer-based theft. In this field, we have given the newest advances. In this research, businessmen wanting to use data mining to discover theft in their organizations, accountants who want to detect theft, and computer administrators who need to defend their systems and software from illegal behavior will be guided. Data mining methods claiming strong categorization and forecasting skills can make the work of theft detection management easier for auditors. This survey aims to examine the utility to detect misrepresentations by utilizing available financial data and compare the performance of three Data Mining methods. Decision trees, neural networks, and the Bayesian belief networks were the approaches used.

References

1. Albrecht, W.S., Albrecht, C.O., Albrecht, C.C., Zimbelman, M.F.: Fraud Examination. Cengage Learning (2015)
2. Bapna, S., Patel, S.C.: Securing computerised models and data against integrity attacks. Int. J. Electron. Finance **4**(4), 343–354 (2010)
3. Bolton, R.J., Hand, D.J.: Statistical fraud detection: a review. Stat. Sci. **17**(3), 235–255 (2002)
4. Chong, H.G.:Does brainstorming help detect and deter fraud? (2011) Available at SSRN 1742702.

5. Gill, N.S., Gupta, R.: Prevention and detection of financial statement fraud: a data mining approach. IUP J. Syst. Manage. **7**(3) (2009)
6. Hormozi, A.M., Giles, S.: Data mining: a competitive weapon for banking and retail industries. Inf. Syst. Manag. **21**(2), 62–71 (2004)
7. Nath, H.V., Mehtre, B.M.: Static malware analysis using machine learning methods. In: International Conference on Security in Computer Networks and Distributed Systems, (pp. 440–450). Springer, Berlin, Heidelberg (2014)
8. Shinde, P.P., Oza, K.S., Kamat, R.K.: Big data predictive analysis: using R analytical tool. In: 2017 International Conference on I-SMAC (IoT in Social, Mobile, Analytics and Cloud) (I-SMAC), pp. 839–842 (2017). https://doi.org/10.1109/I-SMAC.2017.8058297
9. Shinde P (2017) Big data analysis solutions: a review. J. Adv. Scie. Technol. 13
10. Shinde, P., Oza, K., Kamat, R., Katkar, S.: Big data predictive model: towards digital health. AGU Int. J. Eng. Technol. **5**, 25–30 (2017)
11. Ondiek, D.C.O., Onyango, M.A.: An investigation of emerging risks associated with online banking on fraudulent payments among tier one commercial banks in Kenya
12. Patel, S.C., Sanyal, P.: Securing SCADA systems. Inf. Manage. Comput. Secur. (2008)
13. Patel, S., Zaveri, J.: A risk-assessment model for cyber attacks on information systems. J. Comput. **5**(3), 352–359 (2010)
14. Save, P., Tiwarekar, P., Jain, K.N., Mahyavanshi, N.: A novel idea for credit card fraud detection using decision tree. Int. J. Comput. Appl. **161**(13) (2017)
15. Simmers, C.A., Anandarajan, M.: The Internet of People, Things and Services

Chapter 45
An Analytical Approach of Crime Prediction Using Machine Learning

Punya Kapoor

Abstract Crime is considered regarded as our society's maximum critical and serious and growing problem, and stopping it would be an essential duty. On a daily basis, a great number of crimes are perpetrated. This necessitates maintaining note of all crimes and retaining a database for them to be referenced in the destiny. The gift issue is maintaining a dependable crime dataset and analyzing these statistics to aid inside the prediction and backbone of destiny crimes. The goal of this task is to observe a collection including a selection of crimes and forecast the type of crime that could arise within the future based on a diffusion of factors. The science of creating computer systems that make judgments without human intervention is called device getting to know. Machine gaining knowledge has currently been used within the improvement of self-driving vehicles, speech identity, web website seek, and a higher understanding of the human genome. It has also made it feasible to perceive crime using referenced data. This research appears into crime prediction the use of gadget studying. This look examines crime data from India over the last 15 years using critical information-processing methodologies. When predicting crime in India, gadget-mastering predictive models including K-nearest-neighbor and boosted choice tree are used, and crime prediction accuracy levels from thirty-nine % to forty-four%.

45.1 Introduction

Criminality is a big menace to the human race. There are several crimes that can be committed. arise at everyday durations. It is viable that it is growing and spreading at a fast and wide scale. Crime can arise around the globe, from a small village to a huge town. Robbery, homicide, rape, assault, false imprisonment, kidnapping, and homicide are examples of diverse types of crimes. Because crime is on the upward

P. Kapoor (✉)
Vellore Institute of Technology (VIT), Vellore Campus, Tiruvalam Rd, Katpadi, Vellore, Tamil Nadu 632014, India
e-mail: punyakapoor122@gmail.com

S. Yadav et al. (eds.), *Proceedings of Second International Conference in Mechanical and Energy Technology*, Smart Innovation, Systems and Technologies 290,
https://doi.org/10.1007/978-981-19-0108-9_45

push, it is miles important to resolve cases lots more quickly. The price of growth in crook hobby has multiplied, and it is far the task of the police branch to manipulate and decrease criminal interest. [1]. Because there is this sort of big volume of crime information, the police department's biggest issues are crime prediction and criminal identity. There is a call for technology so one can permit faster case decisions. The intention could be to expand a predictive model. The training data set would be used, and the check dataset would be used to validate the education. Depending on the accuracy, a better algorithm could be used to construct the version. For crime prediction, the K-Nearest Neighbor (KNN) classification and different algorithms could be employed. The dataset is visualized so as to analyze any crimes that can have happened inside the United States [2].

Crime is a societal issue that has a negative influence on the quality of life and the economic progress. Depending on the type of civilization and community, the intricacies of how crime is committed differ. Previous studies on crime prediction have discovered that characteristics such as education, poverty, employment, and climate have an impact on the rate of crime. Although India's overall crime rate fell by 1.5% in 2017, significant car break-ins and theft remain a problem. [3]. The Police Department (PD) lately deployed a criminal offense-predictive set of rules to expect crimes related to asset damage-ins, and the town saw a 27% lower in home ruin-ins after it was established. A strategy used by law enforcement to forecast crime is known as crime prediction identifies the most possible crimes based totally on records and statistical analysis. In many regions of the sector, this subject has been the subject of ongoing studies.

Machine learning (Fig. 45.1) is used to forecast the destiny based totally on historic records. Machine learning (ML) is an artificial intelligence (AI) method that allows computer systems to learn while not having to be explicitly programmed. The fundamentals of system getting to know, implementation of easy device mastering algorithm the usage of python [4]. Machine learning focuses on the development of laptop programs that can trade when exposed to new facts and the basics of device mastering, implementation of a simple system mastering algorithm the usage of python. Specialized algorithms are used in the education and prediction method. It feeds the schooling data to set of rules, which then applies the training statistics to new take a look at statistics to make predictions. Machine gaining knowledge of can be divided into three distinct regions.

The technological know-how of creating computers makes judgments without human intervention is referred to as gadget learning. Machine gaining knowledge of has lately been used inside the improvement of self-using motors, speech popularity,

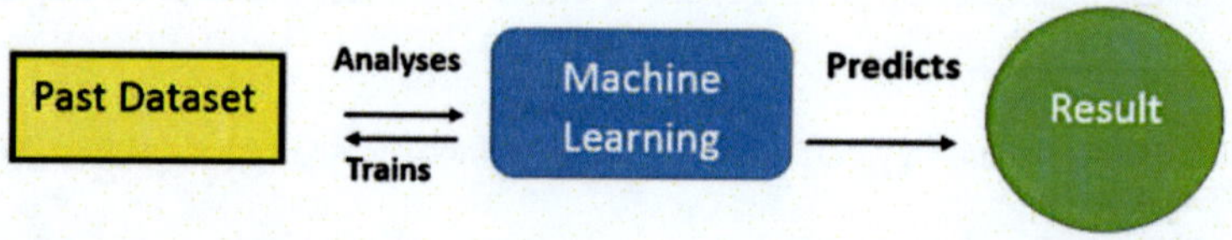

Fig. 45.1 Process of machine learning

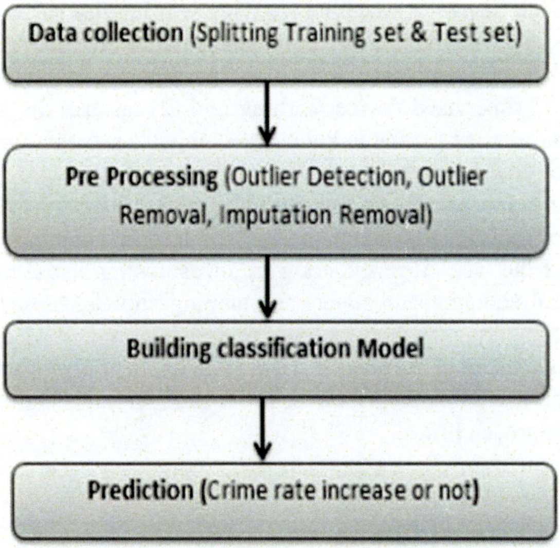

Fig. 45.2 Data flow diagram for machine learning model

net search, and a better expertise of the genome of humans. It has made it as well feasible to assume crime the using referenced facts. Classification is a supervised prediction technique that makes use of nominal elegance labels [5]. Weather forecasting, medical care, finance, and banking are only a few examples of hometown protection, and corporate intelligence is only a few of the areas where class has been carried out.

Machine getting to know is a class of working artificial intelligence with statistical tactics that permit computers to examine from their preceding experiences. The goal of category is to predict a discrete class label, while the undertaking of regression is to are expecting a non-stop amount [6]. This mission aims to forecast the forms of crime so as to occur in a given location. As a result, the type of crime is the have a look act's intention. Some of the techniques that can be used for category include K-Nearest Neighbor (KNN), Support Vector Machine (SVM), Nave Bayesian, Decision Tree, and Ensemble Methods. Every approach has its own advantages and disadvantages in terms of complexity, accuracy, and training time, and distinct effects can be obtained from a single dataset. Figure 45.2 depicts the data waft within the ML working model the usage of a top-down approach.

45.2 Crime Prediction with Machine Learning Algorithms

Machine learning (ML) is a method for a computer system to understand and develop without being explicitly programmed by utilizing previous experiences. After viewing the data, it is not always easy to establish an accurate pattern or information. Machine learning is utilized to interpret the specific pattern and data in such cases [7]. Machine studying is primarily based on the idea that if you give a gadget the right facts, it could learn and resolve each complex mathematical problem and specialized task. In fashionable, there are two sorts of machines getting to know: (1) supervised device learning and (2) unsupervised machine getting to know. In supervised getting to know, the computer is taught the usage of a predetermined set of education examples, while new statistics are provided, it improves its ability to generate accurate and right judgments. Unsupervised studying entails giving a set of statistics to a gadget and asking it to find common styles and relationships most of the facts. Researchers were investigating neural networks, which can be essential equipment in supervised gaining knowledge of because the 1980s [8]. When comparing the random forest and spatiotemporal KDE methods, it became observed that the random woodland is a more efficient algorithm than the same old spatiotemporal KDE approach on the lower time scale and that it outperforms the overall KDE approach [9].

45.2.1 Crime Data Analysis

In the previous decade, crime analysis has emerged as greater spatial The locating of styles or trends the use of strategies from information mining and know-how discovery studies, Combining rule mining, text mining and spatial analysis, and self-organizing maps are only a few examples [10], is any other common method. Various gadget learning algorithms are used to predict crime primarily based on an information set. Machine learning techniques that forecast crime was created using autoregressive integrated moving average models' trends in urban areas [11]. Detecting and analyzing crime patterns is one of the most difficult tasks in the criminal justice system [12]. In this scenario, it is equally vital to understand datasets. We certainly want to be able to predict properly so that we do not spend resources on misleading signals. A strategy for categorizing the crime rate as high, medium, or low was also offered. None of them have classified the types of crimes that could occur and their likelihood of occurring. Crime analysis and prediction are significant activities that can be improved with the use of various approaches and processes [13]. Various researchers in this field have done a great deal of research. The current research is confined to using databases to detect crime hotspots.

To create a reliable prediction version for awaiting crime patterns in metropolitan areas, researchers used a way based on the Auto-Regressive Integrated Moving Average model (ARIMA). The authors advised a probabilistic version of spatial behavior for regarded offenders inside the Metro Vancouver vicinity based on a random-walk-based totally technique to simulate wrongdoer interest [14]. In order to evaluate the role of city indicators in crime prediction in Brazil, the random woodland set of rules changed into applied. A crime-prediction answer for Chilean predominant towns has developed the use of the potential approach, Dumpster-Shafer idea of evidence, and the multi-kernel method [15]. Three strategies were designed, examined, and compared for forecasting crimes within the town of San Francisco: KNN,

Parzen windows, and Neural Networks. The Gradient Boosting Machine (GBM) technology turned into utilized in the beyond to uncover hidden hyperlinks in criminal networks the use of a gadget-mastering prediction model, and the weighted page-rank approach was hired as a powerful tactic to weaken and ruin such networks [16].

45.3 Research Objective

We hired the random woodland algorithm in the research proposed gadget to get higher outcomes and accuracy when compared to the above or current strategies. For accuracy, we appoint random wooded areas. Random forest is a popular and powerful supervised system getting to know set of rules that may perform both type and regression responsibilities. It works by means of education a huge quantity of selection bushes after which. The mode of the teachings (classification) or the mean forecast (regression) of the individual is the output of the class bushes. Random choice forests accurately the tendency of decision trees to overfit their training set. Rainfall, perception, production, and temperature data sets were used to create a random forest, a collection of decision trees based on two-thirds of the records in the datasets. For correct classification, these decision trees are applied to the remaining entries. Random forest algorithm accuracy score is 86.6%.

45.3.1 Random Forest Pros

1. Random Forest uses an Ensemble Learning technique and is based on the bagging algorithm. It grows as many trees as possible from a subset of the data and then aggregates the output of all the trees. As a result, there is less chance of getting into difficulties in decision bushes, as well as a reduction in variance, which enhances accuracy.
2. Random Forest can be used to address problems involving classification and regression..
3. Random Forest can routinely handle lacking values.
4. Four. Random Forest is commonly strong to outliers and might deal with them routinely.
5. Five. Random Forest algorithm may be very solid. Even if a new data point is introduced inside the dataset, the general algorithm is not always affected a great deal since the new statistics can also impact one tree, however, it is far very hard.
6. Random Forest is relatively less impacted through noise. Our proposed system uses random forest which gives accuracy of 85.6%.

45.4 Methodology

45.4.1 Data Source

The source datasets have been accumulated from the city of Vancouver's open information catalog. For this test, two datasets have been used: crime and neighborhood. The VPD has been accumulating crime information considering 2003 and updates it every Sunday morning. Its info the sort of crime dedicated, as well as the date and area of the offense. In the Geographic Information System, the neighborhood dataset contains the borders for the city's 22 neighborhood areas (GIS). The crime dataset is utilized for statistical analysis, while the neighborhood dataset is used to construct maps on this assignment. Table 45.1, which describes the sort of the datasets utilized for this version, gives an in-depth rationalization of the datasets and their attributes.

Statistical Analysis. The crime dataset's distribution is based on the year, month, and day (Fig. 45.3). There are around 31,624 crime events every year, 2720 per month, and 90 per day [17]. As the time intervals grow, the dataset tends to display a normal distribution. The graph of each day, however, contains an unusual maximum value of 650 events, which is thought to be an anomaly - it turns out to be the Stanley Cup riot on June 15, 2011.Trend Analysis.

According to Fig. 45.3, the overall trend shows that the average number of crimes per month decreased from 2003 to 2013 but increased in 2016, and again fell slightly to about 3000 incidents per year in 2018.

The general trend, as proven in the graph, well-known shows that the common number of crimes in line with month reduced from 2003 to 2013, climbed in 2016, and then fell relatively to around 3000 activities according to year in 2018 [18]. The summer season and the middle of each month are the maximum dangerous, in keeping with the time-featured warmness-map plots. On Fridays, Saturdays, and the nights, there also are extra crimes. The warmth map indicated the biggest values near zero hours, which ought to be left out because all the empty records cells had been

Table 45.1 Types of crime

S. No.	Type of crime	No. of records
1	Anti-social behavior	44,070
2	Burglary	22,081
3	Criminal damage and arson	21,333
4	Violent crime	19,673
5	Other theft	19,538
6	Vehicle crime	18,260
7	Drugs	8836
8	Public disorder and weapons	5266
9	Shoplifting	8318
10	Other crime	14,684

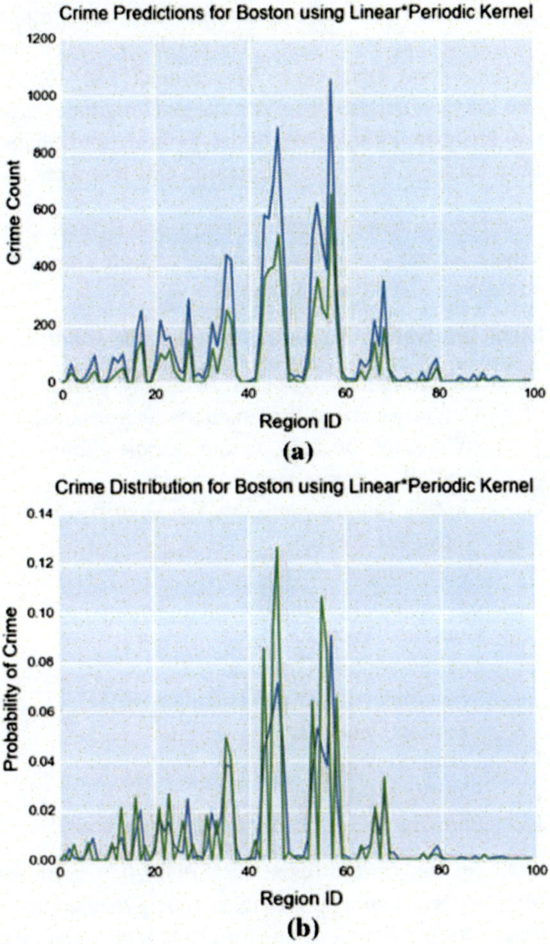

Fig. 45.3 Deviation of predictive data of the proposed model from actual data of the crime **a** Month and year wise. **b** Region-wise

full of 0. The category of theft from cars had the best range of times, observed by way of mischief. However, at the same time as automobile robbery has decreased dramatically in latest years, different forms of robbery have improved in both variety and trend.

Geographical Analysis There are a variety of approaches for mapping hotspots, but choropleth mapping is one of the most used for describing the geographic information of criminal episodes. In a few regions, there was a minor increase in crime, while the remainder of the country was in the typical range (Fig. 45.3b). This makes it simple to spot areas where crime episodes are concentrated, providing insight into criminal behavior [19]

Geographic Information System (GIS) For crime mapping, it has been employed as a strong analytical tool [20]. It displays the series of crime scenes along with diverse geographic data on a single map, assisting police officers in making operational and tactical decisions. The neighborhood boundary dataset was converted from the **Universal Transverse Mercator (UTM)** to the World Geodetic System 1984 (WGS84), often known as latitude and longitude, as the initial step toward geographical analysis. PySal, GeoPandas, Folium, and Shapely are some of the supporting libraries for visualizing geographic data that were utilized to create the map.

Jupyter notebook is the platform, and Python is the programming language [21]. Linear Regression: Nave Baise Gaussian classifier and neural network was the strategy or algorithm I utilized.

Steps involved in Implementation are as follows:

1. Importing packages such as numpy, pandas, pickle, seaborn, matplotlib
2. Defining some utility functions such as normalize_function, randonSplit, rmse, splitLastN, buckets, create Simple Partitions, create Data Matrix, average Prediction, create Heat Map, etc.
3. Plotting function defining(plot Distribution, plot Heat Maps).
4. Normalize the data.
5. Partitioning of data.
6. Split into Train and Test.
7. Gaussian Process
8. Optimization in Gaussian.
9. Train the averages and calculate RMSE.
10. Creating DataMatrix.
11. Count crimes.
12. Plotting histogram.
13. Showing the graph of crime prediction.

In our trials, we utilized CRISPDM (Cross Industry Standard Process for Data Mining) forty-one, a famous facts mining method that become discovered to be nicely proper for predicting duties for crime facts in a comparison of information mining methodologies [22].

45.5 Result Discussions

The data must first be transformed into a readable format before the algorithms can be used. We tested two distinct data-processing procedures and compared the results.

A. **Approach 1**: All categorical variables are converted to binary variables 0 and 1 in the first technique. Every neighborhood and day was turned into a showpiece. Only the correct variable receives a "1," while all other variables receive a "0." All variables having a value of "0" are dummy variables. This provides additional variables for the algorithm to learn and avoids data from skewing to one side. Experiments using skewed data yield a false accuracy rate of 98.9%, which is untrustworthy.

B. **Approach 2:** In the second approach, Numerical variables with unique IDs are created from categorical variables. Different IDs are assigned to different crime kinds and neighborhoods. Vehicle theft, for example, has the crime category ID 10 assigned to it. The work done on Kaggle inspired this technique.

The same algorithms are used in both ways, with the same settings and validation processes. The validity of the classifier algorithm is assessed using fivefold cross-validation. The algorithm is trained on a certain sort of crime. Cross-validation eliminates the problem of overfitting and ensures that the prediction model performs well on new data. The skewed data from approach 1 and the treated dataset using approaches 1 and 2 are shown.

We used official data from the UK Police to develop predictive models for crime frequencies per crime type, per month, and per LSOA code in this article. The analysis' major purpose was to see what kind of performance might be gleaned from 40 months of criminal data. The other purpose was to see how a global model that included all types of crimes compared to a specialized model for a single type of crime. We chose the anti-social behavior crime type for the specialized model since it is the most common of the 16 crime kinds.

45.6 Conclusion

The suggested model aids in the prediction of crime by comparing and compiling data on various crimes that have occurred in recent years, as well as studying and categorizing the data by various contributing elements, which aids in the filtering of the primary cause of the crime. Relationships and patterns have become easier to spot within disparate Thanks to machine learning, we now have access to a plethora of data sets. Present is mostly concentrated on forecasting the type of crime that may occur based on the place where it occurred. We created a model utilizing the principles of machine learning and a training data set that has undergone data cleansing and transformation. The model has a 0.789 accuracy rate in predicting the kind of crime. Crime data from the last 15 years was employed in two separate dataset techniques in this study. The KNN and boosted decision tree machine learning predictive models were utilized to achieve crime-prediction accuracy of 39–44%.

Both the algorithm and the data can be fine-tuned for specific applications to increase prediction accuracy. Although this model's prediction accuracy is low, it does provide a rough framework for additional research. The analysis of a data set is aided by data visualization. Bar, pie, line, and scatter graphs are among the graphs available, each with its unique set of properties. We created a number of graphs and discovered some fascinating statistics that assisted us in better comprehending Chicago crime figures and capturing the aspects that contribute to society's safety.

45.7 Future Scope

The breadth of this development's future is much more than we can anticipate. The project can be rebuilt on a larger scale with technological improvements and more classification models to boost its predicting accuracy which will cause a rise in its profitability.

References

1. Bogomolov, A., Lepri, B., Staiano, J., Oliver, N., Pianesi, F., Pentland, A.: Once upon a crime: towards crime prediction from demographics and mobile data. In: Proceedings of the 16th International. Conference on Multimodal Interaction, pp. 427–434 (2014)
2. Adel, H., Salheen, M., Mahmoud, R.: Crime in relation to urban design. Case study: the greater Cairo region. Ain Shams Eng. J. **7**(3), 925–938 (2016)
3. Overall crime rate in Vancouver went down in 2017, VPD says. CBC News, 15 Feb 2018. Available https://www.cbc.ca/news/canada/british-columbia/crime-rate-vancouver-2017-1.4537831. Accessed:09 Aug 2018
4. Kerr, J.: Vancouver police go high tech to predict and prevent crime before it happens. Vancouver Courier, 23 July 2017 Available https://www.vancourier.com/news/vancouver-police-go-high-tech-to-predict-and-prevent-crime-before-it-happens-1.21295288. Accessed 09-Aug 2018
5. Han, J.: Data Mining: Concepts and Techniques. Morgan Kaufmann, Burlington, Massachusetts, United States (2012)
6. Iqbal, R., Murad, M.A.A., Mustapha, A., ShariatPanahy, P.H., Khanahmadliravi, N.: An experimental study of classification algorithms for crime prediction. Ind. J. Sci. Technol. **6**(3), 4219–4225 (2013)
7. Chen, H., Chung, W., Xu, J.J., Wang, G., Qin, Y., Chau, M.: Crime data mining: a general framework and some examples. IEEE Comput. **37**(4), 50–56 (2004)
8. Beshah, T., Hill, S.: Mining road traffic accident data to improve safety: role of road-related factors on accident severity in Ethiopia. In: Proceedings of Artificial Intellectual for Development (AID 2010), pp. 14–19 (2010)
9. Al Boni, M. Gerber, M.S.: Area-specific crime prediction models. In: 15th IEEE International Conference on Machine Learning and Applications, Anaheim, CA, USA (2016)
10. Zhang, Q., Yuan, P., Zhou, Q., Yang, Z.: Mixed spatial-temporal characteristics based crime hot spots prediction. In: IEEE 20th International Conference on Computer Supported Cooperative Work in Design (CSCWD), Nanchang, China (2016)
11. Mahmud, N., Ibn Zinnah, K., Rahman, Ar, Y., Ahmed, N.: CRIMECAST: a crime prediction and strategy direction service. In: IEEE 19th International Conference on Computing and Information Technology, Dhaka, Bangladesh (2016)
12. Lin, Y.L., Yu, L.C., Chen, T.Y.: Using machine learning to assist crime prevention. In: IEEE 6th International Congress on Advanced Applied Informatics (IIAI-AAI), Hamamatsu, Japan (2017)
13. Bappee,F. K., Júnior, A.S., Matwin, S.: Predicting crime using spatial features, Can. AI 2018: Advances in Artificial Intelligence—Lecture Notes in Computer Science, vol. 10832, pp. 367–373, Springer (2018)
14. Kang, H.W., Kang, H.B.: Prediction of crime occurrence from multi-modal data using deep learning. PLoS ONE **12**(4) (2017)
15. Grover, V., Adderley, R., Bramer, M.: Review of current crime prediction techniques. In: International Conference on Innovative Technology and Applications of Artificial Intelligence, pp. 233–237. Springer, London (2007)

16. Zhao, X., Tang, J. Exploring transfer learning for crime prediction. In: IEEE International Conference on Data Mining Workshop (ICDMW), New Orleans, LA, USA (2017)
17. Marchant, R., Haan, S., Clancey, G., Cripps, S.: Applying machine learning to criminology: semi-parametric spatial demographic Bayesian regression. Secur. Inf. **7**(1) (2018)
18. McClendon, L., Meghanathan, N.: Using machine learning algorithms to analyze crime data. Mach. Learn. Appl.: Int. J. (MLAIJ) **2**(1) (2015)
19. Prabakaran, S., Mitra, S.: Survey of analysis of crime detection techniques using data mining and machine learning. In: National Conference on Mathematical Techniques and its Applications (NCMTA 2018), IOP J. of Physics: Conf. Series, vol. 1000, (2018).
20. Cesario, E., Catlett, C., Talia, D.: Forecasting crimes using autoregressive models. In: IEEE 14th International Conference on Dependable, Autonomic and Secure Compute, Auckland, New Zealand (2016)
21. Tayebi, M.A., Glässer, U., Brantingham, P.L.: Learning where to inspect: location learning for crime prediction. In: IEEE International Conference on Intelligence and Security Informatics (ISI), Baltimore, MD, USA (2015)
22. Alves, L.G.A., Ribeiro, H.V., Rodrigues, F.A.: Crime prediction through urban metrics and statistical learning. Physica A **505**, 435–443 (2018)

Chapter 46
Using Blockchain and Distributed Machine Learning to Evaluate Performance Management and Its Effect on the Construction Industry

Korakod Tongkachok, Asad Ullah, K. V. Pradeep Kumar, Hari Govind Mishra, Harikumar Pallathadka, and Dhiraj Kapila

Abstract Blockchain and machine learning technology are the most important two technologies that help to develop organizational performance in the marketplace. The purpose of this study is evaluation of the performance management and effect of this technology in the construction industry. In the present time, most of the sector has used machine learning and blockchain technology to grow their business. Among these industries, the construction industry is one of the most important sectors that have increased performance through blockchain and machine learning technology. In the construction industry, supply chain management is the most effective aspect. Machine learning and blockchain technology have helped to improve supply chain management. Moreover, the construction industry has improved its performance management through the implementation of 2.0 technology. This research study has selected the secondary qualitative data collection method to get realistic and genuine data for the research study. The teachers have collected the secondary qualitative data from online articles, books, and journals. In this regard, from the secondary qualitative data, the teachers have conducted the thematic analysis to find the authentic result of this study. In this research study, the researcher has analyzed the importance

K. Tongkachok
Faculty of Law, Thaksin University, Khao Rup Chang, Thailand

A. Ullah
Department of Management Studies, Middle East College, Muscat, Oman

K. V. P. Kumar
Department of Mechanical Engineering, Ramaiah Institute of Technology, Bengaluru, India

H. G. Mishra (✉)
School of Business, Shri Mata Vaishno Devi University, Katra, India
e-mail: hgmishra@smvdu.ac.in

H. Pallathadka
Manipur International University, Imphal, Manipur, India

D. Kapila
Department of Computer Science and Engineering, Lovely Professional University, Phagwara, India

S. Yadav et al. (eds.), *Proceedings of Second International Conference in Mechanical and Energy Technology*, Smart Innovation, Systems and Technologies 290,
https://doi.org/10.1007/978-981-19-0108-9_46

of blockchain and machine learning to develop performance management in an organization. After all analysis, this research study concludes that machine learning and blockchain technology have affected the construction industry positively to manage the performance.

46.1 Introduction

Blockchain and machine learning technology play the most effective role to manage the performance of an organization. In the present time, machine learning has helped the organization to develop performance quality and management processes. The blockchain technology has recorded the transaction and tracked the profit of an organization. This technology has collected data in the block system and every block has contained data from the previous block that has created the chain. The construction industry has performed different types of jobs to provide better services to the customers. In this regard, using blockchain and distribution of machine learning has helped this industry to provide better quality services.

46.2 Literature Review

46.2.1 Overview of Blockchain and Machine Learning in an Organization

Blockchain and machine learning are the most effective upgradable techniques that help the organization manage performance. In the present time, every organization has adopted different technologies to improve performance management. The blockchain technology has provided facilities to record data, supply chain management, and analyze organizational performance [1]. The blockchain technology has recorded transaction data in chronological order. In the present time, technological implementation has helped to improve organizational performance. In the business process brass electrode is one of the effective tools that help to improve the business process. In this regard, this tool has provided data security for an organization [2–11]. These upgradable technologies have helped to find issues of the organization. According to the issues the organization can overcome the problems and improve performance management.

46.2.2 Importance of Blockchain and Machine Learning in an Organization

Blockchain technology is the most effective ledger technology that helps to store the data of an organization. Machine learning and blockchain have provided security to the data of an organization. In organizational performance, machine learning and blockchain have impacted positively. Blockchain and machine learning has recording, storage, and updating the data. In organizational management, process transparency is the most effective aspect that helps to manage the organization properly [2]. Blockchain and machine learning has gathered data from different sources to develop the organization. The blockchain and machine learning technology have opened data to all nodes that make the entire system information open and transparent of an organization. In machine learning, MATLAB tool is another most effective machine that helps the business ionization to collect the data and analyze the statistical data to grow the business process. Blockchain technology has some specific rules and protocols that help to formulate the data. In an organization, the blockchain and machine learning has provided a guarantee of correctness and authenticity [12]. The blockchain technology has recorded the data from different nodes and at the same time, it maintains the correctness of data. These all aspects have helped the organization manage performance (Fig. 46.1).

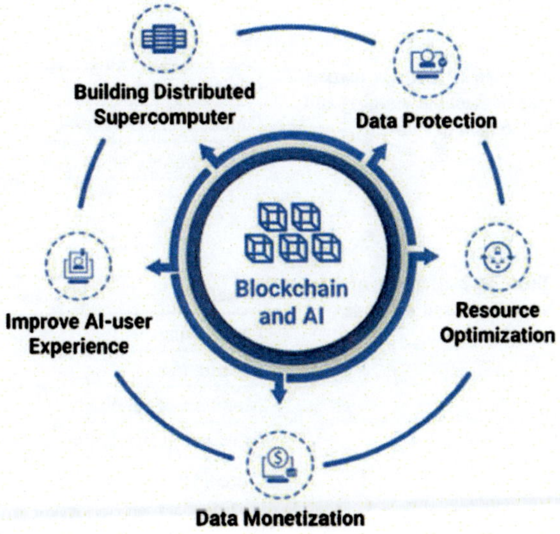

Fig. 46.1 Importance of blockchain and machine learning. *Source* [2]

46.2.3 *Effect of Blockchain and Machine Learning in the Construction Industry*

Blockchain and machine learning technology have developed the organizational function and management process. In the present time, blockchain technology has been used in a wide range to improve organizational performance. Among these industries, the construction industry is one of the most important sectors that need to implement machine learning and blockchain to reduce limitations [12]. The construction industry has adopted the blockchain technology to increase the potentiality to manage performance. In the construction industry, the blockchain has been used mainly for security purposes [3]. In an industry, there are several different types of data and information needed to store important performance. In this regard, blockchain technology has provided security to important data and information. In marketplace the machine learning will increase USD 266.92 Billion and Blockchain will increase USD 2.3 billion in future (Tables 46.1, 46.2).

In the present time, Blockchain and machine learning have provided several benefits to the organization. In an organization, this upgradable technology can save cost of 57% through recording the important information and data. In the marketplace, the organization has faced several risks in its business process. In this regard, the blockchain and machine learning has reduced the risk by 47% [3]. These technological implementations have helped to maintain relevance 42% of an organization. Moreover, it has helped to maintain the standard of an organization.

Table 46.1 Increase market size of machine learning and blockchain in future

Global market	
Blockchain technology	USD 2.3 billion
AI technology or machine learning	USD 266.92 Billion

Source[3]

Table 46.2 Benefits of blockchain and machine learning

Benefits	Percentage (%)
Cost-saving	57
Sharing risk	47
Maintaining relevance	42
Influencing standards	42

Source[3]

46.3 Methodology

In the research study, the methodology is the most effective aspect that helps to get authentic results of the study. The methodology had provided proper systematic techniques to complete this study successfully. In the methodological part, data collection is the most significant aspect that helps to understand the importance of this study. The data collection method is divided into two parts, one is primary and another one is the secondary data collection method [5]. This study has conducted a secondary data collection method to get authentic results. The secondary data collection method has two parts, one is quantitative data collection method and another one is qualitative data collection method.

This study has selected the secondary ***qualitative data collection method*** to complete this study successfully. The researcher has collected the secondary qualitative data from the online sources such as online research articles, books, journals, and authentic websites that were published five years ago. In this research study, from this qualitative data ***thematic analysis*** has been conducted to understand the present condition of this research study [6]. In thematic analysis, the researcher has made some themes to analyze the research topic. These significant themes had helped to analyze the past and present condition of the research topic. This research study has conducted positivism philosophy to increase the significance of this study. Moreover, this research has selected a descriptive research design to describe the research topic and understand the present condition of the research topic.

46.4 Data Analysis

46.4.1 Thematic Analysis

46.4.1.1 Theme 1: Role of Blockchain and Machine Learning

Blockchain technology is the most important decentralized data management technology. In an organization mainly data has been used from different sources. In the other words, in the development context, the organization has used different types of data. In this regard, the blockchain and machine learning has helped to manage the data and give significant information to the organization to develop the performance in the marketplace. In an organization, the blockchain and machine learning are the most effective technologies that help it grow the organization in the marketplace [4]. In the present time, competition in the market has risen rapidly. In this regard, the technological implementation can reduce the competition in the marketplace through the analysis of data. In the organization, the blockchain and machine learning technologies have helped to record the data from different nodes. Machine learning and blockchain have helped to analyze the data and the organization can reduce the cost in the management process [7]. The technological implementation has reduced the cost in an organization.

In an organization, machine learning and blockchain have provided future scope to the organization. The data storage has helped the organization to make outline about the external market. Small diameter holes are the most effective drilling system of business processes. This machine has helped to grow the business in the marketplace through analyzing the business data. In this regard, the organization has made future planning about the organizational performance [13]. Moreover, this data storage and protection increased the performance quality of the organization. In the marketplace, the data of an organization has helped to understand the other organizational performance and capabilities [8]. In this regard, the organization can improve its performance based on the data and achieve more success in the marketplace.

46.4.1.2 Theme 2: Impact of Blockchain and Machine Learning in the Construction Industry

Blockchain and machine learning are the most essential technologies in an organization that can provide beneficial services. In the development context, the organization has used different types of technologies. Among these technologies, the blockchain and machine learning have helped to improve the performance management process. The blockchain technology has gathered data in a systematic way. The blockchain has collected the data from previous blocks of the organization [9]. This technology has helped to exchange information, assets, and digital goods. This technical information is a decentralized network that records information from different nodes and participants. In an organization, the machine learning and blockchain technology have collected information about every participant of the organization. This data has helped to evaluate the performance of participants and can find the issues [4, 10–13]. Based on the data, the organization can provide an effective training process to improve the performance of employees to grow the organization (Fig. 46.2).

The business organization has faced different issues in the marketplace to improve their performance. The non-payment, late payment have created issues for the organizational reputation and performance. In this regard, technological implementation has improved the organizational payment process and increased efficiency [9]. In an organization, big data analysis has played the most significant role to understand organizational performance. In this regard, the organization has adopted different upgradable machines to improve the performance and provide better quality products.

46.4.1.3 Theme 3: Effects of Blockchain and Machine Learning in Performance Management in the Construction Industry

The blockchain and machine learning has developed the organizational performance and in-case efficiency. In the marketplace, most of the sector has used the blockchain

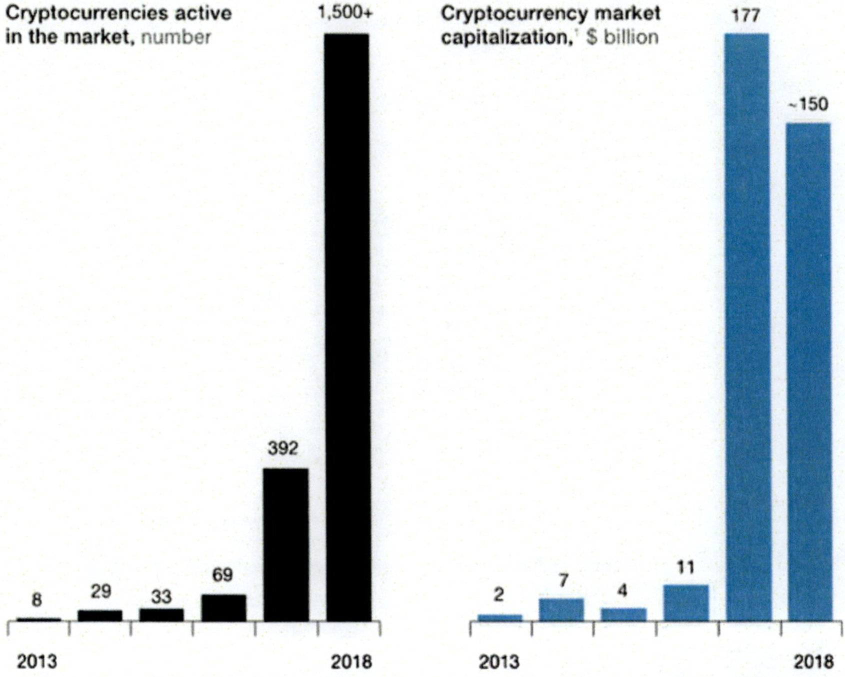

Fig. 46.2 Impact of blockchain and machine learning in the construction industry. *Source* [9]

and machine learning to grow their business. In machine learning, the effective algorithm has helped the business to perform effectively in the marketplace [14]. In this regard, the construction industry is the most effective sector that can improve its performance through the blockchain and machine learning technology [10]. In the construction industry, data security, transparency, efficiency, and cost-effectiveness have helped to develop in the marketplace. At the present time, the construction industry has managed its supply chain management through blockchain and machine learning technology (Fig. 46.3).

The construction industry has collected the data through the blockchain technology and can find the actual issues in the supply chain management that help to improve the function of supply chain management. Moreover, in the construction industry, electronic documents management is the other most effective aspect that helps to improve the performance of the construction industry [4]. At the present time, the construction industry has used 2.0 technology to increase its efficiency in the performance management process. In financial management of the construction industry, the blockchain and machine learning technology play the most effective role [13, 14]. The data storage and data protection has helped to analyze the issues of the construction industry and it has developed the organizational benefit effectively.

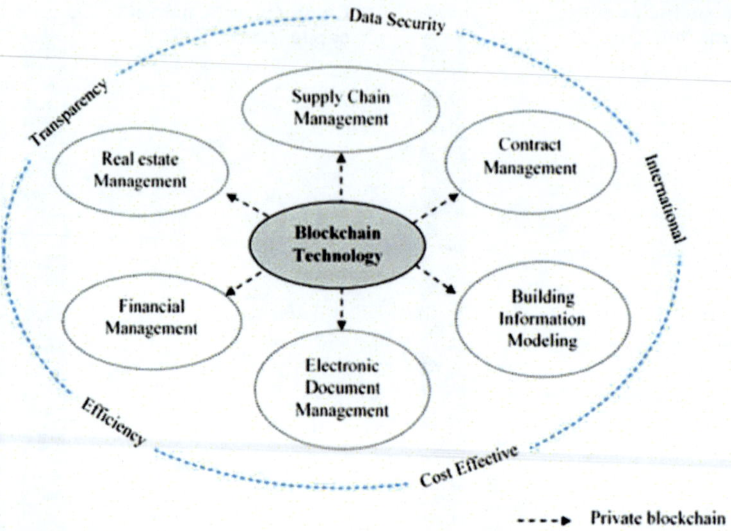

Fig. 46.3 Effects of blockchain and machine learning in the construction industry. *Source* [10]

46.5 Conclusion

After all these discussions it can be concluded that the blockchain and machine learning technology have helped the organization to improve performance management. The construction industry is one of the most effective sectors that use machine learning and blockchain technology to improve the efficiency of the organization. At the present time, the construction industry has used 2.0 technology to improve the supply chain management through storage of the data. Moreover, the construction industry has improved electronic documents management and the use of machine learning and blockchain technology.

References

1. Kaur M, Gupta S (2021) Blockchain technology for convergence: an overview, applications, and challenges. Blockchain AI Technol Industr Internet Things 1–17
2. Abbas, K., Afaq, M., Ahmed Khan, T., Song, W.C.: A blockchain and machine learning-based drug supply chain management and recommendation system for smart pharmaceutical industry. Electronics **9**(5), 852 (2020)
3. Li, J., Greenwood, D., Kassem, M.: Blockchain in the built environment and construction industry: a systematic review, conceptual models and practical use cases. Autom. Constr. **102**, 288–307 (2019)
4. McNamara, A.J., Sepasgozar, S.M.: Intelligent contract adoption in the construction industry: concept development. Autom. Constr. **122**, 103452 (2021)
5. Seo, J., Duque, L., Wacker, J.: Drone-enabled bridge inspection methodology and application. Autom. Constr. **94**, 112–126 (2018)

6. Van der Voet, E., Van Oers, L., Verboon, M., Kuipers, K.: Environmental implications of future demand scenarios for metals: methodology and application to the case of seven major metals. J. Ind. Ecol. **23**(1), 141–155 (2019)
7. Hewa, T., Gür, G., Kalla, A., Ylianttila, M., Bracken, A., Liyanage, M.: The role of blockchain in 6G: challenges, opportunities and research directions. In 2020 2nd 6G Wireless Summit (6G SUMMIT), pp. 1–5. IEEE (2020, March)
8. Liu, Y., Yu, F.R., Li, X., Ji, H., Leung, V.C.: Blockchain and machine learning for communications and networking systems. IEEE Commun. Surv. Tutor. **22**(2), 1392–1431 (2020)
9. Prakash, A., Ambekar, S.: Digital transformation using blockchain technology in the construction industry. J. Inf. Technol. Case Appl. Res. **22**(4), 256–278 (2020)
10. Helo, P., Shamsuzzoha, A.H.M.: Real-time supply chain—a blockchain architecture for project deliveries. Rob. Comput. Integr. Manuf. **63**, 101909 (2020)
11. Jain, A., Yadav, A.K., Shrivastava, Y.: Modeling and optimization of different quality characteristics in electric discharge drilling of titanium alloy sheet. Mater. Today Proc. **21**, 1680–1684 (2019)
12. Jain, A., Pandey, A.K.: Modeling and optimizing of different quality characteristics in electrical discharge drilling of titanium alloy (Grade-5) sheet. Mater. Today Proc. **18**, 182–191 (2019)
13. Jain, A., Pandey, A.K.: multiple quality optimizations in electrical discharge drilling of mild steel sheet. Mater. Today Proc. **8**, 7252–7261 (2019)
14. Panwar, V., Sharma, D.K., Kumar, K.V.P., Jain, A., Thakar, C.: Experimental investigations and optimization of surface roughness in turning of EN 36 alloy steel using response surface methodology and genetic algorithm. Mater. Today Proc. (2021). https://doi.org/10.1016/j.matpr.2021.03.642

Chapter 47
The Role of Artificial Intelligence (AI) in Creating Smart Energy Infrastructure for the Next Generation and Protection Climate Change

Gopal Ramchandra Kulkarni, Manoj Kumar Tamta, Anil Kumar, M. Z. M. Nomani, Charanjeet Singh, and Harikumar Pallathadka

Abstract The application of renewable energy in the world enables in creating of smart energy infrastructure which will help in protecting climate change and prevents the depletion of natural resources. Artificial intelligence in implementing better guidelines for organizing the activities so as to respond to the changing requirements for the next generation. Artificial Intelligence (AI) can enable in overcoming the challenges through adoption of energy-efficient buildings, support in increasing the energy savings, and protect the climate change effectively. The problem related to energy conservation in the building and creation of smart energy infrastructure for the next generation is a multidimensional aspect across countries. It can be stated that AI can enable in offering better solution to the problem through creating smart energy infrastructure which will create comfortable and productive living spaces, it is noted that the complexity of the grids is evolving due to increased energy usage and AI is one of the effective tools which will apply automatic programming which is coupled with the weather conditioning to estimate the full potential of the effective usage of energy and also to take quick action in creating smart energy infrastructure. Through the application of AI, the systems intend to collate and forecast the large

G. R. Kulkarni
Merchant Engineering College, Basna, Mahesana 384380, Gujarat, India

M. K. Tamta (✉)
Department of Geography, Govt. PG College, Bageshwar, Uttarakhand, India
e-mail: manojtamta6@gmail.com

A. Kumar
Department of Botany, DDU Gorakhpur University Gorakhpur, Gorakhpur, India

M. Z. M. Nomani
Faculty of Law, Aligarh Muslim University, Aligarh, India

C. Singh
Electronics and Communication Department, Deenbandhu Chhotu Ram University of Science and Technology, Murthal, Sonipat, India

H. Pallathadka
Manipur International University, Imphal, Manipur, India

S. Yadav et al. (eds.), *Proceedings of Second International Conference in Mechanical and Energy Technology*, Smart Innovation, Systems and Technologies 290,
https://doi.org/10.1007/978-981-19-0108-9_47

volume of information received through the smart grid which is then used to take timely decision making for saving the energy usage and protect climate change. This paper uses secondary data sources in exploring the evolving role of AI in the creation of smart energy infrastructure and provides the necessary steps of implementing such technologies for protecting the climate change in a sustainable manner.

47.1 Introduction

The role of Artificial Intelligence is getting more popular and significant in the energy sector as it possesses more potential in designing efficient energy infrastructure for the energy sector and thereby support in protecting the climate change. The major areas of application of AI in the energy infrastructure revolve around using smart grids, operating the cooling equipments based on the changes in the climate and occupancy levels, electronic trading, and coupling heat and transport. One of the fundamental aspects in the application of AI is the need for distribution of energy and related infrastructure so as to meet the growing energy needs of the individuals and also to protect the climate change thereby safeguarding the environment for the next generation [1]. It has been stated that AI tends to support the energy sector in securing, analyzing, and evaluating the data related to energy, heat, lighting, emission of harmful gases, and forecasting the future energy requirements.

AI is mainly applied in the area of intelligent networking of power and generators across various places, the overall digitization of the power grid has created the opportunity in managing the large number of grids in an effective manner so as to keep the grid in balance, offer better services to the stakeholders, however, this needs more data and information, apply various models in processing the data so as to take quick decisions. Smart grids are considered as the main area of application which enables in transport of electricity and detailed information, with the increase in volatile power generating plants like solar panels, wind turbines, and other modes of renewable sources, hence it is highly critical for power generation to analyze and source the power in an intelligent manner so that the power consumption can be made effectively. Furthermore, it is widely specified that AI can enable in evaluating, analyzing the data from different stakeholders like the manufacturers, energy storage and supply chain companies, consumers, government, etc. so as to connect and analyze their requirements effectively through the smart grid [2]. The main focus of the energy sector is the overall integration of the mobility, moreover, there is an increased focus from government and consumers in using electric vehicles like e-bikes, e-cars, etc. which offers more opportunities for the entire sector in using the available resources to optimize efficiency. Moreover, AI enables in stabilizing the grid through analyzing the anomalies in the power production, transfer, and consumption of the power in rea; time, enable in creating suitable solutions. Moreover, AI tends to coordinate the maintenance work and determine the optimal aspects for the maintenance of networks or the individual systems which supports in reducing the minimizing the cost and enhancing profits.

47.2 Literature Review

AI is considered to be a key factor in the industry as it enables in collating large volume of information in their complex systems the energy sector can be enhanced by stimulating AI in enhancing the overall operational aspects, maintenance, and storage of the energy mainly in the renewable sector and timely delivery of the services. The integration of renewable energy through application of AI is mainly stated as follows:

- Enhancing the grid stability, creating safety in operations
- Managing the supply constraints
- Forecasting the demand and also analyzing weather changes
- Creating better energy storage
- Maintaining better market design

Increased connectivity among the grid components and also better interaction with microgrids for efficient supply of energy (Table 47.1).

The researchers have stated that the implementation of AI can enhance the reliability of renewable energy through analysis of different metrological information, this enables in predicting the weather condition, manage the output of energy, storage, and distribution in an effective manner, furthermore, AI is applied in smart grids so as to make clear balances among them, the role of AI is to analyze the grid before and after the units are absorbed and help in minimizing the congestion in renewable energy generation [3].

Another key role of AI is involved in identification and forecasting of faults in the energy industry, also it supports real-time maintenance of the application, assesses the ideal schedules, and proactively uses the grids for optimal performance. In the energy industry, there is highest form of equipment failure and it impacts the overall production and supply of energy to the stakeholders [4]. AI associated with the key sensors enables overall monitoring of the equipments and also support

Table 47.1 Role of artificial intelligence in creating smart energy infrastructure

Area of application	Critical roles performed
Power grid	Smart grids Monitoring the grid effectively Maintenance and coordination with other grids
Consumption	Creating better and smart homes and offices
Power plant	AI enables virtual power plant Decentralization
Trading	Efficient forecasting and analysis Weather analysis Implementation of algo trading

Source Das et al. [3]

the producers in detection of failures thereby enhancing operational efficiencies, reducing operational cost, and other aspects.

AI supports in creating smart homes through usage of devices like Alex, Sir, etc. which enable the individuals to interact with the energy systems in the home and monitor the energy consumption. This digitization in the home energy allows automatic meters in using AI for optimal usage of energy in the household.

47.3 Research Methodology

The researcher applied descriptive research study in order to understand the role of AI in making smart energy infrastructure for the next generation and also to protect the climate change in a sustainable manner. This paper is a qualitative study in understanding the different AI-enabled practices so as to create smart infrastructure in the energy sector. Furthermore, the researcher focuses on collecting the data from the secondary data sources like published journals, research papers and other related materials. Most recent publications were sourced to understand the relevant practices employed by many countries around the globe.

47.4 Discussion and Analysis

Energy storage supports in solving critical issues related to the intermittent nature of renewable energy, the overall dynamics not only on the sources related to the demand itself. With the overall integration of the AI and other communicating tools, the smart grid tends to integrate with the renewable energy through cyber layers. A highly integrated energy infrastructure is considered as the key determinant in the application of smart grids, this can be achieved through the implementation of AI in the energy sector [5].

The effective usage of AI in the energy sector, mainly in renewable energy industry enables adding more economic value to the projects as it supports in string the power during the peak production and can be used when there is strong demand for the need for energy. In order to enhance the optimal utilization of the system and to enhance the storage system AI enable in providing more possibilities, this will enable in performing the critical analysis which are highly essential in maintaining the production, management of grids, and addressing the latencies effectively.

In order to meet the growing energy needs and also to protect the environment from emission of harmful gases, creating sustainable and intelligent energy infrastructure is the need of the hour. Hence, industry leaders and governments are focusing on implementing AI for creating better and smarter energy infrastructure. The implementation of AI-enabled smart grid focuses on collecting the information and synthesizing them into smart sensors so as to make quick decisions [6]. Moreover, the advances which are made through deep learning and recolonize the demand for energy-efficient economy.

The implementation will enable in creation of micro grids so as to meet the local energy with better resolutions, these aspects can also be coupled with storage technologies so that the future demands can be met effectively. Moreover, studies have stated that the integration of AI with cloud computing intends to increase the robustness and minimize the outages. This enables in creating efficient role in management of smart grids. Furthermore, fog computing supports in analyzing the data locally than processing the information and analysis into cloud, hence through providing on-demand resources, the fog computing offers more advantages in terms of flexibility, scalability of operations, and energy efficiencies.

The smart grids are mainly embedded with the critical information which will offer better communication among different components so that appropriate changes in the energy sector can be achieved. This nature of information layer is mainly created with the installation of sensors, meters for data storage, analysis, etc. [7]. The implementation of the phasor measurement units is essential for creating AI featured smart grids as they provide real-time analysis of the data from different remote points in many grids, this creates a precise and compressive view of the power system and thereby optimizes the production thereby supporting in climate change.

In the coming years, AI coupled with cloud computing and other technologies is expected to increase the operational efficiencies mainly in the energy sector through automation of the operation in the renewable industry, this supports in protecting the climate change, implementing smart infrastructure for enhancing the output. AI further supports framing and creates opportunities in enhancing efficiency based on the technology. These aspects may refer mainly to cost savings, flexibility in operations, etc. for maximizing the production and output.

By integrating advanced computer and communication technology, the smart grid integrates distributed and green energy into the grid, adds a cyber layer to the grid, and provides bidirectional flow and data communication. However, this exposed the smart grid to a number of security issues due to the complexity of smart grid systems and the inherent weakness of communication technology [8]. The most likely consequences of cyber-attacks on smart networks are malfunctions, loss of synchronization, power outages, loss of synchronization, power outages, significant material losses, social damage, data theft, intermittent interruptions, and power failure. AI helps manage energy production and consumption in an ever-changing market environment. The variability of renewable energy sources is one of the biggest challenges in this sector, with increasing importance as the share of AC in total energy production increases [9]. Artificial intelligence is the solution that must be integrated at the microeconomic level to reduce the risk of renewable energy, volatile predictive analysis, standardization, reduce storage costs, provide better network-to-user connectivity, providing stability, reliability, and sustainability. The synergy between renewable energy and artificial intelligence will change the energy sector and improve sustainability nationally and globally. The use of artificial intelligence can increase the efficiency of the renewable energy industry by detecting and predicting patterns, performing specific

Table 47.2 Renewable energy indicators

Particulars	2010 (%)	2019 (%)	% Change
Overall renewable share	14.42	19.73	5.31
Renewable energy related to transport	5.50	8.90	3.40
Renewable electricity generation	21.30	34.10	12.81
Renewable heating and cooling	17.02	22.10	5.08
Greenhouse gas emissions—energy in Mt	1241.855	890.3591	– 28.30

Source DG Energy [11]

tasks without explicit human instructions, optimizing supply, and improving efficiency. It provides a better picture of the process by predicting speed and intelligent connections between vital elements through the rapid development of technology that incorporates artificial intelligence (Table 47.2).

The EU is one of the world's leading players in the renewable energy sector. Environmental aspects have been incorporated into specific regulations at EU and Member State levels. The use of artificial intelligence in this area is important to win the fight against the deterioration and depletion of internal resources but to successfully overcome global competition (China and the United States). Even at the microeconomic level, it is difficult to quantify the use of artificial intelligence (e.g., intelligent storage, micro-networks, central control systems), artificial intelligence from a macroeconomic point of view and the results generated due to a number of factors affecting the outcome of financial processes [9].

Given the development and current efficiency of the analyzed indicators and processes, we conclude that artificial intelligence is widely used within the EU in BR processes, from technology to labor. The prerequisites and requirements for artificial intelligence are the provisions of state-of-the-art technology, as evidenced by the efficiency of transformation processes and structural changes in renewable energy sources from resources associated with a forced and highly skilled workforce [10]. But progress has also been made and the EU as a whole is effectively meeting the goal of renewable energy, set out in the Rules of Procedure for the R & D sector's participation in energy consumption [11]. Raw final energy, a new advance is based on the increasing development of artificial energies. Information is needed to make this sector a leader in the energy sector in all EU Member States (Table 47.3).

AI is involved in detecting and predicting power outages, supporting real-time maintenance, evaluating optimal schedules, and proactively using networks for optimal performance. In the energy sector, there is a large form of equipment failure, which affects the affected total production and energy supply [12]. Artificial intelligence associated with central sensors enables full control of equipment and also helps manufacturers detect faults, which improves operating efficiency, reduces operating costs and other considerations [13]. To meet the growing demand for energy and protect the environment, it is necessary to create a sustainable and smart energy infrastructure against harmful gas emissions [14]. Therefore, industry leaders and the government are focusing on using artificial intelligence to create a better and smarter

Table 47.3 AI-based interactions in energy sector

Specifications	Interactions
Energy generation	Enable in forecasting the weather Optimize performance Manage supply chain effectively
Grid management	Ensuring grid reliability Enhanced connectivity Predict patterns and adapt to it Efficient storage system
Stakeholders' benefits	No latency in supply Low cost Improved efficiency

Source Cui et al. [12]

energy infrastructure. Implementing an AI-enabled smart grid focuses on gathering information and synthesizing it into smart sensors for quick decisions [15]. Through in-depth learning, further progress is made and the demand for an energy-efficient economy is recolonized.

47.5 Conclusion

Artificial intelligence is mainly used in smart grids and generators in various places, the global digitalization of the grid has made it possible to efficiently manage a large number of grids to maintain grid balance, various data processing models are used to streamline decisions. Smart grids are considered to be the main application area, which enables the transmission of electricity and detailed information when the number of power plants such as solar panels, wind turbines, and other renewable energy sources increases, so it is very important for production to analyze energy consumption efficiently. There is no doubt that artificial intelligence can provide a better solution to the problem if it creates a smart energy infrastructure that creates a comfortable and productive space. Efficient tools that implement automated scheduling along with weather conditions to assess the full potential for energy efficiency and take rapid action to build a smart energy infrastructure.

References

1. Puri, V., Jha, S., Kumar, R., Priyadarshini, I., Son, L.H., Abdel-Basset, M., Elhoseny, M., Long, H.V.: A hybrid artificial intelligence and Internet of Things model for generation of renewable resource of energy. IEEE ACCESS **7**, 111181–111191 (2019)
2. Jha, S.K., Bilalovic, J., Jha, A., Patel, N., Zhang, H.: Renewable energy: present research and future scope of artificial intelligence. Renew. Sustain. Energy Rev. **77**, 297–317 (2017)

3. Das, U.K., Tey, K.S., Seyedmahmoudian, M., Mekhilef, S., Idris, M.Y., Van Deventer, W., Horan, B., Stojcevski, A.: Forecasting of photovoltaic power generation and model optimization: a review. Renew. Sustain. Energy Rev. **81**, .912–928 (Part 1) (2018)
4. Mosavi, A., Salimi, M., Ardabili, S.F., Rabczuk, T.: State of the Art of Machine learning models in energy systems, a systematic review. Energies **12**(7), 1301 (2019)
5. Pourbabak, H., Chen, T., Zhang, B., Su, W.: Clean Energy Microgrids, Shin'ya Obara Publisher: The Institution of Engineering and Technology (IET), 2017 (2017)
6. Grecu, E., Aceleanu, M.I., Albulescu, C.T.: The economic, social and environmental impact of shale gas exploitation in Romania: a cost-benefit analysis. Renew. Sustain. Energy Rev. **93**, 691–700 (2018)
7. Arrieta, A.B., Díaz-Rodríguez, N., Del Ser, J., Bennetot, A., Tabik, S., Barbado, A., García, S., Gil-López, S., Molina, D., Benjamins, R., et al.: Explainable Artificial Intelligence (XAI): concepts, taxonomies, opportunities and challenges toward responsible AI. Inf. Fusion **2020**(58), 82–115 (2020)
8. Jain, A., Yadav, A.K., Shrivastava, Y.: Modelling and optimization of different quality characteristics in electric discharge drilling of titanium alloy sheet. Mater. Today Proc. **21**, 1680–1684 (2019)
9. Jain, A., Pandey, A.K.: Modeling and optimizing of different quality characteristics in electrical discharge drilling of titanium alloy (Grade-5) sheet. Mater. Today Proc. **18**, 182–191 (2019)
10. Jain, A., Pandey, A.K.: Multiple quality optimizations in electrical discharge drilling of mild steel sheet. Mater. Today Proc. **8**, 7252–7261 (2019)
11. DG Energy Statistics. Available https://ec.europa.eu/energy/en/data-analysis/energy-statistical-pocketbook#content-heading-1 (2019)
12. Cui, L., Qu, Y., Gao, L., Xie, G., Yu, S.: Detecting false data attacks using machine learning techniques in smart grid: a survey. J. Netw. Comput. Appl. **2020**, 170, 102808 (2020)
13. Sirojan, T., Lu, S., Phung, B., Zhang, D., Ambikairajah, E.: Sustainable deep learning at grid edge for real-time high impedance fault detection. IEEE Trans. Sustain. Comput. **2018** (2018)
14. Panwar, V., Sharma, D. K., Kumar, K. V. P., Jain, A., Thakar, C.: Experimental investigations and optimization of surface roughness in turning of EN 36 alloy steel using response surface methodology and genetic algorithm. Mater. Today Proc. (2021)
15. Amroune, M.: Machine learning techniques applied to on-line voltage stability assessment: a review. Arch. Comput. Methods Eng. **2019**(28), 273–287 (2019)

Chapter 48
Covid-19 Interior Security Tracking System Based on the Artificial Intelligence

Alim Al Ayub Ahmed, Nimit Gupta, Rumi Iqbaldoewes, Muddada Murali Krishna, Rumki Bandyopadhyay, and Md. Khaja Mohiddin

Abstract This article presents an inexpensive artificial intelligence solution aimed at increasing indoor safety of COVID-19, including a number of important aspects: (1) breakdown of the process (2) Method for mask identification (3). Assessment methodology of social distancing The Arduino Uno sensor system uses an infrasound sensor or heat camera, whereas the Raspberry Pi is equipped with computer vision technologies for mask detection and social distance checks. Indoor measures are the most prevalent—people with a high body heat should stay at home, masks should be worn, and their distance should be at least 1.5–2 m. In the first case, the Arduino Uno temperature sensor board is utilized, while we utilize a single-board Pi Raspberry computer coupled with camera for two additional situations, using computer vision techniques. Due to their compact size and cost, we chose to utilize these devices.

A. A. A. Ahmed
School of Accounting, Jiujiang University, 551 Qianjindonglu, Jiujiang, Jiangxi, China

N. Gupta (✉)
School of Management and Liberal Studies, The Northcap University, Gurugram, India
e-mail: nimitgupta1981@rediffmail.com

R. Iqbaldoewes
Faculty of Sport, Universitas Sebelas Maret, Jl. Ir. Sutami, 36A Kentingan, Surakarta, Indonesia

M. M. Krishna
Computer Science and Engineering Department, Dr. Lankapalli Bullayya College of Engineering, Visakhapatnam, India

R. Bandyopadhyay
University of the People, Pasadena, USA

Md. K. Mohiddin
Bhilai Institute of Technology, Raipur, Chhattisgarh, India

S. Yadav et al. (eds.), *Proceedings of Second International Conference in Mechanical and Energy Technology*, Smart Innovation, Systems and Technologies 290,
https://doi.org/10.1007/978-981-19-0108-9_48

48.1 Introduction

Since the final few days of the previous year, SARS-Cov-2 viruses (also known as coronavirus) have impacted nearly every area of human life globally and created new contagious flu-like airborne illness COVID-19. First, it was found in China but in only a few weeks it rapidly spread to other continents. Coronavirus illness is commonly seen in fever, fatigue, sore throat, nasal congestion, taste loss, and odor. It is usually directly (personally and personally) transferred via respiratory outlets and indirectly also via surfaces. Visitation may be fairly long and varying (between 15 and 30 days in extreme cases). Moreover, even asymptomatic individuals (nearly 40% of cases) might spread the illness, exacerbating the issue [1].

The use of facial masks and sanitizers has therefore demonstrated excellent effects in the prevention of disease transmission. The key difficulty, meanwhile, is that the vaccination and medicine are not licensed. Because of this, governments have adopted many safeguards and measures for the reduction of disease spread: compulsory indoor wearing of masks, social distance, quarantine, self-isolation, restriction of the movement of citizens both within and outside country borders, and in many cases along with prohibition and the cancellation of massive public events and meetings. Although at certain moments the pandemic appeared to be weaker, most safety restrictions remain in place because of the unpredictable scenario. Coronavirus illness alters our daily routine, habits, and activities, from job to social connections, athletics, and leisure [2].

This article describes a cost-efficient artificial intelligence solution designed to enable companies to comply with the safety and standards laid out by COVID-19 to limit the spread of the illness. We focus on most popular indoor precautions—people who are at high body temperatures should stay at home, have masks, and should have at least 1,5-2 meters of distance from individuals. For the first case, Arduino Uno is utilized as a 1-board microcontroller with contactless temperature sensor while relying on a 2-board Raspberry Pi 2 computer coupled with camera for another two situations. Because of their compact size and cost, we chose to utilize these devices [3].

48.2 Presentation of Prior Information

Apart from the content itself and associated applications, the function of somatology is to encode the meaning of data, to offer the capacity to comprehend and distribute information, and to make sense of this. In this scenario, knowledge formalization is performed in a way that both human beings and robots can comprehend. The data were presented in relation to ontology inside the concept knowledge bases. Categories, persons, characteristics, and relationships are an ontology. Class is a group of abstract objects of the same kind. Individuals in these groups are examples. Shared hosting to class characteristics. Relationships also define methods to link classes and people. The RDF standardized is utilized for both ontology and fact. It creates triples, which remain in the triple store of meaningful (subject, predicate, object). For the implementation of queries against semantic triple stores, SPARQL is on the other side. Findings are achieved to support various argumentation methods, so that relevant revelations may be derived from the existing set of information [4].

IoT devices frequently utilize ontologies for interoperability and the integration of data from diverse devices and sensors in order to unify their control. In particular, Smart Grid Response Ontology has been created to help rationalize and provide appropriate responses in certain contexts on the events that took place within the artificial intelligence-based system for smart grid tracking. In this article, in the context of the COVID-19 safety control, we apply a similar approach to semantic representation but augment it with the components of abstract awareness [5].

48.3 Open-Source Computer Vision Library

The open-source computer vision library was utilized for the development of mask recognition and methods to verify social distance in the open-source computer vision library [6] (Fig. 48.1).

Face and body recognition methods are built on an existing Viola-Jones Object Detection Framework based on hair cascades from open-source computer vision library. Castor functions are developed from a large number of good and negative pictures. It is a machine learning method. This method is then utilized in new pictures to detect things. The open-source computer vision library includes a trainer as well as a detector. Open-Source Vision Librarians offer preset classifiers, such as visual perception, body, body, and facial sections for the detection of frequent items (both front and back for some of them). Therefore, we use the existing open-source

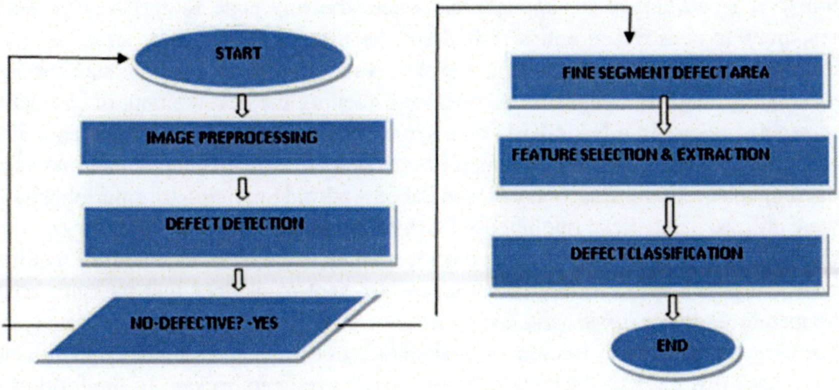

Fig. 48.1 Detection and classification

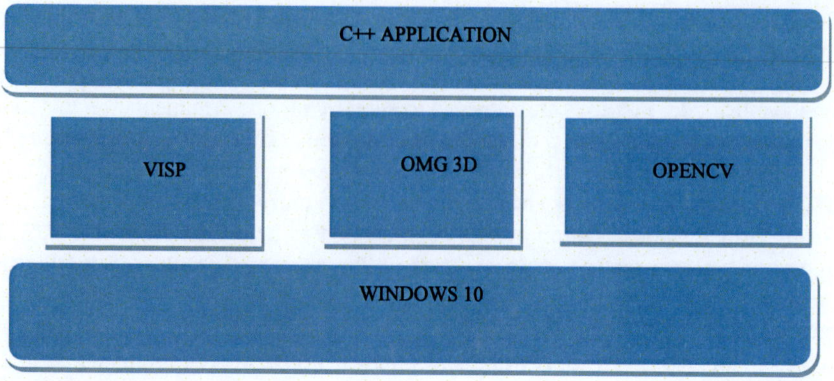

Fig. 48.2 Proposed system VISP

computer vision library classifications in this work since they are sufficient to meet the demands of the solution developed [7].

The open-source computer vision library has been used to operate Raspberry Pi devices within arts and historical heritage booths that demonstrate an adequate result even in instances of real-time. This detector is used to monitor the multimedia reproduction systems [8] (Fig. 48.2).

48.4 Message Queuing Telemetry Transport

Messages queued telemetries transported between Raspberry Pi, Arduino, Edge servers, and smartphones were authorized to figure device communication. It is a lightweight, TCP/IP message subscription protocol. Message queuing telemetry transport is developed for applications where the tiny code footprint or network resources is constrained and is suited for the usage of low-power smart devices in this study. In addition, a message broker is required for publishing subscription messaging methods. We utilize the message queuing the development of telemetry transport brokers in a Node-RED on a server located on the Edge to that end. The associated message queuing customer libraries–utilize for Arduino, Pahoa-message queuing telemetry for Raspberry Pi, and Pahoa Android Servicing for smartphones—were utilized in Artificial Intelligence Devices. Messages involving telemetry confirmation messages to Edge servers in instances when someone does not fulfill requirements to pass specific safety check stages are sent to devices that measure body temperature, detect masks, and social distance. Moreover, the Edge server analyses the message received by the security personnel concerned and transfers it to them to alert them that COVID-19 security regulations have been broken. In the form of a JSON encoded string, each communication is sent [9].

48.5 Breakdown of the Process

The following subsystems are our solution: (1) Arduino Uno-based temperature tracking subsystem (2) Raspberry Pi-based computer vision detecting subsystem and social dissociation control (3) Side of the server (4) Security guards smartphone application. First, all those who try to enter the building must verify the temperature without touch. We use Arduino Uno with an infrared or thermal detection technique to achieve these goals. In addition, WIFI employs Telemetry Message queuing protocol to communicate with the Edge servers. If the body temperature of the individual is greater than usual, the door will be blocked and messages containing both the temperature and the place where the signal was captured, will be transmitted to the computer. This signal is sent to the server, it is analyzed, and forwarded to the security guard smartphone application, in order to ensure that this individual does not try to enter the facility. If not, Arduino sends a signal to open the door if the passenger's temperature is normal. After the checking, travelers move to the next phase. The Raspberry Pi single-board computer equipped with the version 110 Camera Module revision was applied for this assignment. If passengers do not wear a mask or do not wear a nose, safety guards will be notified by telemetry messaging to give a mask or warn that they will depart [10].

However, the door will be opened if the individual under control wears a mask. In addition, Raspberry Pi devices verify whenever they enter the building whether social distance is effectively imposed or not at a certain area. Similarly, a message is delivered to notify safety officials when the social distance in some rooms is not correctly implemented. On the server side, the message queuing transmission broker and checking triple storage is installed, while the analysis of messages, event logging, reasoning, and message transmission is performed. Adjudicate are sent communications, are annotated and rationalized to locate the proper security guard to be informed. A basic mobile Android application for security guards receives messages from the server side and displays data on rule breaches and places within the building. An example of the suggested system based on Artificial Intelligence, which is designed to assure the application of COVID-19 safety directives inside [11].

Overview of the artificial intelligence interior surveillance system COVID-19: 1-Arrival of passengers 2-Value for temperature 3-Door open/close signal 4-Message queue telemetry transit alerting that someone's body temp is greater than normal 5-wears/does not wear a mask 6-Satisfied/unsatisfied with social distancing 7-Telemetry message alerting that somebody attempts to enter without a mask 8-Telescopic message advising passengers that social distance measures are not respected 9-Message queue message notifications of telemetry transfer to the smartphone of security workers regarding breaches of COVID-19 safety precautions in different building [12].

48.6 Method for Mask Identification

We depend on the following open-source vision library classifier to construct the mask detection algorithm: (1) harassed frontal face default—utilized for frontal face detection. (2) Harassed mgs moth—the mouth of man recognizes the image given. (3) nose-detecting moth. The technique described in list 1 is run for each frame coming from the camera stream (Table 48.1).

The first is to transform the camera frame to a gray-scale picture for the face detection, as needed by the hair cascade classification of the open-source computer vision library. Furthermore, a fresh copy of the camera framework is also produced with a black and white version. It was scientifically noticeable that in most cases people with a white mask do not recognize the face correctly using open-source vision library classifiers if the gray image is utilized instead of with black and white pictures. Face detection against pictures is then conducted. Across both situations, if the array length of the identified face is 0, then there is no human being present in the view of the camera. Instead, mouth and nose detection for the appropriate camera frame version is used if a face is identified. If the picture does not include nose and mouth, then the mask is correctly wearing and the door opens. But if the mouth is identified and its coordinates inside the area of the face detected, it is advised to take a mask for action. The user is admonished to put the mask correctly if the

Table 48.1 Mask identification pseudo-code steps

1	Convertor(Image); gray image
2	Concerto(gray image); bw image
3	Detect Faces(gray image); faces
4	Detect Faces (bw image); faces bw
5	if(faces.length = 0, faces bw.length = 0)
6	"No face found!" is the label
7	Other
8	(faces length = 0 and faces bw.length > 0)
9	DetectMouth(bw image); mouths
10	DetectNose(bw image); noses
11	Other
12	DetectMouth(gray image); mouths
13	DetectNose(gray image); noses
14	Break
15	whether(mouths.length = = 0 and noses.length = 0)
16	"Thanks you" = label
17	OpenDoor
18	Other
19	In mouths for every m
20	for each of the f in faces

Table 48.2 Assessment methodology of social distancing

1	Convert Gray(image); gray image
2	Detect Faces gray_image = bodies
3	if(bodies.length ≤ 1)
4	"There aren't enough people for a check!" = text
5	Other
6	each b1 bodies
7	each b2 bodies
8	d = sqrt((b1.x-b2.x) + (b1.y-b2.y)2);
9	ConvertPixelsToMeters = dm
10	(b2 $\neq$ b1 and dm_ threshold)
11	"Social distancing not applied" = label;
12	(social distancing alert, location name); Send MQTT
13	Break if
14	Break if
15	return label;
16	Break

nose is discovered in the facial area. The method works both individually and multi-personally. It is thought that people pass the camera one by one in the single-person mode. In multi-person mode, it is possible for each pair of faces and face noses to determine if the mouth and nose coordinates reside in one face since more than one person may move in the immediate vicinity of the camera [13].

Within the recorded image it uses the haar cascade full-body classifier of the Open-Source Computer Vision Library for human body detection. The previously published social remote control method computer-aided diagnostic is presented in list 2 (Table 48.2).

Each camera frame is transformed into a gray image in the same way as the mask detection method. In addition, the detection of the body is carried out. The length between both individuals is computed and compared with a criterion in meters specified when more than one human body is identified. All distances, depending on the camera properties and the location of the item before comparison, should nonetheless be normalized. The formulation is used for mapping pixels to actual-world distances in *m*.

$$\text{Image Dimension/Focal Length} = \text{Dimension Object/Object Distance}$$

In a particular scenario, the social gap between the two institutions is more or equal to a criterion. If not, the notification is delivered to the server and security operator when this condition is not for at least one pair of bodies.

48.7 Conclusion

The approach suggested is usable for specific performance limits in accordance with the findings attained (such as number of processed frames or measurements per second). It depends both on open hardware and free software to provide such systems with a distinct and desired advantage. In the future, various deep learning and computer vision framework frames to identify objects on Raspberry Pi are intended to be experimented with to obtain a better fps. Furthermore, with environmental sensing methods for adaptive air-conditioning in buildings and airborne ventilation protection, we want to expand this approach to decrease the propagation of coronavirus indoors, particularly in summertime. Eventually, the ultimate idea is to incorporate the system presented in this paper is a framework for efficient resource planning during pandemic crises, so as to provide efficient planning and allocation of masks to security personnel and to evaluate risk based on statistics on safety and air quality guidelines.

References

1. Viola, P., Jones, M.: Rapid object detection using a boosted cascade of simple features. In: Proceedings of the 2001 IEEE Computer Society Conference on Computer Vision and Pattern Recognition. CVPR 2001, vol. 1, pp. I–I. IEEE (2001, December)
2. Davies, J., Studer, R., Warren, P.: (eds.).: Semantic Web Technologies: Trends and Research in Ontology-Based Systems. Wiley (2006)
3. Kings, N.J., Gale, C., Davies, J.: Knowledge sharing on the semantic Web. In: European Semantic Web Conference, pp. 281–295. Springer, Berlin, Heidelberg (2007, June)
4. Davies, J., Lytras, M., Sheth, A.P.: Guest editors' introduction: semantic-web-based knowledge management. IEEE Internet Comput. **11**(5), 14–16 (2007)
5. Davies, J., Studer, R., Sure, Y., Warren, P.W.: Next generation knowledge management. BT Technol. J. **23**(3), 175–190 (2005)
6. Cob, Z.C., Abdullah, R.: Ontology-based Semantic Web services framework for knowledge management system. In: 2008 International Symposium on Information Technology, vol. 2, pp. 1–8. IEEE (2008, August)
7. Sajjad, M., Nasir, M., Muhammad, K., Khan, S., Jan, Z., Sangaiah, A.K., Baik, S.W.: Raspberry Pi assisted face recognition framework for enhanced law-enforcement services in smart cities. Futur. Gener. Comput. Syst. **108**, 995–1007 (2020)
8. Shinde, P.P., Oza, K.S., Kamat, R.K.: Big data predictive analysis: using R analytical tool. In: 2017 International Conference on I-SMAC (IoT in Social, Mobile, Analytics and Cloud) (I-SMAC), pp. 839–842 (2017). https://doi.org/10.1109/I-SMAC.2017.8058297
9. Zhang, Z.F., Gao, Z., Liu, Y.Y., Jiang, F.C., Yang, Y.L., Ren, Y.F., Yang, H.J., Yang, K., Zhang, X.D.: Computer vision based method and system for online measurement of geometric parameters of train wheel sets. Sensors **12**(1), 334–346 (2012)
10. Meenpal, T., Balakrishnan, A., Verma, A.: Facial mask detection using semantic segmentation. In: 2019 4th International Conference on Computing, Communications and Security (ICCCS), pp. 1–5. IEEE (2019, October)
11. Petrović, N., Kocić, Đ.: IoT-based system for COVID-19 indoor safety monitoring. IcETRAN **2020**, 1–6 (2020). (Preprint)

12. Zacharias, J., Narayanan, V.: International journal of basic sciences and applied computing international journal of basic sciences and applied computing. Trans. Power Electron **23**(1), 332–342 (2008)
13. Pramila, M.J., Shewta, P.S.: Wireless temperature detector system using ARDUINO and IOT. Int. J. Comput. Trends Technol. (IJCTT) **67**, 82–83

Chapter 49
Railway Track Monitoring System

Swati Vaishnav and Arti Hadap

Abstract Most of the accidents in railway lines are caused due to the cracks in railway tracks or presence of any object, which causes derailment one of the common reasons behind the accidents. These faults are generally ignored due to manual inspection and uneven maintenance. This results in gross damage to our vigorous property, as well the consequences of these accidents we lose more number of lives each year. The fault detection on rail track is done by vehicle cum robot or we can say it as defense system is powered by solar energy. The suggested robot travels along the path and finds the problem.

49.1 Introduction

The Indian railway network is the world's fourth-largest railway network. It has a total track length of 126,366 km (78,520 miles) and 7349 railway stations, with a route length of 67,956 km (42,226 miles). Train transportation is affordable and convenient mode of transport as well as suitable for long distances and in suburban areas by all kinds of people in India. Train transportation is always limited to railway lines (rails). If one of these tracks fails, it causes a significant problem. According to the newspaper's survey, most of the train accidents are happened due to faults in the railway tracks, which cannot be easily identified. Generally, these railway track faults are occurred by cracks in rail tracks due to natural climates or any other mechanical damage, or presence of any physical object. Also, it takes more time to rectify this problem. Hence, to reduce railway accidents it is needed to take required safety and reliability of railway network. The research work presented in the paper discusses railway track monitoring systems using different technology it is a dynamic

S. Vaishnav (✉) · A. Hadap
Mukesh Patel School of Technology Management and Engineering, NMIMS, Mumbai, India
e-mail: swati.vaishnav@nmims.edu

A. Hadap
e-mail: arti.hadap@nmims.edu

S. Yadav et al. (eds.), *Proceedings of Second International Conference in Mechanical and Energy Technology*, Smart Innovation, Systems and Technologies 290,
https://doi.org/10.1007/978-981-19-0108-9_49

technique that combines solar power and will be resilient, efficient, and steady for both crack identification and object detection in railway tracks.

The rest part of the presented paper is structured as follows—Sect. 49.2—discovers the literature study on earlier developed systems of railway track crack detection by various developers and their research work. While Sect. 49.3—discusses the all the basic facts regarding railway track crack detection system, which will help in the implementation of the ideal system. In Sect. 49.4—the working structure of the proposed system is explained along with the methodology used in the form of flowchart. In the end, Sect. 49.5 concludes the paper review along with the further research work to enhance the system and references.

49.2 Literature Work

In earlier days there is number of railway track monitoring systems are developed by various researchers. Most of the systems are expensive. Hence, to design a smart railway track inspection system, we have summarized formerly developed systems and their respective analysis in this section. The reviews on studied literature work on railway track inspection systems are discussed here.

In [1], author developed solar-based crack detection in railway tracks using Arduino. The accuracy of the sensor for sensing speed is less. Also, sometimes the system is not able to detect the cracks, especially in rainy season. Therefore, overall speed of the developed system is getting reduced.

In [2], author developed a railway track protector system named as Rail-Rakshak. In this system, all data sensed by sensors is stored in cloud. Chatboat is AI-based messaging tool for understanding information. Webhok plays a role of bridge among cloud and Chatbot. For controlling the entire system Ardinuo Uno board is used. This system covers all the requirements excluding the camera.

In [3] author developed railway track fault inspection system using Arduino mega board. The Arduino board is powered by solar energy, used along with the LASER source, Avalanche Photodide [APD], GSM, and GPS module. Also, the system sends an SMS to the concerned authorities along with the link to open a location on Google Maps. MOSFET is used to control the speed of the motors. We cannot see the present scenario due to the absence of camera. Also, the main limitation of LASER is, sensing accuracy can be affected by dust or other materials.

In [4], author established a railway track defense system based on IoT. The system is designed using Raspbery –pi module. The system takes all the required parameters into consideration except only the power supply. If battery is discharged, the system will not work properly because, as the system does not use the solar-powered energy.

In [5], author implemented rail track inspection system based on GSM. For crack detection in rail, tracks system uses ultrasonic and infrared sensors. The system is designed using Arduino board. Camera is not present. There is no provision for uninterrupted power supply.

In [6], author developed a simple railway track monitoring system. The system is built using Arduino uno board. For crack detection in railway, tracks system uses ultrasonic and infrared sensors. Location of the fault is acknowledged using GPS module and regarding message is generated with the help of GSM. There is no facility for continuous power supply. As well as camera is not present.

In [7], author designed a simple monitoring system for railway tracks based on Arduino programming. The system uses ultrasonic sensors for crack detection only. The main limitation of the system is that there is no provision for obstacle detection. Also camera is not present. No backup for battery.

In [8], author implemented a smart fault detection system for railways. The designed system is embedded in the train itself. In case there as any fault is detected, driver of the train get an alarm and results in he will stop the train. But this system is only designed for the detection tracks in rail tracks. There is no provision for obstacle detection.

From this survey, it is come to know that every model supports various functions while other important functions are excluded. Therefore, the main goal of this presented paper is to develop all in one smart railway track monitoring system in such a way that will overcome the boundaries of earlier developed systems. Also, to include as possible as many other essential functions.

49.3 Background of Study

This section highlights all the aspects related to railway track monitoring. At the present, there are so many microcontroller boards are available in the market. Each board was established with a variety of hardware, support, security, improved infrastructure, and communities. Among them Arduino and Raspberry –pi is very popular.

Energy: Energy is the capability of doing work. Energy can be converted in form but not created or destroyed, according to the rule of conservation of energy. There are two sorts of energy: (1) renewable energy and (2) fossil fuel energy.

Renewable Energy: this kind of energy is generated by sunlight, wind, rain, tides, waves, and geothermal heat are examples of renewable resources that replenish themselves spontaneously [9]. Nonrenewable Energy- this type of energy is generated from nonrenewable resources that cannot be easily restocked or renewed itself. Such as coal, oil, natural gas, Nuclear energy.

Benefits of Using Solar Energy

- It is renewable
- Free of cost available in nature
- Miscellaneous Applications
- Maintenance cost is low
- It is Sustainable

- Environment friendly
- Can be generated anywhere, no need for grid connection
- No electricity bills to be paid, thus it saves money
- Is durable.

Limitations of Solar Energy

.

- It is weather dependent

Solar Panel

The conversion of light energy into electric current is done by solar cells or photovoltaic cells. The solar cell or PV works on the phenomenon of photovoltaic effect. Photovoltaic effect states that- is a process that generates voltage or electric current in a photovoltaic cell when it is exposed to sunlight (Fig. 49.1).

Working of Solar Cell

Solar cells are made up of two types of semiconductors, p-type and n-type, which are sandwiched together to form a p–n junction (Fig. 49.2).

- Photons (light particles) attack the upper surface of the cell when sunlight shines on it.
- Photons (yellow blobs) transport energy down the cell membrane.

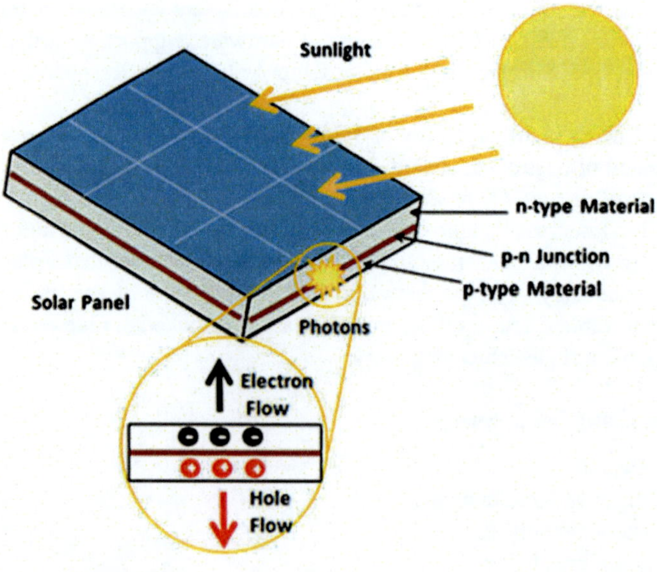

Fig. 49.1 Photovoltaic effect

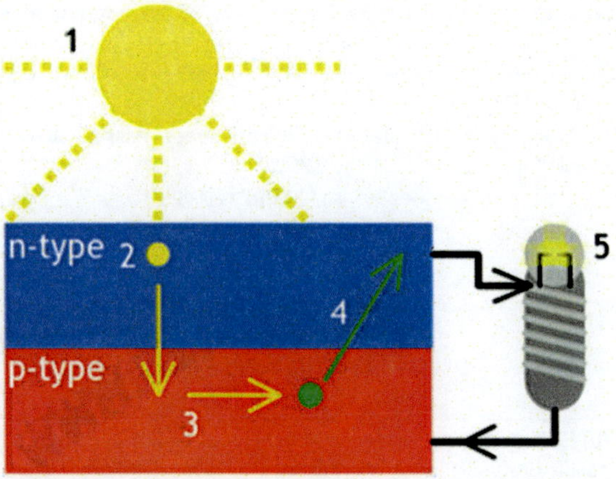

Fig. 49.2 How a solar cell works

- Photons (green blobs) in the lower, p-type layer give up their energy to electrons (green blobs) in the upper, n-type layer, who use it to leap across the barrier into the upper, n-type layer and escape out into the circuit.
- Electrons flow around the circuit, causing the lamp to light up.

The Different Types of Solar Panels

Different types of solar panels attend various requirements and purposes. The detailed classification of solar panel by generation based on the materials used and efficiency is given in Table 49.1.

To achieve larger voltages, currents, and power levels, photovoltaic cells can be electrically coupled in series and/or parallel circuits. The number of solar cells in a panel is formed by joining them together. The number of individual PV panels is connected to make an array. Array with PV panels is generally preferred to be connected in series to produce increased output voltage. When two PV panels are joined in series, for example, the voltage doubles while the current stays the same. Selection criteria of a solar energy system's battery are as follows:

Capacity and power are the first two factors to consider. The total quantity of electricity that a solar battery can store, measured in kilowatt-hours, is known as capacity (kWh).

1. Depth of discharge (DoD) is the measurement of how much of a battery's capacity has been utilized.

   Due to their chemical structure, the most powerful solar batteries must keep a charge at all times. Suppose 100% battery charge is used, the life of battery will be considerably reduced. Thus DoD should be maximum.
2. The quantity of energy that may be used as a fraction of the amount of energy that was needed to store it is referred to as round-trip efficiency. There should be more of it.
3. Warranty and battery life

Table 49.1 The comprehensive overview, of the most common and special types of solar panels

Solar cell type	Advantages	Limitations
First-generation		
(a) Monocrystalline solar panels (mono-SI)	• High efficiency [near about 20%] • Occupy less space • High power output • High lifetime value • When compared to polycrystalline panels, they are less harmed by high temperatures. Suitable for commercial use	• Expensive
(b) Polycrystalline solar panels (Poly-SI)	• Cost is low	• Low efficiency [near about 15%] • They are affected by high degree of temperature
Second generation		
(a) Thin-film solar cells (TFSC)	• Easy for production • Price is low • Flexible • Less affected by high temperature	• More space is required • Shorter lifespan • Not suitable for residential
(b) Amorphous silicon solar cell (A-Si)	• Cost is low	
Third generation		
(a) Cadmium telluride solar cell (CdTe)	• Low cost	• Toxic, if ingested or inhaled
(b) Concentrated PV cell (CVP and HCVP)	• Efficiency rate up to 41%	• A solar tracking system and a cooling unit are required (to reach high-efficiency rate)
(c) Biohybrid solar cell		

The Table 49.2 shows the brief comparative study, about the most common rechargeable batteries available for solar energy systems.

49.4 Proposed System

The basic concept of proposed system diagrammatically is shown in Fig. 49.3 Main building block of the proposed framework will include various sensors, camera for catching images of present situation, microcontroller, motor driver circuit, solar-powered battery system.

Table 49.2 Comparison of rechargeable batteries suitable for solar energy

Type parameters	Battery		
	Lead acid	Lithium-ion	Saltwater
Capacity and power	Small	Largest	Large
Lifespan	Short	Longest	Long
DoD	Low	Medium	High
Cost	Low	High	Medium
Maintenance	Requires frequent maintenance	Little/no maintenance	Little/no maintenance
Safety	Can emit harmful gases if not handled correctly	Low chance of fire	No major concern

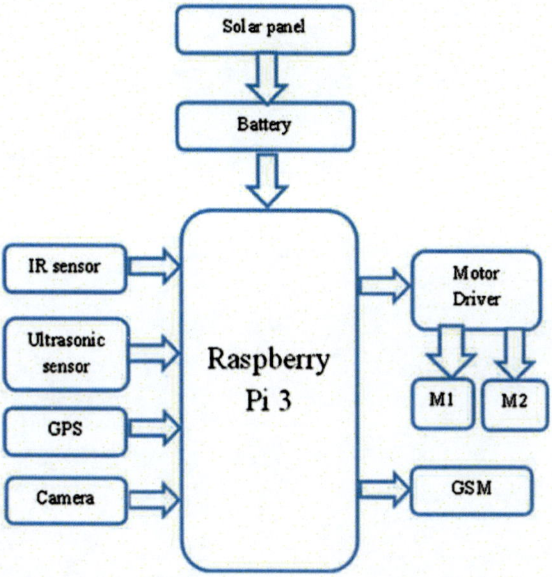

Fig. 49.3 Architecture of proposed solar-based railway track monitoring system

The proposed model will discover the cracks in railway track, detect of presence of any obstacle in railway path as well as, capable to detect the exact locality of crack via the GPS module. The entire system is controlled by the Rpi module. For detection of cracks in railway track, IR sensor is used. At the same time, it is necessary to know the exact location of fault is achieved by GPS. The locality of the fault is determined by the value of longitude and latitude coordinates from GPS module. This location is sent to concerned authorities pre-defined number by using GSM through SMS. The movement of proposed model is controlled by motor driver connected to the Rpi module. Solar panel powered system is used to avoid the power failure.

As shown above, Fig. 49.4 describes the working process of methodology used for monitoring the rack to avoid railway hazards.

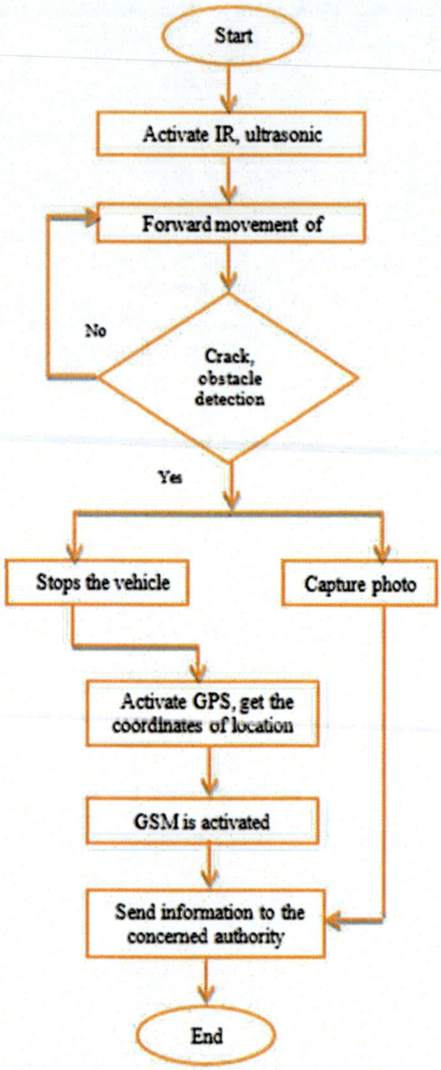

Fig. 49.4 Flowchart of proposed solar-based railway track monitoring system

The list of components required for proposed system are-

IR sensors- is used to detect the crack; Ultrasonic sensor -is used to detect the presence of any obstacle, also provide the distance between the obstacles; GPS- is to detect the exact location of fault; GSM module-is to send the information of latitude and longitude to the concern authorities; Camera for capturing images; Motor Driver-it controls the speed and direction of motor; Dc motors- to ride the vehicle on the tracks; Battery Solar panel.

Proposed system has many advantages as compared with the traditional railway track monitoring technique mentioned below.

Advantages of proposed system include-

System is economical; Model easily portable, reliable; Chances of error are less due to use of GPS system; Minimizes the chances of accidents; Tracking of railway track at any-time is possible; Power consumed by devices is low; Reduces the human efforts; Saves the time; Automatic operation; Prevent accidents; Safety to the peoples; Fault analyze is easy; Easy installation.

49.5 Conclusion

In this work, a brief study of solar energy is discussed. The proposed system will work without any human intervention. The designed model will be helpful for railway maintenance department for regular track inspection. It is more convenient as compared to manual inspection. Due to the use of solar energy, power consumed by device is low. The system will improve accuracy of crack and fault detection in rail tracks. As the consequences of this, it saves the precious lives of travelers and loss of economy. It will also save the time required for the identification of cracks and faults in the railway track.

References

1. Ladola, A., Parekh, C., Patel, D., Bhagatwala, H.: Solar based railway track fault detection system. Int. Res. J. Eng. Technol. (IRJET) **5**(5) (2018)
2. Salvi, S., Shetty, S.: Rail-Rakshak: A solar-powered railway track crack detection and notification system with chatbot support. In: Proceedings of the Third International Conference on I-SMAC-2019 (2019)
3. Mishra, U., Gupta, V., Ahzam, S.M., Tripathi, S.M.: Google Maps based railway track fault detection over internet. Int. J. Appl. Eng. Res. **14**(2) (2019). ISSN 0973-4562
4. Senthil Kumar, R., Kirubakaran, I., Loganathan, A., Janani, K., Janani, K.: Railway trail defense structure based IoT. Int. J. Adv. Res. Ideas Innov. Technol.**5**(2). ISSN 2454-132X
5. Verulkar, S., Dholam, V., Salunkhe, M., Patil, S.: GSM based railway track crack detection using ultrasonic and infrared sensor. JETIR **6**(4) (2019). ISSN-2349-5162
6. Arun Kumar, R., Vanishree, K., Shweta, K., Nandini, C., Shweta, G.: Railway track crack detection. Int. J. Eng. Res. Technol. (IJERT) (2020). ISSN 2278-0181
7. Manoj, V., Bharadwaj, G., Pavan Akhil Eswar, N.R.: Arduino based programmed railway track crack monitoring vehicle. Int. J. Eng. Adv. Technol. (IJEAT) **8**(6) (2019, August). ISSN 2249–8958
8. Mishra, S., Shrivastava, A., Shrivastav, B.: A smart fault detection system for Indian railways. Int. J. Sci. Technol. Res. **8**(10) (2019, October)
9. Basha, I., Bharat, A., Manjunath, N.P., Deshapande, A.: Solar based automatic railway track cracks and discontinuities detection robot. Int. J. Core Eng. Manag. ISSN 2348-9510. (Special Issue, NCETME-2017)

Chapter 50
Determinants of Small and Medium Construction Firms' Financial Performance

F. K. Bondinuba, C. K. Bondinuba, A. Opoku, D. Owusu-Manu, and Manika Mittal

Abstract *Purpose* Whereas construction firms in developed economies often exhibit sound and positive financial performance, the opposite is prevalent in developing economies. It is due to a lack of evidence and consensus on what determines financial performance in the construction context. This paper explores the determinants of Small and Medium Construction Firms' Financial Performance in Ghana through factor analysis. *Study design/methodology/approach* The paper employed a factor analytic approach to validate the financial variables of contractors, involving 125 respondents from two areas in Ghana in a cross-sectional survey using a self-administered structured questionnaire for data collection. Confirmatory factor analysis (CFA) discovered the underlying latent variables that significantly influence the contractor's financial performance were discovered using confirmatory factor analysis (CFA). *Findings* According to the findings, three dimensions relevant to the research setting emerged: funding strategy, management strategy, and prior performance. The developed model's constructs have a high level of reliability and

F. K. Bondinuba
Department of Building Technology, Kumasi Technical University, Kumasi, Ghana
e-mail: francis.kbondinuba@kstu.edu.gh

C. K. Bondinuba
Frontline Limited, KNUST, P.O Box UP 346, Kumasi, Ghana
e-mail: chris@frontlinelimited.com

A. Opoku
UCL Bartlett School of Construction and Project Management, University College London, 1-19 Torrington Place, London WC17HB, UK
e-mail: alex.opoku@ucl.ac.uk

D. Owusu-Manu
Department of Construction Technology and Management, Kwame Nkrumah University of Science and Technology, Kumasi, Ghana
e-mail: d.owusumanu.cap@knust.edu.gh

M. Mittal (✉)
PCJ School of Management, Maharaja Agrasen University, Baddi, Haryana, India
e-mail: info@frontlinelimited.com

S. Yadav et al. (eds.), *Proceedings of Second International Conference in Mechanical and Energy Technology*, Smart Innovation, Systems and Technologies 290,
https://doi.org/10.1007/978-981-19-0108-9_50

validity. *Originality/value* The research adds to the body of knowledge in construction financing models in developing nations. *Research limitations* The findings of this study should be interpreted in the context of Sub-Saharan Africa, particularly the Ghanaian construction industry. Before making broad generalizations, similar research should be conducted in similar circumstances in multiple nations. *Practical implications* Construction companies could easily measure their financial performance using the developed instrument items. *Social implications* The research has a broader social impact in influencing the construction business community's view and acceptance of the findings in managerial decision making aimed at enhancing the company's financial health.

50.1 Introduction

A Contractor is an entity, a business, a person, or an organization who commits with another organization or individual to construct, maintain a building, road, or other construction facility or a project. On the other hand, the contractor's financial performance (F.P.) is simply the financial status and well-being of a construction firm within a period [1]. Whereas in Bhunia et al.'s [12] view, F.P. means the firm's overall financial health, [2] suggested that it represents the leverage, solvency, and cash standing of a construction firm over a given period. It attempts to measure the firm's liquidity, profitability, and other indicators that the business is conducted rationally and typically, ensuring enough returns to shareholders to maintain at least the business market value [2–6]. F.P. is concerned with measuring the results of a firm's policies, operations, and activities over a stated period in monetary terms. The measurement results can either be positive or negative, indicating its soundness or distress. While a sound and positive F.P. is desirable, the opposite is prevalent, as evidenced by the high incidence of construction business failures [9] in the construction industry of many developing markets. Moreover, in the global context, the rate of construction business insolvency has been of great concern to stakeholders. The rate of insolvency is likely to increase further in the wake of the Covid.19 pandemics [6–9].

50.2 Contractor Financial Performance Models in the Construction Industry

The financial capacity of a contractor invariably determines its solvency measure. Therefore, insolvency prediction tools can help assess a firm's financial performance. In doing so, quantitative performance models relied on financial information from firms' balance sheets, while qualitative models are often based on empirical analysis of corporate features and management actions to predict firm performance (Kaplan and Norton 1992). The Altman Z-score model, which is a tool, that uses a combination of both financial ratios and discriminant analyses to assess the likelihood of a firm going bankrupt or not with a 96% accuracy, has its shortcomings (Venkataramana et al. 2012; Hossain et al. 2020). Thus, the model provides a single index known as the Z-score, classifying businesses as failing, at-risk or non-failing. In this model, liquidity and profitability measures, leverage, gearing, and turnover are significant predictors of business failure [6]. However, with the differences in business and industry, such a generic model raises questions about its applicability to all sectors, especially the construction industry [10].

The model combines several ratios into one single performance index to evaluate the firm's financial performance, with a size factor that caters to firms of varying sizes. Another model by Pilateris and McCabe (2003) is the contractor financial evaluation model comprising four inputs and outputs of financial performance benchmarks for Canadian contractors using data envelopment analysis. They used three financial performance measures such as the liquidity, leverage, and profitability. Additionally, Chen (2009) developed a firm-specific model for predicting the financial performance forecasting models for the construction industry [11].

Rajasekhar (2017) further our understanding by developing a financial performance model for contractors in India. In this model, the financial performance of firms was evaluated using four main groups of financial ratios that include liquidity ratios, solvency ratios, profitability ratios, and efficiency measures. There are several mixed variables from contractor selection, financial and performance ratios in all the above models. Many financial ratios that deal with the availability and management of cash of/for construction projects are conspicuously missing in most of them. Surprisingly in the developing world, contractors, in most cases, take on projects that exceed their financial worth or strength for various reasons. These reasons include poor contractor classification arising from inadequate regulations and enforcement, politics, among others. Therefore, it calls for a broader emphasis on understanding the concept of F.P. variables beyond the financial ratios of contraction firms. These variables must be evaluated at the procurement stages of construction projects to prevent bankruptcy, insolvency, and in most cases in the developing world, construction projects abandonments [12].

50.2.1 Conceptualizing Financial Performance in the Context of a Construction Firm

In this paper, Pilateris and McCabe (2003) model is adopted where additional variables from other studies are included to understand better the determining factors of contractors F.P. in the Ghanaian construction industry. These variables not in order of importance range from contractors:- capital structure, debt capital (Donkor 2011), equity capital (Myers 2001), financial ratios, liquidity ratios trends, profitability ratios, financial leverage ratios, management efficiency ratios, cashflow technical capacity (Miller 1962), company size ownership and management structure, contractors credit risk/loan servicing, capital investment (Cuong 2014), past performance (Hong Xiao and Proverbs 2003), industry volatility among others [12–15].

The financial ratio of a firm can also be one of the variables that affect its financial health. Financial ratio provides a quick and effective way of obtaining an insight into a company's operations and financial performance (Kolar 2011; Maria 2015; Rajasekhar 2017). Ihab (2014) contended that the financial ratio analysis is significant for determining its financial performance. On the contrary, Chan and Abdul Aziz (2017) claimed financial ratios alone are not enough but must encompass the contractor's past performance, which is a good indicator of future performance. Beyond the financial ratios debate, liquidity ratio trend has also emerged, which is the ease with which the firm's current assets can be converted into cash to meet the obligation of a firm as a significant predictor of its financial health (Pranowo 2010; Wesa and Otinga 2018). A mismatch between companies' assets and liabilities leads to financial distress and delays projects (Edwards et al. 2017). Bhunia (2010) earlier emphasized that liquidity helps a firm establish its effectiveness in operations hence the need for optimal liquidity levels to meet short-term obligations.

50.2.2 Research Method

Depending on a Small and Medium Construction Firms (SMCF) study conducted in Ghana's two major commercial areas, Greater Accra and Ashanti, the article evaluated the factors of contractor financial performance. These areas are also the most urbanized nation, with over 383 construction companies of all types (MWH 2018). The report uses snowballing as a non-probability sampling strategy, relying on data from SMCF who were accessible and willing to engage in the study via a drop-off style of the survey. According to Dudovskiy (2016), Snowball sampling is very susceptible to selection bias and a significant level of sample error. In a developing country, however, identifying participants based on their workplaces, which might be another sample approach, is difficult because some operate from various hidden locales.

On the other hand, others operate illegally by failing to register with the Ministry of Works, Water, and Housing. As a result, a snowball sampling was feasible since some contractors knew each other because they work in the same area. It also led to the discovery of certain contractors who were previously unknown to the authors.

Between June and November 2019, these companies were sent a standardized questionnaire. It enabled personal contact with contractors' teams after scheduling appointments with the necessary people or firms to acquire responses from their teams. Personnel at the several selected firms responsible for contractors' financial portfolios responded to the pre-written series of questions. Before sending the survey instruments out, the respondent's office locations and instructions were collected by telephone contact, which helped enhance the response rate. The technique resulted in a sample of 200 contractors, with 134 completed surveys, giving a response rate of 67.00%. However, 7.50% ($N = 10$) of the responses were unusable due to errors, inconsistencies, or incompleteness. As a result, 93.30% of the replies ($N = 125$) were used in the study. The years of experience and profession of the person completing the questionnaire were both included in the questionnaire. Table 50.1 shows the 26 measurable variable indicators of financial success that were identified. The literature review provided the basis for these factors. The responses to the different measured variable indicators of financial performance were evaluated on five-point Likert scales, ranging from not significant, less significant, moderately significant, significant, very significant, and coded 1–5, and was assigned a criterion of mean score of 3.0 and above as a significant variable. All 25 variables met the mean score criterion, and one below the mean score was not included in the factor analysis.

50.2.3 *Factor Analyses*

Factor analysis is a multivariate mathematical method that identifies a set of underlying dimensions known as factors to analyze the interrelationship structure among multiple variables. It uses fewer dimensions than the original variables to describe the data and uses latent variables to explain the correlations between a collection of observable variables. The article used a factor analysis strategy that included principal component analysis and the varimax rotation method, similar to earlier investigations. Each F.P. variable in the questionnaire was coded as F.P. 1... F.P. 25 for analytical purposes (see Table 50.1). Cronbach Alpha was used to determine the reliability of all 25 measurement indicator variables, yielding an overall reliability rating of 0.980.

50.2.4 *Survey Results*

50.2.4.1 Characteristics of Respondent Firms

The construction industry in the country is relatively mature, as shown in Table 50.2. Thus, close to about 36% (N = 45) of contractors had operated for less than five years, 16% ($N = 20$) operated between 5 and 10 years, 4% ($N = 5$) between 11 and 15 years, 40% ($N = 50$) between 16 and 20 years and 4% ($N = 5$) above ten years. The profession of respondents was Managing Director 35.20% ($N = 44$), Project Managers 32% ($N = 40$), Site engineers 22.40% ($N = 28$), and cost managers 10.4% ($N = 13$), respectively. In terms of the criticality of the variables using the mean values, industry volatility, although critical, had the least mean value of 3.04. Construction organization cash flow and long-term solvency emerged as the most critical variables with 4.04 mean values, as shown in Table 50.3. It is not surprising as it collaborates with studies such as Alaka et al. [5]. Table 50.4 shows the findings of the exploratory factor analysis, which demonstrate that the Kaiser–Meyer–Olkin

Table 50.1 Variable definitions of SMCF F.P

Coded	Variables	Description	Source
F.P. 1	Capital structure	Combination of debt and equity capital	Bevan and Danbolt (2002), Chiang et al. (2010), Adair and Adaskou [3]
F.P. 2	Liquidity ratio	The ease with which current assets can be converted into cash	Chan and Abdul Aziz (2017), Wesa, et al. (2018), Pranowo (2010)
F.P. 3	Profitability ratio	The measure of a firm's uses of its assets and control of expenses	Chan and Abdul Aziz (2017), Arditi et al. [8]
F.P. 4	Leverage ratio	Measures of the relative contribution of stockholders and creditors and a firm's ability to cover its debt services	Fawzi et al. (2015), Wesa and Otinga (2018)
F.P. 5	Inventory and efficiency	The measure of the company's ability in using its assets to generate optimum returns through management decisions	Alaka et al. [5], Chan et al. [16]
F.P. 6	Cash flow-operating activities	Revenues generated from the core businesses activities	Fawzi et al. (2015), Yap et al. (2012), Ibarra (2009), Rodgers (2011)
F.P. 7	Cash flow-financing activities	The measure of the revenue generated from financing activities	Fawzi et al. (2015)
F.P. 8	Cash flow-investment activities	Measure the cash generated from investing activities to meet its obligation in the long run	Yap et al. (2012), Low et al. (2001), Fawzi et al. (2015)
F.P. 9	Annual turnover	Annual volume of work undertaken	MWH (2001)
F.P. 10	Firm size and image	The scale of the firm operations based on a predefined classification criterion	Chan and Abdul Aziz (2017), Noordin and Kassim (2017)

(continued)

Table 50.1 (continued)

Coded	Variables	Description	Source
F.P. 11	Ownership and Corporate governance	Management and coordinating structure of the business	Dawar (2014), Noordin and Kassim (2017), Petrovic-Lazarevic and Djordjevic (2002)
F.P. 12	Uncertainty and credit risk	Risk of loss through its financial obligations	Nyunja (2011), Wesa and Otinga (2018)
F.P. 13	Subcontractor management	A management function that combines cost, quality, time, health and safety, and operational management	Hong Xiao and Proverbs (2003)
F.P. 14	Industry volatility	A measure of the dispersion of returns in the industry	Hong Xiao and Proverbs (2003)
F.P. 15	Technical capacity	The skill ability of the firm's resources employed in operations	Akomah and Jackson [4], Edwards et al. (2017)
F.P. 16	Past business performance	How previous activities and goals are met efficiently and effectively	Watt et al. (2009)
F.P. 17	Trade union records	Involvement in an organization that looks after your interests	Flanagan and Shaughnessy (2005)
F.P. 18	Industry standing	Industry and corporate Collaborators	Flanagan and Shaughnessy (2005)
F.P. 19	Client/contractor relationship	Client-contractor partnership	Bennett and Jayes (1998)
F.P. 20	Long-term solvency	the long-term financial stability of a company	Lamb and Merna (2004)
F.P. 21	Asset strength	the abilities, skills, knowledge, potential, resilience of the company	Lamb and Merna (2004)
F.P. 22	Payment history	how debt payments are handled over the years	Lamb and Merna (2004)

(continued)

Table 50.1 (continued)

Coded	Variables	Description	Source
F.P. 23	Inflation and interest	Short- and long-term effect of macroeconomic policies	Whitten (2009)
F.P. 24	Below cost tendering	Underpricing of construction works	Scully (2009)
F.P. 25	Project complexity and firm expertise	the measure of the difficulty of executing a project	DBRS (2008)
F.P. 26	Fluctuations in construction volume	The unstable construction job market	Authors' construct (2020)

Source Authors' Construct, 2021

Table 50.2 Firm characteristics

Profession	Frequency	Percent
Managing Director	44	35.20
Project Manager	40	32.00
Site Engineer	28	22.40
Cost Engineer	13	19.40
Total	**125**	**100.0**
Years of experience		
Less than 5years	45	36.0
5–10 years	20	16.0
11–15 years	5	4.0
16–20 years	50	40.0
Above 20 years	5	4.0
Total	**125**	**100.0**

Source Field data, 2019
The 125 was the total sample considered for the study. it consists of SMCFs whose shares given characteristic such as form size among others

measure has a higher value of 0.864, indicating that the data gathered is suitable for factor analysis. The Bartlett's Test of Sphericity (X2: 2882.657, df: 91, Sig.: 0.000) were also significant, showing that the variables chosen are tightly connected and thus effective for structural detection.

Table 50.3 Descriptive statistics

Variables	Mean	S.D	*N*
Firm size	3.84	1.23	125
Trade union	3.56	1.30	125
Profitability ratio	3.68	1.05	125
Leverage ratio	3.24	0.87	125
Efficiency ratio	3.40	0.94	125
Cash flow operations	4.04	1.25	125
Cash flow financing	3.96	1.25	125
Industry standing	3.08	1.06	125
Technical capacity	3.60	1.10	125
Capital structure	3.44	1.17	125
Ownership structure	3.28	1.22	125
Risk and uncertainty	3.64	1.23	125
Subcontractor management	3.20	1.24	125
Industry volatility	3.08	1.10	125
Annual turnover	3.84	1.26	125
Inflation and interest rates	3.44	1.17	125
Asset strength	3.48	1.14	125
Cash flow investment	3.64	1.17	125
Long-term solvency	4.04	1.00	125
Contractor client relations	3.76	1.11	125
Payment history	3.72	1.26	125
Past performance	3.36	1.23	125
Below cost tendering	3.12	1.31	125
Project complexity	3.64	1.10	125
Fluctuation in the volume of work	3.56	1.17	125

Source Field data, 2019

Table 50.4 KMO and Bartlett's test

KMO and Bartlett's test		
Kaiser–Meyer–Olkin measure of sampling adequacy		0.864
Bartlett's test of sphericity	Approx. Chi-Square	2882.657
	df	91
	Sig	0.000

Source Field data, 2019

50.2.4.2 Confirmatory Factor Analysis on the Remaining F.P. Variables

Except for the industrial standing variable, which has a composite reliability value of 0.671, almost all variables have composite reliability values above 0.70. As indicated in Table 50.5, the Cronbach Alpha for all 25 variables is (0.980), suggesting excellent reliability. As shown in Table 50.6, the total variance explained by all variables is 81.88, implying that the principal components account for nearly three-quarters of

Table 50.5 Measurement and scale reliability for F.P. variables

		Communalities extraction
1	Firm size	0.960
2	Trade union	0.865
3	Profitability ratio	0.814
4	Leverage ratio	0.675
5	Efficiency ratio	0.802
6	Cash flow operations	0.891
7	Cash flow financing	0.809
8	Industry standing	0.671
9	Technical capacity	0.889
10	Capital structure	0.814
11	Ownership structure	0.891
12	Risk and uncertainty	0.807
13	Subcontractor management	0.814
14	Industry volatility	0.735
15	Annual turnover	0.905
16	Inflation and interest rates	0.770
17	Asset strength	0.786
18	Cash flow investment	0.843
19	Long-term solvency	0.783
20	Contractor client relations	0.822
21	Payment history	0.890
22	Past performance	0.877
23	Below cost tendering	0.898
24	Project complexity	0.718
25	Fluctuation in volume of work	0.839
	All 25 items (composite Reliability). 0.980	

Source Field data, 2019

The composite reliability in this research is the measure of internal consistency in scale items or variables just like the Cronbach's alpha

Table 50.6 Total variance explained

Component	Initial eigenvalues			Extraction sums of squared loadings			Rotation sums of squared loadings		
	Total	% of variance	Cumulative %	Total	% of variance	Cumulative %	Total	% of variance	Cumulative %
1	17.626	67.791	67.791	17.626	67.791	67.791	9.067	34.873	34.873
2	2.633	10.128	77.919	2.633	10.128	77.919	8.670	33.346	68.219
3	1.029	3.956	81.875	1.029	3.956	81.875	3.551	13.656	81.875
4	0.839	3.227	85.102						
5	0.729	2.802	87.904						
6	0.627	2.410	90.314						
7	0.562	2.161	92.475						
8	0.394	1.514	93.989						
9	0.336	1.291	95.280						
10	0.293	1.128	96.408						
11	0.237	0.912	97.320						
12	0.189	0.726	98.046						
13	0.159	0.612	98.657						
14	0.102	0.391	99.048						
15	0.093	0.359	99.407						
16	0.051	0.197	99.604						
17	0.035	0.134	99.738						
18	0.023	0.089	99.827						
19	0.020	0.078	99.905						
20	0.014	0.055	99.960						

(continued)

Table 50.6 (continued)

Component	Initial eigenvalues			Extraction sums of squared loadings			Rotation sums of squared loadings		
	Total	% of variance	Cumulative %	Total	% of variance	Cumulative %	Total	% of variance	Cumulative %
21	0.006	0.025	99.985						
22	0.002	0.009	99.994						
23	0.002	0.006	100.000						
24	8.250E−016	3.173E−015	100.000						
25	− 6.913E−017	−2.659E−016	100.000						

Extraction Method: Principal Component Analysis
Source Field data, 2019

the F.P. constructs. Because all of the 25 remaining F.P. variables have factor loadings greater than 0.60, as indicated by Kline, they can all be considered vital (2011). Table 50.7 shows that considerable data support the derived dimensions.

50.2.4.3 Validity and Reliability of the Derived Instruments

The main characteristics of construct validity are convergent and discriminant validity, derived from the results shown in Table 50.7. The variables converge strongly to the derived dimensions, and good convergent validity is indicated in the "strong" factor loadings. Additionally, discriminant validity can be deduced because the factor loadings suggest that most variables do not overlap across different dimensions. The reliability results for composite Reliability and average variance extracted are presented in Table 50.8 for each of the derived dimensions of F.P. factors.

50.2.5 Rearrangement and Inferences of the Results

The resulting F.P. variables in Table 50.8 are labeled according to the meaning indicated within the context and notion of contractors F.P. At this level of the examination, liquidity ratios [15] and ownership structure (Petrovic-Lazarevic and Djordjevic, 2002) were not significant, contrary to previous research. The confirmatory factor analysis revealed three dimensions of F.P., which were supported by the results. Their arguments, ramifications, and associated F.P. variables are now explored in the context of contractors' F.P. determinants in Ghana's construction business.

50.2.6 Financing Strategy

The variables that statistically weight on this factor are capital structure (0.840), annual turnover (0.837), asset strength (0.804), profitability ratio (0.802), risk and uncertainty (0.766), long-term solvency (0.741), inflation and interest rates (0.741), cash flow investment (0.715), cash flow financing (0.709), cash flow operations (0.652). SMCFs financing strategy comprises the choice of debt or equity or a combination of which can toll on their financial performance. With the level of capital structure (0.840), respondents emphasized the significance of debt-to-equity ratios in contractor financing, which can be a good proxy to measure its financial performance. Construction projects are capital intensive, especially where conventionally, payments for work done lag several months after execution. This scenario justifies the resort to debt financing options to meet project objectives in the short run. Contractors in Ghana have been disadvantaged with this option due to high-interest rates and strict collateral requirements by financial institutions. Due to high inflation and competition, interest rates often outweigh contractors' profit margins.

Table 50.7 Confirmatory factor analysis of the F.P. Items

Rotated component matrix[a]			
	Components		
	1	2	3
Capital structure	0.840		
Annual turnover	0.837		
Asset strength	0.804		
Profitability ratio	0.802		
Risk and uncertainty	0.766		
Long-term solvency	0.741		
Inflation and interest rates	0.741		
Cash flow investment	0.715		
Cash flow financing	0.709		
Cash flow operations	0.652		
Industry volatility			
Below cost tendering		0.903	
Subcontractor management		0.885	
Firm size		0.781	
Project complexity		0.738	
Contractor client relations		0.730	
Technical capacity		0.712	
Trade union		0.681	
Industry standing		0.670	
Leverage ratio		0.652	
Past performance			0.710
Efficiency ratio			0.709
Payment history			0.606
Eigenvalue	**17.63**	**2.13**	**1.03**
The Percentage of variance explained	**67.79**	**10.13**	**3.96**
Total Variance Explained 81.88			

Source Field data, 2019
Extraction method Principal Component Analysis
Rotation method Varimax with Kaiser Normalization
[a] Rotation converged in 11 iterations
It is the proportion of variance or a part of variance. For example, the total variance in any system is 100%, but there might be many different causes for the total variance
The total variance is the variability that is due to the effects of the true variance plus the variability that is due to error

Table 50.8 Reliability of the labeled derived factors

Emerged factors	Items	Measurement indicator variables	Reliability
Financing strategy	10	Capital structure	0.967
		Annual turnover	
		Asset strength	
		Profitability ratio	
		Risk and uncertainty	
		Long-term solvency	
		Inflation and interest rates	
		Cash flow investment	
		Cash flow financing	
		Cash flow operations	
Construction business management strategy	9	Below cost tendering	0.962
		Subcontractor management	
		Firm size	
		Project complexity	
		Contractor client relations	
		Technical capacity	
		Trade union	
		Industry standing	
		Leverage ratio	
Past performance	3	Past performance	0.917
		Efficiency ratio	
		Payment history	
Cronbach Alpha for all items	22		0.977
Variance extracted for all items	22		78.401

Source Field data, 2019

This situation is further exasperated by the unreliable nature of cash flow projections often anticipated for the debt servicing.

50.2.6.1 Construction Business Management Strategy

The second most important determinant is the Construction Business Management Strategy adopted by SMCFs in the construction market. The variables that statistically load on it are below cost tendering (0.903), subcontractor management (0.885), firm size (0.781), project complexity (0.738), contractor client relations (0.730), technical capacity (0.712), trade union (0.681), industry standing (0.670), leverage ratio (0.652). It also receives a composite reliability value of (0.720). The business structure component has been shown to impact on contractor's financial performance.

Variables including below cost tendering (0.903) strategy rated highest under the component loading (0.938). The management strategy adopted by a business is a significant variable that impacts its financial performance outlook. The business strategy is the preserve of management, who often could lack management knowledge due to owner-manager prominence in the Ghanaian construction market, as shown in the firm characteristics in this paper. Strategic management requires a high level of expertise, most of which are predominantly lacking in many SCMFs in Ghana.

50.2.6.2 Past Performance

Past performance in the construction sector is the third element to evaluate. This influence was not recognized as significant among the factors in the literature review. Three variables load on this factor. These are past performance (0.710), efficiency ratio (0.709), payment history (0.606). Contractors' past performance in the light of project cost overruns can indicate the contractors' financial standing. Contractors with poor financial standing are often associated with "suicide bidding", which is characterized by low-profit margins to win projects. Such contractors turn to capitalize on high claims on variations and other contractual rights to unduly inflate final contract prices at post-contract stages of projects. Therefore, the findings confirmed that SMCFs consider the prevalence of project cost overruns as a significant factor, especially in Ghana. It is equally significant where contractors' actual construction costs record on projects exceeds their contract prices.

50.3 Implication for Policy, Practice, and Research

Financial performance has been widely studied; nonetheless, the results provide more insights into both the existing literature and practice in Ghana. While previous findings suggest that critical variables of financial performance lie within only the financial ratios, financing strategy emerged as the most significant determinant of contractor's financial performance. Financing strategy in this context explains the mix of capital sources the contractor uses to finance the business. Moreover, the capacity and capabilities of the contractor emerged very significantly. Construction projects are typically different in scope and technical requirements. A contractor that lags in a technical capacity; invariably results in a high cost of inputs and poor-quality work performance. It takes a contractor with the requisite capabilities to plan, strategies, and financial resources to profitable use.

50.4 Conclusion

The results point to the fact that the firm's business structure is the third most significant determinant of a contractor's financial performance. To this end, the business structure adopted is the central coordinating framework for decision making, communication, implementation, and monitoring, which is dependent on the type of ownership and management structure of the company. Contractors' past performance has been a long-standing proxy of predicting future performance. This paper reinforces this practice, suggesting that how well a contractor performs in cost, quality, and timeliness significantly impacts their projected future performance. Finally, financial ratios are significant determinants of financial performance. It agrees with the host of previous studies that emphasize that balance sheet ratio variables indicate a company's financial standing. While this can be exceptionally reliable, it is not conclusive evidence to assess firm performance in the construction industry because of the relatively cast variance between cash-outs and cash-ins. This feature is reflected in the results as respondents rate this component least of all extracted. The results generally suggest that attention must be given to financing strategy if contractors want to remain viable and survive in the fluctuating construction industry in Ghana. This results further relieves the quest for determinants of financial performance by providing aggregated measures.

The current study's findings should be evaluated in Sub-Saharan Africa, particularly the Ghanaian construction context. Before making broad generalizations, similar research should be conducted in similar circumstances in multiple nations. Finally, if knowledge evolves and existing regulations and practices change, more items might be categorized under each dimension and the addition of new dimensions. Future research should look at other crucial factors in construction and finance in various developed and growing countries to fully understand the contractor's financial performance characteristics.

References

1. Abidali, A., Harris, F.: A methodology for predicting failure in the construction industry. J. Constr. Manag. Econ. **13**(3), 189–186 (1995)
2. Abinzano, I., Corredor, P. and Martinez, B.: Does family ownership always reduce default risk?. Acc Finance. **61**(3), 4025–4060 (2021)
3. Adair, P., Adaskou, M.: Trade-off theory vs pecking order theory and the determinants of corporate leverage: evidence from a panel data analysis upon french SMEs (2002–2010). Cogent. Econ. Finance. **3**(1), 1–12 (2015)
4. Akomah, B.B., Jackson, E.N.: factors affecting the performance of contractors on building construction projects: central region, Ghana. Int. J. Innov. Res. Devel. **5**(10), 151–158 (2016)
5. Alaka, H. Alaka, H.A., Oyedele, L.O., Owolabi, H.A., Oyedele, A.A., Akinade, O.O., Bilal, M., Ajayi, S.O.: Critical factors for insolvency prediction: towards a theoretical model. For the construction industry. Int. J. Constr. Manag. **17**(1), 25–49 (2017)
6. Altman, E.I.: Predicting financial distress of companies: revisiting the Z-score and ZETA models. Stern School of Business, New York University, pp. 9–12 (2000)

7. Arain, F.M.: Causes of insolvency and unethical practices of contractors in Pakistan construction industry. Women's Career Advancement and Training and Development. p. 1246 (2008)
8. Arditi, D., Koksal, A., Kale, S.: Business failures in the construction industry. Eng. Constru. Architectural Manag. **7**(2), 120–132 (2000)
9. Assaad, R., El-Adaway, I.H.: enhancing the knowledge of construction business failure: a social network analysis approach. J. Constr. Eng. Manag. **146**(6), 04020052 (2020)
10. Baharuddin, N.S.: Determinants of capital structure for listed construction companies in Malaysia. J. Appl. Finan. Bank. **1**(2), 115–132 (2011)
11. Barlow, J.: Innovation and learning in complex offshore construction projects. Res. Policy **29**(78), 973–989 (2000)
12. Bhunia, A., Mukhuti, S.S., Roy, S.G.: financial performance analysis—a case study. Curr. Res. J. Soc. Sci. **3**(3), 269–275 (2011)
13. Blayse, A.M., Manley, K.: Key influences on construction innovation. Constr. Innov. **4**(3), 143–154 (2004)
14. Bondinuba, F.K.: Exploring the challenges and barriers in accessing financial facilities by small and medium construction firms in Ghana. Civ. Environ. Res. **2**(6), 25–35 (2012)
15. Chan, T., Abdul-Rashid, A.: Financial performance and operating strategies of Malaysian property development companies during the global financial crisis. J. Financ. Manag. Prop. Constr. **22**(2), 174–191 (2017)
16. Chan, J.K.W., Tam, C.M., Cheung, R.K.C.: Construction firms at the crossroads in Hong Kong: going insolvency or seeking opportunity. Eng. Constr. Archit. Manag. **12**(2), 111–124 (2005)

Chapter 51
The Impact of Data Mining and Artificial Intelligence on Supply Chain Management and Environmental Performance

Amala Nirmal Doss, Nitu Maurya, K. Guru, Gazala Masood, Sushma Jaiswal, and Mohd Naved

Abstract The desire of practitioners and scholars has risen over the past few years in Data Mining and artificial intelligence technology. Nevertheless, few researchers have empirically examined the value and influence of Data Mining and Artificial Intelligence on the supply chain management process. In order to address this vacuum, we have broadened the idea of organizational information processing to include Data Mining, artificial intelligence, and online education as a stabilizer of the operation of the supply chain. To evaluate a selection of 164 France health facilities using a, respectively. Finally square linear extrapolation structural equation modeling approach, we built a theoretical framework. The research found a major impact on various production processes and on green supply chain management through the application of Data Mining and artificial intelligence technology. The study also examined the considerable influence of environmental performance on environmental convergence and green supply chain cooperation. The outcomes propose the moderating function of green student technologies in connections between large-scale data mining and artificial intelligence and green supply chain cooperation. This

A. N. Doss (✉)
Department of Computing, Muscat College, Muscat, Oman
e-mail: amala@muscatcollege.edu.om

N. Maurya
Institute of Professional Excellent & Management, Ghazibad, Uttar Pradesh, India

K. Guru
SRM Valliammai Engineering College, Chennai, Tamil Nadu, India

G. Masood
Department of Management, Faculty of Commerce and Management, Rama University, Kanpur, India

S. Jaiswal
Department of Computer Science & Information Technology (CSIT), Guru Ghasidas Vishwavidyalaya (A Central University), Bilaspur, India

M. Naved
Department of Business Analytics, Jagannath University, Delhi-NCR, Bahadurgarh, India

S. Yadav et al. (eds.), *Proceedings of Second International Conference in Mechanical and Energy Technology*, Smart Innovation, Systems and Technologies 290,
https://doi.org/10.1007/978-981-19-0108-9_51

paper offers significant insight into the deployment of Data Mining and artificial intelligence technology to achieve supply chain operations and increase sustainable development.

51.1 Introduction

Data mining seems to have become a vital resource for many organizations because of the fast expansion of digital technologies. Data mining consists of diverse representations defined by volume, diversity, speed, and truthfulness. Thanks to information collection, several organizations have developed analytical methods (Data Mining [DM]) to turn data into meaningful information to improve organizational performance and to boost supply chain management. Data Mining on study, particularly for the environmental dimension, remains at an early step of the supply chain. In research, the supply chain (SC) is regarded to promote value chain as organizational aspect. Several studies regard the integration of the environment into the supply chain as a strategic asset for firms. In all sectors, including the healthcare field, which is the main cause of contamination across its supply chain, an improvement in organizations' environmental performance (EP) should be addressed. Healthcare pollution generates many pathogenic, pharmacological, chemical, radioactive, and safety-related hazards [1].

In general, Environmental Health Care efforts are either a reactionary answer to regulatory pressure focused largely on waste reduction or a voluntary strategy confined to a narrow array of environmental practices in the health care supply chain. One of the primary major causes of environmental delays in the healthcare supply chain is the development of healthcare in a suitable liquid with inconsistencies and uncertainty about the implications of healthcare settings and logistical procedures. This becomes time for healthcare to embrace a more realistic data mining technology strategy to increase the accessibility of its SCs. This will help healthcare discover more and more how DM's principles and techniques are applied to build environmental practices [2].

51.2 Theory of the Processing of Organizational Information or Organizational Information-Processing Theory

The OIPT argues that an organization develops inside a system that integrates a number of diverse activities with their complexity and insecurity. Through information-processing mechanisms, theory provides a strong foundation to understand firms' conceptions and organizational conduct. Many causes of uncertainties, Standard operating processes, the instability, and amount of dependency in the supply chain environment. With increasing data volume controlled by enterprises, their use of processing information requires the participation of internally and externally organizations [3].

This amount of information demands more visibility to support effective decisions. An enhanced culture of exchanging generate value inter-organizational information might boost the ability of an organization to process data, boost the creative relationship, and decrease coordination uncertainty. It creates considerable expenses for the companies because of the absence of information-processing culture in an unpredictable environment. The ability to order theory and patient service in the healthcare industry is enhanced on the basis of knowledge and the application of the advanced skills [4].

In this regard, we claim that DM-AI technology may help companies develop and utilize extra knowledge necessary to make choices within and beyond the supply chain. This research validates earlier research which generally investigated firms' effective use of information technology such as DM and enterprise resource planning (ERP) to expand on OIPT methods. The supply chain operations for health care companies are based on numerous unpredictable variables, which fluctuate according to patient demand and the uncertainty of medical activities. These inconsistencies or disruptions might make information processing and decision-making on the adoption of environmental methods more difficult [5].

Healthcare also shares a high level of interconnections between its many components, adding to the complexity of its organization. The use of DM is possibly extremely beneficial in assisting decision-making processes by adjusting measures to enhance healthcare EP [6].

51.3 Data Mining Overview

Considering the recently established prevalence of DM and the complexity of its use, an agreement on its definition is difficult to obtain. Generally speaking, DM is described as the next generation of technology and frameworks meant to gain economic benefit by allowing high-speed acquisition, identification, and/or analysis from very large quantities of a wide range of data. For improved comprehension and decision-making, it demands novel technical forms of information processing. This enables businesses to achieve comparative advantages. These benefits include skill management, supply chain management, and organizational innovation. According to its capability to employ procedures that enable management to make better judgments based on evidence instead of a human sense of judgment or intuition, data mining has gained popularity [7].

It involves the establishment of particular tools for the management of prospective data volumes and the identification of trends, model detection, and worthwhile outcomes. The relevance of Data Mining in the judgment process of organizations has considerably enhanced technological innovation.

Table 51.1 Data mining analysis record

Level of management	Counting	Percentage (%)
Management of Top	69	41.3
Management of middle	98	56
Organizational type		
Education	45	25.5
Hospitals	31	19
Bank	39	22.5
Others	35	19
Experience years		
Less than 5 years	65	37
More than 10 years	55	33
Between 5 and 10 years	50	30
Work domain		
Service to customers	40	24.5
Governance	35	22

In order to achieve useful information, the incorporation, and real-time data collection and analysis of technologies such as smartphones and RFID, Cloud Computing, and internet of things (IoT) offer major advantages. The use of novel technical infrastructures competent for fast analysis of a variety of data mining systems is required in order for this knowledge. Data Mining can help organizations in improving processing power with various technology platforms see Table 51.1.

DM technology sources may also span a number of analytical disciplines, including as predicted, prescriptive and informative analyses. The individualism and insecurity of findings, which may originate from the incorrect sources, is one of the problems faced by businesses in using DM technology. The secrecy and security of common data, which remains a major concern in fields such as healthcare, may also be a further problem. In this case, technology must be mobilized according to regulations, but human skills must also grasp the technical features connected with DM. Data mining can help organizations, by integrating tools, strategies, and procedures, to make effective decisions relating to green operations throughout the supply chain [8].

51.4 In SCM Fields, Application of DM

A significant quantity of data is created in the production line via existing network and interior networks containing sensing devices or instruments on the factory floor. Data mining may enhance the effectiveness of the sales and distribution process and the ongoing tracking of processes and devices to tighten analyses and merging of such datasets. Manufacturers must utilize data mining and analytical approaches to expand their production sector. Forecasting equipment maintenance is an urgent growth-ready industry in this industry. The process of choosing the correct and ideal supplier for the supply chain is challenging, due to the great number of suppliers, as well as the many assessments and decision indications. Cloud services have a major influence on the selection of suppliers. New technologies provide more intuitive access to and exposure of data and users with the power of APIs and integration with contemporary data mining and analysis packages. The research review shows that DM can be applied in several SCM domains. An overview of DM applications in various sectors of the supply chain is discussed in the following subsections [9].

51.5 DM and Network Design of Supply Chain Management

Supply chain management is a reasonable decision that encompasses every strategy on the choice of supply chain partners and establishes corporate policies and programed to accomplish long-term strategic objectives (Fig. 51.1).

The project of supply chain design comprises the physical layout of the supply chain that impacts the majority of business divisions or business functions of a firm. The customer relationship and supply chain efficiency must be determined in the development of the Supply Chain System. The aim of the supply chain design is to develop a group of members which can achieve the company's long-term strategic objectives [10] (Fig. 51.2).

Supplier Relationship Management requires discipline for the development of strategic planning in order to optimize the utility of these connections and to manage

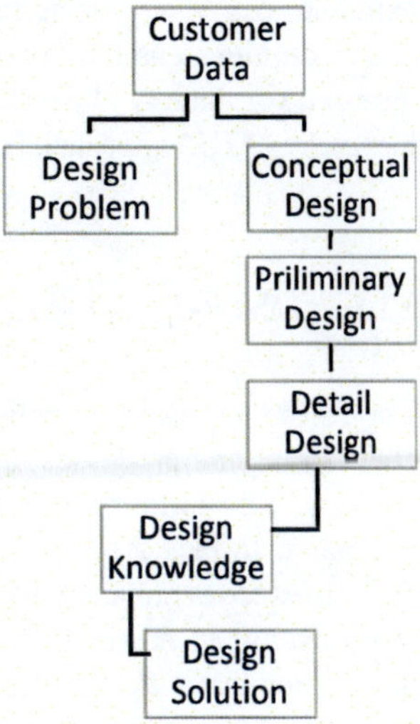

Fig. 51.1 A diagram illustrating the design process

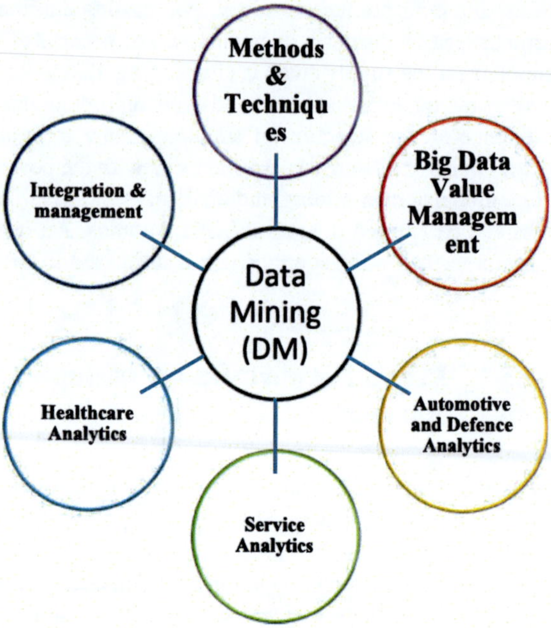

Fig. 51.2 DM and the connection between customers

all relationships with suppliers of businesses. Developing and improving collaboration with key suppliers is crucial in finding and generating new value, and minimizing the risk of SRM failure. The successful elements for the organizations focusing on customer engagement and cooperation include strategy assets and supplier relationship management (SRM). Using DM methods, precise information about corporate spending patterns to support supply chains may be provided. Data Mining, for example, can give precise investment return (ROI) statistics for every investment and in-depth research of possible suppliers. Due to its great capacity as one of the assessed parameters, the study employed a multiple criteria assessment and analytic hierarchy (AHP) approach for the assessment and selection of suppliers [11].

51.6 DM-AI and Healthcare

The results represent significant advantages for management of healthcare supply chain flows and processes in respect of the procurement and delivery of healthcare supplies, transportation and storage, internal production, and waste classification and treatment. In fact, AI-based data mining may be interfaced with programs such as ERP software to enhance decisions and build green practices policy proactively. Our study examined AI technology, which is a healthcare context with established AI procedures that may be utilized with DM. A huge quantity of data is created in the health business to regulate and monitor different treatments, safety, and maintain the medical information of the patients, regulatory standards, and compliance processes. Data Mining in medicine is essential because different forms of data emerge, which are complex, diverse, high-dimension, typically unorganized, and poorly annotated, in current biomedical applications including omics, electronic health records, sensor data, and text. To swiftly organized and evaluate such data, modern and robust technology are necessary. Mining of data All data relating to well enough and patient health care are included in the healthcare sector. Data Mining is designed for the capture, storage, transmission, administration, and analytics of massive amounts of high-speed, complicated, and changeable data requiring advanced technology. The healthcare data Mining covers features such as high-dimensional, diverse, varied, rapid, typically unorganized, badly labeled, and with regard to health care especially, truthfulness.

In a claim that is difficult to recognize using the conventional transaction processing system, DM may help healthcare firms discover fraud and anomalously. Data Mining has numerous healthcare principles including right care, right lifestyle, right development, right supplier, and right value. Data mining may be utilized in future as a new application of Huge Data on population health management and preventative treatment. Assessed the potential of enormous data usage in healthcare, the existing platform is challenging to improve support for the simple package, menu-driven, data analysis, and more real-time solutions. Other issues include data gathering continuity, ownership, consistent data, and data cleaning in the healthcare sector (see Table 51.2).

51.7 Conclusion

This research is the first academic study to emphasize the relationship between the difficulties of management of DM-AI by means of the supply chain, save for a few researches of conceptual character or which fall under medical science research, based on OIPT. In this study, research on the use of DM-AI technology for EP in the sustainable society is becoming increasingly limited. There are also certain restrictions in the document. A full overview of both operating procedural and health care flows utilizing the big sample is provided in the research conceptual framework of the study. Sometimes, when you try to reach exploratory conclusions, it is difficult to apply this viewpoint.

Table 51.2 DM-AI and healthcare Applications

Sensor systems	Main zone	Healthcare applications
Electrocardiograph	Cardiovascular disease monitoring	Mobile phones that are equipped with electrocardiographs are used to support individuals with cardiac disorders in undeveloped regions
Camera	Video and photo collection	Used to detect different kinds of illness, from the point of view of surgical impacts, disease diagnosis, slash observation, skin disease analysis, child health monitoring, etc.
Accelerometer	Measuring velocity	It helps to calculate the orientation of equipment related specifically to the Earth to calculate the velocity. It may be conducted in many activities, such as person tracking, gait control, and monitoring, in different patient monitoring approaches
Location	Tracking of location	gives access through mobile applications to track individuals vulnerable to certain illnesses including Alzheimer's disease and Ebola
Bluetooth	Information exchange and communication	It permits intermediate transmission of data between mobile currencies, many other medical equipment, and monitoring devices
Wi-Fi	Information exchange and interaction	WLAN-based mobile sensors enable the mobile phone for the detection of a disease and its treatment to connect with the physician about healthcare data
Mic	Talk capture	It enables physicians to interact with patients about supporting illness detection and therapy. It also provides a means to evaluate the audio for a patient's feeling with different conditions, such as muscle degeneration
A barometer, a compass, and a gyroscope	Activities physical	Technology and the sensors are used to calculate the static vs the non-stationary activities

(continued)

Table 51.2 (continued)

Sensor systems	Main zone	Healthcare applications
Microphone, GPS, accelerometer, touch interface, and light sensor are all included in the device	Monitoring of sleep schedule	This device arrangement shows the data of fragmented vs consistent sleep patterns in a patient

References

1. Acharya, A., Singh, S.K., Pereira, V., Singh, P.: Big data, knowledge co-creation and decision making in fashion industry. Int. J. Inf. Manage. **42**, 90–101 (2018)
2. Ageron, B., Benzidia, S., Bourlakis, M.: Healthcare logistics and supply chain–issues and future challenges. Supply Chain Forum: An Int. J. **19**(1), 1–3 (2018)
3. Aragón-Correa, J.A., Sharma, S.: A contingent resource-based view of proactive corporate environmental strategy. Acad. Manag. Rev. **28**(1), 71–88 (2003)
4. Bag, S., Pretorius, J.H.C., Gupta, S., Dwivedi, Y. K.: Role of institutional pressures and resources in the adoption of big data analytics powered artificial intelligence, sustainable manufacturing practices and circular economy capabilities. Technol. Forecast. Soc. Change **163**, 120420 (2021)
5. Nobre, G.C., Tavares, E., Lemme, R.P.: Assessing the Role of Big Data and the Internet of Things on the Transition to Circular Economy
6. Alsadi, A.K., Aloulou, W.J.: Impacts of strategic orientations on Saudi firm performance: is supply chain integration a missing link? Int. J. Logistics Manag. (2021)
7. Malik, M.M., Abdallah, S., Hussain, M.: Assessing supplier environmental performance: applying analytical hierarchical process in the United Arab Emirates healthcare chain. Renew. Sustain. Energy Rev. **55**, 1313–1321 (2016)
8. Shinde, P., Oza, K., Kamat, R.: An analysis of data mining techniques in aggregation with real time dataset for the prediction of heart disease. Int. J. Control Theory Appl. **9**, 327–336 (2016)
9. Hussain, M.M., Alam, M.S., Beg, M.S., Malik, H.: A risk averse business model for smart charging of electric vehicles. In: Proceedings of First International Conference on Smart System, Innovations and Computing, pp. 749–759. Springer, Singapore (2018)
10. Benzidia, S., Makaoui, N., & Bentahar, O.: The impact of big data analytics and artificial intelligence on green supply chain process integration and hospital environmental performance. Technol. Forecast. Soc. Change **165**, 120557 (2021)
11. Singh, S.K., El-Kassar, A.N.: Role of big data analytics in developing sustainable capabilities. J. Clean. Prod. **213**, 1264–1273 (2019)

Chapter 52
A Blockchain-Based Decentralized Machine Learning Framework for Performance Management: A Systematic Review

Apoorva Buttan, V. L. Varun, Padmaja Venugopal, P. Balachandra, D. R. Pradeep, and Biswadip Basu Mallik

Abstract In this study, the literature on blockchain applications is assessed in different disciplines systematically. The goal is to study and illustrate how unique characteristics of this composition process might disrupt the business as usual in the current status of blocks chain technology and applications. This study discussed the main basis of numerous studies released in the last ten years in prominent scientific publications and different investigations in gray literature in order to improve our assessment and capture the continuously evolving blockchain sector. The technology of Blockchain (BT) has become one of the most inventive and fashionable technologies. The decentralized BT database highlights data protection and privacy. Moreover, the agreement approach involves safe and authentic information. However, more security issues are still highlighted, including major assaults and double the costs. Blockchain-based safe data requires data collection to handle the difficulties described above. Analysis results of these data show the pertinence of methods for master learning (ML). ML contains a significant amount of detailed decision-making data. In order to improve result accuracy, content validation and availability in ML are highly crucial. The combination results of the two methods (ML and BT) can be precise.

A. Buttan
Amity School of Communication, Amity University, Noida, India

V. L. Varun (✉)
SJB Institute of Technology, Bengaluru, India
e-mail: bhadravathivarun@gmail.com

P. Venugopal · P. Balachandra · D. R. Pradeep
Department of Mathematics, SJB Institute of Technology, Bengaluru, India

B. B. Mallik
Department of Basic Science and Humanities, Institute of Engineering & Management, Kolkata, India

S. Yadav et al. (eds.), *Proceedings of Second International Conference in Mechanical and Energy Technology*, Smart Innovation, Systems and Technologies 290,
https://doi.org/10.1007/978-981-19-0108-9_52

52.1 Introduction

Information has become a vital intelligence source over the last few decades and provides potential for real-life challenges such as wireless, biomedical, agricultural and financial through smart devices. These apps are data-driven and provide user-friendly insights that help users to carry out the intended activity more efficiently. It enhances client interactions, increases efficiency and productivity, and offers the ability to develop new business models. Various smart apps like SG and Smart City make a person's life simpler. These apps create an enormous quantity of data and storing these constantly developing data in the database is a challenge. It consists of a chain of blocks coupled with basic cryptography. Integrity, decentralization, and transparency are the three cornerstones of BT. these three features opened their way to a broad range of applications, such as cryptocurrency presence and smart apps for examination of appropriateness. Those vulnerabilities have also been introduced after installation when BT assures security and privacy concerns. For example, assaults such as large attacks controlling voting or attacks on a false identity to manipulate the consensus become increasingly complex. Sybil attacks a strong IDS system is necessary to deal with the aforementioned problem because the old techniques employ a signature-based approach to identify particular patterns. However, one modern tech known as ML may be utilized to analyze data flow in the detection of intrusions and patterns of assault. There are hence strong needs to manage the blockchain-based intelligent apps to develop efficient and effective algorithms for analyzing the enormous volume of data. ML is therefore very widespread nowadays and utilities without even recognizing it a dozen times a day. ML includes computers for studying, thinking, and acting without human involvement. It is one of the artificial intelligence applications (AI). ML offers computers the ability to learn without being predictive analytics [1].

Blockchains are a permanent collection of records, which are encrypted for audit. It is like a business report. In this case, prior records cannot be altered by the accounting leader, and a trusted party must verify the new data. The main distinction is that fresh blocks with a decentralized node structure with a copy of the ledger are verified. The data are not controlled by a centralized party. The blockchain consists of the linking together of existing blocks; the current block contains the hash of the preceding block, making it traceable and difficult to modify. Older blocks cannot be altered if modified, their hash changed in any way. This emphasis is placed on linking hash to the blockchain network in all following blocks. Each person within the network has a copy of the blockchain; any modifications may be checked from now on by the other users. These blockchain replicas are refreshed by adding a new block [2]. Everybody can then see the block according to the administrative privileges. For the preservation of data integrity inside the block, BT utilities a cryptographic safe hash algorithm (SHA-256) like SHA-512. The hash value of each block is unique. For example, Ethereum employs Keccak-256 and Keccak-512, and Bitcoin uses SHA-256 for double use. This SHA is a confrontation method, which does not generate the same output for two distinct input data. SHA may therefore

Table 52.1 In blockchain-based systems incorporation of machine learning

Properties	Private	Incorporated	Commons
Integrity	Collusion attacks	Collusion attacks	Almost impossible
Transactions acceptance	Millisecond order	Millisecond order	Order of minutes
Mechanisms	Centrally organized	Set of the leader component	Each miner
Confidentiality	Trustworthy	Trustworthy	Dishonest
Consuming	Minimal capacity	Minimal capacity	High capacity
Administration	Whitelist allowed	Allowed component	Lesser permissions
Ownership	Centrally controlled	Semi-centrally controlled	Public
Opinion	Light PoW	Light PoW	Costly PoW
Identification	Users identified	Users identified	Unknown
Treatment Effectiveness	High endurance	High endurance	Low endurance

be used to verify whether or not the data is the same. The NSA and NIST are developing several SHA algorithms belonging to the SHA-2 family. Traditionally, SHA has been designed to create the signature as a segment of the Digital Signature Standard (DSS). Consensus methods are also used to assess the validity of blocks. The choice of methods relies on blockchain types, such as public, public, and private blockchain and consortia. The method used should guarantee agreement between nodes. In the case of assaults, it must be able to efficiently make use of resources and accept some degree of safety. In addition, an intelligent contract is a blockchain software that adds blocks when specific circumstances are fulfilled. It is characterized by the implementation of genuine legal contracts into blockchain applications. It is comparable to storage in complete network [3].

Table 52.1 *shows that In Blockchain-based systems incorporation of machine learning.*

It is added to the blockchain as scripts and runs according to the information it is supplied to generate outputs from the existing deal. It manages transactions that are either carried out in whole or in part using existing inputs. The main objective of an intelligent contract is to ensure greater security and to reduce contractual costs and delays [4] (Fig. 52.1).

ML's learning skills may be used to smarten blockchain technology. The distributed ledger may be enhanced by applying ML security. ML may also be used to improve the time needed to establish agreement by developing better channels for data exchange. It also provides a chance to develop better models by using the decentralized BT design. The smart application collects information from several sources of data, including sensors, intelligent devices, and IoT (Internet of Things), as recommended architecture for ML acceptance in BT-based smart applications. These devices analyze data obtained as part of smart apps. The blockchain works for these intelligent applications as an intrinsic element. In addition, ML may be used to

Fig. 52.1 Smart application

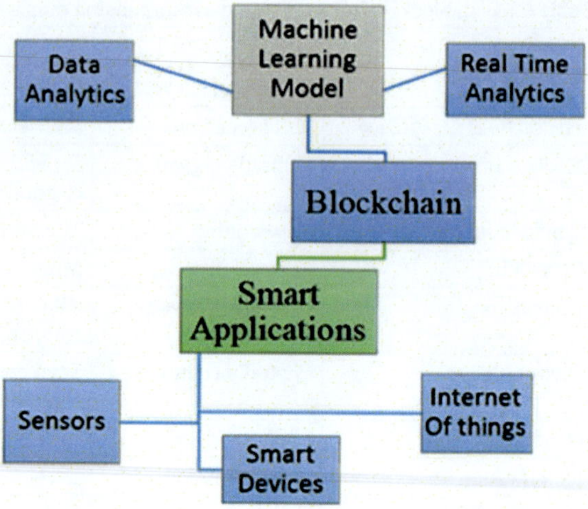

analyze and forecast data of this application (data analysis and real-time analysis). Sets of data may be stored on a blockchain network using ML models. This eliminates data mistakes including duplication, incomplete data, faults, and noises. Blockchain technology is concentrated on data and so ML models may address data-related problems [5].

The ML models might be based rather than the full set of data on selected portions of the network. This may provide individual models for various purposes, including detection of fraud and identity theft. Few of the advantages are given under ML:

- BT provides a high degree of safety and confidence for identity verification as genuine users for requesting and completing any transactions on the blockchain network.
- In order to verify that the requirements and the satisfy those needs before are met, Blockchain incorporates public ML models in blockchain technology.
- BT aims to guarantee that an incentive system is trustworthy; thus, it motivates users/clients to provide data. These enormous data assist to increase the performance of the ML model.
- ML models may be upgraded locally on an individual's personal device without any expenses, with a modest price online BT environment and off-chain.
- Users/consumers may provide reliable data submissions, calculate such data reliably, and the users can be rewarded.
- Payments are made with confidence in a blockchain ecosystem in real-time.
- For example, Ethereum, Blockchains tools deal, around the world, with hundreds of decentralized devices. This ensures that people are never totally unattainable or unavailable.

Concentrate on the objective

Research work from the following aspects was covered in this dimension.

Tracking
Use Self Organization Maps to prototype a blockchain monitoring system (SOMs). The datasets have not been modeled externally. The big vectors here draw on vectors with SOMs to smaller and lower sizes. Blockchain data were analyzed using Kohonen and SOMbrero libraries. This results in the efficient monitoring of the essential properties of the blockchain nodes. In addition, the distributed pattern-identification system with graphic neurons (GN) to monitor the blockchain system was another solution. GNs are scalable and similarities from comparable or incomplete patterns may be recognized. The GN interacted with nearby nodes for the detection of events through input data within the network. Initial reports indicated accurate object recognition by the GN; additional study in the area is still required [6].

Identification
This paragraph detects assaults to deal with data protection problems. Collaborative IDS (CIDS) in a blockchain-based ecosystem with a data-sharing agreement. With the application of BT ML classifiers, data protection problems of the CIDS may be solved. The data holder runs these classifications locally and shares the findings with other users on the network. Manage warnings and data dissemination among members in order to address the confidence calculation problem in CIDS. The data is then checked and an appropriate consensus process is performed to add a new block of data to the blockchain network. These data notifications may be encrypted with keys to specified parties supplied. Another way was to keep them distinct while they are still on the network. A system of anomaly detection (ADS) was also built on the idea that there may be a similar type of assault but on another blockchain entire network. This method does not discard orphans and forks information normally provided by other ADS. This information was exchanged by the attacked nodes with other nodes inside the network. The system effectively avoided the same type of assault in the experimental study with little latency [7].

Predictions
For prediction, the ML models are primarily utilized. The right decision-making and analysis are achieved with a strong prediction model. In addition, an ML model was presented for the prediction of bitcoin values. It employs Bayesian classification and regression trees with various characteristics, including block size, total bitcoins, the peak day, the amount of activities, and the volume of trading. Use log, Z-score, and box-cox normalization methods to normalize the training datasets. In addition, numerous cryptocurrencies such as Ripple, Lite coin, Dash, Bitcoin, and Ethereum coins are subject to a price prediction research. The correlation matrices for function choices utilized and the overall trends in the network are presented here. The model suggested uses a variety of bitcoin regression methods [8].

Reaction
A safe energy exchange technique among an end-user clusters and a virtualized server. This concept employs an architecture based on blockchain to provide a response to needs. Fog-Enabled Intelligent Devices (FEID) on the user side and smart

energy producer contracts on the cloud computing platform have been handled. A group of energy consumers then joined with a Virtualization Unit centralized. The Virtual Nodes have all been predicted and predicted regarding the energy usage. The end-user was taught each time the information was given to the protect the network to enhance its correctness [9].

Prevent the occurrence

The avoidance of a problem is one of the basic uses of ML in a BT-based application, for example, employment verification by a future worker. A technique to decrease the time required to validate a potential employee's employment history information. The formerly operated business must assemble an employee's details and the data will be encrypted using a public key. The employee then creates a new intelligent contract. The smart contract address is then inserted into the registry of the company and then checked by the prospective employer. Job role is taken from the smart employment contract of employees in a future organization and decrypted using the key. Then, compared with the prior company database, each job history entry. The constancy, integrity, and validity of the sender are guaranteed [10].

Blockchain compatibility

Blockchain and its ability to accelerate its digitalization while solving real-life challenges are attractive for businesses in many areas. However, although many IT professionals are considering the use of blockchain in nearly every development, they do not fully comprehend the basic reasons, especially from the point of view of managing data. For example, blockchain will bring no value to already existing technical solutions if no data needs to be ever saved. Likewise, a blockchain will not provide extra guarantee when just one writer in a given system is predicted in comparison to a conventional database which would most likely be a more appropriate solution, especially from a performance viewpoint [11].

We have created a framework to assess the appropriateness of solutions blockchain-based. Specifically, we analyze blockchain potential in four major domain areas against conventional databases: trust assumptions, contextual needs, efficiency, and agreement methods necessary. Figure 52.2 shows that framework for assessing the appropriateness of blockchains the significance of each condition may be measured using an easy three-level scale. The development and public is an extensive tool for professionals to assess whether or not blockchain enhances their systems [12, 13].

52.2 Conclusion

Latest advances have made them breakthrough technologies in Blockchain and ML. The decentralized directory can form the backbone of several smart devices such as intelligent cities, SG, and data trading. Comprehensive BT and ML information

Characteris-tics	Preconditions & Determinants	Block chain Design		
		Permis-sioned	Data-base	Permission less
Faith	Peer-to-peer (P2P) transactions	High	Low	High
	Accountability	High	High	High
	Lack of Trusted Third Parties	High	Low	High
Performanc-es	Costs for mainte-nance	High	Low	High
	Velocity of transac-tion	Medium	High	Low
Opinion	Need for verifiers	High	Low	High
	Commitment guidelines	High	Low	High
Background	Security	High	Low	High
	Transparency of Data	High	Low	High
	Privacy	Medium	Low	High
	Transactions verifi-cation	High	Low	High

Fig. 52.2 Framework for assessing the appropriateness of blockchains

and use in intelligent applications were provided and an ML-BT-based architecture was offered in this document. The ML-BT-based data analysis system may be designed and utilized in this way. There will be a conversation and an analysis of several available studies. Later, we introduced ML-BT surveillance, detection, forecasting, and response. Every dimension is presented with a comparison examination of various methods and approaches. In addition, during ML implementation in BT-based systems, we have identified many research issues which demand answers.

References

1. Viola, P., Jones, M.: Rapid object detection using a boosted cascade of simple features. In: Proceedings of the 2001 IEEE Computer Society Conference on Computer Vision and Pattern Recognition. CVPR 2001, vol. 1, p. I. IEEE (2001)
2. Davies, J., Studer, R., Warren, P. (Eds.): Semantic Web Technologies: Trends and Research in Ontology-Based Systems. Wiley (2006)
3. Kings, N. J., Gale, C., Davies, J.: Knowledge sharing on the semantic web. In: European Semantic Web Conference, pp. 281–295. Springer, Berlin, Heidelberg (2007)
4. Davies, J., Lytras, M., Sheth, A.P.: Guest editors' introduction: semantic-web-based knowledge management. IEEE Internet Comput. **11**(5), 14–16 (2007)
5. Davies, J., Studer, R., Sure, Y., Warren, P.W.: Next generation knowledge management. BT Technol. J. **23**(3), 175–190 (2005)

6. Cob, Z.C., Abdullah, R.:. Ontology-based semantic web services framework for knowledge management system. In: 2008 International Symposium on Information Technology, vol. 2, pp. 1–8. IEEE (2008, August)
7. Sajjad, M., Nasir, M., Muhammad, K., Khan, S., Jan, Z., Sangaiah, A.K., Baik, S.W.: Raspberry Pi assisted face recognition framework for enhanced law-enforcement services in smart cities. Futur. Gener. Comput. Syst. **108**, 995–1007 (2020)
8. Zhang, Z.F., Gao, Z., Liu, Y.Y., Jiang, F.C., Yang, Y.L., Ren, Y.F., ..., Zhang, X.D.: Computer vision based method and system for online measurement of geometric parameters of train wheel sets. Sensors **12**(1), 334–346 (2012)
9. Shinde, P.P., Oza, K.S., Kamat, R.K.: Big data predictive analysis: Using R analytical tool. In: 2017 International Conference on I-SMAC (IoT in Social, Mobile, Analytics and Cloud) (I-SMAC), 2017, pp. 839–842. https://doi.org/10.1109/I-SMAC.2017.8058297
10. Meenpal, T., Balakrishnan, A., Verma, A.: Facial mask detection using semantic segmentation. In: 2019 4th International Conference on Computing, Communications and Security (ICCCS), pp. 1–5. IEEE (2019)
11. Petrović, N., Kocić, Đ: IoT-based system for COVID-19 Indoor safety monitoring. Preprint). IcETRAN **2020**, 1–6 (2020)
12. Zacharias, J., Narayanan, V.: International journal of basic sciences and applied computing international journal of basic sciences and applied computing. Trans. Power Electron **23**(1), 332–342 (2008)
13. Pramila, M.J., Shewta, P.S.: Wireless temperature detector system using ARDUINO and IOT. Int. J. Comput. Trends Technol. (IJCTT) **67**, 82–83

Chapter 53
Exploring the Essentials Elements of Data Analysis and Its Impact on Performance Management Framework

V. L. Varun, M. Poornima, S. Pushpa, A. C. Chaitra, Padmaja Venugopal, and P. M. Sreejith

Abstract The research study sheds light on the impacts of essential elements of data analysis on the performer management frame. Therefore, all the essential elements of data analysis are analyzed in these research articles properly. The purpose of this research article is to investigate the importance and impact of the elements of the data analysis. Furthermore, to analyze the impacts of the essential elements of data analysis on the performance management framework is another purpose of this research article. Apart from that, the secondary data collection method is used in this research article to find out all the basic questions of this research article. Thereafter the secondary data collection method was helpful to collect some proper and accurate data about the elements of data analysis and that was helpful to understand and analyze the concept of the topic. Along with that, the secondary method of data collection was helpful to gain more knowledge about the impacts of the element of data analysis on the performance management framework. On the other hand, the quantitative and qualitative data analysis method is used to analyze the data that are collected by using the mentioned method of data collection. Two specific tables and two specific diagrams are analyzed in this research article that is based on the elements of data analysis. Therefore, some articles and journals were selected that are based on the elements of the data analysis and the impacts of those elements on the performance management frame. The data that is collected from those articles and journals are analyzed in this research article by using several techniques.

V. L. Varun (✉) · M. Poornima · S. Pushpa · A. C. Chaitra · P. Venugopal
Department of Mathematics, SJB Institute of Technology, Bengaluru, India
e-mail: bhadravathivarun@gmail.com

P. M. Sreejith
Cochin University of Science and Technology, Kochi, Kerala, India

S. Yadav et al. (eds.), *Proceedings of Second International Conference in Mechanical and Energy Technology*, Smart Innovation, Systems and Technologies 290,
https://doi.org/10.1007/978-981-19-0108-9_53

53.1 Introduction

Data analysis is an important part of any management frame and there are some elements that are important for data analysis. Therefore, data analysis is a process of analyzing the data that are collected by using some data collection methods. Apart from that, describing and illustrating by using some logical techniques is also called data analysis. The purpose of this research article is to analyze the important elements of data analysis and impact of the elements on the performance management frame. Along with that, there are some objectives or goals of this research study to achieve and the objectives are to investigate the essential elements of the data analysis process, to analyze the importance of statistical techniques for data analysis, and to describe the impact of the elements of data analysis on the performance management frame. According to these objectives of this research article all the data are collected by proper method of data collection and the data are analyzed by using proper data analyzing method.

There are some research questions about the research topic and a proper research method is adopted by the researcher to find out all the answers to these questions. The elements of data analysis are also called the fundamental components and these fundamental components are analyzed in these research articles properly. Apart from that, the performance management frame is an important factor with some steps, and data analysis is essential for a performance management framework. Therefore, the impact of the element of data analysis on the performance management frame is analyzed in this research study significantly. Some research methods are used in this research article for understanding and analyzing the concept of the topic of this research article. Thereafter, the research methods are analyzed in this research study, and the findings are analyzed in the discussion part of this study.

53.2 Essential Elements of Data Analysis

There are a lot of elements of data analysis and all the elements are essential. Different elements of the data analysis process are used for different data analyses. The elements of the data analysis process are very essential and the elements are *code comments, non-data visualization, code, data visualization, data acquisition, tables, statistical models, summary statistics, and computational algorithms*. These elements are most important for data analysis. There are different beneficial sides of these different elements of data analysis and according to those benefits the elements are used for different things [1]. Among all of the elements, statistical models are one of the most important elements of data analysis. Statistical models are based on the summarization of the results of the findings. Therefore, this is helpful for the researcher to analyze the data by using the statistical models. Apart from that statistical models are helpful to understand the data or information strategically and that makes a remarkable data analysis. On the other hand, data visualization is another most important element of data analysis. There are a lot of advantages to using data visualization as an element of data analysis. First of all, the data visualization process is helpful to understand the relationship between the data and that is very important to create a significant data analysis [2]. Apart from that data visualization is a clear and quick process to understand all the information fully stop there for this is helpful for the researcher to save the time and clarify the data analysis. Along with that data, visualization helps the researcher to identify the trends and helps to act easily based on those trends. Thus, the rest elements are also essential just like these two elements (Fig. 53.1).

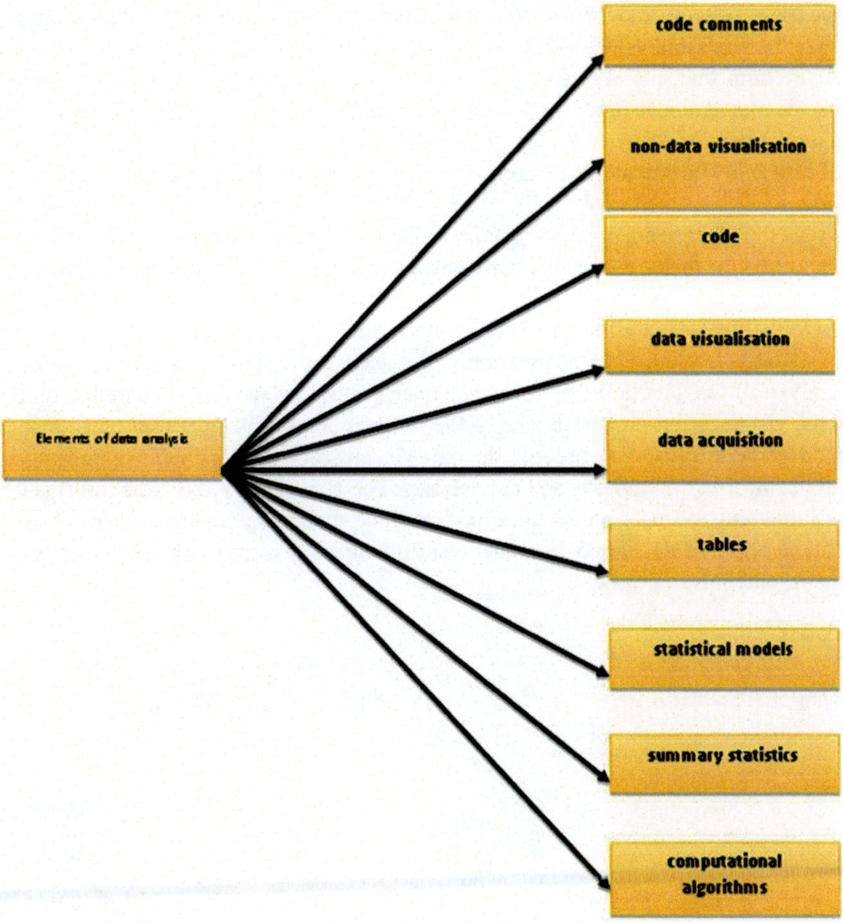

Fig. 53.1 Elements of data analysis

53.3 Impact of Different Elements of Data Analysis on Performance Management Frame

Performance management framework is an important framework that is based on the collected data. There are four steps to create the performance management framework such as *planning* then *activity* therefore *tracking* and then *reviewing*. After collecting the data, analysis of the data is very important for making the performance management framework of an organization. Furthermore, the elements of data analysis are equally important for a performance management framework [3]. First of all, it is important to collect data about the goals and values of the organization. Thereafter, the usage of table and the statistical elements of data analysis can be used to generalize the goals and values of the organization in the performance management form. Apart from that, the decision-making process is important for making the performance management framework of an organization. Thereafter the elements of data analysis help to improve the decision making so it can be saved that the elements of data analysis impact the performance management framework of an organization (Fig. 53.2).

On the other hand, *data acquisition* is another element of data analysis and this is important for making a perfect performance management framework. According to the viewpoint of [4], for making a perfect performance management framework it is important to collect all the necessary data about the performance of the employees of an organization. Data acquisition is the accurate process or key component to analyze all the data in the performance management framework. In addition to that data, visualization process is one of the most important elements for performance management framework because the data visualization process helps to understand all the information quickly and play clearly. The clarification about the information is important to make an accurate performance management framework. Thus, it can be said that the elements of that analysis impact positively on the performance management framework.

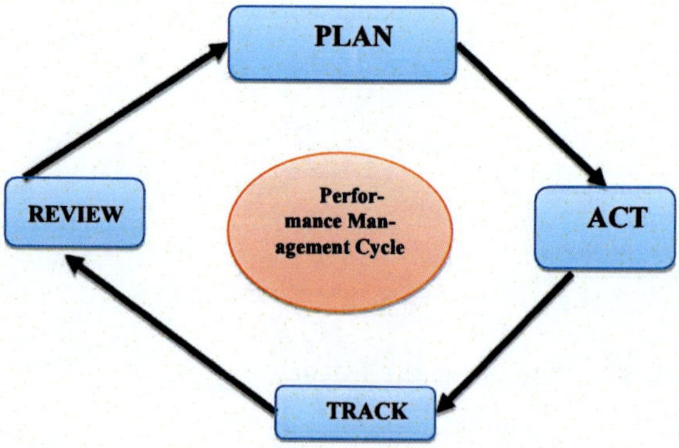

Fig. 53.2 Performance management cycle

53.4 Proposed Methods

Research methods are very important for any research study because research methods help to collect proper data and analyze the data. The researcher has adopted the *positivism research philosophy* for this research study to analyze the concept of the topic. Apart from that, the research has adopted the *descriptive research design* and *deductive research approach* in this research article. These research methods are helpful to create a remarkable research study. On the other hand, the most important research method is the method of data collection because data collection is most important for a research study. Therefore, the research has adopted the "***secondary method of data collection***" gathering data about the research method. The resources of the secondary method of data collection are books, magazines, newspapers, public records, government records, and other published records [5]. The researcher selected some articles and journals from the google scholar that are based on the importance of the elements of data analysis and its impact on performance management framework.

On the other hand, the mixed method of data analysis was adopted by the researcher to analyze the collected data. Therefore, the mixed method of data analysis supports the "***quantitative process of data analysis***" and the "***qualitative process of data analysis***". Quantitative process of data analysis is an important process and the statistical models, tables, and statistical elements of data analysis are used for "quantitative data analysis" in this research article. Apart from that, the thematic, code comments, and other elements of data analysis are used for the "qualitative data analysis" in this research study. According to the point of view of [6], quantitative and qualitative processes of data analysis are efficient and effective to create a significant research study. In addition to that, by this data analysis process, the numerical and statistical data that is collected by using the secondary data collection process, are analyzed in this research article.

53.5 Result and Discussion

The elements of data analysis that are mentioned in this table (Table 53.1) are the most important elements of data analysis. Different types of data about any topic can be collected by using the method of data collection and therefore, different types of elements are used to analyze those data. According to the viewpoint of [11], data visualization is an important element of data analysis and it is used to analyze the data by creating visual formats. Apart from that, statistical models are used to understand the information strategically. In addition to that, tables are also an essential element of data analysis because this element helps to create more presentable data analysis. As per the view of [12], Code element is used to identify one image or a passage in the whole text of data analysis (Table 53.2).

The elements of data analysis are helpful to create the performance management frame. First of all, data visualization is one of the important elements of data analysis

Table 53.1 The elements of data analysis and its usage

Element of data analysis	Usage of the elements
Code comments	This element is used to make the analysis readable for the people
Data visualization	This element is used to create the graph, chart, and several visual formats by using the collected data [7]
Data acquisition	This element is used for sampling the results
Code	This element is used to identify one image or a passage in the whole text of data analysis [8]
Tables	This element is used to make the research study more presentable by putting the collected data
Statistical models	This element is used to understand the data more strategically [9]
Summarize statistics	This element is used to analyze the values of the collected data
Computational algorithms	This element is used to calculate the values of the collected data or information [10]

Table 53.2 The elements of data analysis and its usage in performance management frame

Elements of data analysis	Usage of the elements in performance management frame
Code comments	This element is used to make the performance management framework more presentable [13]
Data visualization	This element is used to analyze the relationship between operational activities and organization
Data acquisition	This element is used to analyze the economic growth of the organization [14]
Code	This element is used to analyze the availability of several factors of the organization
Tables	This element is used to analyze the cash flows, profits, and other data of the organization [15]
Statistical models	This element is used to analyze the strategies of the organization
Summarize statistics	This element is used to identify the loss and profits of the organization [16]
Computational algorithms	This element is used to calculate the values of the organization

for performance management frame and the element is used to analyze the relationship between operational activities and organization. Therefore, data acquisition is also an important element that is used to analyze the relationship between operational activities and organization [17]. Apart from that, tables are also an essential element of data analysis and it is used to analyze the cash flows, profits, and other data of the organization [18]. Along with that, summarizing statistics is also very important for performance management. Thus, the elements of data analysis are essential and effective, and efficient for performance management framework.

53.6 Conclusion

In conclusion, it can be said that the elements of data analysis impact the performance management framework largely. Therefore, statistical models, computational, summarized statistics; code, code comments, tables, data visualization, and data acquisition are the most important elements of data analysis. "Secondary data collection method" played an important role in collecting all the important information about the elements of data analysis. Apart from that, the data collection method was also helpful to collect the data about the impact of the elements of data analysis on the performance management framework. There are two tables that are analyzed with all the data that are collected by using the "secondary method of data collection".

References

1. Al-Htaybat, K., Hutaibat, K., von Alberti-Alhtaybat, L.: Global brain-reflective accounting practices. J. Intell. Capital (2019)
2. Azzone, G., Noci, G.: Seeing ecology and "green" innovations as a source of change. J. Organ. Change Manag. (1998)
3. Branco, M.C., Rodrigues, L.L.: Corporate social responsibility and resource-based perspectives. J. Bus. Ethics **69**(2), 111–132 (2006)
4. Gasevic, D., Rose, C., Siemens, G., Wolff, A., Zdráhal, Z.: Learning analytics and machine learning. In: Proceedings of the Fourth International Conference on Learning Analytics and Knowledge, pp. 287–288 (2014)
5. Grimmer, J.: We are all social scientists now: how big data, machine learning, and causal inference work together. PS, Polit. Sci. Politics **48**(1), 80 (2015)
6. Lokuge, S., Sedera, D.: Is your IT eco-system ready to facilitate organizational innovation? Deriving an IT eco-system readiness measurement model. In: Proceedings of the 37th International Conference on Information Systems (ICIS 2016), pp. 1–22. Association for Information Systems (AIS) (2016)
7. Lytras, M., Visvizi, A., Zhang, X., Aljohani, N.R.: Cognitive computing, Big Data Analytics and Data Driven Industrial Marketing (2020)
8. Manavalan, E., Jayakrishna, K.: A review of Internet of Things (IoT) embedded sustainable supply chain for industry 4.0 requirements. Comput. Ind. Eng. **127**, 925–953 (2019)
9. Mavuri, S., Chavali, K., Kumar, A.: A study on imperative innovation eco system linkages to map Sustainable Development Goal 9. In: 2019 International Conference on Digitization (ICD), pp. 142–147. IEEE (2019)
10. Mergaliyeva, L.: The Nature of innovation eco-system of the Western Kazakh State University. Int. J. High. Educ. **9**(4), 254–260 (2020)
11. Moosavi, S.M., Jablonka, K.M., Smit, B.: The role of machine learning in the understanding and design of materials. J. Am. Chem. Soc. **142**(48), 20273–20287 (2020)
12. Sarif, S.M.: Effects of strategic ta'awun on sustainable, entrepreneurial and urbanised smart society in Muslim world. J. Educ. Soc. Sci. **9**(3), 40–46 (2018)
13. Scharmer, C.O., Kaufer, K.: Leading from the Emerging Future: From Ego-System to Eco-System Economies. Berrett-Koehler Publishers (2013)
14. Scuotto, V., Del Giudice, M., Carayannis, E.G.: The effect of social networking sites and absorptive capacity on SMES' innovation performance. J. Technol. Transf. **42**(2), 409–424 (2017)

15. Sujaya, B. L., Bhaskar, R.S.: A modelling of context-aware elderly healthcare eco-system-(CA-EHS) using signal analysis and "machine learning" approach. Wirel. Pers. Commun., 1–16 (2021)
16. Verhoef, P.C., Broekhuizen, T., Bart, Y., Bhattacharya, A., Dong, J.Q., Fabian, N., Haenlein, M.: Digital transformation: a multidisciplinary reflection and research agenda. J. Bus. Res. **122**, 889–901 (2021)
17. Vidgen, R., Shaw, S., Grant, D.B.: Management challenges in creating value from business analytics. Eur. J. Oper. Res. **261**(2), 626–639 (2017)
18. Katkar, S.V., Dongale, T.D., Shinde, P.P., Kamat, R.K.: Calculation of electrical parameter of solar cell using java framework. AGU Int. J. Sci. Technol. (AGUIJST) 2017, 5, e-ISSN: 2455-0493; p-ISSN: 2455-3808
19. Love, J.H., Roper, S.: SME innovation, exporting and growth: a review of existing evidence. Int. Small Bus. J. **33**(1), 28–48 (2015)
20. Shinde, P.P., Oza, K.S., Kamat, R.K.: Big data predictive analysis: using R analytical tool. In: 2017 International Conference on I-SMAC (IoT in Social, Mobile, Analytics and Cloud) (I-SMAC), 2017, pp. 839–842. https://doi.org/10.1109/I-SMAC.2017.8058297

Chapter 54
Effects of Covid-19 Lockdown: A Perception Survey

Pashmeen Kaur, B. P. Bijay Sankar, and Hemant Bhanawat

Abstract The purpose of the study is to examine the perceptions of people on the effect of lockdown and outbreak of COVID-19 in the National Capital Region (NCR) area of India. In this survey, the two cities of the NCR area, i.e., Delhi and Guru gram are selected for the collection of primary data. The primary data have been collected through the questionnaire. A total of 204 respondents have answered the pre-framed questionnaire. Five-point Likert Scales have been used to judge the degree of agreement with the statements. Descriptive statistics are used to analyze the data. The findings suggest that people pay more attention to personal hygiene during the pandemic time and they feel that working from home is a challenge in India due to bandwidth and technological issues. Job security is a big concern in front of people due to lockdown.

54.1 Introduction

The world is struggling and fighting against the coronavirus pandemic disease. It not only affected human life but also the livelihood of human beings. Various measures have been taken to control the pandemic in India but the situation is still chaotic. The International Committee on a taxonomy of the virus called it as 2019-nCov and syndrome coronavirus 2 (SARS COV-2). It is a new kind of virus found in humans on such a large scale for the very first time.

Coronavirus is a communicable disease and the World Health Organization (WHO) declared it as a pandemic on March 12, 2020. The droplets of saliva or the discharge from the nose spread coronavirus while the infected person is sneezing or coughing. Person having low immunity, older in age, and children less than 10 years

P. Kaur
Chitkara University, Rajpura, Punjab, India

B. P. Bijay Sankar (✉) · H. Bhanawat
Birla Global University, Bhubaneswar, Odisha, India
e-mail: bpbijaysankar@gmail.com

S. Yadav et al. (eds.), *Proceedings of Second International Conference in Mechanical and Energy Technology*, Smart Innovation, Systems and Technologies 290,
https://doi.org/10.1007/978-981-19-0108-9_54

of age are highly at risk of infection from coronavirus. The protection from the infection lies in the proper respiratory etiquette and maintenance of sanitary and hygiene with an alcohol-based sanitizer. It is necessary to cover the mouth and nose with a cloth or mask and frequently touching the face with hands should be avoided.

The first case was reported in Wuhan, the city of China. After that, it is spread into different countries of the world like America, Brazil, Italy, China, Canada, Russia, Peru, Columbia, and almost every country in the world.

Coronavirus has also affected India. The first case in India was reported on 30th January 2020 and there is a substantial rise in the number of infected patients. After that the Indian Government had taken precautionary measures like the 14-h voluntarily "Janta Curfew" was observed on 22nd March 2020. On 24th March 2020 lockdown was observed in the whole country and four phases of lockdown were observed till 31st May 2020. The third phase of lockdown was started from 4th May 2020 to 17th May 2020 with some leniency in the restrictions like the non-essential items and the railway was started.

Due to the lockdown, the shops were closed; there was no transportation in the country. The schools & colleges were closed and jobs of the people were at high risk. Only essential item shops and medical shops were opened during the lockdown. Everybody was restricted to stay at his or her home only. Thus this situation induced us to examine the perception of the people and to know about their worries related to the effects of the lockdown so that immediate measures can be taken to reduce their problems of the people [1–5].

The disease is spreading at a high speed and the public all over the world is scared of being affected by the coronavirus in many ways. The fear among people is related to their businesses, jobs, health, social security, and many more. The cities with high-density populations are highly at risk, where the contact between people is very high and that is the reason the Government has imposed a lockdown in the country. In the COVID-19 pandemic, people in Delhi and Guru gram have their perceptions regarding the lockdown and its effects on them. Perception refers to the way someone thinks about something and the way that someone notices things with his/her senses.

The perception of the effects of lockdown on the people of Delhi and Guru gram is examined through the help of a questionnaire. As the people living in Delhi and Guru gram are earning through jobs in different industries, the study will reflect the effect of lockdown on the people and their jobs & behavior. The paper discusses the financial trouble and job insecurities of the people due to lockdown.

54.2 Conceptual Framework

The number of patient infected from the corona is increasing day by day and the situation due to coronavirus is getting very serious in the world. Various studies have been taken place in such a time in which different views; awareness and perceptions were studied. Perception is the process of receiving the information, collecting it and possessing it, and making apprehension with the mind or senses [6]. Some argue that there is a close relationship between communication and perception. Communication is the perception of verbal and non-verbal behaviors and the assignment of meaning to them [8].

As this study is related to the coronavirus, in February 2003 one more virus, i.e., Severe Acute Respiratory Syndrome (SARS) was recognized. At that time SARS was spread among 28 countries and the cumulative cases were 8096 with 774 deaths. One more related virus was found in Saudi Arabia, in 2012. It was named Middle East Respiratory Syndrome Coronavirus (MERS-COV). It was a zoonotic virus that has entered humans from infected camels. The total number of cases from MARS was 2494 [5].

In the year 2020, the Coronavirus had affected the lives of people all around the globe. One of the studies is conducted on the high socioeconomic status people of China and it was revealed that they were quite knowledgeable, wore masks, and were optimistic about the procedures taken to tackle COVID-19. The women from high socioeconomic status are more knowledgeable and have an optimistic attitude toward the outbreak of the COVID-19 situation in China [10]. Whereas the study was conducted on health care workers in Asia, to know their knowledge and perception related to COVID-19. The research revealed that there was insufficient knowledge among the health care workers but they had a positive perception to prevent the transmission of coronavirus in Asia [1]. The people of US and UK, where it is found that the participants had good knowledge of the main mode of transmission of the coronavirus disease and its common symptoms of it but there are some misconceptions among the few participants like only wearing masks is very effective in safeguarding themselves from the coronavirus and they should be cautious while eating at Chinese restaurants. Participants were also having views that children are at high risk of death due to SARS-CoV-2 [3]. Thus the different studies discussed the mindset of the people belonging to the different areas. Moreover, strong government decisions also play a vital role in the mental health of their people and the country. The weak government resultantly led to high worries and depression for their population. The policymakers not only consider their decisions regarding Covid-19 but how the decision will affect the mental health of the people? [7] The spread of coronavirus is quite fast and its vaccination will take time to develop but there lies a positivity in this situation which thus leads to the development of surveillance techniques and innovative technologies that are introduced to help doctors and healthcare workers such as robots, Toronto based health monitoring A.I (Bluedot), portable lab-on-chip detection kits [2]. Despite taking various steps by the government to minimize the effect of LockDown into day to day life of the general public, still, people are facing a range of problems. Some studies focused on the medical issues and health conditions of people. However, this is to the best of our knowledge that there is no study focused on the perceptions of people on the effect of lockdown in the National Capital Region area of India. The situation is still uncertain all over India, so this study is conducted to know the perception of the people of two cities the NCR area to know the effects of lockdown on them [6–10].

54.3 Research Objectives

The main objective of the study is to examine the perceptions of the people on lockdown due to coronavirus (COVID-19) in Delhi and Guru gram cities of the NCR area this study also aims to analyze the ongoing situation of COVID-19 and to know the personal, financial and social worries of the people in these cities.

54.4 Research Methodology

54.4.1 Data Analysis and Interpretation

This study uses primary data for analysis. The data is collected online mode via emails. The questionnaire was sent to 540 people, out of which 204 answered the full questionnaire. The survey was started on 1st May 2020. We have asked 16 five-point Likert scale questions to the people of Delhi and Gurugram.

The purpose of selecting the Likert scale is that it is used to allow the individual to express his or her view in the degree of Strongly Agree, Agree, Neutral, Disagree, and Strongly Disagree with a particular statement related to the effect of lockdown due to Covid-19. The participants were asked about their current state and future effects on their occupation, social and mental conditions during the lockdown period. This survey was restricted to two cities, i.e., Delhi and Guru gram only because the coronavirus has affected severely in these cities. The Google form was used to prepare the questionnaire for this study. Participants had to answer demographic questions such as age, gender, and educational qualifications. It is mandatory to answer these questions to reach the next questions. The details of the sample are as follows (Table 54.1).

Figure 54.1 represents the age of the respondents, which shows that the major portion of respondents is from the age group of 31–40 years. Figure 54.2 represents the gender of the respondents, which thus reveals that the numbers of male respondents are more than the female respondents (Table 54.2).

Table 54.1 Details segregation of sample size (age, gender, and educational qualification)

Age			Gender		Educational qualification		
20–30	31–40	41–60	Male	Female	Under graduate	Graduate	Post graduate and above
63	120	21	135	69	3	54	147

Source Compiled by the researchers

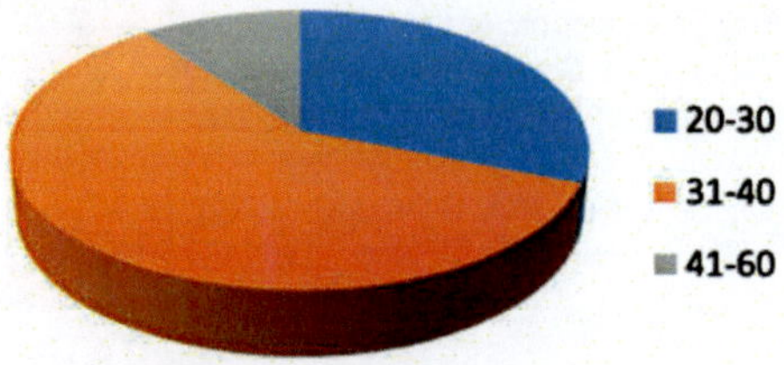

Fig. 54.1 Age of the respondents. *Source* Compiled by the researchers

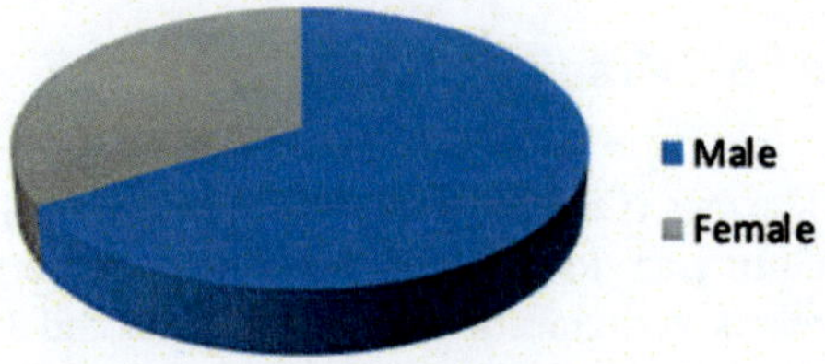

Fig. 54.2 Gender of the respondents. *Source* Compiled by the researchers

Table 54.2 Details segregation of sample size (occupation & income)

Occupation				Income			
Private employee	PSU/Govt. employee	Self employed	Others	0–1 lac	1–3 lac	3–5 lac	More than 5 lacs
132	18	30	24	24	18	30	132

Source Compiled by the researchers

Figure 54.3 represents the occupation of the respondents, which shows that the major portion of respondents is employed in private organizations. Figure 54.4 represents the income of the respondents, which thus reveals that the major portion of the sample includes the respondents who are earning more than 5 lacs.

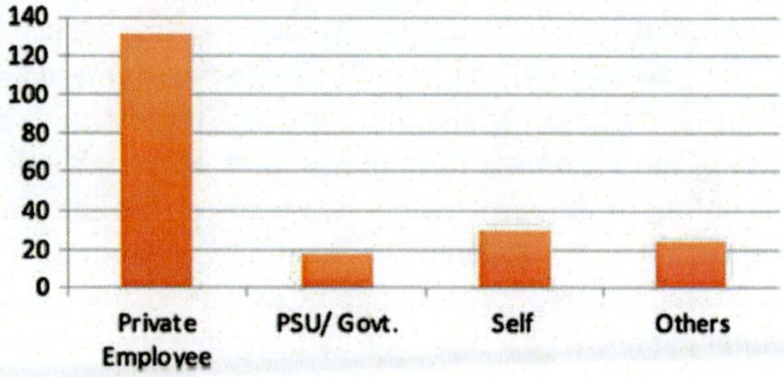

Fig. 54.3 Occupation of the respondents. *Source* Compiled by the researchers

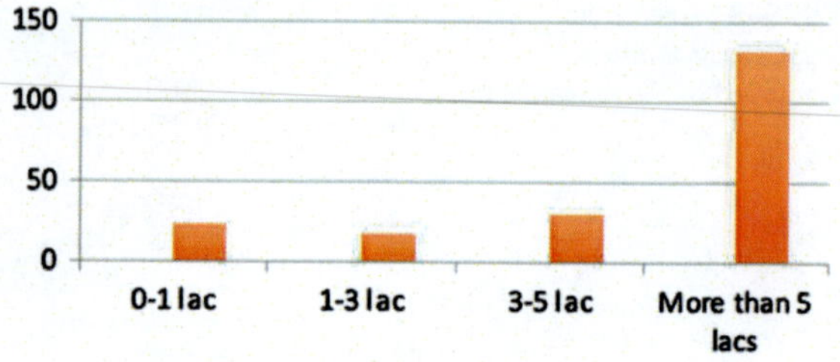

Fig. 54.4 Income of the respondents. *Source* Compiled by the researchers

54.5 Result Analysis

The data is collected after the announcement of lockdown in the area of Delhi and Guru gram to check the perception of the people toward lockdown. The 16 statements were considered as 16 variables and descriptive statistics were used to analyze the data. About 50% of the people strongly agreed that they pay more attention to personal hygiene as compared to the pre-lockdown period. Around 35.3% of the people strongly agreed and approximately 62% of people agreed that there is availability of necessities in the nearby market and there is no problem in the supply of essentials.

Approximately 5.9% of the people strongly agreed and 46% of people agreed that being socially disconnected leads to psychological disorders in the public. 75% of the people agreed that people should not break the coronavirus lockdown. 50% of the people agreed that working from home is another challenge in India due to a lack of network and Internet facilities. 54.4% of the people strongly agreed that job security is also a big reason for worry these days. According to 56% of the people feel that further extension of lockdown will affect the future scope of the business and job. The mean value of Variables in Table 54.3 supports the results. Whereas 34% of people disagreed and 12% of people strongly disagreed that they feel depressed during the lockdown period. 35% of people disagreed that the medical facilities in India are appropriate to fight the pandemic situation in India.

Around 35% of the people also disagree that the pandemic situation will be curbed in the next two months in India. Table 54.3 represents the descriptive analysis and the mean value of variable 1 is 2.9 which represents that people do not feel depressed during the lockdown. The mean value of variable 6 is 2.8 which represents that according to the respondents, the medical facilities in India are not appropriate to fight against the pandemic.

Moreover, respondents support that the pandemic situation will not curb in two months and will go longer than that. It is determined that people disagree that they feel upset during the lockdown and the medical and hospital facilities are not adequate for them. But the respondents are confident that India will control the situation with the help of precautionary measures and support of the people.

Figure 54.5 represents the mean value of the variables in which the values of more than 3 reveal that respondents support the statement by agreeing or strongly agreeing on it. Figure 54.6 represents the frequency distribution of the variables which shows the frequency of strongly agree, agree, strongly disagree, and disagree.

Table 54.3 Descriptive analysis for different variables

Variables	Mean	Std. error	Median	Std. deviation stat	Frequency of strongly agree and agree	Frequency of strongly disagree and disagree
Var 1	2.8529	0.0819	3	1.1694	73	93
Var 2	4.4118	0.0514	4.5	0.7341	196	6
Var 3	4.3088	0.0404	4	0.5768	199	3
Var 4	3.2647	0.0804	3	1.1484	97	60
Var 5	3.3529	0.0646	4	0.9221	106	39
Var 6	2.8088	0.0715	3	1.0208	67	90
Var 7	2.8971	0.0791	3	1.1292	72	91
Var 8	3.5441	0.0664	4	0.9484	126	28
Var 9	4.6567	0.0482	5	0.6831	190	6
Var 10	3.6765	0.0717	4	1.0235	42	39
Var 11	4.4118	0.0555	5	0.7922	187	6
Var 12	4.4559	0.0503	5	0.7178	190	6
Var 13	4.1618	0.0460	4	0.6571	181	3
Var 14	4.3235	0.0473	4	0.6756	187	3
Var 15	3.4265	0.0774	4	1.1051	127	51
Var 16	3.7941	0.0738	4	1.0536	129	28

Source Result formulated from the statistical software

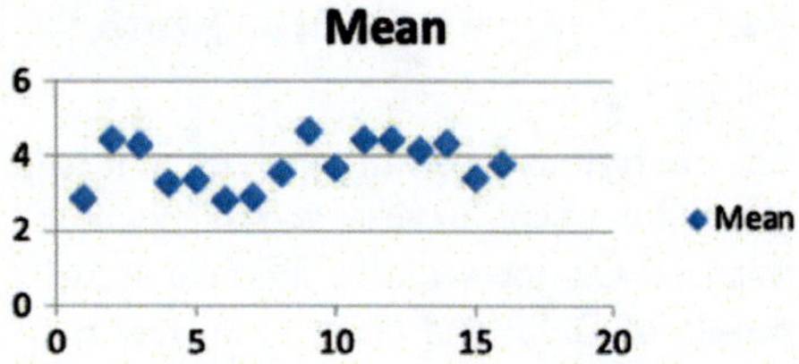

Fig. 54.5 Mean values of the variables. *Source* Compiled by the researchers

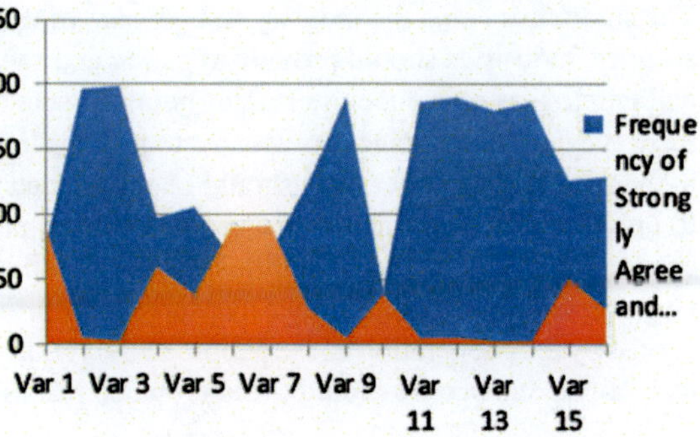

Fig. 54.6 Frequency distribution of the variables. *Source* Compiled by the researchers

54.6 Findings

After studying the primary data it is analyzed that people in the area of Delhi and Guru gram pay more attention to personal hygiene such as frequently washing their hands, wearing masks and social distancing is followed. There is no shortage of necessities in the nearby market. The essentials are easily available in the market. Due to COVID-19, people spend more time on social media and watch television as a source of information and entertainment due to lockdown people cannot visit Movie Theater and go to malls for their enjoyment. But people are satisfied with the banking facilities. They are scared and being confined inside their houses; they agreed that social disconnection due to lockdown will lead to psychological disorders among people. Due to the huge expenditures on health, people agree that everybody should invest in health insurance to save themselves from the unbearable expenditure due to health problems. The lockdown affects the job of the people and they have job security issues. If the pandemic situation will remain like this and lockdown will further be extended, it will affect the prospects of the job and more opportunities in the market. As there is a lack of customers in the market, it is directly affecting the industry. The restriction should remain in the containment zones and those areas that are less affected by the coronavirus should be free from restrictions of the lockdown.

The majority of the respondents are confident that India will control this pandemic in the coming future and the new sense of hygiene will be developed among the public.

54.7 Implication of the Study

The study thus helps the government to prepare for the uncertain situation in the future due to coronavirus and to secure the people from their insecurities regarding financial and job security, medical security, social security, and mental health. The people of Delhi and Guru gram feel that the pandemic should be cured soon by government measures and lockdown will affect the future scope of their business. The study thus helps the industry to take care of their employees and give them job security. Industries should provide hygienic and sanitized conditions to the workers and employees to reduce the risk of being infected by the coronavirus. Moreover, there is a duty of insurance companies to provide low-cost health insurance schemes to the people. Online services should be introduced in almost every field and access to online technology should be available for the public through proper bandwidth connections. The medical facilities should be appropriate for the people and the Government should take further measures to improve the situation. The situation is curbed by the cooperation of the people of the country so that social distancing can be maintained and people should avoid crowded places and use the mask as a protocol.

54.8 Further Scope of Study

This study is an attempt to observe the perception of people National Capital Region area of India during the first lockdown. The researchers can be extended this study by taking more variables from different parts of India. As this is a questionnaire-based study, the research could be possible by one-to-one interviewing the people and data collecting from secondary sources.

54.9 Conclusion

As the second-largest population in the world, India is severely affected by the first outbreak of the COVID-19 pandemic. Various people lost their lives and the negative vibes are all around. The lockdown has impacted the general public in the context of their routine work and Office work. People have suffered lots of technological and financial issues. This research will help in understanding the issues and problems encountered by the people of Gurugram and Delhi. But considering every cloud have a silver line, Indian people are hopeful that India will successfully beat the covid-19 and all will be normal again.

Annexure

Variables	Questions
Var 1	I feel depressed during the lockdown period
Var 2	I pay more attention to personal hygiene as compare to pre-lockdown period
Var 3	There is availability of basic necessities in the nearby market and there is no problem in the supply of essentials
Var 4	I spent more time on social media and watching television during this lockdown period
Var 5	Being socially disconnected leads to psychological disorders in the public
Var 6	The medical facilities in India are appropriate to fight the pandemic situation in India
Var 7	I feel the pandemic situation will be curbed in the next two months in India
Var 8	I am satisfied with the government measures taken to control the coronavirus pandemic
Var 9	People should not break the coronavirus lockdown guidelines
Var 10	Work from home is another challenge in India due to a lack of network and internet facilities
Var 11	Job security is also a big reason of worry these days

(continued)

(continued)

Variables	Questions
Var 12	If the lockdown period will be further extended it will affect the future scope of business and job
Var 13	I am satisfied with the banking services related to hassle-free online services and ATM during the lockdown period
Var 14	I believe everybody should invest in health insurance these days
Var 15	I feel the central government should remove restrictions on a few states those are less affected by coronavirus
Var 16	I am confident that India will control the coronavirus to zero and come up as a role model for the world

References

1. Bhagavathula A.S., Aldhaleei W., Rahmani J., Mahabadi M., Bandari D.: Novel Coronavirus (COVID-19) knowledge and perceptions: a survey on healthcare workers. medRxiv (2020). https://doi.org/10.1101/2020.03.09.20033381
2. Bhagavathula, A.S., Shehab, A.: The story of mysterious pneumonia and the response to deadly novel coronavirus (2019-nCoV): so far! New Emirates Med. J. **1**(1), 7–10 (2019). https://doi.org/10.2174/0250688201901010007
3. Geldsetzer, P.: Use of rapid online surveys to assess people's perceptions during infectious disease outbreaks: a cross- sectional survey on COVID-19. J. Med. Internet Res. **22**(4), e18790 (2020). https://doi.org/10.2196/18790
4. McCloskey, B., Heymann, D.L.: SARS to novel coronavirus-old lessons and new lessons. Epidemiol. Infect. **148**, e22 (2020). https://doi.org/10.1017/S0950268820000254
5. Modi P.D., Nair G., Uppe A., Modi J., Tuppekar B., Gharpure A.S., Langade D.: Covid-19 Awareness among healthcare students and professionals in Mumbai Metropolitan Region: a questionnaire- based survey. Cureus **12**(4), e7514 (2020). https://doi.org/10.7759/cureus.7514
6. Qiong, O.U.: A brief introduction to perception. Stud. Lit. Lang. **15**(4), 18–28 (2017). https://doi.org/10.3968/10055
7. Thiemo, F., Marc, W., Lucas, H., Jon, M.J., Johannes, H., Andriy, I., Stefano, C., Elena, R., Christopher, R., Stefano, F., Margarita, G., Gordon, K.T., Friedrich, M.G., Erez, Y.: Global behaviors and perceptions at the onset of the COVID-19 Pandemic. Working paper (2020). https://doi.org/10.3386/w27082
8. Varner I., Beamer, L.: Intercultural Communication in the Global Workplace, 3rd ed. Shanghai Foreign Language Education Press, Shanghai (2006)
9. World Health Organization: WHO Director-General's opening remarks at the mission briefing on COVID-19-12 March 2020. https://www.who.int/dg/speeches/detail/who-director-general-s-opening-remarks-at-the-mission-briefing-on-covid-19-12-march-2020
10. Zhong B.L., Luo W., Li H.M., Zhang Q.Q., Liu X.G., Li W.T., Li, Y.: Knowledge, attitudes, and practices towards COVID-19 among Chinese residents during the rapid rise period of the COVID-19 outbreak: a quick online cross-sectional survey. Int. J. Biol. Sci. **16**(10), 1745–1752 (2020). https://doi.org/10.7150/ijbs.45221

Chapter 55
A Critical Analysis on the Application of Next-Generation Internet of Things (Iot) for Building Efficient Energy Systems for Improved Temperature Control and Better Energy Consumption

Vivek Veeraiah, Naim Shaikh, A. Narasima Venkatesh, Korakod Thailand, G. Kousalyadevi, and V. Shanmugasundaram

Abstract This research article is focused on analyzing the application of next-generation Internet of Things (IoT) in creating energy-efficient systems so as to improvise the temperature and manage the energy consumption effectively. The term IoT is a new concept that has been growing imminently due to growing needs and requirements of government, individuals, corporations, etc. The IoT is mainly stated as a critical system of computing machines that possess better and enhanced connection so as to automate various aspects of the routine needs of the individuals, it also assists in creating better information approach for evaluation and implementation of different approaches for sustainable living. The IoT system connects the different smart devices and senses through cloud-based controllers, the sensors in the building can enable in collecting and transmitting the real-time data on the environment, this information is then used by the controller's team to offer quick responses on the temperature control ana management. The systems enable in applying predictive

V. Veeraiah
Department of R&D, Computer Science, Adichunchanagiri University, Mandya, Karnataka, India

N. Shaikh (✉)
Global Business School and Research Centre, Dr. D.Y. Patil Vidyapeeth, Tathawade, Pune, India
e-mail: naim.shaikh@dpu.edu.in

A. Narasima Venkatesh
MBA-Department of Human Resource Management, RV Institute of Management, Bangalore, Karnataka, India

K. Thailand
Faculty of Law, Thaksin University, Muang Songkhla District, Thailand

G. Kousalyadevi
Department of Architecture, Mohamed Sathak Engineering College, Kilakarai, India

V. Shanmugasundaram
Department of Electrical and Electronics Engineering, Sona College of Technology, Salem, Tamil Nadu 636005, India

S. Yadav et al. (eds.), *Proceedings of Second International Conference in Mechanical and Energy Technology*, Smart Innovation, Systems and Technologies 290,
https://doi.org/10.1007/978-981-19-0108-9_55

and adaptive algorithms which will enable in executing the real-time operational responses, likc turning on or switching off the lights based on the signals from in-room occupancy and turning on the air conditioning based on the level of occupancy in the building, this enables in optimizing the heating, ventilation and air-conditioning system so as to enhance better energy consumption. This article presents a critical analysis of the usage of IoT in enhancing the energy-efficient system in the building for improved temperature control and enhanced energy consumption.

55.1 Introduction

Based on the available sources it is noted that the conventional host of videos, images, etc. the worldwide web is now enhancing and offering better services to all stakeholders. The IoT system connects different types of sensors and other smart devices to the local or cloud-based. These sensors are mainly responsible for transmitting the real-time information related to the building, environment, etc. so that appropriate controls can be implemented. The emergence of IoT systems enables the government, organizations, and individuals in reacting to energy-efficient systems which will adjust the temperature control in the building thereby supporting efficient usage of power and energy [1]. The application of IoT has enabled in implementation of smart energy systems which have been used to control and automate the energy usage in residential and commercial buildings. The usage of IoT prototypes has significantly supported in preserving the energy consumption and thereby maintains better temperature control in the environment.

It is noted that the overall communication between the command centers and the consumers has enabled in creation of smart energy meters. These aspects have ensured that the overall exchange of messages regarding the consumption of energy and the usage has ensured in creating energy-efficient systems for sustainable growth in the future. The application of IoT on gas meters has enabled in creating and transmitting of real-time data related to the temperature in the building and thereby adjusting the temperature of air condition system for better energy usage.

The IoT possesses smart choice head which received the operational data and is then moved to local nodes stated as smart fusion, this information is monitored and controlled based on the collation and exchange of information from IoT network. It is widely regarded that IoT is considered to be an effective measure in monitoring and controlling the energy usage in the building [2]. The system can be set up so as to measure the overall energy usage, analyze the nature of energy type to be applied like solar energy, natural gas, etc. and also look to meet the energy needs in an effective manner (Table 55.1).

The primary aspect is that it will identify the devices which are responsible for energy usage and make the process more efficient. This enables in understanding the creative actions which enable ineffective usage, analyze the potential in application where the IoT system by monitoring and controlling the usage of power. This is mainly achieved by allowing the IoT to connect the devices grid so as to remove the

Table 55.1 Drivers and challenges of implementing IoT in energy sector

Categories	Features of IoT in energy efficiency
External drivers	Normative pressure Government regulations Need to protect climate change
Internal drivers	Support from the stakeholders Investment allocation Maintenance and cost savings
External challenges	Relative advantage Continuous change in technology Technology compatibility Adoption cost
Internal challenges	Increased operational cost Quality of maintenance Support from employees

Source Saleem et al. [2]

standby power consumption. Lastly, the IoT system can enable accessing the power by limiting the overall consumption which is not critical. If the option tends to exist, the IoT can select the times of the lower grid to power the devices [3]. This enables in more cost-effective and maintains spread the power demand in effective manner. The smart buildings are stated as the creative aspect which enables in application of different sensors which are connected to the different subsystems like lighting, elevators, surveillance cameras, common heating, HVAC, and others.

The IoT implements adaptive and predictive controllers which will help in reducing the consumption of energy, remove wastage and enhance experiences for the tenants and occupants. Moreover, there are enhanced complexities in building the energy environments which are arising due to large variety of datasets and implementation of control mechanisms.

55.2 Literature Review

In the era of globalization, communities around the world are focused on integrating through the application of technologies, this enables in serving different purposes across industries. As stated by the IoT is considered as the overall convergence of virtual, real, and digital aspects with are focused on making smarter cities by applying and using intelligent devices. Which will enhance efficient usage of the energy in the buildings and thereby manage temperature control?

It this been stated that IPV6 is considered as the critical protocol for the concept, this allows the developers in using various physical devices while connected to them and also control them at any point in time [4]. The application of IoT has moved from the traditional virtual and theoretical concept into reality, many experts tend to state the IoT application is highly effective in industrial areas, consumer-related IoT,

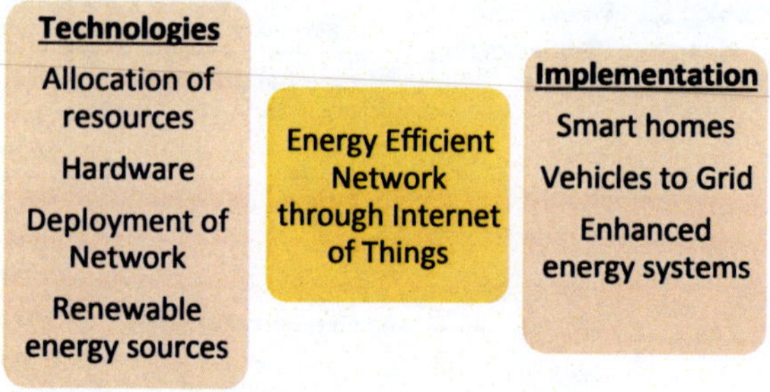

Fig. 55.1 IoT architecture in energy

etc. The usage of smart wearables, smart appliances at home, phones, etc. has made IoT more popular and effective.

Furthermore, the author has stated that the IoT is stated as an effective tool in monitoring and controlling the building energy and its overall usage. The IoT can be set up so as to measure the source power, this focused on isolating each equipment, moreover, IoT can enable in analyzing the power demand for various equipments in different rooms or floors and enable in managing its power consumption in a better manner [5] (Fig. 55.1).

The application of IoT is more focused on collaring the critical data which are highly helpful in different aspects related to application in creating energy-efficient systems for enhanced temperature control and better consumption. The first aspect is that the system will enable in creating for high energy usage and also focus on creating energy-efficient systems. The IoT also supports in understanding the data which are available in each room and equipments in the building forecast the energy usage and also study the results in different energy reduction methods. The next aspect is that the IoT system not only enables monitoring of the temperature but also controls and manages the usage of power based upon occupancy levels and other criteria [6]. This is mainly achieved by allowing the IoT system to disconnect the devices from the grid so that the consumption can be minimized when there is no usage.

Also, the IoT enables in selecting when the devices can access power, for example, need for hot water in the morning and night whereas it is not required during the day, this will limit the consumption of power source, moreover, when the option exists, the IoT can select time of low grid demand to power devices, which will allow more cost-effective for the users, enable in enhancing the power optimization in the communication and use less power, which will reduce the energy bills.

IoT solutions often have limitations and challenges, in many cases, there are solutions and solutions. Common problems can occur due to the network of different types of sensors and devices, each of which performs different functions. This is problematic due to the many problems that arise. In addition, IoT systems need to communicate with each other and wired solutions are not always possible. Connection problems due to structural disruption or an unstable internet connection are common. Another limitation of IoT systems is that they can produce extremely large amounts of data that require efficient handling and storage.

55.3 Methodology

The main aim of the study is to analyze and apprehend the implementation of next-generation IoT in creating energy-efficient systems for enhancing temperature control and energy consumption. This paper is a conceptual analysis of the subject area which stated the role of IoT in creating better energy-efficient systems. The researchers intend to apply the descriptive research so as to understand the role of technology in the energy consumption and effective usage. The researchers have applied secondary data sources in order to analyze the different types of IoT systems that have been implemented in the communities and buildings, the various data sources used are Google scholar, EBSCO, etc. Qualitative analysis is applied in the study so as to provide comprehensive understanding of the research area.

55.4 Implementation of IoT in Building Efficient Energy Systems

Many studies have been explored in analyzing the overall reaction and development of the IoT system related to power-driven appliances recognition. There is an opportunity that the electric appliances detection and monitoring tend to provide greater importance based on the focus on using energy-efficient systems. Furthermore, researchers have supported in application of a monitoring system called Raspberry Pi, which is a low-cost and high reliable IoT system. This enables in using eh Node programming to collate the data from the energy meters and save them for effective analysis. These data can be easily accessible through personal computers or smartphones using various applications [7]. The IoT-enabled energy monitoring system tends to facilitate in analyzing the energy usage and also take adequate steps in conservation measures. The IoT-based system tends to consist of the energy meters, implementation of Raspberry Pi system, cloud-based services, and control devices. It is also noted that there exist other IoT devices apart from Raspberry Pi, which are Arduino, NVIDIA, Nano, etc. (Table 55.2).

Furthermore, the implementation of FX5U IoT enables the device t to possess better communication with the inverter through 3C frames. The implementation of communication devices can enable communicating through 4G/5G networks, Bluetooth devices or Wi-Fi, etc. In another study, it is noted that the IoT system enables in implementation of microcontroller in the temperature sensors which are then connected to modules, this enables in controlling the indoor climate based on the outdoor environment. The system which is used is based on three key aspects, the first is categorized as sensors, the second is the main gateway, and the last part is involved in cloud networking [8]. The overall system flow enables in using the IoT position in the central devices, enable enhanced data flow, the IoT sensor is then situated in the terminal and is linked to the cloud servers. The sensor data are then measured and

Table 55.2 Application of IoT

Categories	Critical description	Intended benefits
Energy regulation	Access to the smart grid for all users	Enhancing the supply chain management system in energy sector, harness the pricing mechanism
Predictive analytics	Analyzing the energy usage	Forecast the energy requirements in the usage, maintain the internal temperature based on the outdoor environment
Smart grids	Operating the grid using IoT	Enhancing energy efficiency and integration of distribution
Microgrids	Operating the smaller grids	Enhancing security of energy supply

Source Rinaldi et al. [7]

stored in the cloud server for effective analysis, this will enable enhanced application of the energy-saving systems.

In order to enhance the critical transfer of data through network various encryption procedures are applied and maintained, this is more powerful and enables better storage of the data and information. The system tends to be found to enhance the cooling capacity by nearly 15% and the performance is optimized by 45% which allows better microclimate control inside the buildings.

The productive temperature control is highly helpful in optimizing the energy usage and overall consumption. The IoT infrastructure is considered to be an effective manner so as to accomplish them, the HVAC system is mainly controlled through the effecting reactive systems. The predictive controller tends to process the data based on patterns relate to heating and cooling based on various factors including weather conditions, operations aspects, occupancy, and related factors [9]. The HVAC system tends to know about the heating of the building and occupancy aspects related to heating process. Moreover, such system tends to detect the overall performance of the HVAC system in enhancing performance. The system is designed to sense the moisture and react in controlling the doors and ventilation aspects to the specified portion of the buildings.

Moreover, IoT data collection system enables in combination with reinforcement to enhance energy usage in the building. The application of IoT systems tends to collect the data, analyze the information and infer energy usage so as to create better HVAC scheduling methods [10]. The IoT system tends to analyze the energy usage, forecast the requirements, and establish the energy-saving policy. The IoT-based architecture enables predictive control of HVAC systems in enhanced situations, the closed-loop control mode tends to control the thermal comfort on the associated energy utilization in the single zone aspects. The IoT system enables connecting the sensors and actuators which are remote database services and actuated in the HVAC system.

55.5 Conclusion

IoT-based solution for predicting indoor temperature. Forecast-based technology such as demand response and demand control is used to reduce energy waste. The solution uses an advanced, non-linear, self-aggressive neural network to make predictions based on data collected by IoT. The forecasts proved to be accurate and reliable. Efficient management of energy consumption of HVAC systems in smart grids with variable energy prices. An energy planning plan was proposed that minimizes the cost of energy consumption over a certain period of time, taking into account the energy price and overall comfort constraints, i.e., temperatures determined by the decisions of the user of a particular living room. The HVAC system took into account user preferences for certain rooms and certain temperatures for maximum comfort. In addition, HVAC system controllers can be remotely monitored and adjusted for "dynamic" changes to maintain the efficiency of the HVAC system. The system used a Thermal Comfort Validator web application that predicted average voice comfort theory based on feedback from passengers in real-time. The system is ready to achieve energy savings based on data collection and processing while offering a high level of convenience.

IoT has an intelligent sampling head that receives operational data and forwards it to the local nodes declared as intelligent fusion, this information is verified and verified based on the collection and exchange of information from the IoT network. It is generally accepted that IoT can be seen as an effective measure to monitor and regulate a building's energy consumption. The system can be configured to measure the total energy consumption, the type of energy used, such as solar, natural gas, etc. To analyze its nature. The IoT system provides access to energy and can also limit its total consumption so that it is not critical. If the option is generally available, IoT can select the shortest network time to operate the devices. This helps to reduce costs and effectively maintain the distribution of energy needs. Smart buildings are presented as a creative aspect that allows the use of different sensors for different subsystems such as lighting, elevators, surveillance cameras, shared heating, HVAC, and more.

IoT implements adaptive and predictable controls that help reduce energy consumption, eliminate waste, and improve tenant and tenant experiences. In addition, the design of energy environments is becoming increasingly complex due to the diversity of data sets and the application of control mechanisms.

Control IoT with cloud computing and intelligence as opposed to controlling smartphones or web apps. A variety of appliances that can be controlled by actuators have been introduced, such as lamps, fans, appliances, and home security. A configuration with three objects was described to create an effective method for an advanced smart home idea and application. a typical IoT application of an intelligent standalone lighting and ventilation system. The system was able to monitor temperature, humidity, CO_2 concentration and from there control an autonomous lighting and ventilation system in energy. The HTTP system stores data on cloud-based servers and can be managed via a smartphone or computer. In addition, the system could record real-time data on the cloud server, where the consumer could also view it, in real-time, anywhere in the world.

References

1. Mahbub, M., Hossain, M., Gazi, M.: IoT-cognizant cloud-assisted energy efficient embedded system for indoor intelligent lighting, air quality monitoring, and ventilation. Internet Things **11**, 1–26 (2020)
2. Saleem, Y., Crespi, N., Rehmani, M.H., Copeland, R.: Internet of Things-aided smart grid: technologies, architectures, applications, prototypes, and future research directions. IEEE Access **2019**(7), 62962–63003 (2019). https://doi.org/10.1109/ACCESS.2913984
3. Lork, C., Choudhary, V., Hassan, N.U., Tushar, W., Yuen, C., Ng, B.K.K., Wang, X., Liu, X.: An ontology-based framework for building energy management with IoT. Electronics **8**, 485 (2019)
4. Hafeez, G., Wadud, Z., Khan, I., Khan, I., Shafiq, Z., Usman, M., Khan, M.: Efficient energy management of IoT-enabled smart homes under price-based demand response program in smart grid. Sensors **20**, 3155 (2020). https://doi.org/10.3390/s20113155
5. Wang, C., Zhu, Y., Shi, W., Chang, V., Vijayakumar, P., Liu, B., Mao, Y., Wang, J., Fan, Y.: A dependable time series analytic framework for cyber-physical systems of IoT based smart grid. ACM Trans. Cyber Phys. Syst. **3**, 7 (2018). https://doi.org/10.1145/3145623
6. Zhao, Y., Genovese, P.V., Li, Z.: Intelligent thermal comfort controlling system for buildings based on IoT and AI. Futur. Internet **12**, 30 (2020). https://doi.org/10.3390/fi12020030
7. Rinaldi, S., Bellagente, P., Ciribini, A.L.C., Tagliabue, L.C., Poli, T., Mainini, A.G., Speroni, A., Cadena, J.D.B., Spagnolo, S.L.: A Cognitive-driven building renovation for improving energy efficiency: the experience of the ELISIR project. Electronics **9**, 666 (2020). https://doi.org/10.3390/electronics9040666
8. Jain, A., Yadav, A.K., Shrivastava, Y.: Modelling and optimization of different quality characteristics in electric discharge drilling of titanium alloy sheet. Mater. Today Proc. **21**, 1680–1684 (2019)
9. Panwar, V., Sharma, D.K., Kumar, K.V.P., Jain, A., Thakar, C.: Experimental investigations and optimization of surface roughness in turning of EN 36 alloy steel using response surface methodology and genetic algorithm Mater. Today: Proc. (2021) https://doi.org/10.1016/j.matpr.2021.03.642
10. Jain, A., Pandey, A.K.: Multiple quality optimizations in electrical discharge drilling of mild steel sheet. Mater. Today Proc. **8**, 7252–7261 (2019)

Chapter 56
Automated Machine Learning-Based Gestational Monitoring Framework in Wearable Internet of Things Environment

K. Nandini, J. Seetha, B. Jegajothi, R. Rajmohan, and D. Nageswari

Abstract Recently, Internet of Things (IoT) and wearable technologies have become popular in diverse areas, and smart health is an important application area. Specifically, IoT and wearable technologies find useful in linking different medicinal gadgets, sensors, and physicians in offering better healthcare services in remote areas. It results in enhanced patient safety, minimum healthcare cost, improved healthcare service accessibility, and high operational efficiency in the healthcare sector. At the same time, high-quality care during pregnancy is essential to determine the probable difficulties earlier and guarantee the healthiness of mother as well as fetus. Several research works have been existed in the literature to monitor the maternal lifetime. But they are developed for a particular health issue and based on short-term data collection approaches. Since maternal monitoring necessitates long-term services, this paper designs an automated machine learning-based gestational monitoring framework in wearable IoT environment. The goal of this paper is to derive a new improved salp swarm optimization (ISSO) with extreme learning machine (ELM), called ISSO-ELM model for continual maternal health monitoring.

K. Nandini
Department of Electronics and Instrumentation Engineering, Easwari Engineering College, Chennai, India

J. Seetha
Department of Computer Science, SRM Institute of Science and Technology, SRMIST, Bharathi Salai, Ramapuram Campus, Chennai 600089, India

B. Jegajothi (✉)
Department of ECE, Saveetha School of Engineering, Saveetha Institute of Medical and Technical Science, Chennai, India
e-mail: jegajothisudhakar@gmail.com

R. Rajmohan
Department of Electronics and Communication Science, Agurchand Manmull Jain College, Meenambakkam, Chennai, India

D. Nageswari
Department of Electrical and Electronics Engineering, RMK College of Engineering and Technology, Chennai, India

S. Yadav et al. (eds.), *Proceedings of Second International Conference in Mechanical and Energy Technology*, Smart Innovation, Systems and Technologies 290,
https://doi.org/10.1007/978-981-19-0108-9_56

The ISSO-ELM model encompasses different sub-processes namely data acquisition, pre-processing, classification, and parameter optimization. The presented model enables IoT devices and wearables to gather healthcare data about pregnant women. Then, the gathered data are pre-processed to remove the unwanted noise. Besides, the ELM model is employed as a classifier to determine the occurrence of difficulties. Moreover, ISSO algorithm is applied as a parameter tuning technique to optimally adjust the parameters that exist in the ELM model. The extensive experimental analysis highlighted the enhanced performance of the ISSO-ELM model over the recent state-of-art methods with the precision, recall, and accuracy of 94.00, 88.00, and 90.07%.

56.1 Introduction

Maternity care's aim is for ensuring the health and wellbeing of the pregnant woman and her fetus. Maternal health has a huge effect on the child at the time of pregnancy but also in the upcoming years. Additionally, health problems in the course of pregnancy, e.g., hypertensive disorders/gestational diabetes, might resonates with equivalent health complications in the pregnant woman's adult age [1]. Henceforth, maternity care is vital for promoting long-term health at the population level and avoiding severe pregnancy problems in individuals. Frequent check-ups at the time of pregnancy are vital for detecting anomalies and to avoid further problems, injuries or mortality. Conventionally, blood glucose urine test and blood pressure were the major concrete considerations to obey in the course of pregnancy and the development of the uterus and maternal gaining weights. To assist a healthier life, maternity care providers should give counseling on other lifestyle and self-management matters, e.g., sleep and physical activities. But, still, these are not monitored systematically [2].

There is a necessity for monitoring ubiquitously pregnant women's health to detect early feasible problems and enhance health factors [3]. Furthermore, frequent observing of distinct health parameters allows fine-grained quantitative data which can offer a well understanding of pregnancy. The advancement of technology in Information and Communication Technology (ICT) is changing the healthcare way is provided. Particularly, the IoT is a developing model in current ICT which utilizes several computing infrastructures, sensing, and communications to provide an innovative network of objects anytime & anywhere [4].

Over 4 decades, ultrasound was the most innovative investigation approaches in the fetal monitoring field. Previously, these measurements can be executed in medical management for a restricted period of time and mainly in a hospital. But, because of the huge amount of unsolved intrauterine fetal mortalities taking place in the 3rd year [5], they should have assessment approaches that can monitor fetal activity, even at home, which has turned into obvious.

With the rapid growth of IoT, wearable devices, and mobile internet, health monitoring has presented an intelligent development recently. Additionally, medical wearable devices like smart ECG devices, 3G blood pressure meters, Bluetooth blood glucose meters were utilized for monitoring ECG, blood sugar, blood pressure, and other physical indications. Finally, the monitoring registers are transmitted to the data environment for real-time diagnoses or for a medical database to keep records [6]. The summary of smart devices for the hospitals could keep the operational cost, improve the medical practice of persons and decrease the labor intensity of medicinal staff [7]. In this study, they have gathered huge amount of big data that have distinct kinds of data like categorical, image, and text data via IoT devices as input data. Now, they have employed a novel ML method to proceed with the learning procedure that maps the data to 2 classifications like "Normal" and "Disease Affected".

This paper designs an automated machine learning-based gestational monitoring framework in wearable IoT environment using improved salp swarm optimization (ISSO) with extreme learning machine (ELM), called ISSO-ELM model. The ISSO-ELM model encompasses different sub-processes namely data acquisition, pre-processing, classification, and parameter optimization. In addition, the ELM model is employed as a classifier to determine the occurrence of difficulties. Furthermore, ISSO algorithm is applied as a parameter tuning technique to optimally adjust the parameters that exist in the ELM model. The ISSO algorithm is derived for balancing the exploration and exploitation with the avoidance of the evolution stagnation. The extensive experimental analysis highlighted the enhanced performance of the ISSO-ELM model over the recent state-of-art methods.

56.2 Literature Review

Zhang et al. [8] a framework for connecting smart things in smart hospitals depending upon NB-IoT, and present edge computing for handling the need of latency in medical procedures. In Kumar et al. [9], a novel systematic method is utilized to the diabetes diseases and the interrelated medical data is created with UCI Repository dataset and the medical sensor for predicting the person who has caused diabetes seriously. Rghioui et al. [10] define a novel scheme to monitor diabetic persons and deliberate prediction analytics with 4 distinct ML methods. The accuracy and performance of the employed methods are deliberated and related to selecting an optimal one regarding various factors.

Rghioui et al. [11] proposed a smart framework to monitor diabetic patients with the help of ML method. The frameworks components include sensors, smartphones, and smart devices for collecting measurements from the body. The smart scheme gathered the data attained from the person and accomplished data classification with the help of ML for making diagnoses. Sarhaddi et al. [12] presented an IoT-based scheme for providing global maternal health monitoring at the time of pregnancy and postpartum. The scheme contains several data gatherers for tracking the mother's condition, with physical activity, stress, and sleep. They executed the whole system execution and accompanied real human subject research on pregnant women in south-western Finland. Later, they estimated the scheme's data reliability, feasibility, and energy efficiency.

56.3 The Proposed ISSO-ELM Model

Primarily, the wearables and IoT devices attached to the pregnant women start collecting healthcare data, which are then transmitted to the cloud server for further processing. Then, the collected healthcare data undergoes pre-processing to remove the unwanted data and improve the quality of the data. Followed by, the ELM model is applied to identify the class labels of the data. At last, the ISSO algorithm is applied to properly tune the parameters involved in the ELM model.

56.3.1 Structure of ELM Model

The ELM is generally made in 1992 [13] and is considered a supervised learning technique able of resolving linear and non-linear classifier problems. If related to other NNs structures, ELM is assumed as single layer FFNN with 1 hidden layer. This classifier was related to CNN classification utilizing the classifier rate to medicinal data classifier. An important simulation of ELM classifier is for separating their classifier data with linear decision surfaces and maximizing the margin among the different types of classes. It is obvious that trained the ELM contains resolving a quadratic optimization problem that needs to utilize of optimization routine in different state-of-art mathematical or heuristic manners. Especially ELM classification is selected to considered application because of the following details: ELM classification gives optimum solutions, even for complex search space and if difficult datasets are utilized during this case, ELM is established has been competitive optimum solutions provider because of their converging features. Also, ELM classifiers decrease the calculation burden and time succeeded from earlier classifications. In ELM attains optimum generalization efficiency at particularly quick learning speeds.

The fundamental ELM classifier technique is provided as follows. Provided N distinct instances $(x_i, y_i) \in R^n \times R^m$, where $x = [x_{i1}, x_{i2}, \ldots, x_{in}]^{\mathrm{T}}$ implies the input vectors and $y_i = [y_{i1}, y_{i2}, \ldots, y_{im}]^{\mathrm{T}}$ represents the target vectors classic SLFNs with M hidden node and activation functions $g(x)$ are expressed as:

$$\sum_{i=1}^{M} \alpha_i f_i(x_j) = \sum_{i=1}^{M} \alpha_i f(w_i \cdot x_j + b_i) = O_j, \; j = 1, \ldots, N, \tag{56.1}$$

where $w = [w_{i1}, w_{i2}, \ldots, w_{in}]^T$ signifies the weight vectors between the input node and ith hidden node, $\alpha_i = [\alpha_{i1}, \alpha_{i2}, \ldots, \alpha_{im}]^{\mathrm{T}}$ represents the weight vectors concerning the ith hidden node and output node, and b_i refers to the bias of ith hidden nodes. Assumed a function for every N instance with zero error; i.e., there exists parameter (w_i, b_i) and α_i as:

$$\sum_{i=1}^{\mathrm{M}} \alpha_i f\left(w_i \cdot x_j + b_i\right) = y_j,\ j = 1, \ldots, N. \tag{56.2}$$

The N formulas are simplified as $H\alpha = T$, where

$$H(w_1, \ldots, w_M, b_1, \ldots, b_M, x_1, \ldots, x_N)$$
$$= \begin{bmatrix} f(w_1 \cdot x_1 + b_1) & \cdots & f(w_{\mathrm{M}} \cdot x_1 + b_{\mathrm{M}}) \\ \vdots & \ddots & \vdots \\ f(w_1 \cdot x_{\mathrm{N}} + b_1) & \cdots & f(w_{\mathrm{M}} \cdot x_{\mathrm{N}} + b_{\mathrm{M}}) \end{bmatrix}_{N\times M},$$

$$\alpha = \left[\alpha_1^{\mathrm{T}}, \ldots, \alpha_M^{\mathrm{T}}\right]_{M\times m}^{\mathrm{T}},\quad T = \left[y_1^{\mathrm{T}}, \ldots, y_N^{\mathrm{T}}\right]_{N\times m}^{\mathrm{T}}. \tag{56.3}$$

The solution of linear model is $\alpha = H^{\dagger}T$, where $H^{\dagger}$ implies the Moore–Penrose generalized inverse of hidden layers output matrix H. The ELM technique is expressed in the subsequent 3 steps.

Input: Trained dataset $\left\{\left(x_j, y_j\right) \middle| x_j \in \middle| R^n,\ t_j \in R^m,\ j = 1, \ldots, N\right\}$, activation function g, and hidden node count M.

Output: it can be the weight of hidden to output layers.

(1) Arbitrarily create the parameter of weights and bias (w_i, b_i), $i = 1, \ldots, M$.
(2) Compute the hidden layers output matrix H.
(3) Compute the weights matrix of output $\alpha = H^{\dagger}T$.

Related to usual ANN, the ELM technique offers a significantly lesser number of parameters to tune.

56.3.2 Design of ISSO Algorithm

In order to optimally tune the parameters involved in the ELM model, ISSO technique is derived. Salp is equivalent to jellyfish from tissues and motion that pushes himself toward pumping water to the body. Its chaining performance is utilized for mathematical module SSA. In SSA, salp population is separated into 2 portions: one follower and leaders. The leader at the front of salp chain searches for the food source F all over the place in the search space, where other is named followers that restart their position according to the position of other partners. In this study, assume as population size of salp is M, the dimensional of search space is d. Afterward, population initiation, the location of salp is recorded from the T matrix, as displayed by Eq. (56.4):

$$T = \left[h_j^m\right]_{M\times d} \tag{56.4}$$

whereas h_j^m denotes value of jth parameter of m th salp, demonstrating the smoothing factors to be enhanced in PNN module, whereas $m \in [1, 2, \ldots, M]$, $j \in [1, 2, \ldots, d]$. The equation of h_j^m is displayed in Eq. (56.5):

$$h_j^m = r_j^m \times [ub(m) - lb(m)] + lb(m) \tag{56.5}$$

whereas r_j^m denotes size $M \times d$ arbitrary matrix with values among zero and one, $ub(m)$ and $lb(m)$ denotes upper & lower boundaries of mth salp, correspondingly. The FF is utilized for computing the fitness value of h_j^m in every optimization iteration of SSA, this equivalent result is kept in a 2D matrix named ST, as shown in Eq. (56.6):

$$ST = \begin{bmatrix} ST^1 \\ ST^2 \\ \vdots \\ ST^{\mathrm{M}} \end{bmatrix} = \begin{Bmatrix} f\left[\left(h_1^1 h_2^1 \ldots h_d^1\right)\right] \\ f\left[\left(h_1^2 h_2^2 \ldots h_d^2\right)\right] \\ \vdots \\ f\left[\left(h_1^M h_2^M \ldots h_d^M\right)\right] \end{Bmatrix} \tag{56.6}$$

In ST matrix, they relate and choose the small fitness value which denotes food source, thus attaining an optimum smoothing factor. The position of leader x_j^{leader} in jth dimension is upgraded by F_j, as evaluated in Eq. (56.7):

$$x_j^{\text{leader}} = \begin{cases} F_j + c_1\left(\left(ub_j - lb_j\right)c_2 + lb_j\right), & c3 \geq 0.5 \\ F_j - c_1\left(\left(ub_j - lb_j\right)c_2 + lb_j\right), & c_3 < 0.5 \end{cases} \tag{56.7}$$

whereas F_j denotes position of food in jth dimensional, that is deliberated as search objective, ub_j and lb_j denotes upper and lower boundaries in jth dimensional correspondingly, c_2 and c_3 indicates 2 arbitrary values of uniform distribution among one and zero, c_1 denotes upgrade coefficient, as provided in Eq. (56.8):

$$c_1 = 2e^{-(4l/l_{\max})^2} \tag{56.8}$$

whereas $l_{\max}$ and l represents maximal and present iteration, correspondingly. They use Newton's Law of motion for inferring the position upgrade equation of followers, as stated in Eq. (56.9):

$$x_j^m = 0.5\left(x_j^{m-1} + x_j^m\right) \tag{56.9}$$

whereas x_j^{m-1} and x_j^m denotes position of $(m-1)$th and mth follower in the jth dimension correspondingly. The novel position upgrade equation of followers and the iterative equation of weight factors are correspondingly defined in Eqs. (56.10) and (56.11):

$$x_j^m = 0.5w(l)\left(x_j^{m-1} + x_j^m\right) \tag{56.10}$$

$$w(l) = w_{\max} r_a - (w_{\max} - w_{\min})(l/l_{\max})^2 \tag{56.11}$$

where $w_{\max}$ and $w_{\min}$ represents upper and lower boundaries of weight factor correspondingly, r_a indicates arbitrary amount that raises the diversity of weight factor. If the technique prematurely converges, the food source at this moment should be locally optimum which would affect the loss of convergence stagnation and population diversity. During this method, the present value could jump out of present search space to a new search region for continuing searching for optimum value. The novel feed source is defined by:

$$F = F(0.5r_m + 1) \tag{56.12}$$

The arbitrary amount t following chaotic map, as displayed in Eq. (56.13):

$$t^{l+1} = [\sin(8\pi t^l) + 1]^2/4 \tag{56.13}$$

The mutation likelihood p where value increases with the augment of iteration, as is denoted in Eq. (56.14), later relates to t for judging either to raise the variations of food source.

$$p = p_{\min} + (p_{\max} - p_{\min})/l_{\max} \tag{56.14}$$

A smaller value of p in the inchoate stage of iteration will not affect the convergence efficiency of the method, while a larger value of p in succeeding iterations would disturb the search trajectory by raising the variations of food source.

56.3.3 ISSO-Based Parameter Optimization Process

The 2 key parameters of presented ELM technique are automatically tuned by utilizing the ISSO approach for simulating the foraging performances of the ISSO method and their connections with the near environments. The presented method is contained 2 important processes comprising inner parameter optimization and outer efficiency estimation. In the inner parameter optimization model, the penalty parameter C and kernel bandwidth γ of ELM is defined dynamically utilizing the ISSO method using fivefold CV analysis. Afterward, the attained optimum parameters pair (C, γ) is inputted as to KELM forecast method for performing the classifier task from outer loop utilizing a tenfold CV approach. The classifier error rate is utilized as FF.

$$\text{fitness} = \frac{\left(\sum_{i=1}^{K} \text{test Error}_i\right)}{k}$$

where test Error_i implies the average test error rates obtained by ELM classification using fivefold CV in the inner parameter optimization model.

56.4 Performance Validation

This section validates the performance of the proposed ISSO-ELM model on the applied PIMA Indians diabetes dataset. The dataset includes a set of 768 instances with 8 features and 2 class labels. The details related to the dataset are given in Table 56.1.

Figure 56.1 investigates the performance of the ISSO-ELM model with existing techniques in terms of prec., rec., and acc. The figure demonstrated that the VP model has accomplished poor results with a prec. of 92%, rec. of 68%, and acc. of 67%. In line with that, the DT model has offered slightly enhanced outcomes with the prec. of 81%, rec. of 79%, and acc. of 74%. Figure 56.2 examines the performance of the ISSO-ELM method with existing algorithms with respect to F-measure, MCC,

Table 56.1 Dataset description

Description	Gestational diabetes
No. of instances	768
No. of attributes	8
No. of class	2
Positive samples (%)	34.90
Negative samples (%)	65.10

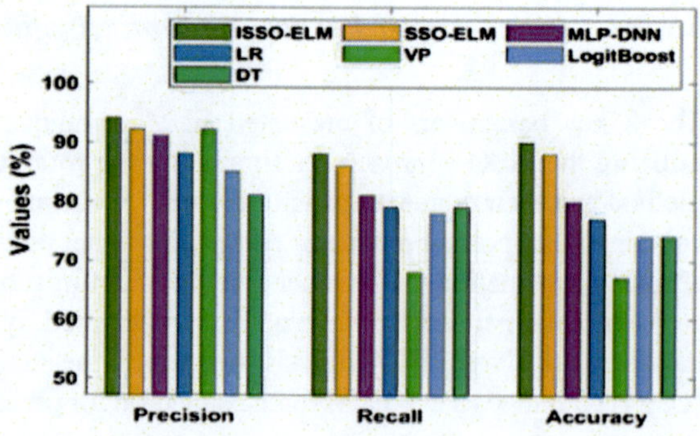

Fig. 56.1 Result analysis of ISSO-ELM model using precision, recall, accuracy

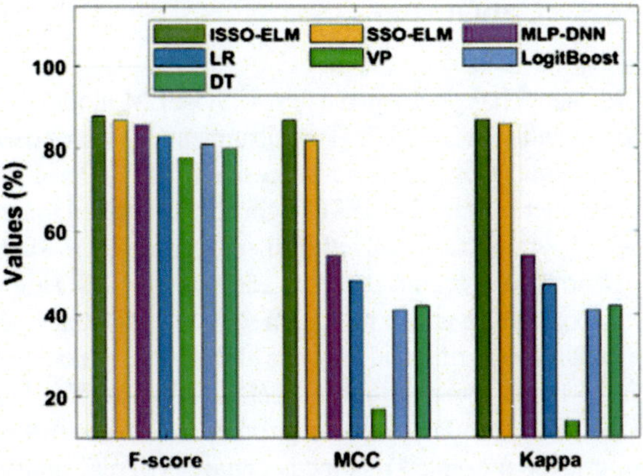

Fig. 56.2 Result analysis of ISSO-ELM model in terms of F-score, MCC, kappa

and kappa. The figure outperformed that the VP method has accomplished worse outcomes with the F-measure of 78%, MCC of 17%, and kappa of 14%.

At the same time, the LR method has resulted in a moderate performance with an F-measure of 83%, MCC of 48%, and kappa of 47%. Concurrently, the MLP-DNN approach has resulted in a manageable result with an F-measure of 86%, MCC of 54%, and kappa of 54%. Eventually, the SSO-ELM algorithm has resulted in a competitive outcome with an F-measure of 87%, MCC of 82%, and kappa of 86%. Finally, the projected ISSO-ELM methodology has accomplished maximum outcomes with the F-measure of 88%, MCC of 87%, and kappa of 87%.

In order to validate the superiority of the presented ISSO-ELM model, a detailed comparative results analysis is made in Fig. 56.3. From the analysis of the detailed results, it is ensured that the ISSO-ELM methodology has obtained superior classification performance with an accuracy of 90.07%. Therefore, it can be employed as an effective tool for gestational monitoring in IoT environment.

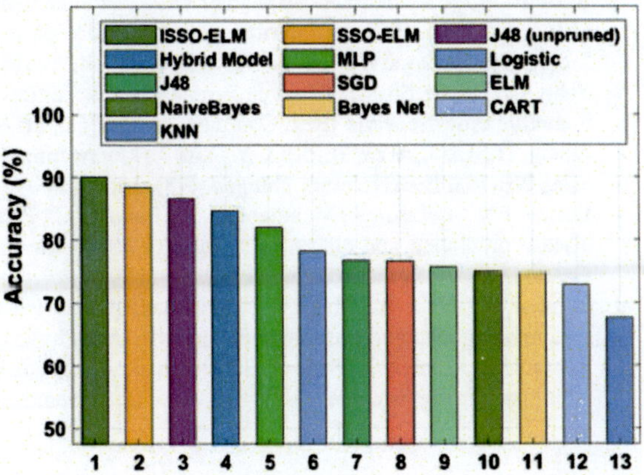

Fig. 56.3 Accuracy analysis of ISSO-ELM model with existing techniques

56.5 Conclusion

This paper has developed a new ISSO-ELM model for continuous gestational monitoring in the wearable IoT environment. The proposed ISSO-ELM model includes data acquisition, data pre-processing, ELM-based classification, and ISSO-based parameter tuning. The ISSO algorithm is applied to appropriately choose an optimal set of parameters involved in the ELM model in such a way that the classification performance gets increased. In addition, the ISSO algorithm is derived for balancing the exploration and exploitation with the avoidance of the evolution stagnation. The extensive experimental analysis highlighted the enhanced performance of the ISSO-ELM model over the recent techniques. The simulation results portrayed that the ISSO-ELM model is superior to other methods with the higher accuracy of 90.07%. In future, the presented method is deployed in real-time environment to assist pregnant women in day to day lives.

References

1. Sharma, S., Sidhu, H., Kaur, S.: Analytical study of intrauterine fetal death cases and associated maternal conditions. Int. J. Appl. Basic Med. Res. **6**(1), 11 (2016)
2. Klemetti, R., Hakulinen-Viitanen, T.: Handbook for Finnish Maternity Health Clinics. Recommendations for practices [in Finnish] Äitiysneuvolaopas. Suosituksia äitiysneuvolatoimintaan. National Institute for Health and Welfare (2013)
3. Grym, K., Niela-Vilén, H., Ekholm, E., Hamari, L., Azimi, I., Rahmani, A., Liljeberg, P., Löyttyniemi, E., Axelin, A.: Feasibility of smart wristbands for continuous monitoring during pregnancy and one month after birth. BMC Pregnancy Childbirth **19**(1), 1–9 (2019)
4. Gubbi, J., Buyya, R., Marusic, S., Palaniswami, M.: Internet of Things (IoT): a vision, architectural elements, and future directions. Futur. Gener. Comput. Syst. **29**(7), 1645–1660 (2013)
5. Goda, M.Á., Telek, T., Kovács, F.: Novel phonography-based measurement for fetal breathing movement in the third trimester. Sensors **21**(1), 211 (2021)
6. Almotiri, S.H., Khan, M.A., Alghamdi, M.A.: Mobile health (m-health) system in the context of IoT. In: 2016 IEEE 4th International Conference on Future Internet of Things and Cloud Workshops (FiCloudW), Vienna, Austria, pp. 39–42 (2016)
7. Yang, G., He, S., Shi, Z., Chen, J.: Promoting cooperation by the social incentive mechanism in mobile crowdsensing. IEEE Commun. Mag. **55**(3), 86–92 (2017)
8. Zhang, H., Li, J., Wen, B., Xun, Y., Liu, J.: Connecting intelligent things in smart hospitals using NB-IoT. IEEE Internet Things J. **5**(3), 1550–1560 (2018)
9. Kumar, P.M., Lokesh, S., Varatharajan, R., Babu, G.C., Parthasarathy, P.: Cloud and IoT based disease prediction and diagnosis system for healthcare using Fuzzy neural classifier. Futur. Gener. Comput. Syst. **86**, 527–534 (2018)
10. Rghioui, A., Naja, A., Mauri, J.L., Oumnad, A.: An IoT Based diabetic patient monitoring system using machine learning and Node MCU. J. Phys.: Conf. Ser. **1743**(1), 012035) (2021)
11. Rghioui, A., Lloret, J., Sendra, S., Oumnad, A.: A smart architecture for diabetic patient monitoring using machine learning algorithms. In Healthcare **8**(3), 348 (2020)

12. Sarhaddi, F., Azimi, I., Labbaf, S., Niela-Vilén, H., Dutt, N., Axelin, A., Liljeberg, P., Rahmani, A.M.: Long-term IoT-based maternal monitoring: system design and evaluation. Sensors **21**(7), 2281 (2021)
13. Huang, G.B., Chen, L.: Enhanced random search based incremental extreme learning machine. Neurocomputing **71**(16–18), 3460–3468 (2008)

Chapter 57
Improving Energy Efficiency in Piezoelectric Effect Based Synchronous Multicast Protocol (PESM) in Wireless Sensor Networks

Tadele Degefa Geleto, C. Suresh Gnana Dhas, Karthikeyan Kaliyaperumal, and S. Noordeen

Abstract The main aim of the proposed system is to optimize the consumption of energy during transmission in Wireless Sensor Networks (WSNs). This is achieved using Piezoelectric Effect based Synchronous Multicast approach (PESM), which is used to conserve energy from sensor nodes and it improves the network lifetime. This technique regenerates electrical effect, which is produced from natural resources to increase the network lifetime. The scavenging process is carried out by piezoelectric crystals in sensor notes. The experiments process that proposed method improves well the network lifetime in WSN more than other methods.

57.1 Introduction

In general, WSN has numerous nodes to measure the physical conditions of the environment. These sensor nodes are employed to monitor such physical environmental conditions. Such sensor nodes are equipped with memory, processor, and better communication capabilities. The sensor devices in WSN are huge in number and it is of heterogeneous one equipped with battery, CPU, sensor, and radio transceiver. The sensor nodes collect data in a periodic way and coverts the data into electrical signals and transmit it to base station. However, it suffers mainly from limited power to process the data [1–5]. The sensor nodes or motes are usually replaceable or battery-powered. The major problem associated with such sensor nodes is their low power sources, low memory, and increased failure rate due to improper node deployment. The lifetime of motes is increased with suitable technique [6–12]. The

T. D. Geleto · C. Suresh Gnana Dhas
Computer Science Department, Ambo University-Hachalu Hundesa Campus, Oromia, Ethiopia

K. Kaliyaperumal
Information Technology Department, Ambo University HH Campus, Oromia, Ethiopia

S. Noordeen (✉)
Department of Computer Science, SAS-AV Campus, VMRF, Chennai, India
e-mail: snoordeen78@gmail.com

S. Yadav et al. (eds.), *Proceedings of Second International Conference in Mechanical and Energy Technology*, Smart Innovation, Systems and Technologies 290,
https://doi.org/10.1007/978-981-19-0108-9_57

mote is a battery-powered one and it is a non-replenishable one. Many applications need hundred motes to understand the event. If a mote dies then the information cannot be retrieved accurately. Once the node dies it is to be replaced with new mote. It is possible to save energy and regenerate the energy from natural resources and this has the potential to increase the lifetime of a mote. Most sensor nodes have non-rechargeable batteries and it is wasted due to poor network design, idle nodes listening for burst nature, and longtime traffic. To attain proper network function, methods like idle listening, congestion control, overhearing, retransmitting, over emitting is deployed. Such mechanism leverages power at a higher rate and hence the power becomes a major challenge [13–15]. To resolve such challenge, the proposed method uses piezoelectric effect to generate power from natural resources. This method provides energy to motes through natural variation in pressure occurring in cloud. The proposed PESM model uses this model to harvest energy in sensor nodes to improve the lifetime. The outline of paper is presented as follows: Sect. 57.2 provides the proposed method. Section 57.3 provides the Energy Balancing Model. Section 57.4 provides the results and Sect. 57.5 concludes the paper [16–20].

57.2 Proposed Approach

This section discusses the proposed PESM method, which is intended to balance and restore the sensor node energy in WSN. The block diagram of proposed method is shown in Fig. 57.1. The PESM process carries out optimization of energy in two different levels, namely, energy balancing level, and energy generation level. The former level delegates the tasks over entire sensor node, so as to avoid overloading and draining of energy. It further ensures that the tasks to sensor nodes are delegates in fair manner. The equal distribution of tasks to nodes processes the required operation with additional wastage of energy. Hence, it is mandatory to use energy regeneration level, since energy balancing cannot maintain the energy level of the network. The non-conventional energy sources are used for regeneration process.

Further, energy balancing is achieved using effective sensing method and proper network design. This reduces the level of congestion in multicasting for data forwarding. It achieves well the sensor nodes synchronization in WSN. The proposed method employs the pressure of vibration to attain energy regeneration in cloud-based WSN.

57.3 Energy Balancing Model

The proper selection of topology reduces well consumption of energy by the sensor nodes. The lifetime of sensor nodes is improved by proper topology selection with better coverage. The proposed system uses peer-to-peer topology to balance well the sensor node energy in large, self-healing, and self-controlled multihop network.

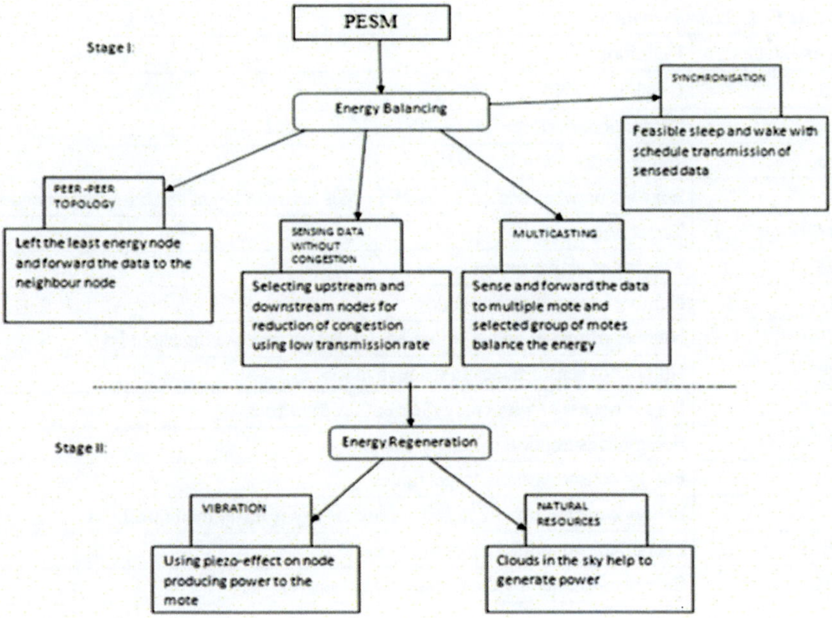

Fig. 57.1 PESM process

The peer-to-peer topology transmits the data between the sensor nodes in a point-to-point manner. During the time of data transmission, the nodes with lesser energy are discarded from exchanging data. Here, nodes will have higher energy is used for data forwarding to attain point-to-point communication and it reduces the hop count (Table 57.1).

The WSN network is assumed in the form of a graph, $G = (\{s_1, s_2, s_3, \ldots, s_i\} \cup ss, E)$.

The base station is reached through target nodes, where the target node-set is represented as, $t = \{t_1, t_2, t_3, \ldots, t_j\}$.The lower threshold sensor energy set is represented as $l = \{l_1, l_2, l_3, \ldots, l_k\}$. Here, the sensor nodes are supplied with renewable energy sources [10]. The consumption of energy by sensor nodes in the network is given in Fig. 57.2. These sensor nodes in energy consumption model operate either in active or inactive mode.

The energy consumed by network (EC_{NT}) is estimated in a time interval (T), which is given as,

$$EC_{NT} = \sum_{i=1}^{m} ECA_i + \sum_{j=1}^{n} ECT_j - \sum_{k=1}^{p} ECLT_k - \sum_{l=1}^{q} ECI_l \qquad (57.1)$$

Table 57.1 List of notations

Notations	Definition
G	Graph in WSN
$s_1, s_2, s_3, \ldots, s_i$	Sensor nodes set
ss	Target nodes set or base station
E	Edge set between sensor nodes and edges between target and base station node
$t_1, t_2, t_3, \ldots, t_j$	Target node-set
M	Number of active sensor nodes
N	Number of target sensor nodes
P	Number of sensor nodes whose energy is less than the threshold level
Q	Number of sensor nodes in inactive mode
$l_1, l_2, l_3, \ldots, l_k$	Sensor nodes set with energy lesser than threshold
ECA	Energy consumption in active nodes
ECT	Energy consumption in target nodes
ECLT	Energy consumption by nodes, whose energy is lesser than threshold
ECI	Energy consumption in inactive nodes
EC_{NT}	Energy consumed by the network w.r.t interval

Initialization

1. Set G = ({s1, s2, s3, ….si} U ss, E) where E is the set of edges
2. Set t = {t1, 2, t3,………..tj} target nodes
3. Set l = (l1, l2, l3,………..lk} below threshold level.

Iteration and Update

1. For each transmission (i,m) such that i and m belongs to energy consumed by active nodes and sum up all the involved active nodes energy value in the network.
2. For each transmission (j,n) such that j and n belongs to energy consumed by target nodes and sum up all the energy values of collected target nodes.
3. For each transmission (k,p) such that k and p belongs to energy consumed by nodes with low level threshold value set and sum up those values.
4. For each transmission (l.q) such that l and q is the values vary with respect to energy consumed by inactive nodes present in the network.
5. Find the transmission with minimum energy sum
6. Update sum of energy in active nodes and target nodes by removing energy in low level threshold nodes and inactive nodes

Fig. 57.2 Energy computation algorithm

a. Data sensing efficiently without congestion

The motes sense the event and forward the desired data to base station. The traffic increases in network when the motes transmit their data simultaneously over the network. The increasing congestion affects the data transmission in reaching the other motes. Hence, there exist packet loss and data retransmission, which increases the consumption of battery power. Therefore, it is necessary to avoid congestion in

Table 57.2 CC2420 (Micaz) specifications

Parameters	CC2420 (Micaz)
Transmit power (P_{tx})	52.2 mW
Receive power (P_{rx})	56.4 mW
Overhear power (P_{Listen})	56.4 mW
Dormant power (P_{sleep})	3 μW
Time to transmit a byte	32 μs
Transmission rate	250 kbps

WSN, which is done by finding the optimal path to transmit the data to the target node and to improve the network lifetime.

The congestion in WSN can be avoided easily, however, the consumption of energy with network congestion is not addressed prominently. Therefore, the congestion control in proposed system is avoided by balancing the network load. This helps to recover the mote quickly from congestion. Further, the retransmission of sensor data leads to increased consumption of energy, which is estimated using congestion ratio. The congestion ratio is defined as the fraction of packets transmitted during congestion and packets received during congestion. Hence, the energy consumption during and after the removal of congestion is given in Eqs. (57.2) and (57.3).

The proposed method uses CC 2240 (Micaz) sensor mote for experimental purposes and the specifications are given in Table 57.2.

The removal of congestion affects the downstream node and latency throughput, which drops the unwanted sensed data transmission.

$$E_{congested} = t_{event} - (t_1 - 1)P_{active} + \big(\tau_{downstreamnode}\big(P_{active} + P_{sleep} - P_{inactive}\big) + (t_{event} - (t_1 - 1) - \tau_{downstreamnode})P_{target}\big) \quad (57.2)$$

$$E_{saved} = t_{event} - (t_1 - 1)P_{active} + [(\tau_{downstreamnode} + t_{event} - (t_1 - 1)]P_{target} - \left[\left(\frac{(P_{active} + P_{sleep})}{2}\right) + t_{event} - (t_1 - 1)\right]P_{sleep} - [t_{event} - (t_1 - 1)]P_{inactive} \quad (57.3)$$

where,

E_{saved} represents the energy saved during congestion,
t_{event} represents the time between two motes,
P_{active} represents the active mode power,
P_{sleep} represents the sleep mode power,
$P_{inactive}$ represents the inactive mode power,
P_{target} represents the target mode power and
$\tau_{downstreamnode}$ represents the downstream node transmission rate. Tables 57.3 and 57.4 shows the consumption of energy in congested and uncongested scenario.

Table 57.3 Energy consumption in congested scenario

P_{active}	P_{target}/P_{rec}	P_{sleep}	$P_{inactive}$	Energy consumed
40.3 mW	22.2 mW	3 μW	1 μW	14.18 J
62.5 mW	50.5 mW	5.5 μW	2.3 μW	22.99 J
80.7 mW	70.3 mW	7.6 μW	4.2 μW	33.35 J
135.2 mW	87.7 mW	10.8 μW	6 μW	53.38 J
190.7 mW	103.4 mW	13.3 μW	5.4 μW	74.35 J

Table 57.4 Energy consumption in uncongested scenario

P_{active}	P_{target}/P_{rec}	P_{sleep}	$P_{inactive}$	Energy consumed
40.3 mW	22.2 mW	3 μW	1 μW	44.07 J
62.5 mW	50.5 mW	5.5 μW	2.3 μW	36.26 J
80.7 mW	70.3 mW	7.6 μW	4.2 μW	30.22 J
135.2 mW	87.7 mW	10.8 μW	6 μW	24.14 J
190.7 mW	103.4 mW	13.3 μW	5.4 μW	18.19 J

The consumption of energy in both scenarios is measured in a time interval of 10 ms, where P_{active}, P_{sleep}, P_{target}, $P_{inactive}$ is measured.

b. **Data forwarding through multicasting**

The proposed method uses multicasting to transmit the data between the motes with less power consumption. This offers selection of multiple destinations from single source nodes. The motes sense the event and forward the sensor data to chosen destination via multicasting. This reduces well the communication overhead and packet overhead reduction attains improved energy consumption efficiency due to selection of multicast communication [3, 18]. Figure 57.3 shows the data building by sensor motes and data acceleration to increase energy.

c. **Energy calculations**

The WSN is assumed as undirected graph with sensor node-set as vertex and link edge set. In multicasting, we have taken a terminal set $M \subseteq \{1 - V\}$ and assign source

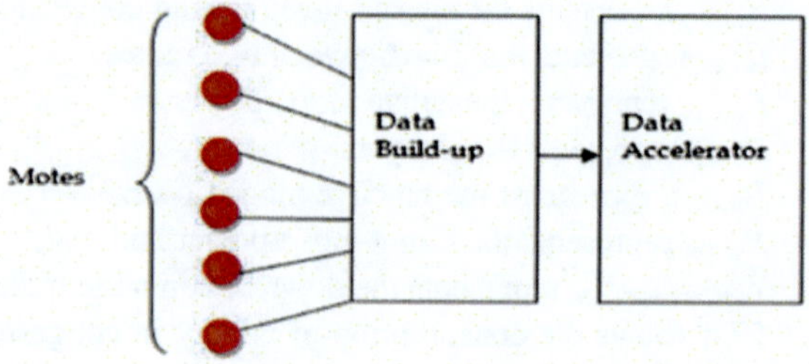

Fig. 57.3 Building up of data

node as s where $s \in M$. Hence receiving nodes are considered as $M - \{s\}$.The poor scheduling policy increases the consumption of energy and vice versa. Hence, planning to schedule the data is needed to optimize well the energy and the energy consumption is estimated using Eq. (57.4)

$$\text{Energy of sensor node} = \text{Transmitting energy} + \text{Receiving energy} \tag{57.4}$$

Here, first-order ratio model and inverse square law are used to acquire the relationship between transmitting and receiving energy. The energy is preserved with less transmit power using multihop transmission. The energy of network is saved by offering sleep mode to the low power node during its idle time. Here, only the target node is made active to receive the data in case of multicast approach to conserve energy. This approach has two groups: join and leave group during routing. The sensor node which is desiring to join the network uses join group and the node which is desiring to leave the network uses leave group. The leave group message is sent to base station when the nodes are dormant. The consumption of energy in leave group is less since the nodes remain dormant. The consumption of energy in join group is high since the nodes remain in transmitting the data. This is the most important principle in multicast routing.

d. **Low power synchronous protocol**

The mote without any transmission or reception will be put in idle state. The dormant and active stage is diagnosed with low power synchronous protocol. The data is retraced from starting stage since at dormant stage the data sensing is stopped abruptly. This requires larger amount of energy and lacks proper synchronization. The time period and relative drift lead to data loss. The synchronization and resynchronization consumes lot of energy and the additional energy for such operation is given as,

$$\text{Clock}_1 = \text{RelativeDrift} * \text{Clock}_2 + \text{offset}$$

```
For each mote in the network
    mote_data=last sensed data
    If (mote_energy ==NULL) Sleep signal;
    else
    Wake signal;
    for (mote_energy mote_data= wake s
    en
    -
```

The congestion-free scheduling mechanism has accurate node transmission and reception facilities. This avoids waiting and reduces delay during transmission. This mechanism consumes lesser energy since the transmission occurs immediately. The data is acknowledged periodically by sensor nodes and this helps to attain reduced

Fig. 57.4 Power generation

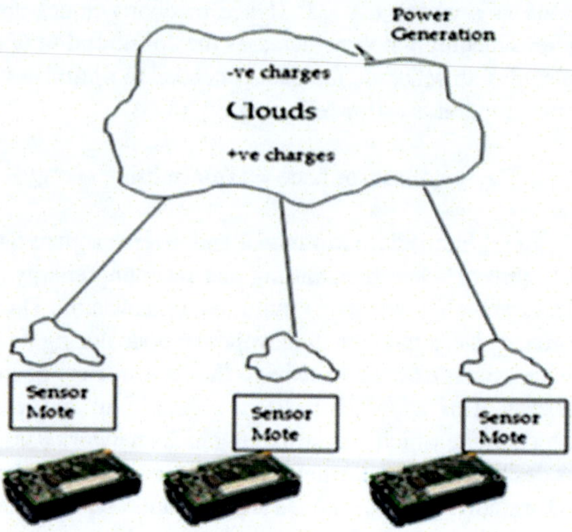

sleep time. The accurate data transmission is determined using low power time-synchronous mechanism. The sensing range is adjusted using power control mechanism and thus power is saved in sensor nodes. The synchronous mechanism avoids data repetition and hence it reduces the consumption of energy and overhead. This eliminates data overhead, node failure, and traffic. Further, it increases the lifetime of network with reduced energy consumption. It further reduces the consumption of resources and reduces usage of links.

e. **Energy Regeneration Vibration**

The application of force regenerates energy in sensor nodes and vice versa. The sensor motes in this study use Piezoelectric crystals, which increase the available power of sensor motes during thunder and lightning. The presence of energy in clouds is used further to generate power in motes, which is shown in Fig. 57.4.

Energy calculation

Once the force is applied, the generation force due to vibration increases the mote lifetime by 30%. Once the energy of mote is zero or the mote is dead, the communication problem occurs. The link breaks usually occur due to death of node and thus the energy consumed in out of coverage area is given by [9],

$$E_G = \sum_{i=1}^{k} T(n_i n_{i+1})$$

where T is the energy consumed by mote during transmission and reception.

$$E_C = E_T + E_R$$

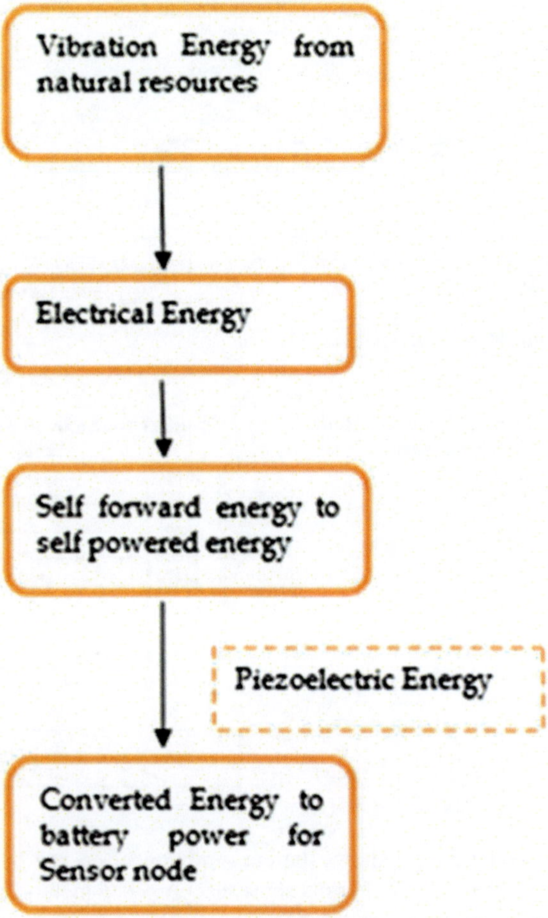

Fig. 57.5 Energy chart diagram

The consumption of energy is increased when there is an increase in total number of motes (see Fig. 57.5).

57.4 Simulation Results

The proposed method simulates a graph for energy with and without regeneration using a Cooja simulator in contiki operating system (see Fig. 57.6). The testing is carried out using Micaz mote (CC2420) type, unit disk graph medium as a distance loss and mote startup delay is 1000 ms.

The mote interface is used to interact the simulated mote with Cooja. The mote battery indication is shown in the form of graph above.

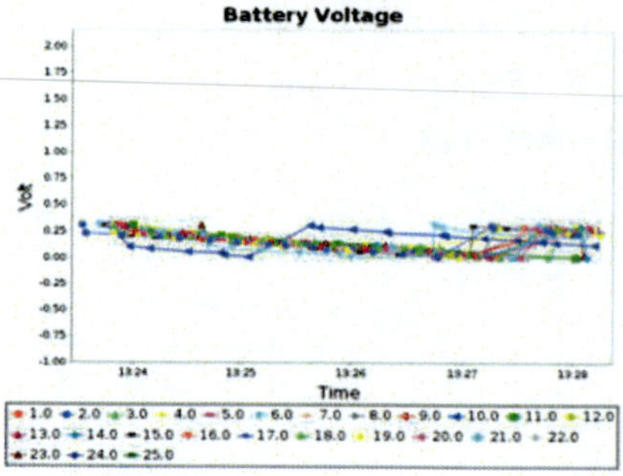

Fig. 57.6 Mote Micaz-battery

Fig. 57.7 Energy consumed Graph using various topologies

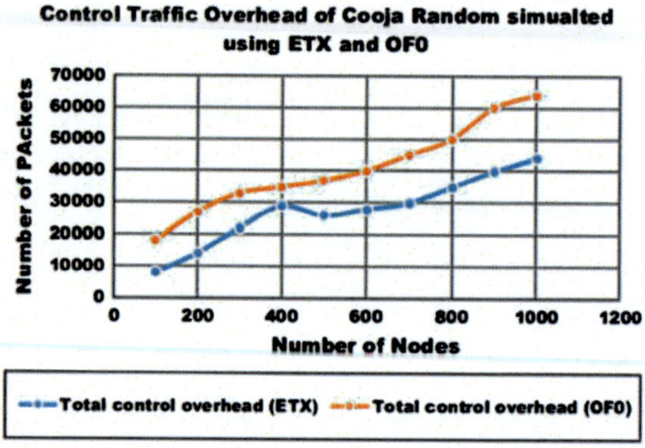

Figure 57.7 shows the consumption of energy by various topologies, which is given below. The results show that proposed method consumes lesser energy during congestion in network than other methods.

The graph simulated at the time of energy consumption with and without congestion is shown in Fig. 57.8.

The average power consumption by the mote is given in Fig. 57.9.

57.5 Conclusions

In this paper, we have proposed PESM method, which balances and restores well the sensor node energy in WSN. The multiple mobile base station model improves well the energy efficiency and increases the lifetime of WSN. The presence of piezoelectric material for harvesting the energy reduces the consumption of energy while transmitting the data across the sensor nodes. The comparison with other methods proves that proposed method attains reduced consumption of power than other methods

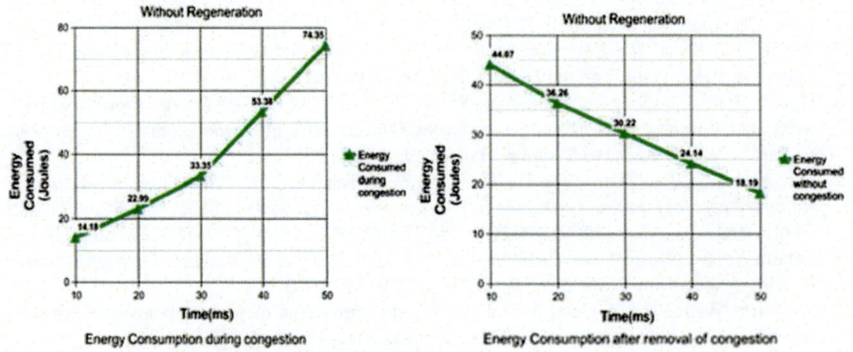

Fig. 57.8 Energy consumed during congestion

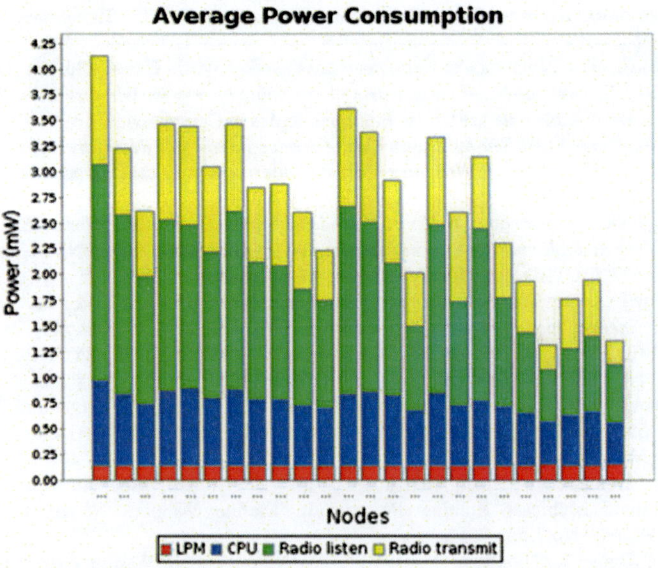

Fig. 57.9 Average power consumption for mote

during and after the congestion in network. Here, the system strictly avoids network congestion, which increases the network energy efficiency.

References

1. Anastasi, G., Coti, M., Frrancesco, M., Passarella, A.: Energy Conservation in Wireless Sensor Networks: A Survey. Elsevier, Adhoc Network (2009)
2. Bachir, A., Dohler, M., Watteyne, T., Leung, K.: MAC essentials for wireless sensor networks.

IEEE Commun. Surv. Tutorials **12** (2010)
3. Zhang, H., Shen, H.: Energy-efficient beaconless geographic routing in wireless sensor networks. IEEE Trans. Parallel Distrib. Syst. **21**(6), (2010)
4. Hongbo Jiang, Shudong Jin, Chonggang Wang," *Prediction or Not? An Energy-Efficient Framework for Clustering-Based Data Collection in Wireless Sensor Networks*", IEEE Transactions on parallel and Distributed Systems", vol.22. No.6 June 2011.
5. Fengyuan Ren, Jiao Zhang, Tao He, Chuang Lin, and Sajal K. Das,"*EBRP: Energy-Balanced Routing Protocol for Data Gathering in Wireless Sensor Networks* ", IEEE TRANSACTIONS ON PARALLEL AND DISTRIBUTED SYSTEMS, VOL. 22, NO. 12, DECEMBER 2011.
6. Samira Yessad,Nassima Tazaeart,Lyes Bakhli,Louiza Medjhoune,"Balanced Energy Efficient Routing Protocol for WSN",ICCIT ",IEEE pp 326–329,2012
7. Zbigniew Watral and Andrzej Michalski"Selected problems of power sources for wireless sensors", IEEE Instrumentation and Measurement Magazine, Feb 2013.
8. Farokh Koroupi,Reza Nourmandipour,Raheleh Hashemzehi et.al,"*Congestion in Wireless Sensor Networks and Mechanisms for controlling Congestion*",IJCSE,Vol.4.No.3.pp 204–207,Jun-July 2013.
9. Ting-yu Lin, Hendro Agus Santoso & Kun-Ru Wu,"*Global Sensor Deployment & Local coverage-aware Recovery Schemes for Smart Environments*", IEEE Transactions on Mobile Computing, 2013.
10. Meng Li,Hiroki Nishiyama,Nei Kato,Yasunoni Owada,Kiyoshi Hamaguihi,"*On the Energy – Efficient of Throughput-based scheme using Renewable Energy for Wireless Mesh Networks in Disaster Area*",IEEE transactions on Emerging topics in Computing ,2013.
11. Tong Shu, Chase Qishi Wu & Daqing Yun,"*Advance Bandwidth Reservation for Energy Efficiency in High performance Networks*", IEEE Conference on Local Computer Networks", pp 541–548, 2013.
12. Pangun Park, Piergiuseppe Di Marco," *Modeling and Optimization of the IEEE 802.15.4 Protocol for Reliable and Timely Communications* ", IEEE TRANSACTIONS ON PARALLEL AND DISTRIBUTED SYSTEMS, VOL. 24, NO. 3, MARCH 2013
13. Hongxing Li, Chuan Wu, Dongxiao Yu, Qiang-Sheng Hua and Francis C.M. Lau, "*Aggregation Latency-Energy Tradeoff in Wireless Sensor Networks with Successive Interference Cancellation*", IEEE TRANSACTIONS ON PARALLEL AND DISTRIBUTED SYSTEMS, VOL. 24, NO. 11, NOVEMBER 2013.
14. Zhao Han,Jie Wu,Jie Zhang,Liefeng Liu,Kaiyun Tian, "A General Self-Organized Tree-Based Energy-Balance Routing protocol for Wireless Sensor Networks",IEEE transactions on Nuclear Science,Vol.61,No.2 April 2014.
15. Shaojie Wen,Zhenyu Zhang,Wenzhong Yang & Siquan Flou,"An Energy-efficient opportunistic Multicast Routing Protocol in Mobile Wireless Networks",Journal of Networks,Vol.9,No.7,July 2014.
16. Jie Tang ,Daniel K.C.So,Emad Alsusa & Khairi Ashour Hamdi,"*Resource Efficiency: A new paradigm on Energy Efficiency & Spectral Efficiency Tradeoff*",IEEE Transactions on Wireless Communications Vol 13.No.8,pp. 4656–4669,Aug 2014
17. Songtao Guo, Cong Wang, and Yuanyuan Yang, IEEE, "*Joint Mobile Data Gathering and Energy Provisioning in Wireless Rechargeable Sensor Networks*",IEEE TRANSACTIONS ON MOBILE COMPUTING, VOL. 13, NO. 12, DECEMBER 2014.
18. Daehee Kim, Sejun Song, Baek-Young Choi," *APCP: Adaptive Path Control Protocol for efficient branch based Multicast Routing in Wireless Sensor networks*", 2014, 10th International conference on Mobile Adhoc & Sensor network, IEEE Computer society.
19. Luo, J., Hu, J., Wu, D., Li, R.: Opportunistic routing algorithm for relay node selection in wireless sensor networks. IEEE Trans. Ind. Inf. **11**(1) (2015)
20. Ren, X., Liang, W., Xu, W.: Quality–aware target coverage in energy harvesting sensor networks. IEEE Trans. Emerg. Topics Comput. **3**(1), (2015)

Chapter 58
An Advanced Security Framework Scheme Based on SCSR for Hierarchical Wireless Sensor Network

Anwaar Ahmad Wani, Razeef Mohammad, and Idrees Ahmad Bhat

Abstract Wireless sensor networks play a judicious role in the collection of data from the city to deliver smart services. The transfer of data over WSN's must be secured for the protection of data. However, the threat to the security of WSN's is a major challenge in smart cities. To address the problems and security challenges of WSNs this paper presents an advanced security framework scheme for smart hierarchical wireless sensor networks. Furthermore, the action of handout of security to the network after dynamic computing may result in the exploitation of the data by a malicious node. Thus the paper proposes dynamic key administration technique SCSR for heterogeneous mobile wireless sensor networks to reduce the risk of sensor-based attacks and computation overhead. The results and the simulations of the proposed scheme validate its flexibility and efficiency over the conventional schemes.

58.1 Introduction

Mobile sensor nodes require separate services from one location to the further, wherever they are, in a mobile wireless sensor network (MWSNs). Thanks to the movement of network nodes, fire reply, target tracking, detection of dairy cattle health and healthcare surveillance can be used in structuring emergency response. Other large scale applications like military surveillance, high level security for health care monitoring, SMART city etc. require huge number of nodes with different configurations need to be deployed in the field along with mobility support. Every wireless sensor node could be mobile in the future omnipresent environments. When nodes pass from place to place the security of nodes must be guaranteed, confidentiality of communication and information integrity. The authenticity and privacy of all communications exchanged during the cross-cluster routing should also be maintained [1–3].

A. A. Wani (✉) · I. A. Bhat
Mewar University, Chittorgarh, Rajasthan, India
e-mail: anwaar@mewaruniversity.co.in

R. Mohammad
Sher-I-Kashmir University of Agriculture Science & Technology, Srinagar, J&K, India

S. Yadav et al. (eds.), *Proceedings of Second International Conference in Mechanical and Energy Technology*, Smart Innovation, Systems and Technologies 290,
https://doi.org/10.1007/978-981-19-0108-9_58

For secure communication in WSNs, symmetric key-based security architecture was introduced. This mechanism uses three keys which are the password, the lock key and the lock key. The key is exchanged with the BS, the cluster key is used in a cluster, and the cluster key is used in certain clusters between all the nodes [4, 5]. This framework, however, does not attain the least overhead storage. The nodes closest to the BS are easily exhausted in most wireless sensory networks. A topology training scheme has been proposed to overcome this problem. The backbone tree is designed using strong nodes. This solves the problem of the hot spot.

58.2 Network Architecture

To meet the necessities of the large scale HWSN, the network architecture should be designed hierarchically. Also different types of key management should be provided at various levels to apprehend the critical requirements. The architecture proposed is composed of BS and sink, BN, CH and n number of nodes (Nj), where j = 1 to n. These include BS and BNs, which are known to be healthier than CHs and member nodes. The capabilities of the CHs are very high in energy and computing power compared to the member nodes [6, 7]. It has been noted that the BN is mainly considered as the constant and hence it may be not critically possible to forecast the stated topology. Hence it is regarded that architecture suggested is outlined below (Fig. 58.1).

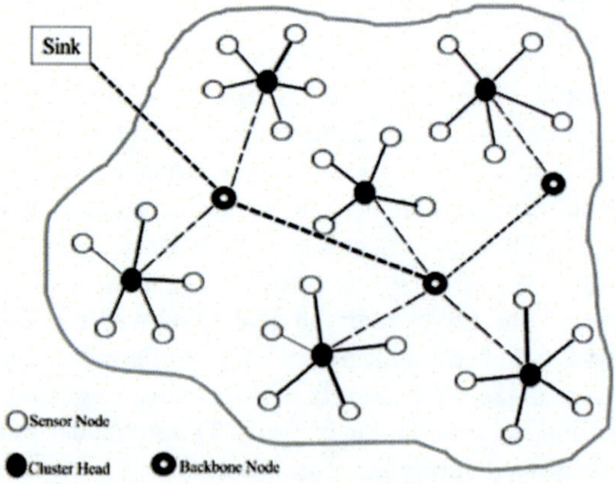

Fig. 58.1 Deployment of BNs on PSO technology

58.3 Dynamic Key Management

We use the dynamic button, which changes for each session, in the projected design for intra-cluster message. The remainder of the battery power is increasing after each operation is carried out in the node. The current battery power cannot be correctly measured with the source and the endpoint nodes. Thus, the source and destination nodes could not use this value for dynamic key generation. Therefore, VBP is considered a commodity used instead of actual battery power [8, 9]. It is presumed that when it is initially deployed, every sensor node will have a digital battery power cost. The switch in VBP is used to dynamically measure the keys for communications within the cluster.

Forward privacy: The member node which has left the group may not able access the stated credentials after being removed from the stated group, hence they are regarded as forward mode.

Backward privacy: When the new person enters a party, the previous qualifications should not be seen. It is called the secrecy behind it. The dynamic keys are generated from VBP costs automatically. There are no existing or new keys as the VBP shifts by transaction. Consequently, the technology preserves privacy [10].

58.4 Secure Clustering

BNs deployed in the HWSN were part of the proposed architecture. BN has high capacity for transmission. The number of BNs to implement is rational straight to the number of network nodes. We also conclude that the BN nodes information has been disappearing.

58.4.1 Secure Clustering Phase

This is being created after the deployment of the nodes in the network after a certain period of time (our estimation is 120 s). This method is carried out to form network clusters.

Notation Description

NjjthNode
KN is a Network Key
BNiith is a Backbone Node
PKBiMj is a Pair wise key among *ith considered as* Mainstay and *jth is the* Associate Node
PKBij is Pairwise key among the mainstay
HPtx is a High transmission power
R_Mes is a response related Communication

A_Mes, Alive
NDj is a Node Degree of *jth* Node
E{x} K Encoded communication *using the required K.*

Initialization

[Pre-conditions: the BS key pool is KN, PKBij and PKBiMj,] [Post-conditions: the BS key pool has consistently been dispersed to all BNs and nodes of membership: KN, PKBij and PKBiMj].

1. Start
2. Provide the information into the related nodes, assigned by *BS,*
3. Through on *BS, EBS,* or assigned other set of equation
4. End

Cluster Formation

[Post-conditions: (1) Clusters were safely formed [Pre-conditions: Initialization has been done and trusting is given to all nodes]. (2) All nodes of members are aware of their CH]

1. Start
2. *BNi* broadcasts *{A_Mes}:* inside *BNi'sHPtx* range,
3. *Nj*ℵ assigned supportive nodel which will be covered through the appropriate message
4. *BNi* calculates *ND, BNi*ℵ *Nj: E{ID* of *BNi}* ||*ND*|| *{PKBiMj} KN*
5. Each *Nj* computes *Wi* as

$$\begin{cases} if S_{av} is 0, \; set S_{av} = 0.1 = 0.1 \\ else S_{av} \quad = \quad current value \end{cases}$$

The c1, c2, c3 and c4 values shall be considered respectively 0.20, 0.45, 0.25 and 0.45. Preferably the Sav, advanced VBP and fewer DBN node with lower mobility should be selected as CH.

6. *NjBNi: E{ID of Nj* || *Wi} PKBiMj*
7. *BNi* selects *CH* consuming the function *Min (Wj),* somewhere $i = 1...n$ in its transmission area.
8. *BNi* transmission *CH* information → *Ni Nj*
9. End

When node systems are ready to deploy in the network, a number of pairs of keys between BNs and sensor nodes just between BNs using the KN are securely loaded by the BS. The secure clustering process was initiated after the network was deployed. Each of the specified transmissions intends to share messages to different nodes inside the broadcast range of specified Secure Clustering phase. The nodes will send R Mes to KN encrypted BNi after receiving A Mes, containing the node ID and location of that node. BNi calculates the Member nodes and communicates it together with the pair key encrypted by KN after getting R Mes from every of its neighbors. Each NJ then calculates weight according to parameters including node rate, BN length, average battery speed and digital power [11, 12]. Each Nj detects its value to BNi, coded between itself and the BN by a parallel key. BNi selects the CH node with the smallest weight value when gathering the encrypted weight values from the entire NJ and transmits them to its members.

58.5 Simulation Study

The implementation of the proposed framework SCSR is done in C++ and verified in NS2 with 2.32 TCL simulation scripts. The performance of SCSR and EDDK are tested in a number of simulated scenarios. The simulation scenarios have been created by changing the number of nodes and the topology of the network.

58.6 Network Threat Model

Our objective is to ensure clustering and routing safety. The size of the network comprises 100, 200, 300, 400 and 500 clusters and BN nodes. Because the nodes in BN use the PSO system, only 2% of the overall network size is needed for the required BNs. Use of the safe clustering process to form clusters after deployment of BNs. If the cluster Member or CH inclines to leave the network or to join, a safe maintaining stage of the cluster is called for.

Nodes	Clusters	BN nodes
100	10	2
200	20	4
300	30	6
400	40	8
500	50	10

The concept of risk encompasses both external and internal attackers. The external attacker launches different categories of attacks. The RREQ package is one of the attacks to choose from. The internal attacker initiates a flood attack and play attack of MovConf during the maintenance phase of the cluster.

58.7 Simulation Parameters

Simulators NS2 version 2.32 is used to assess the SCSR proposed. The sink nodes are supposed to be located 100 m from the above location. The imitation is carried out over 20 test runs for different scenarios and the average for each value is taken. As seen below, the simulation variables.

The total number of specified nodes ranging between 100 and 500
The nature of the size is 500 × 500 m
Traffic Source is CBR
Specified major model
MAC IEEE 802.11
Mobility Model BN: No Mobility.

58.8 Results and Discussions

In accordance with the EDDK system, the SCSR output is. The quality is measured mainly by the following measures.

Energy: It is the normal energy disbursed for the information transmission.

Network Resilience: Now we will quantify how attackers disturb the future of network resilience. It is determined by calculating the fraction of messages cooperated by collecting x-nodes among non-compromised nodes.

Packet Delivery Ratio (PDR): This reflects the proportion of the number of packets effectively delivered to the entire number of packets distributed and restrained when attacks take place.

No. of Alive Nodes: The remainder of the nodes remaining alive through the simulation rounds in the network is calculated.

Results

In a 500 node example, the numbers of attackers vary between 5, 10, 15, 20 and 25. The figure displays the fraction of SCSR and EDDK methods' compromised communications for various scenarios for the attackers. When the numbers of assailants are increased from 5 to 25, the number of attacks will increase as shown in the figure. Throughout intra-cluster and inter-cluster routing, SCSR recognizes malicious nodes and separates them from the network. We therefore cannot engage further in the communications. Therefore, node capturing attacks will be challenging for the attackers.

Figure 58.2 signifies the energy consumption for the various attacker situations for SCSR and EDDK techniques. Because EDDK frequently performs the local key and the cluster key, it uses further resources, while SCSR upgrades dynamically with less energy consumption.

The PDR for EDDK is shown in Fig. 58.3. In this system the data cannot be retrieved when one or several packets are attacked or discarded even if there are multiple paths entering the destination. This means that the PDR is less than the SCSR (Fig. 58.4).

In contrast to dispersal technology, multipath routing is used in SCSR. With this dispersion process, the data is regenerated by the destination if minimum t shares between q shares directed by the origin are received. The flood messages from MovConf are filtered on the BN stage because they are not authenticated.

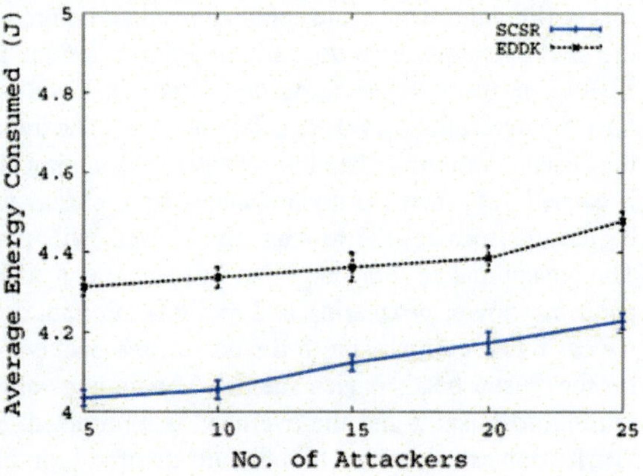

Fig. 58.2 Average energy consumption in the presence of attackers

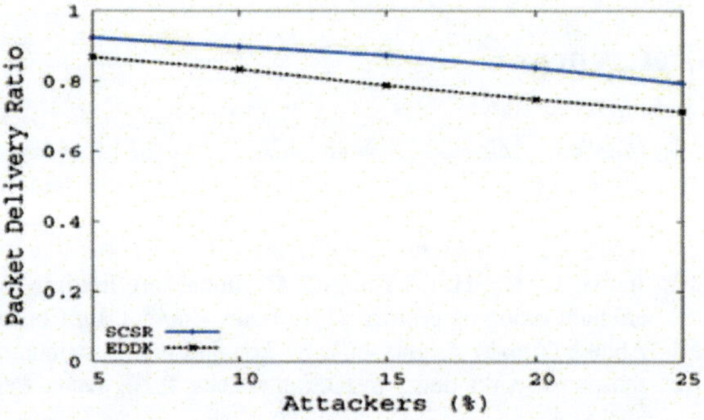

Fig. 58.3 Packet delivery ratio for EDDK

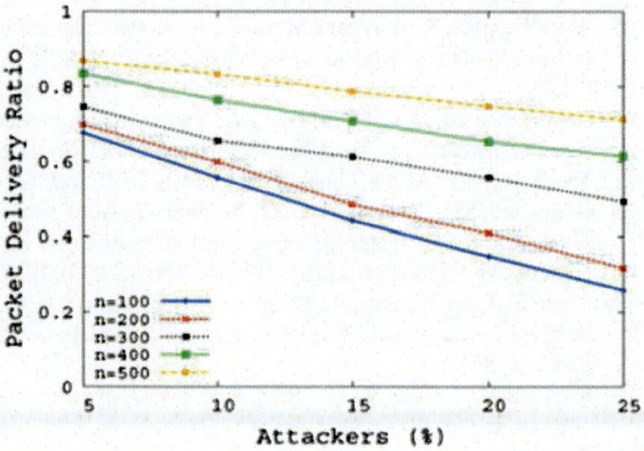

Fig. 58.4 Packet delivery ratio for SCSR and EDDK with 500 nodes

58.9 Conclusion

This section summarizes the dynamic key administration technique SCSR for heterogeneous mobile wireless sensor networks. This security framework can be applied for medium and large scale applications that require security in all aspects. Includes the heterogeneous BNs used by the PSO technique in the network. BN nodes use EBS to perform stable clustering. Then, CCV values based on location, ND and VBP are determined by the cluster members. The dynamic keys are created by CCV, then used to communicate intra-cluster information. The stable cluster maintenance is carried out when the node changes from cluster to cluster. When the data must be passed from source to sink, the clusters will perform secure route discovery. The conceptual results show that in comparison with existing technologies SCSR consumes lower processing and requires minimal communication costs for sensor nodes. When comparison to the cost of the backbone nodes, the connectivity cost for the cluster heads is also smaller. The findings of the test show that the software introduced is safer and the overhead is minimized. SCSR provides a good packet distribution ratio because of the multi-dispersal routing technology.

References

1. Mir, I.A., Mir, G.M., Wani, A.A.: A novel procedure to bring forth security for static homogeneous WSN's using secret keying protocol. Res. Rev. Int. J. Multi. **03**(12) (2018)
2. Sahingoz, O.K.: Large scale wireless sensor networks with multilevel dynamic key management scheme. J. Syst Architect **59**(9), 801–807 (2013)
3. Jiun-J, L., Lin-H, C., Hung-C, C.: Improving lifetime in heterogeneous WSN with energy efficient grouping protocol. Int J Innov Comput, Inf Control **8**(9), 6037–6047 (2012)
4. Abdelhakim, M., Lightfoot, L.E., Ren, J., Li, T.: Distributed detection in mobile access wireless sensor networks under byzantine attacks. IEEE Trans. Parallel Distrib. Syst. **25**(4), 950–959 (2013)
5. Alrajeh, N.A., Khan, S., Shams, B.: Intrusion detection systems in wireless sensor networks: a review. Int. J. Distrib. Sens. Netw. 2013, 7p (2013)
6. Azarderskhsh, R., Reyhani-Masoleh, A.: Secure clustering and symmetric key establishment in heterogeneous wireless sensor networks. EURASIP J. Wirel. Commun. Netw. 2011:12p (2011)
7. Banihashemian, S., Bafghi, A.G.: A new key management scheme in heterogeneous wireless sensor networks. In: 2010 the 12th International Conference on Advanced Communication Technology (ICACT) (Phoenix Park 2010), IEEE, pp. 141–146 (2010)
8. Boujelben, M., Cheikhrouhou, O., Youssef, H., Abid, M.: A pairing identity based key management protocol for heterogeneous wireless sensor networks. In: International Conference on Network and Service Security, N2S'09. (Paris 2009), IEEE, pp. 1–5 (2009)
9. Chen, S., Liao, X., Shu, R., Shen, X., Xu, X., Zheng, X.: Dynamic key management scheme in wireless sensor networks. In: High Performance Networking, Computing, and Communication Systems, 2011
10. Erfani, S.H., Javadi, H.H.S., Rahmani, AM.: A dynamic key management scheme for dynamic wireless sensor networks. Secur. Commun. Netw. **8**(6), 1040–1049 (2015)
11. Goyal, P., Kumar, M., Sharma, R.: A novel and efficient dynamic key management technique in wireless sensor network. Int. J. Adv. Netw. Appl. **4**(1), 1462–1466 (2012)
12. Anwaar Ahmad, W., Dixit, S.: Enhanced Frame Aggregation Scheduler (EFAS) for data Transmission over IEEE802.11ac. Res. Rev. Int. J. Multi. **4**(05), 05 (2019)

Chapter 59
Influential Effects of Process Parameters of Fused Deposition Modelling on Wear of a PLA Specimen: A Comprehensive Review

Shivam Gupta, Sonali Gupta, Kundan Kumar Pandey, Kritik Subodh Dwivedi, Shahroz Akhtar Khan, and P. K. Arora

Abstract Additive manufacturing (AM) was commonly used for rapid prototyping, to visualize, test and authenticate designs, and then the production begins. The ability to build complex parts layer by layer, minimizing cycle time of production is only possible with the help of Additive Manufacturing's fused deposition modelling (FDM). Thickness of material-layers, orientation, width, angle of raster and the air gap have significant effect on mechanical properties and thus surface quality of parts. Optimization and selection of process parameters for FDM are very important in having enhanced mechanical properties for printed parts under wear. Taguchi's signal to noise (S/N) ratio and Artificial Neural Network (ANN) are utilized in order to find a set of parameters having great results for their respective response characteristics. Impact of each parameter is found using analysis of variance (ANOVA). Material under consideration is Poly lactic acid (PLA). This study intends to analyze the wear characteristics of the PLA specimen by finding optimal process parameters of 3-dimensional (3D) printing with FDM process. Thickness of layer, angle orientation, printing speeds are the process variables considered for this study.

59.1 Introduction

AM construct parts layer by layer directly from data available from CAD software. In conventional manufacturing techniques, the material need to be removed to obtain required objects. AM creates 3D parts by applying material in layer-by-layer manner. This enables AM very flexible to print all kind of geometry with any level of complexity with lower time and cost of production [1]. Almost all AM techniques have similar working principle, i.e., layering a material in predefined track for

S. Gupta · S. Gupta · K. K. Pandey · K. S. Dwivedi · S. A. Khan (✉) · P. K. Arora
Department of Mechanical Engineering, Galgotias College of Engineering and Technology, Greater Noida, UP 210310, India
e-mail: Shahroz.khan@galgotiacollege.edu

S. Yadav et al. (eds.), *Proceedings of Second International Conference in Mechanical and Energy Technology*, Smart Innovation, Systems and Technologies 290,
https://doi.org/10.1007/978-981-19-0108-9_59

creating a desired part. The process binds together the 2-dimensional layers of material to create a 3D object. AM is currently compatible for a wide range of materials from precious metals to low grade polymers [2, 3]. One of the widely used consumer focused 3D printing technologies is FDM which offers ease to set up and print in comparison to other AM technologies [4]. The most researched and used biodegradable as well as renewable polyester (aliphatic) is poly lactic acid or polylactide. It has proven potential in order to replace traditional petrochemical based polymers for application in industries or Biomaterial for medical purposes [5, 6].

59.1.1 Fused Deposition Modelling (FDM)

Stratasys Inc. was the developer of FDM process in USA in the 1990s. It was introduced as a rapid prototyping technique. FDM uses thermoplastics like acrylonitrile–butadiene–styrene (ABS) and PLA as its materials [7]. Material spools are installed in the machine from where material is then extracted using drive wheels into the liquefier section of machine in which the material melts and gets spread upon the bed with the help of a nozzle moving on a predefine track. The nozzle temperature has been found affecting the mechanical properties of the material significantly and therefore design of nozzle for FDM require serious considerations [8]. Depending on the requirement and geometry of parts to be printed more than one material can be used in FDM (see Fig. 59.1) [9].

FDM has the ability to produce parts made of thermoplastics having a complex geometry with high dimensional accuracy. FDM allows the manufacturer to freely control the characteristics of printed parts with the help of a wide range of changeable parameters called process parameters. Many research have been conducted on FDM printed parts to study their physical properties on changing the process parameters.

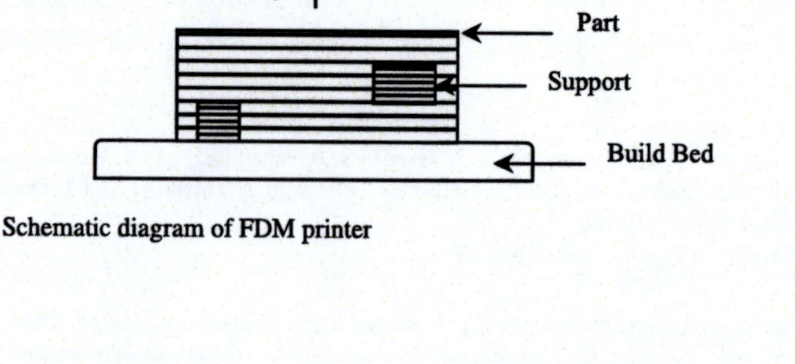

Fig. 59.1 Schematic diagram of FDM printer

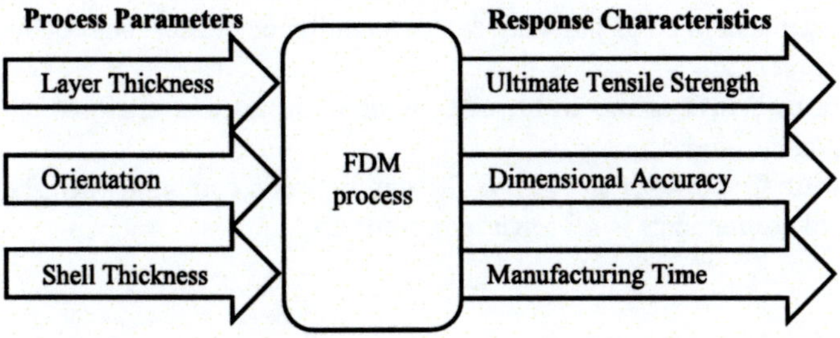

Fig. 59.2 Experimental process layout

Figure 59.2 can be referred to achieve better response characteristics on printed parts by controlling the significant process parameters.

59.1.2 *Process Parameters*

Process parameters are the variables which can be altered during any process so that to obtain significant change in the result. The FDM machine offers control over various machining process parameters which have significant impact on efficiency of production and characteristics of parts. Common process parameters are orientation of build, density infill, thickness of layer, printing speed, width and orientation of raster, air gap, temperature of extruder, pattern infill, shells number and heat treatment temperature. The details of some important process parameters are [10–17]:

1. **Layer Thickness** is the height of layer poured by the tip of nozzle as seen in the Fig. 59.3. The nozzle diameter and material have significant effect on the amount of layer thickness.
2. **Raster Angle** is the angle subtended between raster model with respect to X-axis on underside part layer. The usual raster angle varies from 0° to 90° (see Fig. 59.3).
3. **Raster Width** is breadth of material bead used for raster. The higher value of raster width, the component will be built more vigorous interior. The nozzle tip diameter has the effect on amount of raster width.
4. **Orientation Build** is the angle subtended by the part with the build platform with respect to *x*, *y* and *z*-axis.

Fig. 59.3 Layer thickness and orientation

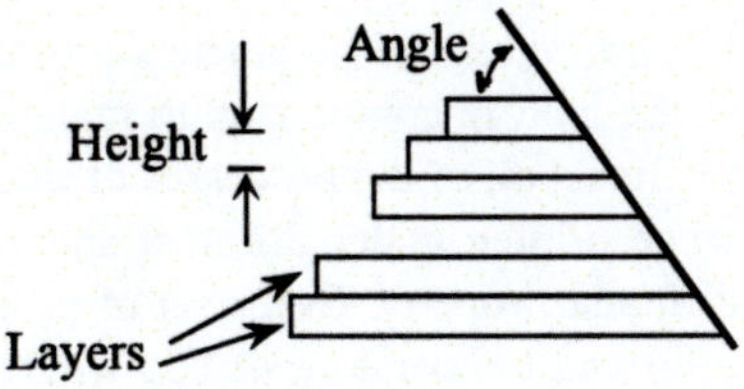

5. **Air Gap** is defined as gap between two contiguous raster tool conduits on the same layer.
6. **Contour Width** is the width of contour tool part that encloses coils of the component.
7. **Contour to Contour air gap** is the gap between contours when the packing style of components is set to various contours.

59.1.3 Various Responses Characteristics of Interest

1. **Surface Roughness** [18–20]

This property is used to define the texture of the outer surface of the part. Surface roughness plays a very important role in processes like moistening and wetting. Rougher surface means it retains moisture on itself and in case of metals it means faster corrosion.

2. **Dimensional Accuracy** [21, 22]

Whenever the thermoplastics settle on the bed and cools down, they show shrinkage in size. Also, since the FDM printed parts are printed layer wise so they don't exactly have a curved surface instead they have a stair like structure. These factors lead to inaccuracies in dimensions of the part after printing.

3. **Tensile Strength** [23]

The maximum load a material can support without yielding while being stretched. An important property which determines if the printed part can withstand the forces, it is being designed to handle.

4. **Build Time** [19, 24]

Time required by the printer to print a single part is called the "Build time" of the part. Various researches have been made to decrease the time for build to increase amount of part production.

59.2 Research Findings on the Effects of FDM Process Parameters

In AM, values of parameters are always given higher interest by the researchers in the field in comparison to material properties of the part and this is in contrary to almost all of the processes of manufacturing. Identical parts from same material when printed under different sets of parameters have been seen to offer entirely different property. Every set of process parameters like thickness of layer, pattern infill and temperature of bed would generate a unique part structure and leads to varying values of mechanical properties [25]. Sood et al. conducted experiments with five parameters, namely thickness of layer, width of raster, angle of raster, air gap and orientation to analyze the changes in tensile, impact and flexural strengths of the specimen under test [21]. They observed shrinkage along the length of the part whereas increase in the thickness of the printed parts. Lanzotti et al. have performed tests to measure variations in ultimate tensile strength (UTS) and nominal strain on parts printed out of 3D printing process [26]. Gorski et al. performed flexural, impact and tensile tests on the specimens made of ABS having unique orientations [27].

Chacon et al. performed the analysis on bending strength and tensile strength of PLA part under different sets of orientation build, thickness of layer and rate of feed [28]. They observe increase in ductility with rise in thickness of layer and rate of feed. Apart from changing the process parameters the physical properties of printed part can be enhanced by improving the interface bonding of the adjacent layers by reheating the last layer so that it can bond well with next layer. Seth Collins P. used hot air flow to heat the layer [29], Ravi et al. employed lasers to heat layers [30] and Kishore et al. used infrared radiations to reheat the deposited layer [31].

In the year 2001, Anitha et al. analyzed effects of FDM parameters on surface roughness with ABS printed parts using Taguchi and ANOVA [18]. In the year 2004, Thrimurthulu et al. utilized a technique called genetic algorithm (GA) for the prediction of optimal orientation [19]. In year 2007, Wang et al. used statistics to study effect of process parameters like thickness of layer, deposition style, structures for support and orientation [20]. They concluded that orientation and thickness of layers play a crucial role for dimensional accuracy of printed parts.

In this research optimum parameters were different for different conditions so no optimum parameters were concluded in the research. So, in later work all the different responses were combined into one response to obtain optimum parameters. Nancharaiah et al. in their research resulted in thickness of layer and air density of infill were two process variables which affect the accuracy for dimensions the most [22]. Ahn et al. focused on tensile and compressive strengths of the printed parts in their research where it was concluded that optimum values of thickness of layer and orientation can lead to higher tensile and compressive strengths [23].

Thrimurthulu et al. in the year 2004 used a mathematical model for predicting and optimizing time for the build [19]. This model was also found successful in predicting the build time of the parts printed by other publishers in their respective researches. In the year 2011 in another research Nancharaiah et al. used Taguchi L9 technique to determine the relation between air gap, thickness of layer and time for build of parts. Thickness of layer affected time for build by around 66.57% and air gap around 30.77%. A thickness of layer of 0.330 mm and an air gap of 0.020 mm were found optimum in their research [24].

59.3 Wear Property

Surface property of the material to get worn out due to continuous exposure to other surfaces is called wear. Wear in a part gives the time for the part to get used in a particular environment. Wear can be caused through surroundings and touching parts. Long duration of exposure to wind and flowing water can also cause a part to wear off.

59.3.1 Recent Research Findings on Effect of FDM Process Parameters on Wear Property

In the year 2010 Srinivasan et al. conducted their research on effect of FDM process variables on wear strength of carbon-fibre infused PLA material. Parameters included thickness of layers, percentage infill and pattern in infill [32]. It is found that thickness of layer and infill percentages were the parameters which affected wear the most. Increased density of infill and minimal thickness of layers provided the best results. For similar values of thickness of layer and percentage of infill, parts printed with grid pattern as pattern of infill showed lesser wear in them.

In the year 2020, Mohamad Nordin et al. used response surface modelling (RSM) to optimize the parameters for ABS material. They conducted tests on three parameters; thickness of layer, pattern infill and temperature of nozzle. They found in their research that thinner layers result in less wear rate. Using ANOVA and RSM the optimum value for layer thickness was 0.10 mm with triangle pattern and 234 °C nozzle temperature. This research did not include the grid pattern [33]. Researchers in 2021 use Box Behnken-based design of experiment to conduct tests to achieve optimize input parameters for wear. RSM is used to set up a mathematical model between combinations of input parameters and obtained output. As a result of ANOVA, it was concluded that thickness of layer and the infill density were more significant in comparison to print speed [34].

Kumar et al. in 2021 studied the effect of sliding wear on the specimen made of PLA and ABS material. They found that wear rate reaches to a maximum for 90° build orientation [35].

There are certain tools and techniques, like Taguchi and ANN, known to optimize the constraints and provide the results which can help in saving time, money, material and can lead to better optimized results.

Table 59.1 Steps followed in Taguchi methodology

S. No.	Steps followed
1	Defining the problem and the objective
2	Identify all factors and their levels
3	Designing the required orthogonal array
4	Conduction the set of experiments
5	Data analysis (S/N ratio, ANOVA, response plot)
6	Determination of optimum levels for factors
7	Conducting the confirmation experiment

59.4 Taguchi Method

Taguchi method can be easily applied with the software MINITAB; this helps in generating plots and can obtain results under given optimized conditions. Table 59.1 shows the sequence of steps followed in Taguchi methodology:

59.4.1 Orthogonal Array

There are a lot of possible combination of factors and their levels. Taguchi helps in selecting the special arrangement of combinations of levels of key factors known as orthogonal array [36].

59.5 S/N ratio and ANOVA

S/N ratio is utilized to calculate robustness, signal indicates the controllable factors while the noise depicts the uncontrollable factors [37, 38]. There are various options available to select for plotting data points as per the need. For example, in wear testing "smaller is better" is used while for tensile strength "larger is better" can be used.

The analysis of variance (ANOVA) technique is utilized to find linear plot between input and output factors [37]. It also helps in determination of p-values and F-values that determine which factors are actually affecting the output, and which factor(s) have no effect on the output result. It also predicts the r-square values (0–100) which determine the closeness of the derived curve's fitness with the obtained data. The response table also gives delta values and rank of dominance of factor.

There are certain limitations with Taguchi design [37] like

- Difficulty in accounting interactions between various parameters.
- Method is not appropriate for dynamically changing process.

Thus, ANN is used to obtain better validations.

ANN (Artificial-Neural-Network)
ANN is human simulation of neurological system. The diverse connection between these elements helps in better approximation of any non-linear functions and thus called universal function approximators [39]. The ANN model is trained between input and output parameters [40]. The neurons from one layer can interact with neurons from another layer; there interconnections are assigned with weights and biases [41]. In ANN 70% of data is used for training while rest of the data is being used for the purpose of testing [42]. The flow chart, given in Fig. 59.4 below, depicts the proposed working of the ANN system [37] using MATLAB tool. Different methods

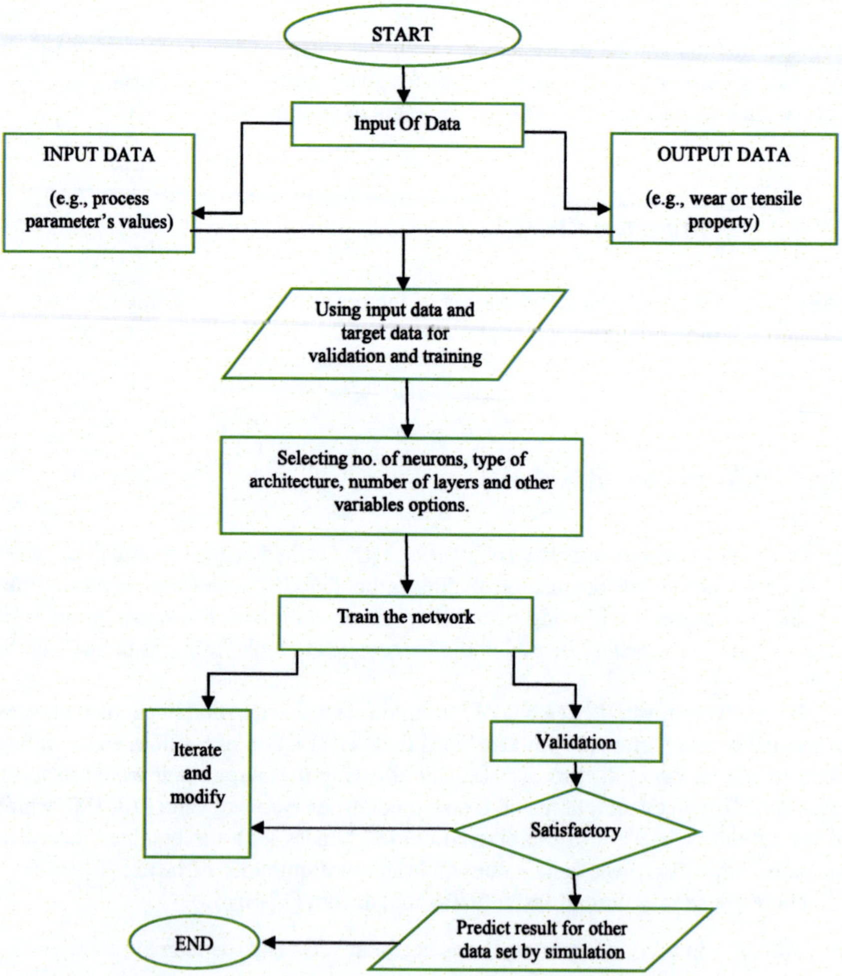

Fig. 59.4 The flow chart

such as feed forward backdrop, feed forward delay, Hopfield, generalized regression, learning vector quantization (LVQ), etc. are available. The most commonly used is feed forward backdrop method [37], there are options in transfer functions as well namely PURELIN, TANSIG, LOGSIS.

59.6 Conclusions

It is observed that in certain conditions where dynamically changes take place, Taguchi method fails to give better predictions. For same data set ANN gives more appropriate results in comparison to Taguchi. In ANN more data is required for the purpose of training and validation and testing while in Taguchi predictions can be made with relatively lesser data. ANN is capable of predicting results for larger data set easily as compared to Taguchi. Taguchi method provides a variety of options as compared to ANN, like information regarding rank of factors, their effectiveness and ineffectiveness, tells which factor is more significant thus Taguchi method is informative, while for working on larger data set and better accuracy in result, ANN dominates Taguchi.

59.6.1 Future Aspects in the Research

- Using more process parameters can give a more accurate answer for the optimum values for least wear rate.
- Since the use of different infill patterns showed different results in above researches it can be inferred that using a wide range of levels within the parameter can also help to narrow down the optimum parameters.
- Applying more than single optimizing techniques can help to determine which techniques have a better prediction rate for the wear rate. Results from different techniques and models can be compared to each other.
- For same data results can be predicted and compared using different transfer functions to know at which transfer function the error is least.
- Regression equation in Taguchi method can be obtained from ANOVA analysis and prediction, these equations are different and analysis can be done so as to compare which equation gives more accurate results.

References

1. Mohamed, O.A., Masood, S.H., Bhowmik, J.L.: Optimization of fused deposition modelling process parameters: a review of current research and future prospects. J. Adv. Manuf. **3**, 42–53 (2015)
2. Mellor, S., Hao, L., Zhang, D.: Additive manufacturing: a framework for implementation. Int J Prod Econ. **14**, 194–201 (2014)
3. Ngo, T.D., Kashani, A., Imbalzano, G., Nguyen, K.T.: Additive manufacturing (3D printing): a review of materials, methods, applications and challenges. Compos. B Eng. **143**, 172–196 (2018)
4. Jackiewicz, J.: Manufacturing of instructional aids for students at low cost by means of 3D printing. Mater. Manuf. Process. **32**, 1116–1130 (2017)
5. Savioli Lopes, M., Jardini, A.L. MacielFilho, R.: Procedia Eng. **42**, 1402–1413 (2012)
6. Yadav, A., Srivastav, A., Singh, A., Mushtaque, M.D., Khan, S.A., Kumar, H., Arora, P.K.: Investigation on the materials used in additive manufacturing: a study. Mater. Today: Proc. **43**, 154–157 (2021). https://doi.org/10.1016/j.matpr.2020.10.975
7. Kumar, H., Khan, S.A., Arora, P.K.:Retrospective investigation on emergence and development of additive manufacturing. Indian J. Eng. Mater. Sci. (IJEMS), **28**(2), 115–124 (2021)
8. Kedare, P.K., Khan, S.A., Kumar, H.: 3D Printer nozzle design and Its parameters: a systematic review. In: Yadav S., Singh D., Arora P., Kumar H. (eds) Proceedings of International Conference in Mechanical and Energy Technology. Smart Innovation, Systems and Technologies, vol. 174. Springer, Singapore (2020). https://doi.org/10.1007/978-981-15-2647-3_73
9. Alafaghani, A., Qattawi, A., Alrawi, B., Guzman, A.: Experimental Optimization of Fused Deposition Modelling Processing Parameters: A Design-for-Manufacturing Approach. Procedia Manufacturing **10**, 791–803 (2017). https://doi.org/10.1016/j.promfg.2017.07.079
10. PLA vs ABS Filament: Plastic Strength, Flexibility Compared! Which Is Better For 3D Printing? Available online: https://www.allthat3d.com/pla-vs-abs/ (accessed on 21 June 2019).
11. Panda, S.K., Padhee, S., Anoop Kumar, S., Mahapatra, S.S.: Optimization of fused deposition modelling (FDM) process parameters using bacterial foraging technique. Intell. Inf. Manag. **1**, 89–97 (2009)
12. Akande, S.O.: Dimensional accuracy and surface finish optimization of fused deposition modelling parts using desirability function analysis. Int. J. Eng. Res. Technol **4**, 196–202 (2015)
13. Gurrala, P.K., Regalla, S.P.: Multi-objective optimisation of strength and volumetric shrinkage of FDM parts: a multi-objective optimization scheme is used to optimize the strength and volumetric shrinkage of FDM parts considering different process parameters. Virtual Phys. Prototyp. **9**, 127–138 (2014)
14. Rao, R.V., Rai, D.P.: Optimization of fused deposition modeling process using teaching-learning-based optimization algorithm. Eng. Sci. Technol. Int. J. **19**, 587–603 (2016)
15. Mohamed, O.A., Masood, S.H., Bhowmik, J.L.: Optimization of fused deposition modeling process parameters: A review of current research and future prospects. Adv. Manuf. **3**, 42–53 (2015)
16. Popescu, D., Zapciu, A., Amza, C., Baciu, F., Marinescu, R.: FDM process parameters influence over the mechanical properties of polymer specimens: A review. Polym. Test. **69**, 157–166 (2018)
17. Raju, M., Gupta, M.K., Bhanot, N., Sharma, V.S.: A hybrid PSO–BFO evolutionary algorithm for optimization of fused deposition modelling process parameters. J. Intell. Manuf. 2018, 1–16
18. Anitha, R., Arunachalam, S., Radhakrishnan, P.: Critical parameters influencing the quality of prototypes in fused deposition modelling. J Mater Process Technol **118**(1–3), 385–388 (2001)
19. Thrimurthulu, K., Pandey, P.M., Reddy, N.V.: Optimum part deposition orientation in fused deposition modeling. Int J Mach Tools Manuf **44**(6), 585–594 (2004)
20. Wang, C.C., Lin, T.W., Hu, S.S.: Optimizing the rapid prototyping process by integrating the Taguchi method with the gray relational analysis. Rapid Prototyp J **13**(5), 304–315 (2007)

21. Sood, A.K., Ohdar, R.K., Mahapatra, S.S.: Parametric appraisal of mechanical property of fused deposition modelling processed parts. Mater. Des. **31**(1), 287–295 (2010)
22. Nancharaiah, T., Raju, D.R., Raju, V.R.: An experimental investigation on surface quality and dimensional accuracy of FDM components. Int J EmergTechnol **1**(2), 106–111 (2010)
23. Ahn, S.H., Montero, M., Odell, D., et al.: Anisotropic material properties of fused deposition modeling ABS. Rapid Prototyp J **8**(4), 248–257 (2002)
24. Nancharaiah, T.: Optimization of process parameters in FDM process using design of experiments. Int J EmergTechnol **2**(1), 100–102 (2011)
25. Gorski, F., Kuczko, W., Wichniarek, R.: Influence of process parameters on dimensional accuracy of parts manufactured using fused deposition modelling technology. Adv. Sci. Technol. Res. J. **7**(19), 27–35 (2013)
26. Lanzotti, A., Grasso, M., Staiano, G., Martorelli, M.: The impact of process parameters on mechanical properties of parts fabricated in PLA with an open-source 3-D printer. Rapid Prototyping Journal **21**(5), 604–617 (2015)
27. Gorski, F., Wichniarek, R., Kuczko, W., Zawadzki, P., Bun, P.: Strength of ABS parts produced by fused deposition modelling technology – a critical orientation problem. Advances in Science and Technology Research Journal **9**(26), 12–19 (2015)
28. Chacon, J.M., Caminero, M.A., Garcia-Plaza, E., Núñez, P.J.: Additive manufacturing of PLA structures using fused deposition modelling: effect of process parameters on mechanical properties and their optimal selection. Mater. Des. **124**, 143–157 (2017)
29. Seth Collins, P.: Fused deposition modeling with localized pre-deposition heating using forced air. Montana State Univ 41(10):1975–1975 (1985)
30. Ravi, A.K., Deshpande, A., Hsu, K.H.: An in-process laser localized pre-deposition heating approach to inter-layer bond strengthening in extrusion based polymer additive manufacturing. J Manuf Process **24**, 179–185 (2016)
31. Kishore V, Ajinjeru C, Nycz A, Post BK, Lindahl JM, Kunc V, Duty CE (2016) Infrared preheating to improve interlayer strength of big area additive manufacturing (BAAM) components. AdditManuf 14:321–324
32. Srinivasan, R., Aravindkumar, N., Aravind Krishna, S., Aadhishwaran, S., & George, J. (2020). Influence of fused deposition modelling process parameters on wear strength of carbon fibre PLA. Materials Today: Proceedings.doi:https://doi.org/10.1016/j.matpr.2020.03.738
33. Norani, M. N. M., Abdollah, M. F. B., Abdullah, M. I. H. C., Amiruddin, H., Ramli, F. R., Tamaldin, N.: 3D printing parameters of acrylonitrile butadiene styrene polymer for friction and wear analysis using response surface methodology. In: Proceedings of the Institution of Mechanical Engineers, Part J: Journal of Engineering Tribology (2020). 135065012092560. https://doi.org/10.1177/1350650120925601
34. Kumar, H., Sharma, A., Shrivastava, Y., Khan, S. A., & Arora, P. K. (2021). Optimization of process parameters of pin on disc wear set up for 3D printed specimens. Journal of Engineering Research, (EMSME Special Issue), pp.133–145. https://doi.org/10.36909/jer.emsme.13899
35. Kumar, A., Choudhary, A., Tiwari, A., James, C., Kumar, H., Kumar Arora, P., Akhtar Khan, S.: An investigation on wear characteristics of additive manufacturing materials. Materials Today: Proceedings **47**, 3654–3660 (2021). https://doi.org/10.1016/j.matpr.2021.01.263
36. Srivastava, M., Rathee, S.: Optimization of FDM process parameters by Taguchi method for imparting customized properties to componentsVirtual and Physical Prototyping **13**, 203–210 (2018)
37. Pant, M., Singari, R.M., Arora, P.K., Moona, G., Kumar, H.: Wear assessment of 3–D printed parts of PLA (polylactic acid) using Taguchi design and artificial neural network (ANN) technique. Mater. Res. Express **7**, 115307 (2020). https://doi.org/10.1088/2053-1591/abc8bd
38. Darbar, R., Patel, D.M., Patel, J.: International Journal of Engineering, Process Parameters Optimization Of FDM Process And Application Of Taguchi Approach And ANN – The Review, Research and Applications (IJERA) ISSN: 2248-9622 www.ijera.com Vol. 3, Issue 1, January -February 2013, pp 743–746
39. Deshwal, S., et al.: Exercising hybrid statistical tools GA-RSM, GA-ANN and GA-ANFIS to optimize FDM process parameters for tensile strength improvement. NULL (2020). https://doi.org/10.1016/j.cirpj.2020.05.009

40. Yadav, D., Chhabra, D., Kumar Garg R., et al.: Optimization of FDM 3D printing process parameters for multi-material using artificial neural network, Mater. Today: Proc. https://doi.org/10.1016/j.matpr.2019.11.225
41. Weake, N., Pant, M., Sheoran, A., Haleem, A.: Optimising parameters of fused filament fabrication process to achieve optimum tensile strength Using artificial neural network pp. 373–381 (2020) http://hdl.handle.net/2324/4068614
42. Kustrin, S.A., Beresford, R.: Basic concepts of artificial neural network (ANN) modeling and its application in pharmaceutical research. J. Pharm. Biomed. Anal. **22**(5), 717–727 (2002)

Author Index

S. Yadav et al. (eds.), *Proceedings of Second International Conference in Mechanical and Energy Technology*, Smart Innovation, Systems and Technologies 290,
https://doi.org/10.1007/978-981-19-0108-9